Mathematical Ideas

THIRD EDITION

Mathematical Ideas

THIRD EDITION

Charles D. Miller   Vern E. Heeren
American River College, Sacramento, California

Scott, Foresman and Company   Glenview, Illinois
Dallas, Tex.   Oakland, N.J.   Palo Alto, Cal.   Tucker, Ga.   London, England

The cover: Euclid surrounded by students, from Raphael's mural *The School of Athens.* The entire mural is shown on the preceding pages, and some comments are on page 289. Two other figures from the mural are featured in this book: Aristotle (page 36) and Pythagoras (page 118).

Library of Congress Cataloging in Publication Data
Miller, Charles David, 1942–
Mathematical Ideas.
Includes bibliographies and index.
1. Mathematics—1961–
I. Heeren, Vern E., joint author.
II. Title.
QA39.2.M55 1978   510   77-26071
ISBN 0-673-15090-9

Credits appearing on pages 506–508 represent
an extension of the copyright page.

2 3 4 5 6-RRC-84 83 82 81 80 79 78

Preface

This book is written for non-mathematics and non-science majors who need an understanding of the key ideas of contemporary mathematics. Such students include liberal arts students, prospective and in-service elementary school teachers, and those in business and social science.

Often, many or all of these students are enrolled in the same sections. For this reason, the book offers a large variety of different applications, so that students from all these areas will find much of interest.

The only prerequisite that we assume is arithmetic. The book contains enough algebra and geometry to review the necessary basic mathematics and to teach elementary problem solving skills.

Mathematical Ideas can be used in a variety of courses. Previous editions have been used in survey courses, introduction to mathematics, mathematics for liberal arts students, mathematics for elementary school teachers, topics in contemporary mathematics, applications of mathematics, and finite mathematics. The book can be used for a full year course; the rich variety of topics presented makes the book very flexible for shorter courses. Complete course outlines for several different courses are included in the Instructor's Guide.

This Guide is especially complete. In addition to the answers for the even-numbered exercises, the Guide also includes two forms of a test for each chapter, and a collection of test items for constructing a final examination. Many supplementary topics are included as work sheets in the Guide; these may be reproduced and supplied to students as local curriculum needs or instructor interests dictate.

The Guide also includes an extensive computer supplement. The many complete computer programs (in BASIC) are keyed to sections of the text.

In preparing this third edition, we have been guided by the comments of the many people acknowledged after this preface. Their comments helped us shape the following characteristics:

Topics There is a broad selection of topics in the book. The chapters on sets and logic are carefully paced and include an "everyday" discussion of these topics. There is added material on the real number system; and in the probability chapter, the difficult ideas of permutations and combinations are minimized. Optional topics are identified by use of an asterisk.

Exercises Each exercise set begins with a large number of routine numerical problems covering the ideas of the section. They are followed by two types of additional problems — applications that show the usefulness of the topics, and more challenging problems. We have indicated the more challenging problems with an asterisk.

Appeal to Students We have drawn examples and exercises from areas of common interest, as well as included a large number of relevant photographs and drawings. Extensive captions relate the illustrations to the material of the text itself.

We have tried to write in a way that shows students that mathematics can be enjoyed and is not something to be feared. Today, however, it is probably more important to show that mathematics is *useful*. We try to do this by emphasizing applications. We include much material on consumer mathematics (more than just the one section indicated in the Table of Contents). In the logic chapter, we analyze the logic of a rental contract. We also discuss the problems of "consumers" of statistics in Chapter 10. And many of the exercise sets include brief applications of the topics of the section that each such exercise set follows.

The book includes chapter summaries and chapter tests.

The text has been designed for ease of reading. We use a type face that is large and clear. A second color is used for both pedagogical and decorative purposes.

The publication of this book marks the tenth anniversary of our association with Scott, Foresman. Throughout this time, we have always found the company dedicated to the production of quality textbooks. In particular, the last few years with Robert Runck, Executive Editor for Mathematics, have been a real pleasure for us.

Nat Weintraub and Jill Reschke, editors, spent much time on the book. Donna Sharp typed much of the manuscript. Matthew J. Pittner did the bulk of the work on the index. Our thanks go to all these people for their valued contributions.

Charles D. Miller

Vern E. Heeren

Acknowledgments

Rebecca M. Baum, *Lincoln Land Community College*
Phillip W. Bean, *Mercer University*
Sandra A. Bollinger, *Longwood College*
Harold W. Brockman, *Capital University*
Martin Byrer, *Lehigh County Community College*
Charles A. Church, Jr., *University of North Carolina*
David A. Cusick, *Marshall University*
Thomas B. Dellaquila, *Monroe Community College*
Patricia B. Dyer, *Broward Community College*
Richard C. Enstad, *University of Wisconsin–Whitewater*
Howard W. Eves, *University of Maine*
David Fontana, *American River College*
Judith L. Gersting, *Indiana University–Purdue University at Indianapolis*
Joan M. Golliday, *Santa Fe Community College*
Joseph A. Gore, *Valdosta State College*
Robert J. Greger, *Harford Community College*
James L. Hall, *Isothermal Community College*
Adelaide T. Harmon, *California Polytechnic State University*
Edward Harper, *American River College*
Maurice Holtfrerich, *Phoenix College*
Adam J. Hulin, *Louisiana State University*
Barbara C. Jones, *Northeast Louisiana University*
Alfred Mudrich, *Allegany Community College*
Charles A. Nichol, *University of South Carolina*
Maryjo M. Nichols, *University of Detroit*
Norman Nystrom, *American River College*
Theron D. Oxley, Jr., *Drake University*
Chester I. Palmer, *Auburn University*
John W. Petro, *Western Michigan University*
Kenneth L. Pothoven, *University of South Florida*
David Presser, *DePaul University*
Mike Reagan, *Northeastern Oklahoma State University*
Mignon S. Riley, *Northeast Louisiana University*
Darrell Ropp, *Rock Valley College*
Albert Schild, *Temple University*
Charles E. Schwartz, *Glassboro State College*
Margaret S. Scott, *Harper College*
Lawrence A. Sher, *Manhattan Community College*
Jesse L. Smith, *Middle Tennessee State University*
Thomas Spradley, *American River College*
Leonard J. Swiatkowski, *Wilbur Wright College*
Howard E. Taylor, *West Georgia College*
Lowell T. Van Tassel, *San Diego City College*
Waunetta Walling, *Tarrant County Junior College (Northeast)*
Peter R. Weidner, *Edinboro State College*

Contents

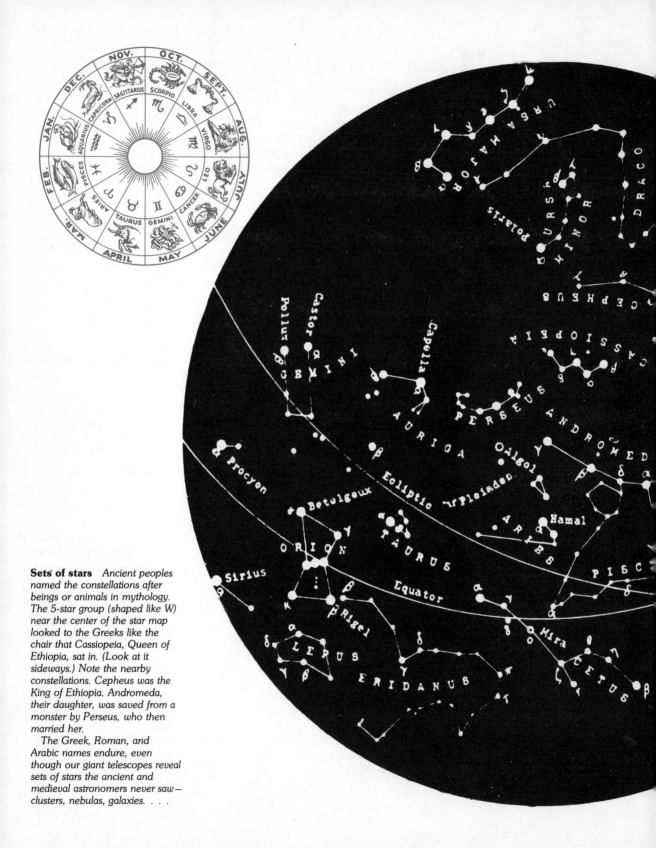

Sets of stars *Ancient peoples named the constellations after beings or animals in mythology. The 5-star group (shaped like W) near the center of the star map looked to the Greeks like the chair that Cassiopeia, Queen of Ethiopia, sat in. (Look at it sideways.) Note the nearby constellations. Cepheus was the King of Ethiopia. Andromeda, their daughter, was saved from a monster by Perseus, who then married her.*

The Greek, Roman, and Arabic names endure, even though our giant telescopes reveal sets of stars the ancient and medieval astronomers never saw — clusters, nebulas, galaxies. . . .

Chapter 1 Sets

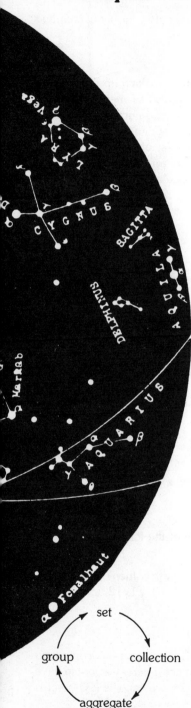

Psychologists tell us that the human mind likes to create collections. Instead of seeing a group of five stars as five separate items, we tend to see them as *one* group of stars. The mind always tries to find order and patterns.

This tendency is represented in mathematics with the idea of a *set*, or collection of objects. We form sets by collecting together all the solutions to a given problem, or perhaps all the numbers that have meaning in a given problem.

We will encounter sets throughout this book. For example, we study sets of numbers in Chapters 4 and 5, mathematical systems (which require sets of objects) in Chapter 6, and sets of outcomes for probability in Chapter 9.

The basic ideas of set theory were developed by the German mathematician Georg Cantor (1845–1918) about the year 1875. Some of the things he proved, such as "The whole is not always greater than its parts," flew in the face of the then-accepted mathematical beliefs. Controversial ideas are seldom well received, and this was especially so in the 1870s. We shall discuss Cantor's ideas in more detail in Section 1.6.

1.1 DESCRIBING SETS

In beginning our study of sets, we come right away to a difficulty. It is really not possible to mathematically define the word **set.** We can explain what a set is: a set is a collection or group or assemblage of objects. A dictionary definition does not help. If we look up *set*, we find the meaning *collection.* The word *collection* is given the meaning *aggregate,* which in turn is given the meaning *group.* Finally, if we look up *group*, we find the meaning *set.* We have gone in a circle.

1

Infinity   *Close-up of a camera lens shows the infinity sign ∞, defined as any distance greater than 1000 times the focal length of a lens.*

The sign was invented by the mathematician John Wallis in 1655. Wallis used 1/∞ to represent an infinitely small quantity. The philosopher Voltaire described the ∞ as a loveknot, and was skeptical about the sign making the idea of infinity any clearer. Infinity is still a matter of controversy in mathematics.

Any attempt to define the word *set* leads to a circular definition. We start with *set* and we end with *set*.

Even though we don't define the word *set* itself, we use sets to define many other mathematical terms. For example, we define the set of *counting numbers* as

$$\{1, 2, 3, 4, 5, 6, \ldots\}.$$

In listing a set, we use braces to enclose the objects making up the set. The objects belonging to the set are called the **elements,** or *members,* of the set. For example, 4 is an element of the set of counting numbers. The three dots show that the elements continue in the same way but that the others are not actually listed; there is no last number in the set of counting numbers.

The set of counting numbers is an example of an **infinite set;** there is no end to the elements in the set. If the number of elements of a set is 0 or a counting number, the set is a **finite set.** Thus, $\{1, 2, 3, 4\}$, read "the set of elements 1, 2, 3, and 4," contains four elements, and so it is a finite set.

For convenience, sets usually are named—generally with capital letters. If we choose F, for example, to name $\{1, 2, 3, 4\}$, then we can refer to the set $\{1, 2, 3, 4\}$ simply as set F.

If we let E be the set of all letters of our alphabet, then

$$E = \{a, b, c, d, e, f, g, h, i, j, k, l, m, n, o, p, q, r, s, t, u, v, w, x, y, z\}.$$

Since set E is finite, all of its elements can be listed. However, because there are 26 letters in the alphabet, it is a lot of work to write every element of set E. To get around this difficulty, we write only some of the letters. Just as for an infinite set, three dots would indicate that some elements are left out. However, enough elements should be included to show a pattern so that we know exactly which elements are left out. Elements from the beginning and the end usually are enough. Using three dots, we can write set E, for example, as

$$E = \{a, b, c, d, \ldots, x, y, z\}$$

or

$$E = \{a, b, c, d, e, \ldots, y, z\}.$$

Example 1   Write out all the elements of each of the following sets: (a) $\{5, 6, 7, \ldots, 14\}$ (b) $\{5, 6, 7, \ldots\}$.

(a) The elements of the set $\{5, 6, 7, \ldots, 14\}$ start with 5, then 6, and then 7. To continue the pattern, we need to include 8, 9, 10, 11, 12, and 13. We stop at 14. This finite set is completely described as

$$\{5, 6, 7, 8, 9, 10, 11, 12, 13, 14\}.$$

(b) The set $\{5, 6, 7, \ldots\}$ starts with 5, 6, 7. The elements that come next are 8, 9, 10, 11, and so on. There is no stopping point here. We would have to keep listing elements of the set forever, which we can't do. This set is infinite.

Example 2   List the elements of the set of counting numbers between 10 and 11.

Solution.   There is no counting number between 10 and 11. The set of counting numbers between 10 and 11 contains no elements.

     A set, such as the one of Example 2, that contains no elements, is called the **empty set,** or *null set.* The symbol $\emptyset$, or sometimes { }, is used for this set. With this symbol, the answer to Example 2 could be written

$$\{\text{all counting numbers between 10 and 11}\} = \emptyset.$$

Empty sets   *Some Zen Buddhists meditate facing a blank wall, symbol of the universal void.*

     Zen koans are questions intended to spark enlightenment. Example: You have heard the sound of two hands clapping; tell me, what is the sound of one hand clapping?

     Sylvia Porter illustrates the empty set in one of her newspaper columns "Your Money's Worth." She advises consumers to check out the legal name of a company. If you sue a company under the name it advertises with or the "monogram" of its logo, you may be "going after a ghost." The popular name, unless it agrees exactly with the legal name, won't stand up in court.

     The number of elements of a set is called the **cardinal number** of the set. The symbol $n(A)$ represents the cardinal number of set A.

Example 3   Find the cardinal number of each of the following sets: (a) $K = \{10, 12, 14, 16\}$ (b) $M = \{0\}$ (c) $\emptyset$.

(a) Set $K = \{10, 12, 14, 16\}$ contains four elements. Thus, the cardinal number of set K is 4, and $n(K) = 4$.

(b) Set M contains only one element, zero, so that $n(M) = 1$.

(c) The empty set, $\emptyset$, contains no elements, and $n(\emptyset) = 0$.

     For a set to be useful, it must be **well-defined.** This means that if we are given a particular set and some particular element, it must be possible to tell whether or not the element belongs to the set. For example, the set E of letters of our alphabet is well-defined. If someone gives us the letter q, we know that q is an element of E. If someone gives us the Greek letter θ, we know that it is not in set E.

Are you there, Herman?

However, given the set C of all fat chickens, and a particular chicken, Herman, it is not possible to say whether

Herman is an element of C

or   Herman is *not* an element of C.

The problem is the word "fat"; how fat is fat? Since we cannot necessarily decide whether or not a given chicken belongs to set C, we say that set C is not well-defined.

We know that q is an element of set E, where E is defined as the set of all the letters of our alphabet. To show this, we use the symbol ϵ to replace the words "is an element of," and write

$$q \in E,$$

which is read "q is an element of set E." The Greek letter θ is not an element of E; write this with ϵ and a cancel mark:

$$\theta \notin E.$$

Example 4 Decide whether each statement is true or false: (a) $3 \in \{5, 7, 9, 10, 13\}$ (b) $0 \in \{-3, -2, 0, 1, 2, 3\}$ (c) $5 \notin \{6, 8, 10\}$.

(a) The statement $3 \in \{5, 7, 9, 10, 13\}$ claims that the number 3 is an element of the set $\{5, 7, 9, 10, 13\}$, which is false.

(b) Since 0 is indeed an element of the set $\{-3, -2, 0, 1, 2, 3\}$, the statement is true.

(c) This statement says that 5 is *not* an element of the set $\{6, 8, 10\}$. This is true.

Two sets are **equal** if they contain exactly the same elements, regardless of order. By this definition,

$$\{a, b, c, d\} = \{a, c, d, b\}$$

since both sets contain exactly the same elements.

Example 5 Are $\{3, 2, 5\}$ and $\{0, 3, 2, 5\}$ equal sets?

Solution. The first set contains three elements: 3, 2, and 5. The other contains four elements: 0, 3, 2, and 5. Thus, the sets do not contain exactly the same elements, and are not equal. This is written

$$\{3, 2, 5\} \neq \{0, 3, 2, 5\}.$$

All the sets thus far in this chapter have been identified by listing their elements. By this method, we would write the set of even counting numbers less than 11 as

$$\{2, 4, 6, 8, 10\}.$$

Sometimes a description of the common property of the elements is more useful than a list of the elements themselves. In that case, we use **set-builder notation.** With set-builder notation, the set of all even counting numbers less than 11, for example, would be written

$$\{x \mid x \text{ is an even counting number less than } 11\}.$$

This is read "the set of all elements x such that x is an even counting number less than 11." Here the *variable* x is a sort of place-holder; the set contains all possible elements x that satisfy the given condition. Set-builder notation often is used to reduce the possibility of ambiguity. For example,

$$\{x \mid x \text{ is an odd counting number between 2 and } 810\}$$

is preferable to $\{3, 5, 7, 9, \ldots, 809\}$.

1.1 EXERCISES

List the elements of the sets in Exercises 1–14.

1. $\{12, 13, 14, \ldots, 20\}$ **4.** $\{3, 9, 27, \ldots, 729\}$

2. $\{8, 9, 10, \ldots, 17\}$ **5.** $\{17, 22, 27, \ldots, 47\}$

3. $\{1, 1/2, 1/4, \ldots, 1/32\}$ **6.** $\{74, 68, 62, \ldots, 38\}$

7. {all counting numbers less than 6}

8. {all counting numbers greater than 7 and less than 15}

9. {all counting numbers not greater than 4}

10. {all counting numbers not less than 3 or greater than 10}

11. $\{x \mid x \text{ is an even counting number larger than 3}\}$

12. $\{x \mid x \text{ is a counting number less than 13}\}$

13. $\{x \mid x \text{ is an odd counting number between 8 and 16}\}$

14. $\{x \mid x \text{ is an odd counting number not greater than 15}\}$

Identify the sets in Exercises 15–22 as *finite* or *infinite*.

15. $\{4, 5, 6, \ldots, 15\}$ **17.** $\{1, 1/2, 1/4, 1/8, \ldots\}$

16. $\{4, 5, 6, \ldots\}$ **18.** $\{0, 1, 2, 3, 4, 5, \ldots, 75\}$

19. $\{x \mid x \text{ is a counting number larger than 5}\}$

20. $\{x \mid x \text{ is a person alive on the earth now}\}$

21. $\{x \mid x \text{ is a fraction between 0 and 1}\}$

22. $\{x \mid x \text{ is a positive even counting number}\}$

Find n(A) for the sets in Exercises 23–28

23. $A = \{0, 1, 2, 3, 4\}$ **25.** $A = \{1, 2, 3, \ldots, 1000\}$

24. $A = \{-1, 0, 1, 2, 3, 4, 5\}$

26. A = {1, 2, 3, 4, . . . , 2000}   **28.** A = {x|x is a letter in our

27. A = {x|x is a state in the U.S.}   alphabet}

Identify the sets in Exercises 29–36 as *well-defined* or *not well-defined*.

29. {x|x is a counting number}   **33.** {x|x is friendly}

30. {x|x is a positive number}   **34.** {x|x is good to its mother}

31. {x|x is a good movie}   *35.** {x|x is a mouse weighing over 300 pounds}

32. {x|x is a lousy mathematics teacher}   *36.** {x|x is a counting number less than 1}

Complete the blanks in Exercises 37–44 with either $\in$ or $\notin$ so that the resulting statement is true.

37. 6 _____ {3, 4, 5, 6}   **41.** 0 _____ {−2, 0, 3, 4}

38. 9 _____ {3, −2, 5, 9, 8}   **42.** 0 _____ {5, 6, 7, 8, 10}

39. −4 _____ {4, 6, 8, 10}   *43.** {3} _____ {2, 3, 4, 5}

40. −12 _____ {3, 5, 12, 14}   *44.** {5} _____ {3, 4, 5, 6, 7}

Write *true* or *false* for each of Exercises 45–56.

45. 3 $\in$ {2, 5, 6, 8}   **49.** 9 $\notin$ {2, 1, 5, 8}

46. 6 $\in$ {−2, 5, 8, 9}   **50.** 3 $\notin$ {7, 6, 5, 4}

47. 1 $\in$ {3, 4, 5, 11, 1}   **51.** {2, 5, 8, 9} = {2, 5, 9, 8}

48. 12 $\in$ {18, 17, 15, 13, 12}   **52.** {3, 0, 9, 6, 2} = {2, 9, 0, 3, 6}

53. {5, 8, 9} = {5, 8, 9, 0}

54. {3, 7, 12, 14} = {3, 7, 12, 14, 0}

55. {x|x is a counting number less than 3} = {1, 2}

56. {x|x is a counting number greater than 10} = {11, 12, 13, . . .}

Write *true* or *false* for each of Exercises 57–64.

Let A = {2, 4, 6, 8, 10, 12}, B = {2, 4, 8, 10}, C = {4, 10, 12}.

57. 4 $\in$ A   **59.** 4 $\notin$ C   **61.** 10 $\notin$ A

58. 8 $\in$ B   **60.** 8 $\notin$ B   **62.** 6 $\notin$ A

63. Every element of C is also an element of A.

64. Every element of C is also an element of B.

In the study of probability, the set of all possible outcomes for an experiment is called the *sample space* for the experiment. (See Chapter 9.) For example, the sample space for tossing an honest coin is the set {h, t}. Find the sample space for each of Exercises 65–68.

*65.** Rolling an honest die.   *67.** Tossing a coin three times.

*66.** Tossing a coin twice.   *68.** Having three children.

{BBB . . . GGG}

_____

*An asterisk identifies a more challenging problem.

Tycho's star   *In 1572 the Danish astronomer Tycho Brahe observed what we call today a supernova—a star flaring up millions of times more brilliant than before. The star he watched was near the constellation Cassiopeia (see opposite page 1), and it flared for 16 months. Tycho was famous for his accurate calculations of star positions. His assistant, mathematician-astronomer Johannes Kepler, was even more famous (see Chapter 8).*

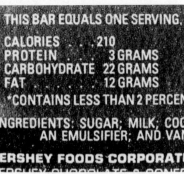

THIS BAR EQUALS ONE SERVING.

CALORIES . . . 210
PROTEIN . . . 3 GRAMS
CARBOHYDRATE 22 GRAMS
FAT . . . . . 12 GRAMS
*CONTAINS LESS THAN 2 PERCEN

NGREDIENTS: SUGAR; MILK; COO
AN EMULSIFIER; AND VAN

ERSHEY FOODS CORPORAT

Sorry about that! *An astronomer proposed in 1975 the existence of a comparatively small galaxy near our own Milky Way. It was nicknamed Snickers because by comparison to Milky Way, it's "only peanuts."*

A Hershey bar of a certain size contains 210 calories. Suppose you eat two of these candy bars, and then decide to exercise and get rid of the calories. A list of possible exercises shows the following information.

Exercise	Abbreviation	Calories per Hour
Sitting around	si	100
Light exercise	li	170
Moderate exercise	mo	300
Severe exercise	se	450
Very severe exercise	ve	600

List all possible sets of exercises (with no exercise repeated) that will burn off the calories from the two candy bars in:

69.   one hour   **70.**   two hours

1.2 SUBSETS

When we are working a particular problem from a particular field of study, we usually have an idea about the type of answer that we might expect. For example, if we are working a problem about money, we would expect our answer to be in dollars and cents. We would not get an answer involving names, animals, or pencils.

In every problem that we work, there is either a stated or an implied *universe of discourse*. The universe of discourse includes all things under discussion at a given time. For example, if we were studying reaction to a proposal that a certain campus begin the sale of beer, the universe of discourse might be all the students at the school, the nearby members of the public, the board of trustees of the school, or perhaps all of these groups of people.

When using sets, the universe of discourse is called the **universal set.** We reserve the letter U for the universal set. The universal set might well change from problem to problem. In one problem the universal set might be the set of all whole numbers, while in another problem the universal set might be the set of all females over 25 years of age who have two or more children.

The idea of a universal set comes from Augustus De Morgan, whose laws of logic will be discussed in the exercises for Section 2.4. The idea was then taken up by the logician John Venn (1834–1923), who invented diagrams such as those used below. Venn compared the universal set to our field of vision. It holds what we look at. We ignore everything we cannot see.

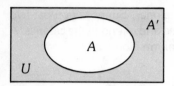

Figure 1.1 The region in color represents set A'

Suppose that for a particular discussion, the universal set is

$$U = \{1, 2, 3, 4, 5\}, \qquad \text{while} \qquad A = \{2, 4, 5\}.$$

The set of all elements of U that are not elements of A is called the **complement** of A. Here the complement of A contains the elements 1 and 3. The symbol A' (read "A prime") represents the complement of A. With this symbol, we have

$$A' = \{1, 3\}.$$

In Figure 1.1, the region bounded by the rectangle represents the universal set U, while that bounded by the circle represents set A. The region inside U and outside the circle represents A'. Set diagrams like Figure 1.1 are called **Venn diagrams.**

Example 1 Let $U = \{a, b, c, d, e, f, g, h\}$,
$\qquad\qquad\quad M = \{a, b, e, f\}$.
$\qquad\qquad\quad N = \{b, d, e, g, h\}$.

Find each of the following sets: (a) M' (b) N'.

(a) Set M' contains all the elements of set U that are not in set M. Since set M contains the elements a, b, e, and f, set M' will contain c, d, g, and h, or

$$M' = \{c, d, g, h\}.$$

(b) Set N' contains all the elements of U that are not in set N. Hence, $N' = \{a, c, f\}$.

If set $U = \{1, 2, 3, 4, 5\}$, what is the set U' ? To find U', we need all the elements of U that do not belong to U. There can't be any such elements, so there can be no elements in set U'. Thus,

$$U' = \varnothing$$

for any universal set U. (In the same way, $\varnothing' = U$.)

Suppose that in a particular discussion, the universal set is $U = \{1, 2, 3, 4, 5\}$, while one of the sets we are discussing is $A = \{1, 2, 3\}$. Every element of set A is also an element of set U. Because of this, set A is called a **subset** of set U, written in set notation as

$$A \subset U.$$

A Venn diagram showing that set M is a subset of set N is in Figure 1.2

Example 2 Write $\subset$ or $\not\subset$ (not a subset) in each blank.

(a) $\{3, 4, 5, 6\}$ _____ $\{3, 4, 5, 6, 8\}$. Every element of $\{3, 4, 5, 6\}$ is also an element of $\{3, 4, 5, 6, 8\}$. Therefore, the first set is a subset of the second, so that $\subset$ goes in the blank.

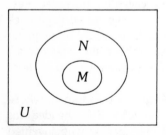

Figure 1.2 Set M is a subset of set N

(b) $\{1, 2, 3\}$ ____ $\{2, 4, 6, 8\}$. The element 1 belongs to $\{1, 2, 3\}$ but not to $\{2, 4, 6, 8\}$. Thus, $\not\subset$ goes in the blank.

(c) $\{5, 6, 7, 8\}$ ____ $\{5, 6, 7, 8\}$. Every element of $\{5, 6, 7, 8\}$ is also an element of $\{5, 6, 7, 8\}$. Place $\subset$ in the blank.

As Example 2(c) illustrates, every set is a subset of itself; for example,

$$B \subset B \quad \text{for any set B.}$$

Set A is a subset of set B if *every* element of set A is also an element of set B. We can reword this definition, and say that set A is a subset of set B if there are no elements of A that are not also elements of B. From the second definition, we see that the empty set is a subset of any set, or

$$\varnothing \subset B, \quad \text{for any set B.}$$

This is true since it is not possible to find any elements of $\varnothing$ that are not also in B. (There are no elements in $\varnothing$.)

By now, we know that every set (except $\varnothing$) has at least two subsets, $\varnothing$ and the set itself. Let us see if we can find a rule to tell *how many* subsets a given set has. We begin by looking at an example.

Example 3 Find all possible subsets of each set: (a) $\{7, 8\}$ (b) $\{a, b, c\}$.

(a) By trial and error, we find 4 subsets for $\{7, 8\}$:

$$\varnothing, \quad \{7\}, \quad \{8\}, \quad \{7, 8\}.$$

(b) Here, trial and error leads to 8 subsets for $\{a, b, c\}$:

$$\varnothing, \{a\}, \{b\}, \{c\}, \{a, b\}, \{a, c\}, \{b, c\}, \{a, b, c\}.$$

In Example 3 we found all subsets of $\{7, 8\}$ and all subsets of $\{a, b, c\}$ by trial and error. An alternative method involves drawing a **tree diagram.** A tree diagram is a systematic way of listing all the subsets of a given set. Figure 1.3 (a) and (b) shows tree diagrams for $\{7, 8\}$ and $\{a, b, c\}$.

Figure 1.3 (a)
Tree diagram listing subsets of the set $\{7, 8\}$

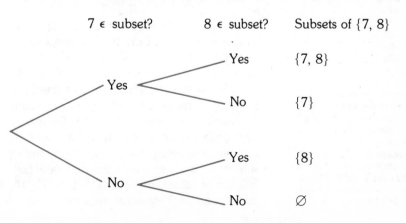

| 7 ϵ subset? | 8 ϵ subset? | Subsets of $\{7, 8\}$ |

Figure 1.3(b)
Subsets of {a, b, c}

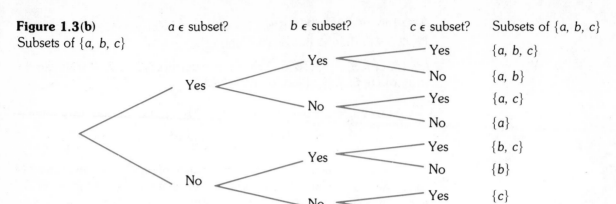

For a set containing a very large number of elements, it would be virtually impossible to find the number of subsets by either trial and error or by drawing a tree diagram. Perhaps we can find a formula.

We begin with the set containing the least number of elements possible— the empty set. This set has only itself as a subset. Next, a set with one element has only itself and $\varnothing$ as subsets. Tree diagrams show that a set of 2 elements has 4 subsets, and a set of 3 elements has 8 subsets. We can place this information in a chart:

Number of elements	0	1	2	3
Number of subsets	1	2	4	8

We notice that as the number of elements of the set increases by 1, the number of subsets doubles. This suggests that the number of subsets in each case might be a power of 2. Looking at the second row of the chart we see that each number is indeed a power of 2. We can add this information to the chart:

Number of elements	0	1	2	3
Number of subsets	$1 = 2^0$	$2 = 2^1$	$4 = 2^2$	$8 = 2^3$

This chart shows that the number of elements in each case is the same as the exponent on the 2. This suggests the following generalization: The number of subsets of a set with n elements is 2^n. Of course, we have shown this to be true *only* for sets with 0, 1, 2, or 3 elements.

The method by which we discovered the formula 2^n for $n = 0, 1, 2,$ or 3 is the same method by which many mathematicians have made important discoveries. However, once a mathematician has made such a discovery from observations, it must then be proved that it is true in the general case.

Mathematicians *have* proved that *any* set containing n elements has 2^n subsets, so we will accept this generalization as true. Now we can find the number of subsets of, say, the set {5, 12, 14, 17, 19, 20, 27} by first noting that it contains 7 elements; thus the set has $2^7 = 128$ subsets.

Example 4 Find the number of subsets of each set: (a) {3, 4, 5, 6, 7} (b) {−1, 2, 3, 4, 5, 9, 12, 14}.

(a) This set has 5 elements. Thus it has $2^5 = 2 \times 2 \times 2 \times 2 \times 2 = 32$ subsets.

(b) The set {−1, 2, 3, 4, 5, 9, 12, 14} has 8 elements, and so it has $2^8 = 256$ subsets.

Using more extensive mathematical tables, we can find that a set with 100 elements has 2^{100} subsets, where

$$2^{100} = 1,267,650,600,228,229,401,496,703,205,376.$$

1.2 EXERCISES

Insert $\subset$ or $\not\subset$ in each blank so that the resulting statement is true.

1. {2, 4, 6} _____ {3, 2, 5, 4, 6}
2. {1, 5} _____ {0, −1, 2, 3, 1, 5}
3. {0, 1, 2} _____ {1, 2, 3, 4, 5}
4. {5, 6, 7, 8} _____ {1, 2, 3, 4, 5, 6, 7}
5. ∅ _____ {1, 4, 6, 8}
6. ∅ _____ ∅
7. {1, 2} _____ {x|x is a counting number}
8. {3, 4, 5, 6} _____ {x|x is an odd counting number}

For exercises 9–30, tell whether each statement is *true* or *false*. Let

$$A = \{2, 4, 6, 8, 10, 12\} \qquad D = \{2, 10\}$$
$$B = \{2, 4, 8, 10\} \qquad U = \{2, 4, 6, 8, 10, 12, 14\}.$$
$$C = \{4, 10, 12\}$$

9. $A \subset U$
10. $C \subset U$
11. $D \subset B$
12. $D \subset A$
13. $A \subset B$

14. $B \subset C$
15. $\emptyset \subset A$
16. $\emptyset \subset D$
17. $\emptyset \subset \emptyset$
18. $D \subset C$

19. $\{4, 8, 10\} \subset B$
20. $\{0, 2\} \subset D$
21. $D \not\subset B$
22. $A \not\subset C$

23. There are exactly 32 subsets of A.
24. There are exactly 16 subsets of B.
25. There are exactly 6 subsets of C.
26. There are exactly 4 subsets of D.
27. There is exactly one subset of ∅.
28. The drawing correctly represents the relationship among sets A, C, and U.

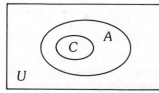

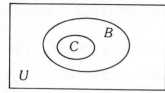

29. The drawing here correctly represents the relationship between sets B, C, and U.

*30. The symbol {∅} represents the empty set.

Find the number of subsets of each set in Exercises 31–38.

31. {4, 5, 6}
32. {3, 7, 9, 10}
33. {5, 8, 10, 15, 17}
34. {6, 9, 1, 4, 3, 2}
35. ∅
36. {2, 5, 9, 15, 17, 18}
37. {x|x is an even number between 7 and 17}
38. {x|x is an odd number between 8 and 14}

Let U = {1, 2, 3, 4, 5, 6, 7, 8, 9, 10} and find the complement of each set in Exercises 39–44.

39. {1, 4, 6, 8}
40. {2, 5, 7, 9, 10}
41. {1, 3, 4, 5, 6, 7, 8, 9, 10}
42. {3, 5, 7, 9}
43. ∅
44. U

Find all possible subsets of:

*45. ∅   ***46.** {∅}   ***47.** {0}   ***48.** {{0}}

Term insurance

Low cost
Pays off at death
No retirement benefits

There are two kinds of life insurance in common use today, term insurance and whole life. A few of the characteristics of each of these types of insurance are listed in the margin.

49. Find a universal set U that contains all the characteristics of both term insurance and whole life.

Whole life

High cost
Pays off at death
Retirement benefits

Let *T* represent the set of characteristics of term insurance, and let *W* represent the set of characteristics of whole life. Use the universal set from Exercise 49 to find each of the following:

50. T′   **51.** W′

Find the set of all elements common to both of the given sets, where T and W are defined as above.

52. T and W   **53.** T′ and W   **54.** T and W′

Suppose a family consists of a mother and father, Donna and Lance, and three children, Anna, Kerry, and Shamus. List all the possible ways that the given number of family members can be together in the living room.

55. 5   **56.** 4   **57.** 3   **58.** 2   **59.** 1   **60.** 0

*61. Find the total number of ways that the family members can be together in the living room. (*Hint:* Add your answers to Exercises 55–60.)

*62. How does your answer in Exercise 61 compare with the number of subsets of a set of 5 elements? How can you interpret the answer in Exercise 61 in terms of subsets?

1.3 INTERSECTION AND UNION OF SETS

After comparing the campaign promises of two candidates for sheriff, a voter made the following lists of promises made by the two people running for the office. Each promise is given a code letter.

Honest Tom Weintraub	Tough Lynn Lawson
Spend less money, *m*	Spend less money, *m*
Emphasize traffic law enforcement, *t*	Crack down on crooked politicians, *p*
Increase service to suburban areas, *s*	Increase service to the city, *c*

Tangram figures. See page 19.

The only promise common to both lists is promise *m*, to spend less money. Suppose we take each list of promises to be a set. The promises of candidate Weintraub give the set $\{m, t, s\}$, while the promises of Lawson give $\{m, p, c\}$. The only element common to both sets is *m*; this element is the **intersection** of the two sets $\{m, t, s\}$ and $\{m, p, c\}$. In symbols,

$$\{m, t, s\} \cap \{m, p, c\} = \{m\}.$$

The cap-shaped symbol $\cap$ represents intersection. Note that the intersection of the original two sets is itself a set. A Venn diagram showing the intersection is given in Figure 1.4.

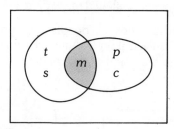 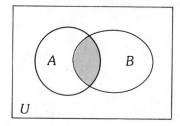

Figure 1.4   The intersection is set $\{m\}$ $A \cap B$ is in color

In general, the intersection of sets A and B is defined by

$$A \cap B = \{x \mid x \in A \text{ and } x \in B\}.$$

Example 1   Find the intersection of the given sets in (a) – (d).

(a) $\{3, 4, 5, 6, 7\}$ and $\{4, 6, 8, 10\}$. The elements common to both sets are 4 and 6. Thus,

$$\{3, 4, 5, 6, 7\} \cap \{4, 6, 8, 10\} = \{4, 6\}.$$

(b) $\{5, 7, 9, 11, 18, 19\}$ and $\{7, 11, 18, 19, 20\}$. The common elements are 7, 11, 18, and 19, so

$$\{5, 7, 9, 11, 18, 19\} \cap \{7, 11, 18, 19, 20\} = \{7, 11, 18, 19\}.$$

(c) {5, 9, 11} and ∅. There are no elements in ∅, so that there can be no elements belonging to both {5, 9, 11} and ∅. Because of this,

$${5, 9, 11} \cap \varnothing = \varnothing.$$

(d) {9, 14, 25, 30} and {10, 17, 19, 38, 52}. These two sets have no elements in common, so that

$${9, 14, 25, 30} \cap {10, 17, 19, 38, 52} = \varnothing.$$

In Example 1(d), we saw two sets that have no elements in common. Sets with no elements in common are called **disjoint sets.** A set of men and a set of women would be disjoint sets. In mathematical language, sets A and B are disjoint if A ∩ B = ∅. Two disjoint sets are shown in the Venn diagram of Figure 1.5.

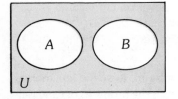

Figure 1.5 Sets A and B are disjoint

At the beginning of this section, we saw lists of campaign promises of two candidates running for sheriff. We found the intersection of those lists above. Suppose we are taking a political science class and have to write a paper on the types of promises made by candidates for public office. We would need to study *all* the promises made by *either* candidate. This would be the set

$${m, t, s, p, c}.$$

This set is the **union** of the sets of promises made by the two candidates. In symbols,

$${m, t, s} \cup {m, p, c} = {m, t, s, p, c}.$$

The cup-shaped symbol ∪ denotes set union—but don't confuse this symbol with the universal set U. Again, the union of two sets is a set. A Venn diagram showing the union of the candidates' sets is in Figure 1.6.

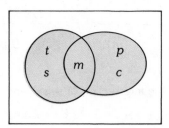 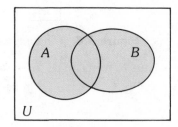

Figure 1.6 The union is set {m, t, s, p. c}   A ∪ B is in color

In general, the union of sets A and B is written A ∪ B, and is defined by

$$A \cup B = {x | x \in A \text{ or } x \in B}.$$

Form the union of sets A and B by taking all those elements of set A and then adding the elements of set B that are not already included.

Example 2   Find the union of the given sets.

(a) {2, 4, 6} and {4, 6, 8, 10, 12}. Start by using all the elements from the first set, 2, 4, and 6. Then use all the elements from the second set that are not in the first set, 8, 10, and 12 here. The union is made up of all these elements, or

$$\{2, 4, 6\} \cup \{4, 6, 8, 10, 12\} = \{2, 4, 6, 8, 10, 12\}.$$

(b) {a, b, d, f, g, h} and {c, f, g, h, k}. Their union is

$$\{a, b, d, f, g, h\} \cup \{c, f, g, h, k\} = \{a, b, c, d, f, g, h, k\}.$$

(c) {3, 4, 5} and ∅. There are no elements in ∅. Thus the union of {3, 4, 5} and ∅ contains only the elements 3, 4, and 5, or

$$\{3, 4, 5\} \cup \varnothing = \{3, 4, 5\}.$$

Recall from the previous section that A′ represents the complement of set A. To form A′ we take all the elements of the universal set U that are not in A. This idea is shown in the next example.

Example 3   Let A = {1, 2, 3, 4}   C = {1, 3, 6, 9}
               B = {2, 4, 6}   U = {1, 2, 3, 4, 5, 6, 9}

Find each set: (a) $A' \cap B$ (b) $B' \cup C'$ (c) $A \cap (B \cup C')$.

(a) Here we need the set of all elements belonging to both set A′ and set B. We first identify the elements of set A′. To find A′, we list all the elements of U that are not in set A:

$$A' = \{5, 6, 9\}.$$

Now we can find A′ ∩ B:

$$A' \cap B = \{5, 6, 9\} \cap \{2, 4, 6\} = \{6\}.$$

(b) $B' \cup C' = \{1, 3, 5, 9\} \cup \{2, 4, 5\} = \{1, 2, 3, 4, 5, 9\}.$

(c) To find the elements in a more complicated set such as A ∩ (B ∪ C′), work first inside the parentheses:

$$B \cup C' = \{2, 4, 6\} \cup \{2, 4, 5\} = \{2, 4, 5, 6\}.$$

Now finish the problem:

$$A \cap (B \cup C') = A \cap \{2, 4, 5, 6\}$$
$$= \{1, 2, 3, 4\} \cap \{2, 4, 5, 6\}$$
$$= \{2, 4\}.$$

Intersection and union of sets are examples of *set operations*. In general, an **operation** is a rule or procedure by which one or more objects are used to obtain another object. The objects involved in an operation are usually sets or numbers. The most common operations on numbers are addition, subtraction, multiplication, and division. For example, if we start with the numbers 5 and 7, the addition operation produces the number $5 + 7 = 12$. With the same two numbers, 5 and 7, the multiplication operation would produce $5 \times 7 = 35$.

While intersection and union are the most common operations for sets, there are others. Forming the complement of a set is a set operation discussed in Section 1.2. Finding the difference of two sets is discussed in Exercises 67–72 below.

1.3 EXERCISES

Write *true* or *false* for Exercises 1–12.

1. $\{5, 7, 9, 10\} \cap \{7, 9, 11, 15\} = \{7, 9\}$
2. $\{8, 11, 15\} \cap \{8, 11, 19, 20\} = \{8, 11\}$
3. $\{2, 1, 7\} \cup \{1, 5, 9\} = \{1\}$
4. $\{6, 12, 14, 16\} \cup \{6, 14, 19\} = \{6, 14\}$
5. $\{3, 2, 5, 9\} \cup \{2, 7, 8, 10\} = \{2, 3, 5, 7, 8, 9, 10\}$
6. $\{8, 9, 6\} \cup \{9, 8, 6\} = \{8, 9\}$
7. $\{3, 5, 9, 10\} \cap \varnothing = \{3, 5, 9, 10\}$
8. $\{3, 5, 9, 10\} \cup \varnothing = \{3, 5, 9, 10\}$
9. $\{1, 2, 4\} \cup \{1, 2, 4\} = \{1, 2, 4\}$
10. $\{1, 2, 4\} \cap \{1, 2, 4\} = \varnothing$
11. $\varnothing \cup \varnothing = \varnothing$
12. $\varnothing \cap \varnothing = \varnothing$

In Exercises 13–26, list the elements in each set. Let

$$X = \{2, 3, 4, 5\} \qquad Z = \{2, 4, 5, 7, 9\}$$
$$Y = \{3, 5, 7, 9\} \qquad U = \{2, 3, 4, 5, 7, 9\}.$$

13. $X \cap Y$
14. $X \cup Y$
15. $Y \cup Z$
16. $Y \cap Z$
17. $X \cup U$
18. $Y \cap U$
19. X'
20. Y'
21. $X' \cap Y'$
22. $X' \cap Z$
23. $Z' \cap \varnothing$
24. $Y' \cup \varnothing$
25. $X \cup (Y \cap Z)$
26. $Y \cap (X \cup Z)$

Write *true* or *false* for each of Exercises 27–30. Assume that A and B represent *any* two sets.

27. $A \subset (A \cup B)$
28. $A \subset (A \cap B)$
29. $(A \cap B) \subset A$
30. $(A \cup B) \subset A$

The lists in the margin show some symptoms of an overactive thyroid and an underactive thyroid.

31. Find the smallest possible universal set U that includes all of the symptoms listed.

Let N be the set of symptoms for an underactive thyroid, and O be the set of symptoms for an overactive thyroid. Find each set.

32. O' 33. N' 34. $N \cap O$ 35. $N \cup O$ 36. $N \cap O'$

Describe in words each set in Exercises 37–44. Let
A = {all students studying math}   D = {all students on the G.I. Bill}
B = {all students studying history}   U = {all students}.
C = {all students over 25}

Underactive thyroid

Sleepiness, *s*
Dry hands, *d*
Intolerance of cold, *c*
Goiter, *g*

Overactive thyroid

Insomnia, *i*
Moist hands, *m*
Intolerance of heat, *h*
Goiter, *g*

---

37. $A \cap B$ **40.** $A \cup C$ **43.** $A' \cap (B \cap C)$
38. $B \cap C$ **41.** $A' \cap C$ **44.** $(A \cup B) \cap C'$
39. $A \cap D$ **42.** $A' \cap B'$

Exercises 45–56 use sets X, Y, Z, and U as defined for Exercises 13–26. Work out the set operation on the left of the equals sign, and then work out the set operation on the right of the equals sign. If the two resulting sets are the same, write *true;* if the two sets are different, write *false.*

Example: $X \cap Y = Y \cap X$. True or false?
   $X \cap Y$, on the left of the equals sign, is
   $X \cap Y = \{2, 3, 4, 5\} \cap \{3, 5, 7, 9\} = \{3, 5\}$;
   $Y \cap X$, on the right, is
   $Y \cap X = \{3, 5, 7, 9\} \cap \{2, 3, 4, 5\} = \{3, 5\}$.

Since the sets on the right and left sides of the equals sign are the same, write *true.* (Remember from the definition of equal sets that the order in which the elements of a set are listed makes no difference.)

45. $X \cup Y = Y \cup X$ **51.** $X \cup (Y \cup Z) = (X \cup Y) \cup Z$
46. $(X')' = X$ **52.** $X \cap (Y \cap Z) = (X \cap Y) \cap Z$
47. $X \cup \emptyset = \emptyset$ **53.** $X' \cap Y' = (X \cap Y)'$
48. $Y \cup \emptyset = Y$ **54.** $X' \cup Y' = (X \cup Y)'$
49. $X \cap \emptyset = \emptyset$ **55.** $X' \cap Y' = (X \cup Y)'$
50. $X \cup \emptyset = X$ **56.** $X' \cup Y' = (X \cap Y)'$

To prepare for exercises 57–62, draw a square. Select any two points inside it. Label one point A and the other point B. Draw line segment AB. Is all of AB inside the square? Would it be possible to choose points A and B (inside the square) so that part of segment AB would be outside the square? Try this same thing with a circle, a triangle, and the two figures here.

A figure is called *convex* if for any two points A and B inside the figure, the line segment AB is always completely inside the figure. A square, circle, triangle, and the top figure in the margin are convex; the bottom figure in the margin is not. A convex figure, together with the points of its interior is often called a **convex set.**

Write *convex set* or *not convex set* for each figure in Exercises 57–62.

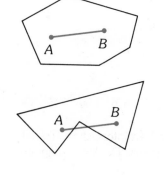

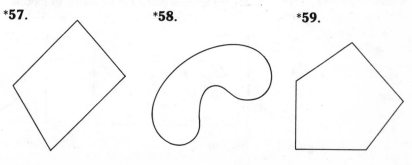

*57.   ***58.**   ***59.**

*60.   ***61.**   ***62.**

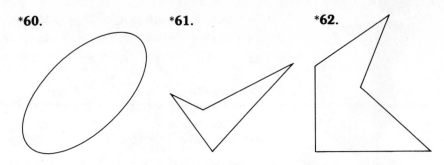

The figure in the margin shows convex set A and convex set B.

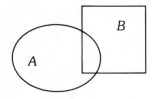

*63.   Is A ∩ B convex?

*64.   Do you think the intersection of two convex sets is always convex?

*65.   Is A ∪ B convex?

*66.   Do you think the union of two convex sets is ever convex?

The **difference of two sets** A and B is defined as

$$A - B = \{x \mid x \in A \text{ and } x \notin B\}.$$

For Exercises 67–72, let M = {3, 7, 9, 11}, N = {7, 11, 15, 17}, and U = {3, 7, 9, 11, 15, 17, 18}. Perform the operations:

*67.   M − N   ***69.**   N′ − M   ***71.**   M ∩ (M − N)

*68.   N − M   ***70.**   M′ − N   ***72.**   N ∪ (N − M)

Suppose A is any set. Describe the conditions under which each statement would be true.

*73.   A = A − B   ***74.**   A = B − A

1.4 VENN DIAGRAMS FOR MORE THAN TWO SETS

In the previous section we looked at set union and set intersection. To help explain these two operations, we used Venn diagrams. The Venn diagrams for these operations are repeated in Figure 1.7.

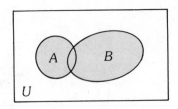

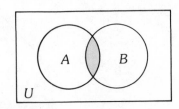

Figure 1.7   Set union, A ∪ B   Set intersection, A ∩ B

The rectangular region in a Venn diagram represents the universal set, *U*. If we include only the single set A inside the universal set, as shown in Figure 1.8(a), we divide the total region of U into two parts. One part represents set *A*, while the other part represents the elements not in A. The elements not in A make up the complement of A, set A'.

If we include two sets A and B inside U, we get the Venn diagram of Figure 1.8(b). Notice that two sets divide the universal set into four regions. Region 1 includes those elements outside of both set A and set B. Region 2 includes the elements belonging to A and not to B. Region 3 includes those elements belonging to both A and B. How would you describe the elements of region 4?

Tangram *This puzzle-game comes from China, where it has been a popular amusement for centuries. The above figure and those on page 13 are tangrams. Each is composed of the same seven tans (the pieces making up the square below). Note that each tan is convex, but there are only 13 convex tangrams possible. All others, like the figure above, are concave. To "solve" tangrams you must work out how the tans fit together (see page 25).*

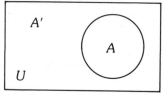

Figure 1.8(a)
One set, two regions

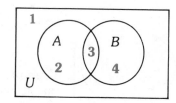

Figure 1.8(b)
Two sets, four regions

Example 1 Draw a Venn diagram similar to Figure 1.8(b) and shade the region or regions representing the sets: (a) A' ∩ B (b) A' ∪ B'.

(a) Refer to the labeling in Figure 1.8(b). Set A' contains all the elements outside of set A, in other words, the elements in regions 1 and 4. Set B is made up of the elements in regions 3 and 4. The intersection of sets A' and B, the set A' ∩ B, is made up of the elements in the region common to 1 and 4 and 3 and 4, that is, region 4. Thus, A' ∩ B is represented by region 4, which is shaded in Figure 1.9.

(b) Again set A' is represented by regions 1 and 4, while B' is made up of regions 1 and 2. To form the union of A' and B', the set A' ∪ B', we need the elements belonging either to regions 1 and 4 or to regions 1 and 2. This union is composed of regions 1, 2, and 4, which are shaded in Figure 1.10.

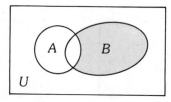

Figure 1.9
A' ∩ B is in color

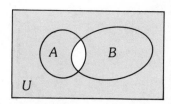

Figure 1.10
A' ∪ B' is in color

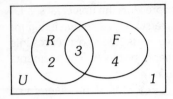

Figure 1.11

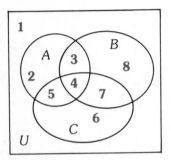

Figure 1.12
Three sets, eight regions

Example 2   Suppose set *R* contains all the red cards in an ordinary deck of cards, while set *F* contains all the face cards. (A deck of cards is pictured in Chapter 9). Identify which cards go in each region of Figure 1.11.

Solution.   Region 1 contains the cards that are not red and not face cards — the black number cards and black aces. Region 2 contains the cards that are red but not face cards. Thus, region 2 contains the red number cards, and the red aces. Region 3 contains the cards that are both red and face cards, or the red face cards. Region 4 contains the face cards that are not red, or the black face cards.

We can draw Venn diagrams with three sets inside U. Three sets divide the universal set into eight regions, which are numbered in Figure 1.12.

Example 3   Shade the set $(A' \cap B') \cap C$ in a Venn diagram similar to the one in Figure 1.12.

Solution.   Work first inside the parentheses. As shown in Figure 1.13, set A' is made up of the regions outside set A, or regions 1, 6, 7, and 8. Set B' is made up of regions 1, 2, 5, and 6. The intersection of these sets is given by the overlap of regions 1, 6, 7, 8 and 1, 2, 5, 6, or regions 1 and 6. We must now find the intersection of regions 1 and 6 with set C. From Figure 1.13, set C is made up of regions 4, 5, 6, and 7. The overlap of regions 1, 6 and 4, 5, 6, 7 is region 6 only. Region 6 is shaded in Figure 1.13.

Example 4   Shade $A' \cup (B \cap C')$ in a Venn diagram.

Solution.   Here we first find $B \cap C'$. Set B contains regions 3, 4, 7, and 8, while set C' contains regions 1, 2, 3, and 8. The overlap of these regions, the set $B \cap C'$, is made up of regions 3 and 8. Set A' is made up of regions 1, 6, 7, and 8. The union of regions 3, 8 and 1, 6, 7, 8 is regions 1, 3, 6, 7, 8, shaded in Figure 1.14.

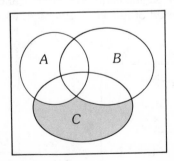

Figure 1.13
$(A' \cap B') \cap C$ is in color

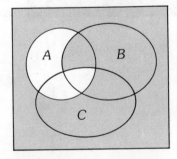

Figure 1.14
$A' \cup (B \cap C')$ is in color

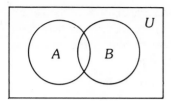

1.4 EXERCISES

1. Suppose U = {a, b, c, d, e, f, g}
$$A = \{b, d, f, g\}$$
$$B = \{a, b, d, e, g\}.$$

Place the elements of these sets in the proper location on the Venn diagram.

2. Let U = {5, 6, 7, 8, 9, 10, 11, 12, 13}
$$M = \{5, 8, 10, 11\}$$
$$N = \{5, 6, 7, 9, 10\}.$$

Place the elements of these sets in the proper location on the Venn diagram.

Use a Venn diagram similar to the one at the right to shade each of the following sets.

3. B ∩ A' **9.** B' ∩ B
4. A ∪ B **10.** A ∩ B'
5. A' ∪ B **11.** B' ∪ (A' ∩ B')
6. A' ∩ B' **12.** (A ∩ B) ∪ B
7. B' ∪ A **13.** U'
8. A' ∪ A **14.** ∅'

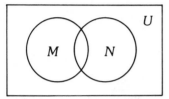

15. Let U = {m, n, o, p, q, r, s, t, u, v, w}
$$A = \{m, n, p, q, r, t\}$$
$$B = \{m, o, p, q, s, u\}$$
$$C = \{m, o, p, r, s, t, u, v\}.$$

Place the elements of these sets in the proper location on a Venn diagram similar to the one in the margin.

16. Let U = {1, 2, 3, 4, 5, 6, 7, 8, 9}
$$A = \{1, 3, 5, 7\}$$
$$B = \{1, 3, 4, 6, 8\}$$
$$C = \{1, 4, 5, 6, 7, 9\}.$$

Place the elements of these sets in the proper location on a Venn diagram.

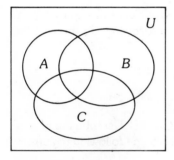

Exercises 15–26

Use a Venn diagram to shade each of the following sets.

17. (A ∩ B) ∩ C **22.** (A ∪ B) ∪ C
18. (A ∩ C') ∪ B **23.** (A ∩ B') ∪ C
19. (A ∩ B) ∪ C' **24.** (A ∩ C') ∩ B
20. (A' ∩ B) ∩ C **25.** (A ∩ B') ∩ C'
21. (A' ∩ B') ∩ C **26.** (A' ∩ B') ∪ C

For each of Exercises 27–36, draw two Venn diagrams: one for the set on the left of the equals sign, and one for the set on the right. If the two Venn diagrams are the same, mark the original statement *true*. If the two Venn diagrams are different, mark the original statement *false*.

27. $A \cap B = B \cap A$ **30.** $A' \cup B' = (A \cup B)'$

28. $A \cup B = B \cup A$ **31.** $A \cap A' = \varnothing$

29. $A' \cap B' = (A \cap B)'$ **32.** $A \cap A' = U$

33. $(A \cap B) \cap C = A \cap (B \cap C)$

34. $(A \cup B) \cup C = A \cup (B \cup C)$

35. $A \cap (B \cup C) = (A \cap B) \cup (A \cap C)$

36. $A \cup (B \cap C) = (A \cup B) \cap (A \cup C)$

For each of Exercises 37 and 38, draw two Venn diagrams: one for the set on the left of the $\subset$ symbol, and one for the set on the right. If the given subset relationship is true, write *true;* otherwise *false.*

37. $(A \cap B) \subset A$ **38.** $(A \cup B) \subset A$

For the following, use two diagrams that show the given subset relationship. In one of them, shade the set on the left of the equals sign; in the other, shade the set on the right of the equals sign. Decide if the stated equality is true.

39. If $A \subset B$, then $A \cup B = A$.   **40.** If $A \subset B$, then $A \cap B = B$.

The electrocardiograph (EKG) is used to monitor electrical impulses from the heart. When a physician reads an EKG, the following are three of the items that are checked:

    N: normal heart rate (60–99 beats per minute)
    R: regular heart rhythm
    P: p wave precedes r wave

Draw a Venn diagram showing regions for *N, R,* and *P.* Describe the type of EKG that would be expected for patients in each of the following regions.

Example: $N \cap R \cap P'$. Which type of EKG?

Solution. The patient has a normal heart rate, a regular rhythm, and the r wave precedes the p wave.

41. $N \cap R \cap P$ **43.** $N' \cap R \cap P$ **45.** $(N \cap R) \cap P$

42. $N \cap R' \cap P$ **44.** $N' \cap R' \cap P$ **46.** $(N \cup P) \cap R$

*47. In this section, we have looked at Venn diagrams containing one, two, or three sets. Use this information to complete the following table. (*Hint:* Predict for 4 sets from the pattern of 1, 2, and 3 sets.)

Number of sets	1	2	3	4
Number of regions dividing U	2	___	___	___

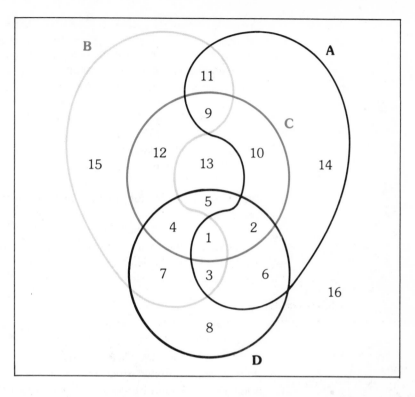

The figure shows U divided into 16 regions by 4 sets, A, B, C, and D. Find the numbers of the regions belonging to each set in Exercises 48–51.

Dark constellations *Detail from Louise Nevelson's Night Rhythm I, 1968. These are the two bottom "subsets" of an 8-box assemblage (opposite) about two feet square. Nevelson often constructs from odd pieces of wood she paints a uniform color. Her sculptures range from small assemblages to wall-size "total environments."*

*48. A ∩ B ∩ C ∩ D   ***50.** (A ∩ B) ∪ (C ∩ D)

*49. A ∪ B ∪ C ∪ D   ***51.** (A′ ∩ B′) ∩ (C ∪ D)

*52. If we placed 5 most generally related sets inside U, how many regions should result?

*53. Make a general formula. If we placed *n* most generally related sets inside U, we would get how many regions?

1.5 SURVEYS: AN APPLICATION OF SETS

We can use Venn diagrams to solve problems in surveying groups of people. For example, suppose we question a group of college students about their favorite singers, with the following results:

22	like Johnny Cash	17	like Johnny and Mick
25	like Helen Reddy	20	like Helen and Mick
39	like Mick Jagger	6	like all three
9	like Helen and Johnny	4	like none of these performers

Let us now use this information to find the number of people who were questioned.

At first glance, it might seem reasonable to add up all the numbers in the list above, but this won't work since there is some overlapping. For example, the 17 students who like Johnny Cash and Mick Jagger are also counted as part of the 22 who like Johnny Cash and the 39 who like Mick Jagger. Figure 1.15 shows a Venn diagram for the three performers in our survey.

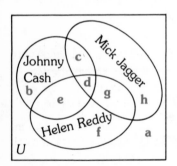

Figure 1.15

The universal set contains all the students who were surveyed. From the survey results we know that 22 students like Johnny Cash. The set of students who like Johnny Cash includes regions b, c, d, and e. Thus, these 22 students are distributed among the four regions. We must now decide how to distribute the students among the regions so that we avoid overlapping.

The smallest total listed is the 4 students who do not like any of these performers. Since these 4 students are outside any of the circles for Johnny Cash, Mick Jagger, or Helen Reddy, they must go in region a. The number 4 has been placed in region a of Figure 1.16.

There are 6 students who like all three of these performers. Since region d is the intersection of all three circles, place the number 6 in region d.

The 20 students who like Helen Reddy and Mick Jagger are represented by regions d and g. Thus, a total of 20 students goes in regions d and g. However, region d represents 6 students already, so that $20 - 6$, or 14, students must be placed in region g. Region g represents the students who like Helen Reddy and Mick Jagger, but not Johnny Cash. By the same method, region c represents 11 students, while region e represents 3.

A total of 22 students like Johnny Cash. The students who like Johnny Cash go in regions b, c, d, and e. We have seen that regions c, d, and e account for a total of $11 + 6 + 3 = 20$ students. Thus, region b must represent $22 - 20 = 2$ students. In the same way, region f represents 2 students, while region h represents 8.

Using all of this information, we can complete Figure 1.16. No students are counted twice on this diagram, so that we can find the total number of students surveyed by adding all the numbers from the regions:

$$4 + 2 + 14 + 6 + 3 + 2 + 11 + 8 = 50.$$

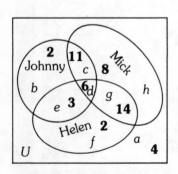

Figure 1.16

Fifty students were questioned.

More on tangrams *Here is the "solution" to the figure on page 19. Mathematicians have described various properties of tangrams. "Proper" tangrams have tans joined only by sides, not by corners. (Is the above figure a proper tangram?) The number of proper tangrams is infinite. "Snug" tangrams are a subset of proper— "snug" means that the sides of the small right triangles match. Are convex tangrams necessarily snug? Are other types of tangrams ever snug?*

Example 1 Using the data from the survey on student preferences for performers, answer the following questions.

(a) How many students like Helen Reddy only?

Solution. A student who likes Helen Reddy *only* does not like Johnny Cash and does not like Mick Jagger. These students will be inside the Helen Reddy region, and outside the regions for Mick Jagger and Johnny Cash. The region we need is region f (see Figure 1.15). Region f contains 2 students; these two students like Helen Reddy but neither of the other two performers.

(b) How many students like exactly two performers?

Solution. The students in regions c, e, and g like only two performers. The total number of students who like only two performers is thus

$$11 + 3 + 14 = 28.$$

Example 2 Bill Bradkin is a section chief for an electric utility company. The employees in his section cut down tall trees, climb poles, and splice wire. Bradkin reported the following information to the management of the utility:

Out of the 100 employees in my section,

45	can cut tall trees	20	can climb poles and splice wire
50	can climb poles	25	can cut trees and splice wire
57	can splice wire	11	can do all three
28	can cut trees and climb poles	9	can't do any of the three

From the data supplied by Mr. Bradkin, we can find the numbers shown in Figure 1.17. By adding all the numbers from the regions, we find the total number of Bradkin's employees to be

$$9 + 3 + 14 + 23 + 11 + 9 + 17 + 13 = 99.$$

Bradkin claimed to have 100 employees, but his data indicates only 99. The management decided that Bradkin didn't qualify as section chief, and reassigned him as night-shift information operator in Guam. (The moral: He should have taken this course.)

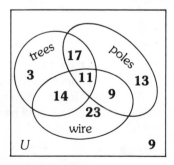

Figure 1.17

1.5 EXERCISES

Use Venn diagrams to answer each of the following questions

1. A survey of 80 sophomores at a certain western college showed that

   36 take English
   32 take History
   32 take Political Science
   16 take Political Science and History
   16 take History and English
   14 take Political Science and English
    6 take all three

   How many students:
   (a) take English and neither of the other two? 12
   (b) take none of the three courses? 20
   (c) take History, but neither of the other two? 6
   (d) take Political Science and History but not English? 10
   (e) do not take Political Science? 48

2. The following data shows the preferences of 102 people at a wine tasting party.

   99 like Spañada
   96 like Ripple
   99 like Boone's Farm Apple Wine
   95 like Spañada and Ripple
   94 like Ripple and Boone's
   96 like Spañada and Boone's
   93 like all three

   How many people prefer:
   (a) none of the three?
   (b) Spañada, but not Ripple?
   (c) anything but Boone's Farm?
   (d) only Ripple?
   (e) exactly two kinds of wine?

3. Bill Bradkin (Example 2 in the text) was again reassigned, this time to the Home Economics department of the electric utility. He interviewed 140 people in a suburban shopping center to find out some of their cooking habits. He obtained the following results. Should he be reassigned yet one more time?

   58 use microwave ovens
   63 use electric ranges
   58 use gas ranges
   19 use microwave ovens and electric ranges
   17 use microwave ovens and gas ranges
    4 use both gas and electric ranges
    1 uses all three
    2 cook only with solar energy

4. Toward the middle of the harvesting season, peaches for canning come in three types: earlies, lates, and extra lates, depending on the expected date of ripening. During a certain week, the following data was recorded at a small fruit receiving station.

   34 trucks went out carrying early peaches
   61 had late peaches
   50 had extra lates
   25 had earlies and lates
   30 had lates and extra lates
    8 had earlies and extra lates
    6 had all three
    9 had only figs (no peaches at all)

   (a) How many trucks had only late variety peaches?
   (b) How many had only extra lates?
   (c) How many had only one variety of peaches?
   (d) How many trucks in all went out during the week?

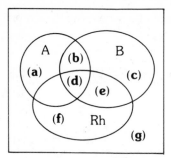

Human blood can contain either no antigens, the A antigen, the B antigen, or both the A and B antigen. A third antigen, called the Rh antigen, is important in human reproduction, and again may or may not be present in an individual. Blood is called type A-Positive if the individual has the A and Rh, but not the B antigen. A person having only the A and B antigens is said to have type AB-Negative blood. A person having only the Rh antigen has type O-Positive blood. Other blood types are defined in a similar manner.

5. Identify the blood type of the individuals in each region of the Venn diagram.

6. (Use the diagram for Exercise 5.) In a certain hospital, the following data on patients was recorded:

    25 patients had the A antigen
    17 had the A and the B antigens
    27 had the B antigen
    22 had the B and the Rh antigens
    30 had the Rh antigen
    12 had none of the antigens
    16 had the A and the Rh antigens
    15 had all three antigens

How many patients:
(a) are represented here?
(b) have exactly one antigen?
(c) have exactly two antigens?
(d) have O-Positive blood?
(e) have AB-Positive blood?
(f) have B-Negative blood?
(g) have O-Negative blood?
(h) have A-Positive blood?

7. A certain chicken farmer in Wappingers Falls, New York, surveyed his flock with the following results. The farmer has:

    9 fat red roosters
    2 fat red hens
    37 fat chickens
    26 fat roosters
    7 thin brown hens
    18 thin brown roosters
    6 thin red roosters ✓
    5 thin red hens

Answer the following questions about the flock. (*Hint:* You need a Venn diagram with circles for fat, for male (a rooster is a male, a hen is a female) and for red (assume that brown and red are opposites in the chicken world).) How many chickens are:
(a) fat?   (d) fat, but not male?
(b) red?   (e) brown, but not fat?
(c) male?   (f) red and fat?

8. Country-Western songs emphasize three basic themes, love, prison, and trucks. A survey of the local Country-Western radio station produced the following data:

    ① 12 about a truck driver who is in love while in prison
    ② 13 about a prisoner in love
    ⑦ 28 about a person in love
    ③ 18 about a truck driver in love
    ④ 3   about a truck driver in prison who is not in love
    ⑤ 2   about people in prison who are not in love and do not drive trucks
    LAST 8 about people who are out of jail, are not in love, and do not drive a truck
    ⑥ 16 about truck drivers who are not in prison

51 (a) How many songs were surveyed?
Find the number of songs about:
31 (b) truck drivers
18 (c) prisoners
15 (d) truck drivers in prison
33 (e) people not in prison
23 (f) people not in love

The following table shows the number of people in a certain city in Georgia who fit into the given categories.

Age Drink	Vodka (V)	Bourbon (B)	Gin (G)	Totals
21–25 (Y)	40	15	15	70
26–35 (M)	30	30	20	80
over 35 (O)	10	50	10	70
	80	95	45	220

*9. Using the letters given in the table, find the number of people in each of the following sets.

(a) $Y \cap V$
(b) $M \cap B$
(c) $M \cup (B \cap Y)$
(d) $Y' \cap (B \cup G)$
(e) $O' \cup G$
(f) $M' \cap (V' \cap G')$

The following table shows the results of a survey in a medium-sized town in Tennessee. The survey asked questions about the investment habits of local citizens.

Age	Stocks (S)	Bonds (B)	Savings accounts (A)	Totals
18–29 (Y)	6	2	15	23
30–49 (M)	14	5	14	33
50 or over (O)	32	20	12	64
	52	27	41	120

*10. Using the letters given in the table, find the number of people in each of the following sets.

(a) $Y \cap B$
(b) $M \cup A$
(c) $Y \cap (S \cup B)$
(d) $O' \cup (S \cup A)$
(e) $(M' \cup O') \cap B$

*1.6 CARDINAL NUMBERS OF INFINITE SETS

As we mentioned in the introduction to this chapter, most of the early work in set theory was done by Georg Cantor. He devoted much of his life to a study of the cardinal numbers of sets. Recall that the *cardinal number* of a finite set is the number of elements that it contains. For example, the set {5, 9, 15} contains 3 elements, and has a cardinal number of 3. The cardinal number of ∅ is 0. (Why?)

Cantor proved many results about the cardinal numbers of infinite sets. The proofs of Cantor are quite different from the type of proofs you may have seen in an algebra course. Because of the novelty of Cantor's methods, they were not universally accepted by the mathematicians of his day. (In fact, some other aspects of Cantor's theory lead to paradoxes.) The results we will discuss here, however, are commonly accepted today.

Before we can find the cardinal number of an infinite set, we need the idea of 1-to-1 correspondence. For example, each of the sets $\{1, 2, 3, 4\}$ and $\{9, 10, 11, 12\}$ has four elements. We could pair off corresponding elements of the two sets in the following manner:

$$\{1, \quad 2, \quad 3, \quad 4\}$$
$$\updownarrow \quad \updownarrow \quad \updownarrow \quad \updownarrow$$
$$\{9, 10, 11, 12\}.$$

Such a pairing is a **1-to-1 correspondence** between the two sets. The "1-to-1" refers to the fact that to each element of the first set is paired exactly one element of the second set. Also, to each element of the second set is paired exactly one element of the first set.

The following correspondence between sets $\{1, 8, 12\}$ and $\{6, 11\}$

$$\{1, \quad 8, \qquad 12\}$$
$$\{6, \qquad 11\}$$

is not 1-to-1 since the elements 8 and 12 from the first set are both paired with the element 11 from the second set.

It seems reasonable to say that if two non-empty sets have the same cardinal number, then we can establish a 1-to-1 correspondence between them. Also, if we can establish a 1-to-1 correspondence between two sets, then the two sets must have the same cardinal number. These two facts are fundamental in discussing the cardinal numbers of infinite sets.

The basic set used in discussing infinite sets is the set of counting numbers, $\{1, 2, 3, 4, 5, \ldots\}$. The set of counting numbers is said to have the infinite cardinal number $\aleph_0$ (the first Hebrew letter, aleph, with a zero subscript, read "aleph-null"). Think of $\aleph_0$ as being the "smallest" infinite cardinal number. To the question "How many counting numbers are there?" we would answer "There are $\aleph_0$ of them."

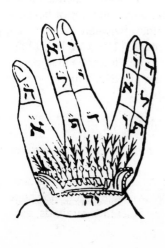

From what we said above, it follows that any set that can be placed in 1-to-1 correspondence with the counting numbers will have the same cardinal number as the counting numbers, namely, $\aleph_0$. It turns out that many sets of numbers have cardinal number $\aleph_0$. The next few examples show some of them.

Example 1 Show that the set of whole numbers $\{0, 1, 2, 3, \ldots\}$ has cardinal number $\aleph_0$.

Solution. The set of whole numbers has one more element (the number 0) than the set of counting numbers, and thus should have cardinal number

$\aleph_0 + 1$. However, the correspondence

$$\begin{array}{cc} \{1, \ 2, \ 3, \ 4, \ 5, \ 6, \ \ldots, \quad n, \quad \ldots\} & \text{Counting numbers} \\ \updownarrow \ \updownarrow \ \updownarrow \ \updownarrow \ \updownarrow \ \updownarrow \qquad\quad \updownarrow & \\ \{0, \ 1, \ 2, \ 3, \ 4, \ 5, \ \ldots, n-1, \ \ldots\} & \text{Whole numbers} \end{array}$$

is a 1-to-1 correspondence between the set of counting numbers and the set of whole numbers. Since we have a 1-to-1 correspondence, we conclude that the cardinal number of the set of counting numbers is the same as the cardinal number of the set of whole numbers, or

$$\aleph_0 + 1 = \aleph_0 \, .$$

This result shows that intuition is a poor guide for dealing with infinite sets. Intuitively, it is "obvious" that there are more whole numbers than counting numbers. However, since the sets can be placed in 1-to-1 correspondence, the two sets have the same cardinal number.

Example 2   Show that the set of integers $\{\ldots, -3, -2, -1, 0, 1, 2, 3, \ldots\}$ has cardinal number $\aleph_0$.

Solution.   Every counting number has a corresponding negative; the negative of 8, for example, is -8. Therefore, we might expect the cardinal number of the set of integers to be $\aleph_0 + \aleph_0$, or $2\aleph_0$. However, we can set up a 1-to-1 correspondence between the set of integers and the set of counting numbers, as follows:

$$\begin{array}{c} \{1, \ 2, \quad 3, \ 4, \quad 5, \ 6, \quad 7, \ldots, 2n, \ 2n+1, \ \ldots\} \\ \updownarrow \ \updownarrow \quad \updownarrow \ \updownarrow \quad \updownarrow \ \updownarrow \quad \updownarrow \qquad\quad \updownarrow \qquad \updownarrow \\ \{0, \ 1, \ -1, \ 2, \ -2, \ 3, \ -3, \ldots, \ n, \quad -n, \quad \ldots\} \end{array}$$

Because of this 1-to-1 correspondence, the cardinal number of the set of integers is the same as the cardinal number of the set of counting numbers, or

$$2\aleph_0 = \aleph_0 \, .$$

As shown by Example 2, there are just as many integers as there are counting numbers. This result is not at all intuitive. However, the next result is even less intuitive. We know that there is an infinite number of fractions between any two counting numbers. For example, between the counting numbers 1 and 2, we have the infinite set of fractions, $\{1\frac{1}{2}, \ 1\frac{3}{4}, \ 1\frac{7}{8}, \ 1\frac{15}{16}, \ 1\frac{31}{32}, \ \ldots\}$. This should imply that there are "more" fractions than counting numbers. It turns out, however, that there are just as many fractions as counting numbers, as shown by the next example.

Example 3   Show that the cardinal number of the set of rational numbers* is $\aleph_0$.

---

*A rational number is a quotient of integers, with the exception that zero is never a divisor. The set of integers is $\{\ldots, -3, -2, -1, 0, 1, 2, 3, \ldots\}$. These sets are discussed in Chapter 4.

Paradox   *The word in Greek originally meant "wrong opinion" as opposed to orthodox ("right opinion"). Note that a right angle is a translation of the more formal term "orthogonal." Anyhow, the word came to mean self-contradiction, for example*
"This sentence is false."
Assume it to be true; assume it false—yes, it's paradoxical.
*   The famous paradoxes of Zeno (born about 496 BC in southern Italy) demonstrate that motion is impossible. Since an object must move through space, which can be divided infinitely, motion cannot begin. One of Zeno's paradoxes concerns Achilles, who cannot overtake a tortoise even though he runs faster. Tortoise has a head start of one meter, say, and goes one-tenth as fast as Achilles. A reaches the point where T was, but T is one-tenth of a meter ahead. A goes that one-tenth, but now T is ahead by one-hundredth; and so on and on.*

Solution. To show that the cardinal number of the set of rational numbers is $\aleph_0$, we must set up a 1-to-1 correspondence between the set of rational numbers and the set of counting numbers. This is done by the following ingenious scheme, devised by Georg Cantor. Look at the chart of Figure 1.18. The non-negative rational numbers whose denominator is 1 are written in the first row; those whose denominator is 2 are written in the second row, and so on. Every non-negative rational number appears in this list sooner or later. For example, $^{327}/_{189}$ is in row 189 and column 327.

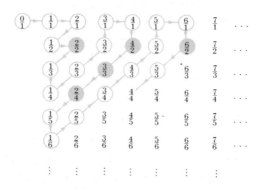

Figure 1.18

To set up a 1-to-1 correspondence between the set of non-negative rationals and the set of counting numbers, follow the path drawn in Figure 1.18. Let $^0/_1$ correspond to 1, let $^1/_1$ correspond to 2, $^2/_1$ to 3, $^1/_2$ to 4 (skip $^2/_2$, since $^2/_2 = ^1/_1$) $^1/_3$ to 5, $^1/_4$ to 6, and so on. The numbers under the colored disks in Figure 1.18 are omitted, since they can be reduced to lowest terms.

Since we can set up a 1-to-1 correspondence between the set of non-negative rationals and the counting numbers, both of these sets must have the same cardinal number, $\aleph_0$. If we use the method of Example 2, we can extend this correspondence to include the negative rational numbers as well. Therefore, the set of all rational numbers has cardinal number $\aleph_0$.

A set is called **countable,** if it is finite, or if it has cardinal number $\aleph_0$. All the sets we have discussed so far—the counting numbers, the whole numbers, the integers, and the rational numbers—are countable. It seems now that *every* set is countable, but this is not true. The next example shows that the set of real numbers is not countable.

Example 4 The set of all numbers used in arithmetic is called the set of *real numbers.** The real numbers include all rational numbers, square roots such as $\sqrt{2}, \sqrt{5}, -\sqrt{3}$, and numbers such as π. Show that the set of real numbers does *not* have cardinal number $\aleph_0$.

Solution. There are two possibilities:

_____

*The real numbers are discussed in Chapter 5.

(1)   The set of real numbers has cardinal number $\aleph_0$.
(2)   The set of real numbers does not have cardinal number $\aleph_0$.

Let us assume for the time being that (1) is true. If (1) is true, then we can set up a 1-to-1 correspondence between the set of real numbers and the set of counting numbers. We do not know what sort of correspondence we might have, but we assume it can be done.

It is a fact that every real number can be written as a decimal. Thus, in the 1-to-1 correspondence we are assuming, some decimal corresponds to the counting number 1, some decimal corresponds to 2, and so on. Suppose the correspondence is as follows.

$$1 \leftrightarrow .68458429006 \ldots$$
$$2 \leftrightarrow .13479201038 \ldots$$
$$3 \leftrightarrow .37291568341 \ldots$$
$$4 \leftrightarrow .935223671611 \ldots$$

We are assuming the existence of a 1-to-1 correspondence between the counting numbers and the real numbers. Thus, we are claiming that every decimal is in the list above. Let's construct a new decimal K as follows. The first decimal in the above list has 6 as its first digit; let K start as $K = .4. \ldots$ We picked 4 since $4 \neq 6$; we could have used any other digit except 6. Since the second digit of the second decimal in the list is 3, we let $K = .45 \ldots$ (since $5 \neq 3$). The third digit of the third decimal is 2, so let $K = .457 \ldots$ (since $7 \neq 2$). The fourth digit of the fourth decimal is 2, so let $K = .4573 \ldots$ (since $3 \neq 2$). Continue K in this way.

Is K in the list that we assumed to contain all decimals? The first decimal in the list differs from K in at least the first position (K starts with 4, and the first decimal in the list starts with 6). The second decimal in the list differs from K in at least the second position, and the n-th decimal in the list differs from K in at least the n-th position. Every decimal in the list differs from K in at least one position, so that K cannot possibly be in the list. In summary:

(1)   We assume every decimal is in the list above.
(2)   The decimal K is not in the list.

Since both these statements cannot be true, our original assumption has lead to a contradiction. Thus, we must accept the only possible alternative to our original assumption. It is not possible to set up a 1-to-1 correspondence between the set of reals and the set of counting numbers. Thus, the cardinal number of the set of reals is not equal to $\aleph_0$.

The set of counting numbers is a subset of the set of real numbers. Thus, it would seem reasonable to say that the cardinal number of the set of reals, commonly written as c, is greater than $\aleph_0$. Other, even larger infinite cardinal numbers can be constructed. For example, the set of all subsets of the set of real numbers has a cardinal number larger than c. By continuing this process of finding cardinal numbers of sets of subsets, we can produce more and more, larger and larger infinite cardinal numbers.

We have seen that c is a larger cardinal number than $\aleph_0$. Is there a cardinal number between $\aleph_0$ and c? The person who began the study of set theory, Georg Cantor, did not think so, but he was unable to prove his guess. Cantor's *Hypothesis of the Continuum* was long considered one of the major unsolved problems of mathematics. However, in the early 1960s, the American mathematician Paul J. Cohen solved the problem in an unusual way. He showed that the assumption that no such cardinal number exists leads to a valid, consistent body of mathematical results. He also showed that if it is assumed that such a cardinal number does exist, equally valid mathematical results will be obtained. (Compare this with the problems caused by Euclid's Fifth Postulate. See Chapter 8.)

1.6 EXERCISES

Set up a 1-to-1 correspondence between the elements of each pair of sets, where possible.

1. {1, 4, 9, 12} and {8, 12, 16, 20}

2. {Shamus, Anna, Kerry} and {□, ○, ☆}

3. {0, 4, 5, 9} and {4, 5, 9}

4. {∅} and {2}

Find the cardinal number of each set in Exercises 5–14.

5. {9, 10, 11, 12, 13}   **8.** {v, a, c, u, u, m}

6. {1, 2, 3, . . . , 52}   **9.** {0}

7. {ϵ, h, u, c, k}   **10.** ∅

11. {1001, 1002, 1003, . . .}

12. {1/2, 1/4, 1/8, 1/16, 1/32, . . .}

13. the set of points on a circle

14. the set of all decimals between 0 and 1

Show that each set in Exercises 15–22 has cardinal number $\aleph_0$, by setting up a 1-to-1 correspondence between the counting numbers and the given set.

15. {5, 10, 15, 20, 25, 30, . . .}
(*Hint:* $10 = 5 \times 2$, $15 = 5 \times 3$, and so on)

16. {1,000,000, 2,000,000, 3,000,000, . . .}

17. {positive even numbers}   **19.** {even numbers}

18. {positive odd numbers}   **20.** {odd numbers}

21. {2, 4, 8, 16, 32, 64, 128, . . .}
(*Hint:* $4 = 2^2$, $8 = 2^3$, $16 = 2^4$, and so on)

22. {1/3, 1/9, 1/27, 1/81, 1/243, . . .}

*23. Show how a 1-to-1 correspondence can be set up between the points of a one-inch line segment and the points on a line of infinite length. (*Hint:* Bend the segment.)

Work each of the following problems.

*24. $\aleph_0 + 15$ ***28.** $c + 55$

*25. $\aleph_0 + 8{,}000{,}000$ ***29.** $c + c$

*26. $\aleph_0 + \aleph_0 + \aleph_0$ ***30.** $c + \aleph_0$

*27. $\aleph_0 - 18{,}000{,}000{,}000$

A hotel contains a countably infinite number of rooms, numbered 1, 2, 3, 4, 5, ... Each room is occupied by a guest.

*31. If one more guest arrives, show how a room can be made available for the guest. (*Hint:* Each of the current guests must move.)

*32. Suppose $\aleph_0$ new guests arrive. Show how rooms can be made available for all of them.

CHAPTER 1 SUMMARY

To find meanings of these words, look on the page given in the Index.

Keywords

Set	Universal set	Disjoint sets
Finite set	Complement of a set	Equal sets
Infinite set	Empty (null) set	Convex set
Well-defined	Operation	Cardinal number
Element (member) of a set	Union	1-to-1 correspondence
Set-builder notation	Intersection	Countable
Variable	Venn diagram	Counting numbers
Subset	Tree diagram	Real numbers

Symbol	Meaning	Example
$\{\ \ \}$	Set braces	$\{3, 4, 5\}$ is a set
$\varnothing$	Empty set	
$n(A)$	Cardinal number of set A	$n(\{3, 4, 5, 6\}) = 4$
$\in$	Element of	$4 \in \{5, 6, 8, 4, 3\}$
$\{x \mid x \text{ has property } P\}$	Set-builder notation	
U	Universal set	
A'	Complement	If $U = \{3, 6, 8, 9\}$ and $A = \{3, 9\}$, then $A' = \{6, 8\}$.
$\subset$	Subset	$\{2, 4, 7\} \subset \{2, 3, 4, 5, 6, 7, 8\}$
$\cap$	Intersection	$\{8, 9, 11, 12\} \cap \{7, 8, 10, 12\} = \{8, 12\}$
$\cup$	Union	$\{8, 9, 11, 12\} \cup \{7, 8, 10, 12\}$ $= \{7, 8, 9, 10, 11, 12\}$
$-$	Set difference	$\{3, 4, 5, 6\} - \{4, 6, 8, 10\} = \{3, 5\}$
c	Cardinal number of the set of real numbers	
$\aleph_0$	Cardinal number of the set of counting numbers	

CHAPTER 1 TEST

Let $U = \{1, 2, 3, 4, 5, 6, 7\}$, $A = \{1, 3, 5, 7\}$, $B = \{3, 6, 7\}$, and $C = \{1, 3\}$. Find each of the following sets.

1. A' **3.** $B \cup C$

2. $A \cap B$ **4.** $A' \cap (B \cup C')$

Use sets A, B, C and U given above and decide whether the following statements are *true* or *false*.

5. $B \subset A$ **9.** $7 \in A$

6. $C \subset A$ **10.** $3 \notin C$

7. $A \not\subset U$ **11.** $A \cup B' = A' \cup B$

8. $\varnothing \subset C$ **12.** $(A \cap C)' = A' \cup C'$

Draw Venn diagrams for each of the following sets.

13. $M \cap N'$ **15.** $P \cap (Q \cup R')$ *answer incorrect in book* }

14. $M' \cup N$ **16.** $(P' \cup Q') \cap R$ *answer incorrect in book* }

The following list shows some household appliances, and the typical daily cost of energy to operate them in an area of fairly low rates.

Appliance	Cost	Appliance	Cost
Outdoor gas light	8¢	Toaster	¼¢
Coffee maker	½¢	Color television	5¢
Dishwasher	2½¢	Clothes dryer	6¢
Frostless refrigerator	15¢	Central air conditioner	72¢
Manual refrigerator	5¢	Waterbed heater	10¢

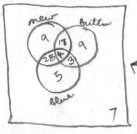

Let K be the set of those appliances normally found only in the kitchen. (Clothes dryers do not belong in K.) Let C be the set of those appliances that operate on less than $5\frac{1}{2}$¢ worth of energy per day. Find the appliances in each of the following sets.

17. K' **18.** C' **19.** $K \cap C$ **20.** $K \cup C'$

The local department store inventoried its Levi's department as follows:

59 pairs were new   7 had a button fly and were blue

34 had a button fly   32 were new and blue

40 were blue   4 were new and blue, with a button fly

22 were new and had a button fly   7 were none of these

Find the number of pairs of Levi's which were

21. blue, but not new   **24.** new and blue, but no button fly

22. new, but not blue   **25.** not blue

23. not new, and not blue

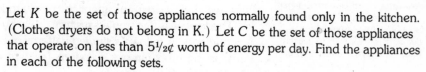

Are you there, Achilles?

Chapter 2 Logic

Aristotle *could be a rock singer as Raphael painted him in* The School of Athens *mural. He is debating a point with his teacher Plato; your eye is directed to the two of them by a series of archways. Indeed, Platonism and Aristotelianism seem to be the dominant world views in the West. Aristotle's logic and analysis of how things work, it is said, are inseparably woven into the ways we are taught to think. People often look to the East for "new" answers—but the questions were asked first by Aristotle, in his* Ethics, *say, the book he is holding in the mural.*

See **null-A** *on page 50.*

Logic, logical thinking, and correct reasoning are the subject of this chapter. These ideas have wide application in many fields, in areas as diverse as law, psychology, rhetoric, science, and mathematics. In our daily life, logic, the science of correct reasoning, certainly plays a vital role, but it is perhaps not quite as important as emotions in guiding our actions.

A good example of the "battle" between logic and emotion is given in the television series *Star Trek*. Mr. Spock, the Vulcan, claims to let logic rule his life totally. On the other hand, Dr. McCoy, an earth resident, has scant use for logic except as it applies to his field of medicine.

While an interesting study can certainly be made of logic in human lives, we shall restrict our attention mainly to logic as it is used in mathematics. This logic was first studied systematically by Aristotle (384 BC – 322 BC). Aristotle and his followers studied patterns of correct and incorrect reasoning. Perhaps the most famous *syllogism* (pattern of logical reasoning) to come from Aristotle is the following:

> All men are mortal.
> <u>Socrates is a man.</u>
> Socrates is mortal.

The work of Aristotle was carried forward by medieval philosophers and theologians, who made an intimate study of logical arguments. A big advance in the study of mathematical logic came with the work of Gottfried Wilhelm von Leibniz (1646 – 1716), one of the inventors of calculus. Leibniz introduced *symbols* to represent ideas in logic—letters for statements and other symbols for the relations between statements. Leibniz hoped that logic would become a *universal characteristic* and unify all of mathematics.

Leibniz (*or Leibnitz in French*) *was a wide-ranging philosopher and a universalist who tried to patch up Catholic-Protestant conflicts. He promoted cultural exchange between Europe and the East. Chinese ideograms led him to search for a universal symbolism. (See also page 39.)*

For two hundred years after Leibniz, mathematicians had little interest in his universal characteristic. In fact, no further work in symbolic logic was done until George Boole (1815–1864) began his studies. Boole's major contribution came when he pointed out the connection between logic and sets. His rules for sets make up the subject of *Boolean algebra,* which is a key idea in the development of modern computers and calculators. Boole's work is summarized in his most famous book, *An Investigation of the Laws of Thought,* published in 1854.

Logic became a favorite subject of study by British mathematicians after Boole. John Venn (1834–1923) developed diagrams such as we saw in Chapter 1. Augustus De Morgan (1806–1872) proved certain laws that we shall see in this chapter. Charles Dodgson (1832–1898) wrote several popular textbooks on logic. Dodgson, a professor at Oxford University in England, wrote *Alice's Adventures in Wonderland* under the pen name Lewis Carroll. We see some of his logic exercises in Section 6 of this chapter.

The work of all these mathematicians was summarized and expanded in *Principia Mathematica,* a 2000-page, three-volume work which appeared between 1910 and 1913. The book begins with a small number of basic assumptions, or *axioms,* and a few undefined terms of logic. Then, with this foundation, all the theory of arithmetic is developed in terms of the definitions and assumptions of logic. *Principia Mathematica* was written by Bertrand Russell (1872–1970) and Alfred North Whitehead (1861–1947).

2.1 INDUCTION, DEDUCTION AND VALID ARGUMENTS

George Boole *grew up in poverty. His father, a London tradesman, gave him his first math lessons and taught him to make optical instruments. Boole was largely self-educated. At 16 he worked in an elementary school and by 20 had opened his own school. He studied mathematics in his spare time. Lung disease killed him at age 49.*

Boole's ideas have been used in the design of computers and telephone systems.

Look at the following list of numbers:

$$2, 9, 16, 23, 30.$$

What is the next number of this list? Most people would say that the next number is 37. Why? They probably reason something like this: What have 2 and 9 and 16 in common? What is the pattern?

After looking at the numbers for awhile, we might see that $2 + 7 = 9$, and $9 + 7 = 16$. Is something similar true for the other numbers in this list? Do you add 7 and 16 to get 23? Do you add 7 and 23 to get 30? Yes; any number in the given list can be found by adding 7 to the preceding number. Thus, the next number in the list should be $30 + 7 = 37$. The next number after 37 would be $37 + 7 = 44$.

We won't go into the question of why people assume that a pattern exists and then study the numbers to find it. This question is studied a lot in psychology classes. Many puzzles or "brain teasers" require you to "see" the pattern in a list of numbers. There is some evidence that the ability to see these patterns is related to native intelligence, so that find-the-next-number questions appear on standard tests, such as those given for employment or college admission.

Long life—new directions
During the decade that Whitehead and Russell worked together on Principia Mathematica, *Whitehead (above) was teaching mathematics at Cambridge University, and had written* Universal Algebra. *In 1910 he went to the University of London, exploring science—not only the philosophical basis but the "aims of education" (as he called one of his books). It was as philosopher that he was invited to Harvard University in 1924. In the U.S. he completed his philosophy of organism in three unique works:* Science and the Modern World; Adventures in Ideas; Process and Reality. *His* Process *book was a cosmology, evolving from a novel conception of "experience," and Whitehead intended to replace Newton's universe-as-mechanism with his own organic view.*
Whitehead died at the age of 86 in Cambridge, Massachusetts. (*See* **Russell** *on page 42.*)

In any case, you set out to find the "next number" by reasoning from your observation of the numbers in the list. You may have jumped from the facts given in the list above ($2 + 7 = 9, 9 + 7 = 16$, and so on) to the general statement that any number in the list is 7 more than the preceding number. This reasoning process is called *inductive reasoning,* or **induction.** When you use inductive reasoning, you reason from specific examples to a general statement.

By using inductive reasoning, we concluded that 37 was the next number in the list. But this is wrong. You were set up. You've been tricked into using logic to get a wrong answer. Not that your logic was faulty; but the person making up the list has another answer in mind. The list of numbers

$$2, 9, 16, 23, 30$$

actually gives the dates of Mondays in June if June 1 is on a Sunday. The next Monday after June 30 is July 7. With this pattern, the list continues as

$$2, 9, 16, 23, 30, 7, 14, 21, 28, \ldots.$$

The process you may have used to obtain the rule "add 7" in the list above reveals one main flaw of inductive reasoning. You can never be sure that what is true in a specific case will be true in general. Even a larger number of cases may not be enough. For example, when medical scientists were looking for an improvement in a drug used to treat malaria, they tested different substances. None of them was better than the drug then in use. By inductive reasoning, the scientists might well have concluded that there was *no* drug better than the drug in use. However, a drug tested later proved to be much superior to the one in use. Here inductive reasoning would have led to a false conclusion, with unfortunate consequences for malaria sufferers. (In another case, scientists tested over 14,000 possible drugs before finding one that was better than the drug in use.)

Inductive reasoning does not guarantee a result, but it does have the advantage of giving you a *hypothesis* (an educated guess) that can then be tested. In testing a general rule that you obtained by inductive reasoning, it takes only one example that doesn't work to prove the rule false. Such an example that proves a rule false is called a *counterexample.*

As we have seen, a result arrived at by inductive reasoning may well be wrong—what is true in many examples may not be true in general. In mathematics, this difficulty is overcome by insisting that all hypotheses receive a *proof* before being accepted. This proof must be based on previously known facts, and must follow the rules of *deductive reasoning,* or **deduction.**

Deductive reasoning goes from general statements to specific situations. For example, suppose you want to show that the area of your living room is 300 square feet. You measure the living room and find it to be 15 feet by 20 feet. You then use the general formula for the area of a rectangle, Area = length × width (a formula which is known to be true), and find the area of your living room: Area = $20 \times 15 = 300$ square feet. Here you reasoned from a general formula to a specific situation.

Principia Mathematica *The title chosen by Whitehead and Russell was a deliberate reference to Philosophiae naturalis principia mathematica, or "mathematical principles of the philosophy of nature," Isaac Newton's epochal work of 1687. Newton's* Principia *pictured a kind of "clockwork universe" that ran via his Law of Gravitation. Newton independently invented the calculus, unaware that Leibniz had published his own formulation of it earlier. A controversy over their priority continued into the 18th century.*

Madam du Châtelet (*1706– 1749) is pictured between Leibniz and Newton, and she participated in the scientific activity of the generation following them. She was educated in science, music, and literature, and was studying mathematics at the time (1733) she began a long intellectual relationship and affair with the philospher Voltaire (1694– 1778). Her chateau was equipped with a physics laboratory. She and Voltaire competed independently in 1738 for a prize offered by the French Academy on the subject of fire. Although the Mme. de Châtelet did not win, her dissertation was published by the academy in 1744. (Euler won— see page 43.) By this time she had published* Institutions of Physics *(expounding in part some ideas of Leibniz) and a work on Vital Forces. During the last four years of her life she translated Newton's* Principia *from Latin into French—the only French translation to date.*

Arguments   When we reason through a problem, we usually work from certain *premises*. A premise can be an assumption, law, rule, widely held idea, or observation. We reason inductively or deductively from the premises to obtain a *conclusion*. The premises and conclusion make up a **logical argument.**

Example 1   Identify each premise and the conclusion in each of the following arguments. Then tell whether each argument is an example of inductive or deductive reasoning.

(a) My mother's Chihuahua is nasty. My neighbor's Chihuahua is nasty. Therefore, all Chihuahuas are nasty.

Solution.   The premises are "My mother's Chihuahua is nasty" and "My neighbor's Chihuahua is nasty." The conclusion is "Therefore, all Chihuahuas are nasty." Since we are reasoning from specific examples to a general statement, the argument is an example of inductive reasoning.

(b) All typewriters will type the letter x. I have a typewriter. I can type the letter x.

Solution.   Here the premises are "All typewriters will type the letter x" and "I have a typewriter." The conclusion is "I can type the letter x." We are reasoning from general to specific, so that deductive reasoning is being used.

(c) Today is Wednesday. Tomorrow will be Thursday.

Solution.   There is only one premise here, "Today is Wednesday." The conclusion is "Tomorrow will be Thursday." We are using the fact that Thursday follows Wednesday, even though this fact is not explicitly stated. Since the conclusion comes from general facts that apply to this special case, we have used deductive reasoning.

Validity   When we reason from the premises of an argument to obtain a conclusion, we want the argument to be valid. The argument is **valid** if the fact that all the premises are true forces the conclusion to be true. An argument that is not valid is *invalid,* or a **fallacy.**

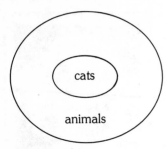

Figure 2.1

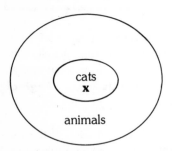

Figure 2.2

Let **x** represent "Tom"

There are several techniques that can be used to check whether an argument is valid. One of these is the visual technique based on *Euler diagrams,* illustrated by the following examples.

Example 2   Is the following argument valid?

> All cats are animals.
> Tom is a cat.
> _____
> Tom is an animal.

Solution.   Here we use the common method of placing one premise over another, with the conclusion below a line. To begin, we draw a rough circle to represent the first premise. This is the region for "animals." Since all cats are animals, the region for "cat" goes inside the region for "animals," as in Figure 2.1.

The second premise, "Tom is a cat," suggests that "Tom" would go inside the region representing "cats." Let **x** represent "Tom."

From the final diagram, Figure 2.2, we see that "Tom" is inside the region for "animals." Therefore, if both premises are true, then the conclusion that Tom is an animal must be true also. The argument is valid, as checked by Euler diagrams.

Example 3   Is the following argument valid?

> All rainy days are cloudy.
> Today is not cloudy.
> _____
> Today is not rainy.

Solution.   In Figure 2.3 we have drawn the region for "rainy days" entirely inside the region for "cloudy days." Since "Today is *not* cloudy," we place an **x** for "today" *outside* the region for "cloudy days." (See Figure 2.4.) Placing the **x** outside the region for "cloudy days" forces it to be also outside the region for "rainy days." Thus, if the first two premises are true, then it is also true that today is not rainy. The argument is valid.

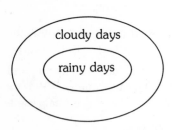

Figure 2.3

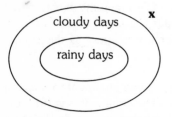

Figure 2.4

Let **x** represent "today"

Example 4   Is the following argument valid?

> All banana trees have green leaves.
> That plant has green leaves.
> _____
> That plant is a banana tree.

Solution.   The region for "banana trees" goes entirely inside the region for "plants having green leaves." (See Figure 2.5.) We have a choice for locating the **x** that represents "that plant." The **x** must go inside the region for "plants having green leaves," but can go either inside or outside the region for "banana trees." Even if the premises are true, we are not forced to conclude that the conclusion is true. This argument is not valid; it is a fallacy.

It is important to note that the validity of an argument is not the same as the truth of its conclusion. The argument in Example 4 was not valid, but the conclusion "That plant is a banana tree" may or may not be true—we don't know.

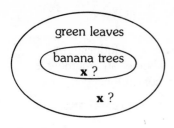

Figure 2.5
We have a choice

Example 5 Is the following argument valid?

All expensive things are desirable.
All desirable things make you feel good.
All things that make you feel good make you live longer.

Expensive things make you live longer.

Solution. A diagram for the argument is given in Figure 2.6. If each premise is true, then the conclusion must be true, making the argument valid.

Figure 2.6

Arguments with the word "some" can be a little tricky. Work them out as shown in the final example of this section.

Example 6 Is the following argument valid?

Some tenants get ripped off by the landlord.
I am a tenant.

I get ripped off.

Solution. The first premise is sketched in Figure 2.7. As the sketch shows, *some* (but not necessarily *all*) tenants get ripped off. Where do we put **I**? There are two possibilities, as shown in Figure 2.8.

One possibility is that **I** get ripped off; the other is that **I** don't. Thus, the truth of the premises does not force the conclusion to be true, so the argument is invalid.

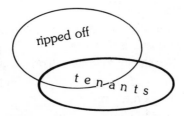

Figure 2.7
Some but not all tenants

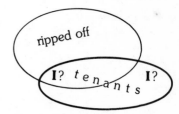

Figure 2.8
Several possibilities

The logic of a rental contract
on pages 78–79 is a practical application of what's happening in this chapter.

Bertrand Russell *was a student of Whitehead's before they wrote the* Principia. *Like his teacher, Russell turned toward philosophy. His works include a critique of Leibniz, analyses of mind and of matter, and a history of Western thought.*

Russell became a public figure because of his involvement in social issues. About 1901 he experienced something like a religious conversion, becoming deeply aware of human loneliness. He was "passionately desirous of finding ways of diminishing this tragic isolation." During World War I he was an anti-war crusader, and was even briefly imprisoned. Again in the 1960s he championed peace. He wrote many books on social issues, winning the Nobel Prize for Literature in 1950.

2.1 EXERCISES

Decide whether each of Exercises 1–8 is an example of inductive or deductive reasoning. (Don't worry here about whether an argument is valid or invalid.) Identify all premises and each conclusion.

1. I must go to class today. Every time I go, I need to take notes. I will have to take notes today.

2. Textbooks always cost too much. This book is a textbook. It costs too much.

3. The last four Presidents of the United States have been men. Every President has been a man.

4. I had 12 apples. I gave 5 away at the Crafts Faire. Therefore, I have 7 left.

5. All the merchants on this street raised the price of raisins. Ms. Carter is a merchant. She raised the price of raisins.

6. Roosters do not lay eggs. Therefore, my pet rooster Roger will not lay an egg.

7. Prudent people never buy $25,000 cars. Cheryl Bradkin is prudent. Cheryl would never buy a $25,000 car.

8. This year's winner of the flower show has a greenhouse. Therefore, a greenhouse is necessary if you want to win.

Decide which of the arguments in Exercises 9–24 are valid.

9. All politicians make promises.
   Ella is a politician.
   Ella makes promises.

10. All mathematics majors have trouble with their checkbooks.
    Chuck is a mathematics major.
    Chuck has trouble with his checkbook.

11. All football players have agents to make demands.
    Kung Wong has an agent.
    Kung Wong is a football player.

12. All car buyers make payments.
    I make payments.
    I am a car buyer.

13. All chickens love to scratch on the ground.
    Roger Rooster is a chicken.
    Roger Rooster loves to scratch on the ground.

14. All residents of Harris County live on farms.
    Norm lives on a farm.
    Norm lives in Harris County.

15. All nursing students need mathematics.
    Fred does not need mathematics.
    Fred is not a nursing student.

16. All members of the Consumer Co-Op save money.
    Lynn saved no money.
    _____
    Lynn is not a member of the Consumer Co-Op.

17. All people using small claims court must appear before the judge.
    Robin appeared before the judge.
    _____
    Robin is using small claims court.

18. All residents of El Dorado Hills have a huge heating bill in winter.
    Pam and Fred have a huge heating bill in winter.
    _____
    Pam and Fred live in El Dorado Hills.

19. Some animals devour their mates.
    Irma Spyder is an animal.
    _____
    Irma Spyder devours her mate.

20. Some vegetarians are healthy.
    Peter Plant is a vegetarian.
    _____
    Peter Plant is healthy.

21. Some roommates make a lot of noise.
    Robert is my roommate.
    _____
    Robert makes a lot of noise.

22. Some cars have power steering.
    Some cars are black.
    _____
    Some black cars have power steering.

23. Some boring people have bad breath.
    Some boring people eat turnips.
    _____
    No turnip eater has bad breath.

24. Some people who play blackjack in Reno lose.
    Some people who play Keno in Reno win.
    _____
    People who lose at blackjack always win at Keno.

The premises marked A, B, and C are followed by several possible conclusions (Exercises 25–30). Take each conclusion in turn, and check whether the resulting argument is valid.

A. All people who drive contribute to air pollution.
B. All people who contribute to air pollution make life a little worse.
C. Some people who live in a suburb make life a little worse.

25. Some people who live in a suburb drive.

26. Some people who live in a suburb contribute to air pollution.

27. Some people who contribute to air pollution live in a suburb.

28. Suburban residents never drive.

29. All people who drive make life a little worse.

30. Some people who make life a little worse live in a suburb.

*31. Find examples of arguments in magazine ads. Check them for validity.

*32. Find examples of arguments on television commercials. Check them for validity.

Leonhard Euler (*1707–1783*) *won the academy prize and edged out Mme. du Châtelet and Voltaire. That was a minor achievement, as was the invention of "Euler circles" (which antedated Venn diagrams). Euler was the most prolific mathematician of his generation, perhaps the greatest. This was despite blindness, which forced him to dictate from memory.*

Euler was born in Switzerland (Helvetia); the stamp below shows one of various "Euler equations," this one from trigonometry. Euler wrote in all mathematical fields, creating new results as well as organizing several fields, above all calculus and analysis. He applied calculus to explain the motion of objects and the forces acting on them. His pioneer work in topology is discussed in Chapter 8.

Exercises 33–38 list several classic puzzles involving logic. We have not tried to explain how to work these puzzles, since there really is no precise method—just keep thinking until you come up with the answer.

*33. A bus leaves Miami for Tampa. An hour later, a person starts walking from Tampa to Miami. When the bus and the walker meet, which is farther from Miami?   *(Same distance)*

*34. Which arithmetic symbol can be inserted between 7 and 8 to make a number larger than 7 and less than 8?   *7.8*

*35. Several soldiers must cross a deep river at a point where there is no bridge. The soldiers spot two children playing in a small rowboat. The rowboat can hold only two children or one soldier. All the soldiers got across the river. How?

*36. A person must take a wolf, a goat, and some cabbage across a river. The rowboat to be used has room for the person plus either the wolf, the goat, or the cabbage. If the person takes the cabbage in the boat, the wolf will eat the goat. While the wolf crosses in the boat, the cabbage will be eaten by the goat. The goat and cabbage are safe only when the person is present. Even so, the person gets everything across the river. Explain how. (This problem dates back to around the year 750.)

*37. You have eight coins. Seven are genuine and one is a fake, which weighs a little less than the other seven. You have a balance scale, which you may use only three times. Tell how to locate the bad coin in three weighings.

*38. Three women, Ms. Thompson, Ms. Johnson, and Ms. Andersen, are sitting side by side at a meeting of the neighborhood improvement group. Ms. Thompson always tells the truth. Ms. Johnson sometimes tells the truth, and Ms. Andersen never tells the truth. The woman on the left says, "Ms. Thompson is in the middle." The woman in the middle says, "I'm Ms. Johnson," while the woman on the right says, "Ms. Andersen is in the middle." What are the correct positions of the women?

2.2 STATEMENTS

In this section, we begin our study of *symbolic logic,* which uses letters to represent statements and symbols for words such as *and, or, not.* One of the main applications of logic is in the study of the *truth value* (that is, the truth or falsity) of statements with many parts. The truth value of these statements depends on the components that make them up.

Many kinds of sentences occur in ordinary language, including sentences that state facts, sentences of opinion, commands, and questions. In symbolic logic we study only the first type of sentence, the kind that involves facts. We define a **statement** as a sentence which is either true or false. For example, both of the following are statements:

> Water runs downhill.
> $3+1=5$.

Each one either is true or is false. On the other hand, the following sentences are not statements under the definition:

> Close the door.
> Will you eat your spinach?
> The Fonz is Number One.
> This sentence is false.

We cannot say that they are either true or false. The first sentence is a command, and the second is a question. The third is an opinion. "This sentence is false" is a paradox; if we assume it is true, then it is false, and if we assume it is false, then it is true.

   A **compound statement** is formed by connecting one or more component statements with a **connective**. Common connectives include *and, or, not,* and *if . . . then.*

Example 1   Identify each of the following that are compound.

(a) The Ming Tree restaurant serves flaming duck and the Gold Mine Bar serves flaming drinks.

Solution.   This statement is compound; it is made up of the component statements "The Ming Tree restaurant serves flaming duck" and "The Gold Mine Bar serves flaming drinks." The connective is *and.*

(b) No person is an island.

Here the connective *no* makes the statement compound.

(c) If I go fly a kite, then I may find electricity.

The connective here is *if . . . then.* We shall discuss this connective in more detail in Section 2.4.

Negation   The sentence "Tom Jones wears tight pants" is a statement; the *negation* of this statement is "Tom Jones does not wear tight pants." The negation of a true statement is false, and the negation of a false statement is true.

Example 2   Form the negation of each statement.

(a) Every town has a mayor.

Introduce *not* into the sentence to form the negation:

> Not every town has a mayor.

(b) The shirt is washable.   The negation is: The shirt is not washable.

Traffic symbol *The importance of symbols was emphasized by the American philosopher-logician Charles Sanders Peirce (1839–1914), who asserted the nature of humans as symbol-using or sign-using organisms. Symbolic notation is half of mathematics, Bertrand Russell once said.*

The symbols for connectives in logic are introduced below. A comparison between these and the symbols of set notation is given on page 48.

By the way, do the words "No U Turn" make a statement as defined in this chapter?

Be careful when forming the negation of a statement involving the words *all, some,* or *none.* For example, let us write the negation of

All dogs have fleas.

This statement is false. Its negation must be true. Many people would write the negation as "No dogs have fleas" or "All dogs do not have fleas." However, these last statements are also false, and cannot be negations of the given statement. The correct negation of "All dogs have fleas" is actually

Not all dogs have fleas

or   It is not the case that all dogs have fleas

or   At least one dog does not have fleas

or   Some dogs do not have fleas.

The word *some* means "at least one." *possibly all*

Example 3 Write the negation of each statement.

(a) Some dogs have fleas.

Solution. Since *some* means "at least one," the statement "Some dogs have fleas" is really the same as "At least one dog has fleas." The negation of this is   No dog has fleas.

(b) Some dogs do not have fleas.

Solution. This statement claims that at least one dog, somewhere, does not have fleas. The negation of this is   All dogs have fleas.

(c) No dogs have fleas.   The negation is: Some dogs have fleas.

The following table can be used to find the negation of a statement involving *all, some,* or *no.*

Original statement	Negation
All dogs have fleas.	Some dogs do not have fleas.
Some dogs have fleas.	No dogs have fleas.
Some dogs do not have fleas.	All dogs have fleas.
No dogs have fleas.	Some dogs have fleas.

Symbols To simplify our work with logic, we introduce some standard symbols. Statements are represented with letters, such as p, q, or r. Symbols for some connectives are shown in the table. The table also gives the type of compound statement having the given connective.

Connective	Symbol	Type of statement
and	$\wedge$	Conjunction
or	$\vee$	Disjunction
not	$\sim$	Negation

Example 4 Let p represent "It is hot today," and let q represent "It is windy today." Write each symbolic statement in words:
(a) p ∨ q (b) ~p ∧ q.

(**a**) From the table above, ∨ symbolizes *or;* thus, p ∨ q represents

It is hot today or it is windy today.

(**b**) It is *not* hot today *and* it is windy today.

The symbol ~ represents the connective *not.* If p represents the statement "Each student has a book," then ~p represents the negation of this statement, that is, "Not every student has a book."

2.2 EXERCISES

Decide whether each of Exercises 1–16 is a statement.

1. July 4, 1976, was a Monday.

2. The area code of Chicago is 312.

3. Bring the paper.

4. This book costs $1.

5. $8 + 7 = 15$

6. $9 + 12 = 21$

7. $4 - 2 = 1$

8. $17 + 3 = 22$

9. Not all numbers equal 0.

10. Millard Fillmore was President in 1850.

11. On the average, women outlive men.

12. Smoking is delightful.

13. I can't stand the mayor.

14. Your Honda makes too much noise.

15. A parrot at Busch Gardens in Tampa can recite the U.S. Constitution, including Amendments.

16. One cubic foot of gold weighs less than twelve pounds.

Name the connective used in each of Exercises 17–24 that is compound. Write *not compound* otherwise.

17. I like Joyce or I like Fred.

18. Not all men are immortal.

19. Today is Saturday or today is Sunday.

20. Jose is over 30, and so is Lupe.

21. $3 + 1 \neq 4 - 8$

22. $6 \neq 2 + 5$

23. All chickens eat chicken feed.

24. Some cowboys wear levis.

Let p represent the statement "He has brown hair," and let q represent "He is tall." Convert each compound statement into words.

25.	~p	**29.**	~p ∨ q	***33.**	~(~p ∧ q)
26.	~q	**30.**	p ∧ ~q	***34.**	~(p ∨ ~q)
27.	p ∧ q	**31.**	~p ∨ ~q		
28.	p ∨ q	**32.**	~p ∧ ~q		

Let p represent the statement "Laura is tall." and let q represent "Freddie is tall." Convert each of the following compound statements into symbols. Assume that "tall" and "short" are opposites.

35. Both Laura and Freddie are tall.

36. Laura is tall and Freddie is short.

37. Laura is short and so is Freddie.

38. Laura is short, or Freddie is tall.

*39. Either Laura is tall or Freddie is tall, but not both.

*40. Neither is tall.

Write a negation for each statement in Exercises 41–53.

41. His mother is tall.

42. The grass needs to be cut.

43. All cows were once calves.

44. No dew forms in the desert.

45. Some people eat pancakes.

46. All stars are yellow.

47. All tax must be paid by April 15.

48. Some sentences have stupid negations.

49. Some ground squirrels are not happy.

50. No telephone installer can play poker.

*51. Nobody doesn't like Sara Lee.

*52. Everybody doesn't like me.

*53. Everybody loves somebody sometime.

Errors in forming negations of statements are very common in everyday language. For each of Exercises 54–58, decide what is literally meant by an exact reading of the statement. Then decide what is probably *really* meant.

*54. No one may enter the White House.

*55. Everyone can't be an expert in mathematics.

*56. All people are not on the committee.

*57. I have tried all week to find a book on how to play the tuba without success.

*58. Do not open this door.

*59. "Albany Kills Bill to Repeal Law Against Birth Control." Does this headline state that action was taken for or against it?

*60. What is the final status of the law here? "Supreme Court refuses to hear challenge to lower court decision approving trial judge's refusal to allow defendant to refuse to speak."

Brain teaser *True statements may have opposites, which are usually false. But consider the following statement:*
"The number of words in this sentence is nine."
That's a true statement. Now think up an opposite statement that is also true! (One answer is on page 53.)

Sets	Logic
A'	~p
A ∩ B	p ∧ q
A ∪ B	p ∨ q
U	T
∅	F

The above table compares symbols in set theory and in logic.

Negations of statements are analogous to complements of sets. Intersection of sets plays the same role that the connective and plays in logic; similarly, union and or. The universal and empty sets compare with the logical ideas of T and F respectively.

2.3 TRUTH TABLES

In this section, we look at the truth values (true or false) of compound statements. To begin, let us decide on the truth values of the conjunction p *and* q, symbolized p $\wedge$ q.

In everyday language, the connective *and* implies the idea of "both." Thus,

A quarter is round and a dollar bill is rectangular

is true, since both component statements are true. On the other hand,

A quarter is round and a dollar bill is round

is false, even though part of the statement (a quarter is round) is true. For the conjunction p $\wedge$ q to be true, both p and q must be true. This is summarized by a table, called a **truth table,** which shows all four of the possible combinations of truth values for the conjunction p *and* q.

AND

p	q	p $\wedge$ q
T	T	T
T	F	F
F	T	F
F	F	F

Conjunction truth table

Example 1 Let p represent "Austin is the capital of Texas," and let q represent "Charleston is the capital of South Carolina." Here p is true and q is false, so looking in the second row of the conjunction truth table we see that p $\wedge$ q is false.

In ordinary language, the word *or* is ambiguous. The expression "this or that" can mean either "this or that or both," or "this or that but not both." For example,

I will paint the wall or I will paint the ceiling

has the following meaning: "I will paint the wall or I will paint the ceiling or I will paint both." On the other hand,

I will take Linda or Mary to the film festival

probably means "I will take Linda, or I will take Mary, but I shall not take both."

Let us agree that the symbol $\vee$ represents the first *or* described. That is, p $\vee$ q means "p or q or both." With this meaning of *or*, p $\vee$ q is called the *inclusive disjunction,* or just the **disjunction** of p and q.

In everyday language, the disjunction implies the idea of "either." For example, the disjunction

I have a quarter or I have a dime.

is true whenever I have either a quarter, a dime or both. The only way this disjunction could be false would be if I had neither coin: A disjunction is false only if both component statements are false.

OR

p	q	p $\vee$ q
T	T	T
T	F	T
F	T	T
F	F	F

Disjunction truth table

We know that the negation of a statement has an opposite truth value from the statement itself. This leads to the truth table for *negation.*

NOT

p	~p
T	F
F	T

Negation truth table

Example 2   Suppose p is false, q is true, and r is false. Is the compound statement ~p ∧ (q ∨ ~r) true or false?

Solution.   Here parentheses are used to group q and ~r together. Work first inside the parentheses. Since r is false, ~r will be true. We know q is true. Thus, to find the truth value of q ∨ ~r, we look in the first row of the *or* truth table. From this row, we find the result T. Since p is false, ~p is true, and the final truth value of ~p ∧ (q ∨ ~r) is found in the top row of the *and* truth table. From the *and* truth table, we see that when p is false, q is true, and r is false, then the statement ~p ∧ (q ∨ ~r) is true.

In Example 2 we found the truth value for a given statement by going back to the basic truth tables. In the long run, it is easier if we first create a complete truth table for the given statement itself. Then final truth values can be read directly from this table. The procedure for making new truth tables is shown in the next few examples.

Example 3   Construct a truth table for (~p ∧ q) ∨ ~q.

Solution.   We begin by listing all possible combinations of truth values for p and q, as we did above. We then find the truth values of ~p ∧ q. Do this by listing the truth values of ~p, which are the opposite of those of p.

p	q	~p
T	T	F
T	F	F
F	T	T
F	F	T

Use only the "~p" column and the "q" column, along with the *and* truth table, to find the truth values of ~p ∧ q. List them in a separate column:

p	q	~p	~p ∧ q
T	T	F	F
T	F	F	F
F	T	T	T
F	F	T	F

Next include a column for ~q.

p	q	~p	~p ∧ q	~q
T	T	F	F	F
T	F	F	F	T
F	T	T	T	F
F	F	T	F	T

Finally, make a column for the entire compound statement. To find the truth values, use *or* to combine ~p ∧ q with ~q.

p	q	~p	~p ∧ q	~q	(~p ∧ q) ∨ ~q
T	T	F	F	F	F
T	F	F	F	T	T
F	T	T	T	F	T
F	F	T	F	T	T

Example 4 Suppose both p and q are true. Find the truth value of (~p ∧ q) ∨ ~q.

Solution. Look in the first row of the final truth table for Example 3, where both p and q have truth value T. Read across the row to find that the compound statement is false.

Example 5 Find the truth table for p ∧ (~p ∨ ~q).

Solution. Proceed as shown.

p	q	~p	~q	~p ∨ ~q	p ∧ (~p ∨ ~q)
T	T	F	F	F	F
T	F	F	T	T	T
F	T	T	F	T	F
F	F	T	T	T	F

Example 6 Find the truth table for (~p ∧ r) ∨ (~q ∧ ~p).

Solution. This statement has three component statements, p, q, and r. The truth table thus requires eight lines to list all possible combinations of truth values of p, q, and r. The final truth table, however, can be found in much the same way that we found the ones above.

p	q	r	~p	~p ∧ r	~q	~q ∧ ~p	(~p ∧ r) ∨ (~q ∧ ~p)
T	T	T	F	F	F	F	F
T	T	F	F	F	F	F	F
T	F	T	F	F	T	F	F
T	F	F	F	F	T	F	F
F	T	T	T	T	F	F	T
F	T	F	T	F	F	F	F
F	F	T	T	T	T	T	T
F	F	F	T	F	T	T	T

A statement with four different letters would require a truth table with sixteen lines, five letters would require thirty-two lines, and so on. The moral: If you are on a quiz show don't choose "truth tables of complicated statements" as a category.

2.3 EXERCISES

For Exercises 1–10, let p represent a true statement, and let q represent a false statement. Find the truth value of the entire statement.

1. ~p   **5.** p ∨ ~q   **9.** ~(p ∧ ~q)

2. ~q   **6.** ~p ∧ q   **10.** ~(~p ∨ ~q)

3. p ∨ q   **7.** ~p ∨ ~q

4. p ∧ q   **8.** p ∧ ~q

Let p represent a false statement, while q and r represent true statements. Find the truth value of each compound statement in Exercises 11–18.

11. p ∧ r   **15.** ~(p ∧ q) ∧ (r ∨ ~q)

12. q ∨ ~r   **16.** (~r ∧ ~q) ∨ (~r ∧ q)

13. p ∧ (q ∨ r)   **17.** ~[(~p ∧ q) ∨ r]

14. (~p ∧ q) ∨ ~r   **18.** ~[r ∨ (~q ∧ ~p)]

Make a truth table for each compound statement in Exercises 19–34.

19. ~p ∨ ~q   **27.** (~p ∧ ~q) ∨ (~p ∨ q)

20. ~p ∧ q   **28.** (p ∨ ~q) ∧ (p ∧ q)

21. p ∨ ~q   **29.** r ∨ (p ∧ ~q)

22. ~(p ∧ q)   **30.** (~p ∧ q) ∧ r

23. (p ∧ ~q) ∧ p   **31.** (~r ∨ ~p) ∧ (~p ∨ ~q)

24. (q ∨ ~p) ∨ ~q   **32.** (~p ∧ ~q) ∨ (~r ∨ ~p)

25. ~p ∨ (~q ∧ ~p)   ***33.** (~r ∨ s) ∧ (~p ∧ q)

26. ~q ∧ (~p ∨ q)   ***34.** ~(~p ∧ ~q) ∨ (~r ∨ ~s)

Tell how many rows would be in the truth tables for each compound statement.

35. (~p ∧ q) ∨ (~r ∨ ~s)

36. [(p ∨q) ∧ (r ∧ s)] ∧ t

37. [(~p ∧ ~q) ∧ (~r ∧ s ∧ ~t)] ∧ (~u ∨ ~v)

38. [(~p ∧ ~q) ∨ (~r ∨ ~s)] ∨ [(~m ∧ ~n) ∧ (~p ∧ ~r)]

39. Lawyers sometimes use the phrase "and / or." This phrase corresponds to which usage of the word *or* as discussed in the text?

Decide whether the following compound statements are true or false. Here, *or* is the exclusive disjunction; p or q means "either p or q, but not both."

40. 3 + 1 = 4 or 2 + 5 = 7   **42.** 3 + 1 = 7 or 2 + 5 = 7

41. 3 + 1 = 4 or 2 + 5 = 9   **43.** 3 + 1 = 7 or 2 + 5 = 9

44. Complete the truth table for *exclusive disjunction*. The symbol ⊻ represents "one or the other, but not both."

p	q	p ⊻ q
T	T	F
T	F	T
F	T	T
F	F	F

Exclusive disjunction

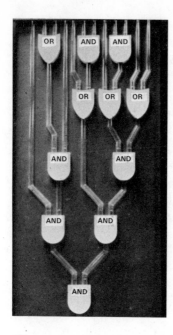

Make truth tables for the following.

*45. p $\underline{\vee}$ ~q

*46. ~p $\underline{\vee}$ q

*47. (~p ∧ q) $\underline{\vee}$ (~p)

*48. (~p $\underline{\vee}$ ~q) ∧ q

*49. ~(~p ∧ q) $\underline{\vee}$ (~p $\underline{\vee}$ ~q)

*50. (~p $\underline{\vee}$ q) $\underline{\vee}$ (p ∧ ~q)

The photograph shows the AND and OR game at the Ontario Science Center in Toronto. To win, you must release six of the twelve Ping-Pong balls at the top, with exactly one Ping-Pong ball making it to the bottom. A ball will pass through an AND gate only when both tubes entering the gate contain a ball. A ball passes through the OR gate if either or both tubes are filled.

51. Pick six balls so that one makes it to the bottom.

52. Can you pick a different set of six balls so that one gets through?

Exercises 53–56 come from "Solving Whodunits by Symbolic Logic," by Lawrence Sher, in *The Two-Year College Mathematics Journal*, December 1975, p. 36.

When Charlie Chan neared the end of a case, he would gather all the suspects together in a room. As he recited the clues that he had unearthed, the circle of guilt would narrow until it contained only the suspect. Chan's method was derived from the greatest of all fictional detectives, Sherlock Holmes. Holmes' technique was summarized as follows: "To solve a mystery, I simply eliminate all that is impossible. Whatever remains, however improbable, is the solution." We can apply the Holmes method to solve whodunits by the use of truth tables. Consider the case in which we know

He did it or she did it, and he didn't do it.   (*)

53. Let H represent "He did it," and let S represent "She did it." Write the statement labeled (*) in symbols.

There are two people involved, so that there are four possibilities: (1) Both he and she did it, (2) He did it and she didn't, (3) She did it and he didn't, (4) Neither did it. Each line in a truth table for the clue represents one of the possibilities. Holmes' method is to eliminate the impossible. The truth table shows what is impossible by having the row for those cases turn out to be false. (Another set of clues is given at the end of Section 2.4.)

54. Form the truth table for (H ∨ S) ∧ ~H.

55. Which row of the truth table leads to a true statement?

56. Who did it?

An answer to the brain teaser: *"The number of words in this sentence is not nine."*

2.4 THE CONDITIONAL

One of the most common connectives in everyday use is the **conditional,** or *implication,* with the connective *if . . . then.* For example,

If I buy a new shirt, *then* I will want new pants.
If you get a Big Mac, *then* you will want fries.
If it doesn't rain soon, *then* I will have to water the lawn.

In each of these conditional statements, the component coming after the word *if* gives a condition under which the statement coming after *then* will be true. Thus, "If it is over 90°, then I'll go to the mountains" tells us one condition under which I will go to the mountains—if the temperature is over 90°.

The conditional is written with an arrow, so that "if p, then q" is symbolized as

$$p \rightarrow q.$$

Read $p \rightarrow q$ as "p implies q" or "p arrow q."

The conditional connective is "hidden" in many everyday expressions. For example, the statement

Tom always buys a Ford

can be written in the *if . . . then* form as

If Tom buys a car, then the car is a Ford.

IF . . . THEN

p	q	p → q
T	T	T
T	F	F
F	T	T
F	F	T

Conditional truth table;
ARROW truth table

Also, "She buys a toy when she is in Tampa" can be written

> If she is in Tampa, then she buys a toy.

The arrow truth table is a little harder to construct than were the tables in the previous section. To see how to find the arrow truth table, let us analyze a statement made by a politician, Senator Spend:

> If I am elected, then taxes will go down.

As before, there are four possible combinations of truth values for the two simple statements. (We let p represent "I am elected," and let q represent "Taxes go down."

(1) She was elected, and taxes did go down.   (p is T, q is T)

(2) She was elected, and taxes did not go down.   (p is T, q is F)

(3) She was defeated, and taxes went down.   (p is F, q is T)

(4) She was defeated, and taxes did not go down.   (p is F, q is F)

(1)   For the first possibility we assume that the Senator was elected and taxes did go down (p is T, q is T). The Senator told the truth, and we place T in the first row of the truth table. We do not claim that taxes went down *because* she was elected; it is possible that she had nothing to do with it at all.

(2)   Next, if the Senator was elected and taxes did not go down (p is T, q is F), then she did not tell the truth. Thus, we place F in the second row of the truth table.

(3)   In the third case, we assume the Senator was defeated, but taxes went down anyway (p is F, q is T). Senator Spend did not lie; she only promised a tax reduction if she were elected. She said nothing about what would happen if she were not elected. In fact, her campaign promise gives no information about what would happen if she lost. Since we cannot say that the Senator lied, we place T in the third row of the truth table.

(4)   Also, if the Senator is defeated and taxes do not go down (p is F, q is F) we cannot blame her, since she only promised to reduce taxes if elected. Thus, T goes in the last row also. The complete truth table for the conditional is shown here.

We emphasize that the use of an arrow connective in no way implies a cause and effect relationship. Any two statements may have an arrow placed between them to create a compound statement. For example

> If I pass mathematics, then the sun will rise the next day

is true, since the second part is true. (Why does this make the entire statement true?) There is, however, no cause and effect connection between my passing mathematics and the sun's rising. The sun will rise no matter what grade I get in a course.

Logic circuit board   (*opposite*) *from Bell Telephone Laboratories has "spaghetti" connectives. Symbolic logic shows how to simplify circuits (see pages 71– 73). Special logic circuits in computers are necessary for step-by-step performance of calculations.*

Many devices and contraptions are described by Martin Gardner in Logic Machines and Diagrams *(McGraw-Hill, 1958). One early electrical logic machine was hooked to a phonograph so that the machine could "answer" as follows: "My name is Leibniz Boole De Morgan Quine. I am nothing but a complicated and slightly neurotic robot, but it gives me great pleasure to announce that your statement is absolutely false." (Professor Willard V. Quine is a logician.)*

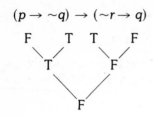

$(p \to \sim q) \to (\sim r \to q)$

Example 1   Find the truth value of the statement $(p \to \sim q) \to (\sim r \to q)$ if p, q, and r are all false.

Solution.   As before, we first work inside the parentheses. Since q and r are both false, $\sim q$ and $\sim r$ are both true. The final truth value can be found through the steps shown in the margin.

The statement $(p \to \sim q) \to (\sim r \to q)$ is false when p, q, and r are all false.

We can work out truth tables with arrows just as we did in the previous section. This procedure is shown in the next examples.

Example 2   Make a truth table for $(\sim p \to \sim q) \to (\sim p \land q)$.

Solution.   First insert the truth values of $\sim p$ and of $\sim q$. Then find the truth value of $\sim p \to \sim q$.

p	q	$\sim p$	$\sim q$	$\sim p \to \sim q$
T	T	F	F	T
T	F	F	T	T
F	T	T	F	F
F	F	T	T	T

Next use $\sim p$ and q to find the truth values of $\sim p \land q$.

p	q	$\sim p$	$\sim q$	$\sim p \to \sim q$	$\sim p \land q$
T	T	F	F	T	F
T	F	F	T	T	F
F	T	T	F	F	T
F	F	T	T	T	F

Now use the arrow truth table to combine the truth values of $\sim p \to \sim q$ and $\sim p \land q$.

p	q	$\sim p$	$\sim q$	$\sim p \to \sim q$	$\sim p \land q$	$(\sim p \to \sim q) \to (\sim p \land q)$
T	T	F	F	T	F	F
T	F	F	T	T	F	F
F	T	T	F	F	T	T
F	F	T	T	T	F	F

Example 3   Make a truth table for $(p \to q) \to (\sim p \lor q)$.

Solution.   Go through steps similar to the ones above.

p	q	$p \to q$	$\sim p$	$\sim p \lor q$	$(p \to q) \to (\sim p \lor q)$
T	T	T	F	T	T
T	F	F	F	F	T
F	T	T	T	T	T
F	F	T	T	T	T

As the truth table shows, the statement $(p \rightarrow q) \rightarrow (\sim p \vee q)$ is always true, no matter what the truth values of the components. Such a statement is called a **tautology**. Other examples of tautologies (as can be checked by forming truth tables) include $p \vee \sim p$, $p \rightarrow p$, $(\sim p \vee \sim q) \rightarrow \sim (q \wedge p)$, and so on.

2.4 EXERCISES

Rewrite the following statements in Exercises 1–12 using the *if . . . then* connective. Rearrange the wording or add words as necessary.

1. If it's Tuesday, this must be Belgium.
2. If you watch too much television, you'll get a headache.
3. It's a bird if it flies.
4. You can believe it if it's in *The Wall Street Journal*.
5. Sally goes downtown every Saturday.
6. Snoopy likes dog food.
7. All people have heads.
8. All biologists love formaldehyde.
9. No chickens are teetotalers.
10. No candy comes from Bolivia.
11. Europe trembles when Napoleon shouts.
12. A banana tree won't grow here.

Let p represent the statement "I study hard," let q represent the statement "I pass this lousy course," and let r represent "My instructor is weird." Express each compound statement in words.

13. $p \rightarrow q$	17. $\sim p \rightarrow \sim r$	21. $(p \wedge q) \rightarrow r$
14. $q \rightarrow r$	18. $\sim q \rightarrow \sim p$	22. $(\sim p \vee \sim r) \rightarrow \sim q$
15. $\sim r \rightarrow p$	19. $p \rightarrow (r \wedge q)$	
16. $\sim p \rightarrow q$	20. $\sim q \rightarrow (\sim r \vee p)$	

Let p represent "I go to town," let q represent "It rains," and let r represent "I'll buy an umbrella." Write each compound statement in symbols.

23. If I go to town, then I'll buy an umbrella.
24. If it rains, then I'll go to town.
25. If I buy an umbrella, then it won't rain.
26. If I don't go to town, then it won't rain.
27. I'll buy an umbrella, and if it rains, then I won't go to town.
28. I'll go to town, or if I buy an umbrella, then it will rain.
29. It'll rain if I buy an umbrella.
30. I'll go to town if it doesn't rain.

Find the truth values of the statements in Exercises 31–42. Assume that p and r are true, with q false.

31. $p \rightarrow \sim q$		**37.** $\sim q \rightarrow (p \wedge r)$	
32. $\sim r \rightarrow q$		**38.** $p \rightarrow (q \vee \sim r)$	
33. $\sim p \rightarrow \sim r$		**39.** $(\sim r \vee p) \rightarrow (q \wedge p)$	
34. $\sim q \rightarrow r$		**40.** $(p \rightarrow \sim q) \wedge (p \rightarrow r)$	
35. $r \rightarrow \sim p$		**41.** $(\sim p \wedge \sim q) \rightarrow (p \wedge \sim r)$	
36. $p \rightarrow q$		**42.** $(p \rightarrow \sim q) \rightarrow (\sim p \wedge \sim r)$	

Make a truth table for each statement in Exercises 43–54. Identify any tautologies.

43. $p \rightarrow \sim q$	**49.** $(\sim p \rightarrow \sim q) \rightarrow (p \wedge q)$
44. $\sim q \rightarrow p$	**50.** $(p \vee q) \rightarrow (q \vee p)$
45. $p \rightarrow (p \vee q)$	**51.** $[(r \vee p) \wedge \sim q] \rightarrow p$
46. $\sim q \rightarrow (p \wedge \sim q)$	**52.** $r \rightarrow (p \wedge \sim q)$
47. $(\sim p \rightarrow q) \rightarrow p$	***53.** $(\sim r \rightarrow s) \vee (p \rightarrow \sim q)$
48. $(\sim q \rightarrow \sim p) \rightarrow \sim q$	***54.** $(\sim p \wedge \sim q) \rightarrow (\sim r \rightarrow \sim s)$

Two statements are *equivalent* if they have exactly the same truth tables. Decide which pairs of statements in Exercises 55–62 are equivalent.

Example: $\sim p \wedge \sim q$ and $\sim(p \vee q)$. Are they equivalent statements?

Solution.   Make a truth table for each statement, with results as follows.

p	q	$\sim p \wedge \sim q$
T	T	F
T	F	F
F	T	F
F	F	T

p	q	$\sim(p \vee q)$
T	T	F
T	F	F
F	T	F
F	F	T

Since both truth tables are the same, then $\sim p \wedge \sim q$ and $\sim(p \vee q)$ are equivalent. Equivalence is written with a three-bar sign, $\equiv$, in distinction to the equals sign. Thus, $\sim p \wedge \sim q \equiv \sim(p \vee q)$.

55. $\sim p \vee \sim q$ and $\sim(p \wedge q)$	**59.** $p \rightarrow q$ and $q \rightarrow p$
56. $p \rightarrow q$ and $\sim p \vee q$	**60.** $\sim(\sim p)$ and p
57. $q \rightarrow p$ and $\sim p \rightarrow \sim q$	**61.** $\sim p \wedge q$ and $\sim p \rightarrow q$
58. $p \rightarrow q$ and $\sim q \rightarrow \sim p$	**62.** $p \wedge \sim q$ and $\sim q \rightarrow \sim p$

The two equivalences

$$\sim p \wedge \sim q \equiv \sim(p \vee q) \qquad \text{[in the example above]}$$

and $\qquad \sim p \vee \sim q \equiv \sim(p \wedge q) \qquad$ [Exercise 55]

are called *De Morgan's Laws,* after the British logician Augustus De Morgan (1806–1871).

The connectives *unless* and *either . . . or* are common in everyday language. Define *p unless q* by p unless $q \equiv \sim q \rightarrow p$. (If q does not occur, then p will, or, as long as q does not occur, p will occur.)

*63. Write a truth table for *unless*.

Define *either p or q* by either p or $q \equiv (p \vee q) \wedge \sim(p \wedge q)$.

*64. Write a truth table for *either . . . or*.

Decide on the truth value for each of the following. Assume p and r are true, but q is false.

*65. either p, or (q unless r)

*66. $\sim p$ unless (either $\sim p$ or $\sim r$)

*67. $p \wedge$ (either $\sim p$ or q)

*68. $(\sim q \wedge \sim p) \rightarrow (\sim p$ unless $r)$

*69. $(r \wedge \sim p) \rightarrow$ (either $\sim p$ or $\sim q$)

*70. $(\sim p \rightarrow \sim q) \rightarrow (\sim r$ unless $p)$

2.5 MORE ON THE CONDITIONAL

The connective *if . . . then* is perhaps the most useful of the various connectives that we have studied. This is because of the many ways that this connective can be translated into English. For example,

> If you go to the shopping center,
> then you will find a place to park

can also be written

> Going to the shopping center is *sufficient*
> for finding a place to park.

According to this statement, going to the shopping center is enough to guarantee finding a place to park. Going to other places, such as schools or office buildings, might also guarantee a place to park, but at least we know that going to the shopping center does. Thus, $p \rightarrow q$ can be written "p is sufficient for q." Knowing that p has occurred is sufficient to guarantee that q will also occur. *On the other hand,*

> Going to the shopping center is necessary
> for finding a place to park $\hspace{2cm}$ (*)

has a different meaning. Here, we are saying that one condition that is necessary for finding a parking place is that you go to the shopping center. This may not be enough; there may be no parking spaces during the Christmas season, for example. We could rewrite the statement labeled with (*) as

> If you find a place to park, then you went to the shopping center.

CB Static Watch what questions you ask over CB radio. You may get an answer like the following, which a friend of ours got when he asked for the time.

If I tell you the time, then we'll start chatting. If we start chatting, then you'll want to meet me at a truck stop. If we meet at a truck stop, then we'll discuss my family. If we discuss my family, then you'll find out that my daughter is available for marriage. If you find out that she is available for marriage, then you'll want to marry her. If you want to marry her, then my life will be miserable since I don't want my daughter married to some fool who can't afford a $10 watch.

If I tell you the time,
then my life will be miserable.

As this example shows, $p \rightarrow q$ is the same as "q is necessary for p." In other words, if q doesn't happen, then neither will p.

The statement "If you get a parking place, then you went to the shopping center" can also be written "You get a parking place *only if* you went to the shopping center."

In summary, $p \rightarrow q$ can be translated into words in any of the following ways:

if p, then q

q, if p

p implies q

p, only if q

p is sufficient for q

q is necessary for p.

Traffic symbol *Does this stand for a conditional statement?*

Example 1 The statement

If you are 18, then you can vote

can be written in any of the following ways (among others).

You can vote if you are 18.
You are 18 only if you can vote.
Being able to vote is necessary for you to be 18.
Being 18 is sufficient for being able to vote.

Example 2 Let p represent "A triangle is equilateral," and let q represent "A triangle has three equal sides." Write each of the following in symbols.

(a) A triangle is equilateral if it has three equal sides.

$$q \rightarrow p$$

(b) A triangle is equilateral only if it has three equal sides.

$$p \rightarrow q$$

If we form the conjunction of the statements in Example 2, we get

p if q *and* p only if q

which is often written *p if and only if q*. We use the *biconditional* symbol $\leftrightarrow$ to represent "if and only if," and write

p if and only if q   or $p \leftrightarrow q$.

If we form the conjunction of the answers we found in Example 2, we get

$$(p \rightarrow q) \wedge (q \rightarrow p)$$

which we use as the definition of the **biconditional statement** $p \leftrightarrow q$.

Example 3 Make a truth table for $p \leftrightarrow q$.

Solution. Use the definition of $p \leftrightarrow q \equiv (p \rightarrow q) \wedge (q \rightarrow p)$.

$$p \leftrightarrow q$$

p	q	$(p \rightarrow q) \wedge (q \rightarrow p)$
T	T	T
T	F	F
F	T	F
F	F	T

Biconditional truth table

Example 4 Tell whether each compound statement is true or false.

(a) $6 + 9 = 15$ if and only if $12 + 4 = 16$.

Both $6 + 9 = 15$ and $12 + 4 = 16$ are true. By the truth table for the biconditional, the compound statement is true.

(b) $5 + 2 = 10$ and $17 + 19 = 36$.

Since the first component is false, $(5 + 2 = 10)$ and the second is true, the entire compound statement is false.

Given a conditional statement $p \rightarrow q$ (called the **direct statement**) there are three new statements that we may form:

The **converse** of $p \rightarrow q$ is $q \rightarrow p$.

The **inverse** of $p \rightarrow q$ is $\sim p \rightarrow \sim q$.

The **contrapositive** of $p \rightarrow q$ is $\sim q \rightarrow \sim p$.

Example 5 Write the converse, inverse, and contrapositive of the statement

If I have a dollar, then I can afford to mail a letter.

Solution. Let p represent "I have a dollar," and q represent "I can afford to mail a letter." Then the direct statement may be written $p \rightarrow q$. The converse, $q \rightarrow p$, is

If I can afford to mail a letter, then I have a dollar.

The converse is not true, even though the direct statement is. The inverse, $\sim p \rightarrow \sim q$, is

If I don't have a dollar,
then I cannot afford to mail a letter,

which is again not necessarily true. The contrapositive, $\sim q \rightarrow \sim p$, is true:

If I cannot afford to mail a letter,
then I do not have a dollar.

Element tables and truth tables

Given two sets A and B, there are four possibilities for x being an element (member) of the sets:

$x \in A$ and $x \in B$
$x \in A$ and $x \notin B$
$x \notin A$ and $x \in B$
$x \notin A$ and $x \notin B$

List this in a table:

A	B
$x \in A$	$x \in B$
$x \in A$	$x \notin B$
$x \notin A$	$x \in B$
$x \notin A$	$x \notin B$

To simplify, write only $\in$ and $\notin$ to make an element table. How does it compare with the truth table alongside?

A	B		p	q
$\in$	$\in$		T	T
$\in$	$\notin$		T	F
$\notin$	$\in$		F	T
$\notin$	$\notin$		F	F

This example shows that the converse and inverse of a true statement need not be true. They *can* be true, but need not be. The relationship between the truth values of the direct statement, converse, inverse, and contrapositive, are shown in the next example.

Example 6 Make a truth table of the statements $p \to q$, $q \to p$, $\sim p \to \sim q$, and $\sim q \to \sim p$.

Solution.

		Direct	Converse	Inverse	Contrapositive
p	q	$p \to q$	$q \to p$	$\sim p \to \sim q$	$\sim q \to \sim p$
T	T	T	T	T	T
T	F	F	T	T	F
F	T	T	F	F	T
F	F	T	T	T	T

As this truth table shows, the direct statement and the contrapositive always have the same truth values. Thus, any statement may be replaced by its contrapositive without affecting its logical meaning. (We shall use this idea in the next section.) Also, the converse and inverse always have the same truth values.

Example 7 Write the converse, inverse, and contrapositive of $\sim p \to \sim q$.

Solution.   The converse is $\sim q \to \sim p$. The inverse is $p \to q$, and the contrapositive is $q \to p$.

2.5 EXERCISES

Write each of the following in the form *if . . . then.*

1. I will stay home if it rains.
2. I will go skiing only if it snows.
3. Doing homework is necessary for passing this class.
4. Watering the grass is sufficient to make it grow.
5. Being in Raleigh is sufficient for being in North Carolina.
6. I drive fast only if I have a Porsche.
7. Being against pollution is necessary to be elected.
8. Long grass is necessary before I will cut it.
9. The city will redevelop Eastlake only if the government approves.
10. Doing logic is (very) sufficient for driving me crazy.
11. I feel bad only if I lose on the slot machines.
12. Going from Boardwalk to Oriental, it is necessary that I pass Go.
13. I can live in West Virginia only if I like coal.
14. Having a CB radio is necessary for going 65 on the highway.

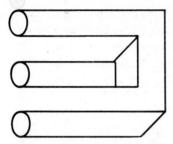

Write the converse, inverse, and contrapositive of each statement in Exercises 15–30.

15. If I buy a Ford, I will need a garage.

16. If she comes, then I go.

17. Cigarette smoking endangers health.

18. Orange juice contains Vitamin C.

19. All dogs have fleas.

20. People who live in big houses have big payments.

21. Birds of a feather flock together.

22. A rolling stone gathers no moss.

23. Moss grows on the north side of a tree.

*24. All that glitters is not gold. (Is this old saying true?)

25. $\sim p \to q$ **27.** $\sim p \to \sim q$ ***29.** $p \to (q \lor r)$

26. $p \to \sim q$ **28.** $\sim q \to \sim p$ ***30.** $(r \lor \sim q) \to p$

Find the truth value of each statement.

31. $3 + 1 = 6$ if and only if $9 = 8$

32. $5 = 9 - 4$ if and only if $8 + 2 = 10$

33. $6 \times 2 = 14$ if and only if $9 + 7 = 16$

34. $8 + 7 = 15$ if and only if $3 \times 5 = 9$

35. Truman was President if and only if Eisenhower was President.

36. Some money is green if and only if pennies are copper.

37. McDonalds sells Big Macs if and only if Exxon only sells trucks.

38. Topeka is the capital of Kansas if and only if Harrisburg is the capital of New York.

Two statements which can both be true about the same object are called *consistent.* For example, "It is green" and "It is small" are consistent statements. Statements which cannot both be true about the same object are called *contrary;* "It is a Ford" and "It is a Chevrolet" are contrary. Label the following pairs of statements as either *contrary* or *consistent.*

39. That is Tom Jones. That is Harry Smith.

40. That animal has four legs. That animal is a cow.

41. That is a pig. That is a chicken.

42. She is rich. She is poor.

43. He has power. He has wealth.

44. That clock is green. That clock is slow.

Let us take another example from the article by Lawrence Sher mentioned at the end of Section 2.3. As we saw, when a row in the truth table for a group of clues is false, the case is impossible and we eliminate it.

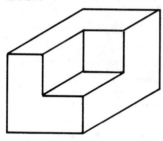

Visual paradoxes   *Here are two examples of optical illusions that change back and forth as you look at them.*

However, when a row ends in T, this does not mean that the case is the truth. It means that it is *possible*. The way to solve the case is to eliminate all but one possibility. The last remaining case is the truth.

Sometimes it takes more than one clue to eliminate all but one possibility. Consider the mystery for which we have the following clues.

(1)   If the butler did it, then the maid didn't.
(2)   The butler or the maid did it.
(3)   If the maid did it, then the butler did it.

45.   Write these clues in symbols. Use *b* for "The butler did it," and use *m* for "The maid did it."

46.   Make a truth table for the first clue.

47.   Which possibility can now be eliminated?

48.   We now know that the butler and the maid could not both have done it. Test the remaining three possibilities by completing a truth table for the second clue, $b \lor m$.

b	m	$b \lor m$
T	F	
F	T	
F	F	

49.   Which row is eliminated by this table? This now leaves only two possibilities to be tested by the last clue. Test these two cases, the butler did it alone and the maid did it alone, by completing a truth table for the last clue, $m \to b$.

b	m	$m \to b$
T	F	
F	T	

50.   Who did it?

Logic and problem solving
Logic forms the basis for problem solving behavior, and such behavior is of interest and concern to psychologists and other social scientists. For instance, do we have to lose our mental powers as we grow older? From recent studies, it appears that older people become less alert and lose their powers of memory mostly because they think they should: they are victims of social learning and expectation.

The National Institute of Aging is currently conducting long-range studies of problem-solving performance. The volunteer below is recording his responses to a light-box testing apparatus at the Gerontology Research Center in Baltimore.

2.6 VALID ARGUMENTS AND TRUTH TABLES

In Section 1 of this chapter we looked at methods of using diagrams to test the validity of arguments. While diagrams work fine for simple arguments, we run into problems with more complex arguments. These problems come from the fact that we test for valid arguments with diagrams by drawing a sketch showing every possible case. In more complex arguments, we are never sure that all cases have been considered.

Truth tables give us a way out of this difficulty. We can test the validity of arguments with a truth table, since the table ensures that we have considered every possible case. Let us look at an example:

If the floor is dirty, then I must mop it.

The floor is dirty.

I must mop it.

As mentioned in Section 2, this argument is made up of the two *premises* "If the floor is dirty, then I must mop it" and "The floor is dirty." The *conclusion* is "I must mop it."

To find out if this argument is valid, let us write the argument in symbols. If we let *p* represent "The floor is dirty," while *q* represents "I must mop it," the argument becomes

$$p \rightarrow q$$
$$\underline{p}$$
$$q$$

An argument is **valid** if the conclusion is true whenever all the premises are true. To test for validity of an argument using truth tables, proceed as follows:

Step 1. Form the conjunction of the premises.

In our argument, the conjunction of the premises is

$$(p \rightarrow q) \wedge p.$$

Step 2. Let the result from Step 1 imply the conclusion.

We have:

$$[(p \rightarrow q) \wedge p] \rightarrow q.$$

Step 3. Form the truth table from the result of Step 2. If this truth table is a tautology, the argument is valid. The following truth table shows this argument to be a tautology, showing that the argument is valid.

p	*q*	$p \rightarrow q$	$(p \rightarrow q) \wedge p$	$[(p \rightarrow q) \wedge p] \rightarrow q$
T	T	T	T	T
T	F	F	F	T
F	T	T	F	T
F	F	T	F	T

The pattern of argument used here,

$$p \rightarrow q$$
$$\underline{p}$$
$$q$$

is called *modus ponens,* or the **Law of detachment.**

Example 1 Is the following argument valid?

If diamonds are expensive, then coal is expensive.

Coal is not expensive.

Diamonds are not expensive.

$p \rightarrow q$
$\sim q$
$\sim p$

Solution. If p represents "Diamonds are expensive," and q represents "Coal is expensive," the argument can be written as in the margin.

Form the statement $[(p \rightarrow q) \wedge \sim q] \rightarrow \sim p$. The truth table for this statement is a tautology, so the argument is valid.

p	q	$p \rightarrow q$	$\sim q$	$(p \rightarrow q) \wedge \sim q$	$\sim p$	$[(p \rightarrow q) \wedge \sim q] \rightarrow \sim p$
T	T	T	F	F	F	T
T	F	F	T	F	F	T
F	T	T	F	F	T	T
F	F	T	T	T	T	T

The pattern of reasoning used here is called *modus tollens,* or the **Law of contraposition.**

The argument of Example 1 is valid even though the conclusion is false. A valid argument can have either a true conclusion or a false conclusion, depending on the truth values of the premises. If the premises are all true and the argument is valid, then the conclusion is true.

Example 2   Is the following argument valid?

If she buys a coat, then she will buy a hat.
She buys a hat.
She buys a coat.

Solution.   This argument can be written in the form

$p \rightarrow q$
q FALLACY
p

A truth table shows that the statement

$$[(p \rightarrow q) \wedge q] \rightarrow p$$

is not a tautology. The argument is therefore invalid. An invalid argument is called a **fallacy.**

If $p \rightarrow q$ is replaced above by $q \rightarrow p$, the argument becomes valid. The argument would be valid if the direct statement and the converse had the same truth values, which they do not. For this reason, this argument is sometimes called the **fallacy of the converse.**

In the same way,

$p \rightarrow q$
$\sim p$ FALLACY
$\sim q$

$p \rightarrow q$
$q \rightarrow r$
$p \rightarrow r$

is often called the **fallacy of the inverse.**

Example 3   Is the argument in the margin valid?

Solution. To find out, write the statement $[(p \rightarrow q) \land (q \rightarrow r)] \rightarrow (p \rightarrow r)$. If the truth table for this statement is a tautology, then the argument is valid. The truth table is found as follows.

p	q	r	$p \rightarrow q$	$q \rightarrow r$	$(p \rightarrow q) \land (q \rightarrow r)$	$p \rightarrow r$	$[(p \rightarrow q) \land (q \rightarrow r)] \rightarrow (p \rightarrow r)$
T	T	T	T	T	T	T	T
T	T	F	T	F	F	F	T
T	F	T	F	T	F	T	T
T	F	F	F	T	F	F	T
F	T	T	T	T	T	T	T
F	T	F	T	F	F	T	T
F	F	T	T	T	T	T	T
F	F	F	T	T	T	T	T

The argument is valid.

The form of logical argument in Example 3 can be used to solve the problems written by Lewis Carroll, author of *Alice's Adventures in Wonderland.* Lewis Carroll was really the Reverend Charles Lutwidge Dodgson, a mathematics lecturer at Oxford University in England. Queen Victoria told Dodgson how much she enjoyed *Alice* and how much she wanted to read his next book; he is said to have sent her *Symbolic Logic,* his most famous mathematical work. (Don't laugh—there are worse gifts.) In this book, Dodgson makes some mighty strong claims for the subject: "It [symbolic logic] will give you clearness of thought—the ability to *see your way* through a puzzle—the habit of arranging your ideas in an orderly and get-at-able form—and, more valuable than all, the power to detect *fallacies,* and to tear to pieces the flimsy illogical arguments which you will so continually encounter in books, in newspapers, in speeches, and even in sermons. . . ."

Queen Alice *Near the end of* Through the Looking Glass, *the sequel to* Alice's Adventures in Wonderland, *Alice finds a crown on her head. She scolds herself in the straightest Victorian manner: ". . . it'll never do for you to be lolling about on the grass like that! Queens have to be dignified, you know!" After walking around awhile with the crown on her head, she decides: ". . . if I really am a Queen, I shall be able to manage it quite well in time."*

Queen Victoria had the time. She assumed the crown in 1837, at the age of 18, and reigned over the British Empire until her death in 1901.

Tweedlogic *"I know what you're thinking about," said Tweedledum, "but it isn't so, nohow." "Contrariwise," continued Tweedledee, "if it was so, it might be; and if it were so, it would be, but as it isn't, it ain't. That's logic."*

The next example is from Lewis Carroll.

Example 4   Supply a valid conclusion for the following premises.

Babies are illogical.
Nobody is despised who can manage a crocodile.
Illogical persons are despised.

Solution.   First, write each premise in the form *if . . . then.*

If you are a baby, then you are illogical.
If you can manage a crocodile, then you are not despised.
If you are illogical, then you are despised.

Let p be "You are a baby," let q be "You are logical," let r be "You can manage a crocodile," and let s be "You are despised." With these letters, the statements can be written symbolically as

$$p \rightarrow \sim q$$
$$r \rightarrow \sim s$$
$$\sim q \rightarrow s.$$

Now choose any letter that appears only once. Here p appears only once. Using the contrapositive of $r \rightarrow \sim s$, which is $s \rightarrow \sim r$, we rearrange the three statements as follows:

$$p \rightarrow \sim q$$
$$\sim q \rightarrow s$$
$$s \rightarrow \sim r.$$

From the three statements, and the form of valid argument from Example 3 above, we get the valid conclusion

$$p \rightarrow \sim r.$$

In words, "If you are a baby, then you cannot manage a crocodile," or, as Lewis Carroll would have written it, "Babies cannot manage crocodiles."

Charles Dodgson (*1832–1898*) *began using the pen name* **Lewis Carroll** *to write humorous pieces for magazines. The Alice books made Carroll famous. Late in life, however, Dodgson shunned attention and denied that he and Carroll were the same person— though he gave away hundreds of signed copies to children and children's hospitals. Dodgson-Carroll merged again only in A Tangled Tale, ten short stories, each with a mathematical problem, and in The Game of Logic, designed to make learning logic fun.*

2.6 EXERCISES

Use truth tables to test the validity of each argument in Exercises 1–12.

1. $p \lor q$
$\underline{p }$
$\sim q$

5. $\sim p \to \sim q$
$\underline{q }$
p

9. $(p \to q) \land (q \to p)$
$\underline{p }$
$p \lor q$

2. $p \land \sim q$
$\underline{p }$
$\sim q$

6. $p \to \sim q$
$\underline{\sim p }$
$\sim q$

10. $(\sim p \lor q) \land (\sim p \to q)$
$\underline{p }$
$\sim q$

3. $p \to \sim q$
$\underline{q }$
$\sim p$

7. $p \to q$
$\underline{q \to p}$
$p \land q$

*11. $(r \land p) \to (r \lor q)$
$\underline{(q \land p) }$
$r \lor p$

4. $\sim p \to q$
$\underline{p }$
$\sim q$

8. $p \lor \sim q$
$\underline{p }$
$\sim q$

*12. $(\sim p \land r) \to (p \lor q)$
$\underline{(\sim r \to p) }$
$q \to r$

Test the validity of the arguments in Exercises 13–20.

13. If I buy a Honda, then I'll need a helmet.
I buy a Honda.
I'll need a helmet.

14. If my GI check comes in time, then I'll go to summer school.
I'm not going to summer school.
My GI check didn't come in time.

15. If I get ripped off, then I'll go to small claims court.
I'll go to small claims court.
I get ripped off.

16. If Helen Reddy comes to town, then I will go to the theatre.
Helen Reddy isn't coming to town.
I'm not going to the theatre.

17. If you like Tums, then you'll like Rolaids.
I don't like Rolaids.
I don't like Tums.

18. If I didn't have to do homework, then I'd have a beer.
I'm having a beer.
I didn't have to do homework.

19. If math were interesting, then I'd get an A.
Math isn't interesting.
I'm not getting an A.

20. If my calculator batteries die, then I will flunk math.
My batteries die.
I will flunk math.

Alice and the Cheshire Cat

"*In that direction,*" the Cat said, waving its right paw round, "*lives a Hatter: and in that direction,*" waving the other paw, "*lives a March Hare. Visit either you like: they're both mad.*"

"*But I don't want to go among mad people,*" Alice remarked.

"*Oh, you can't help that,*" said the Cat: "*we're all mad here. I'm mad. You're mad.*"

"*How do you know I'm mad?*" said Alice.

"*You must be,*" said the Cat, "*or you wouldn't have come here.*"

Alice didn't think that proved it at all: however, she went on: "*And how do you know that you're mad?*"

"*To begin with,*" said the Cat, "*a dog's not mad. You grant that?*"

"*I suppose so,*" said Alice.

"*Well, then,*" the Cat went on, "*you see a dog growls when it's angry, and wags its tail when it's pleased. Now I growl when I'm pleased, and wag my tail when I'm angry. Therefore I'm mad.*"

"*I call it purring, not growling,*" said Alice.

"*Call it what you like,*" said the Cat. "*Do you play croquet with the Queen today?*"

"*I should like it very much,*" said Alice, "*but I haven't been invited yet.*"

"*You'll see me there,*" said the Cat, and vanished.

The following premises come from Lewis Carroll. Write each premise in symbols, and then give a valid conclusion. For Exercises 21–23, let p be "ducks," q be "my poultry," r be "officers," and s be "willing to waltz."

21. No ducks waltz.   **23.** All my poultry are ducks.

22. No officers ever decline to waltz.   **24.** Give a valid conclusion.

In Exercises 25–27, let p be "able to do logic," q be "fit to serve on a jury," r be "sane," and s be "your sons."

25. Everyone who is sane can do logic.

26. No lunatics are fit to serve on a jury.

27. None of your sons can do logic.

28. Give a valid conclusion.

In Exercises 29–31, let p be "guinea pigs," q be "hopelessly ignorant," r be "keep silent," and s be "appreciates Beethoven."

29. Nobody who really appreciates Beethoven, fails to keep silent while the Moonlight Sonata is being played.

30. Guinea pigs are hopelessly ignorant of music.

31. No one who is hopelessly ignorant of music, ever keeps silent while the Moonlight Sonata is being played.

32. Give a valid conclusion.

In Exercises 33–37, let p be "honest," q be "pawnbrokers," r be "promise-breakers," s be "trustworthy," t be "very communicative," and u be "wine-drinkers."

33. Promise-breakers are untrustworthy.

34. Wine-drinkers are very communicative.

35. A person who keeps a promise is honest.

36. No teetotalers are pawnbrokers. (*Hint:* Assume "teetotaler" is the opposite of "wine-drinker.")

37. One can always trust a very communicative person.

38. Give a valid conclusion.

In Exercises 39–43, let p be "loves fish," q be "green eyes," r be "tail," s be "teachable," t be "whiskers," and u be "willing to play with a gorilla."

39. No kitten that loves fish, is unteachable.

40. No kitten without a tail, will play with a gorilla.

41. Kittens with whiskers always love fish.

42. No teachable kitten has green eyes.

43. No kittens have tails unless they have whiskers.

44. Give a valid conclusion.

Handwritten margin work:

(24.)
$$p \to \sim s$$
$$r \to \sim s$$
$$q \to p$$
$$q \to p$$
$$p \to \sim s$$
$$\sim s \to \sim r$$
$$q \to \sim r$$

(28)
$$r \to p$$
$$\sim r \to \sim q$$
$$s \to \sim p$$
$$s \to \sim p$$
$$\sim p \to \sim r$$
$$\sim r \to \sim q$$
$$s \to \sim q$$

(32)
$$s \to r$$
$$p \to q$$
$$q \to \sim r$$
$$p \to q$$
$$q \to \sim r$$
$$\sim r \to \sim s$$
$$p \to \sim s$$

(38.)
$$r \to \sim s$$
$$u \to t$$
$$\sim r \to p$$
$$\sim u \to \sim q$$
$$t \to s$$
$$q \to u$$
$$u \to t$$
$$t \to s$$
$$s \to \sim r$$
$$\sim r \to p$$
$$q \to p$$

(44)
$$p \to s$$
$$\sim p \to \sim u$$
$$t \to p$$
$$s \to \sim q$$
$$\sim t \to \sim r$$

$$q \to \sim s$$
$$\sim s \to \sim p$$
$$\sim p \to \sim t$$
$$\sim t \to \sim r$$
$$\sim r \to \sim u$$

Pg. 59

In Exercises 45–50, let *p* be "going to a party," *q* be "brushed hair," *r* be "self-command," *s* be "looks fascinating," *t* be "opium-eater," *u* be "tidy," and *v* be "wears white kid gloves."

45. No one who is going to a party ever fails to brush his hair.

46. No one looks fascinating if he is untidy.

47. Opium-eaters have no self-command.

48. Everyone who has brushed his hair, looks fascinating.

49. No one wears white kid gloves, unless he is going to a party.

50. A man is always untidy, if he has no self-command.

51. Give a valid conclusion.

In Exercises 52–60, let *p* be "begin with 'Dear Sir'," *q* be "crossed," *r* be "dated," *s* be "filed," *t* be "in black ink," *u* be "in third person," *v* be "I can read," *w* be "on blue paper," *x* be "on one sheet," and *y* be "written by Brown."

*52. All the dated letters in this room are written on blue paper.

*53. None of them are in black ink, except those that are written in the third person.

*54. I have not filed any of them that I can read.

*55. None of them that are written on one sheet are undated.

*56. All of them that are not crossed are in black ink.

*57. All of them, written by Brown, begin with "Dear Sir."

*58. All of them written on blue paper, are filed.

*59. None of them written on more than one sheet, are crossed.

*60. None of them that begin with "Dear Sir" are written in the third person.

*61. Give a valid conclusion.

*2.7 CIRCUITS–AN APPLICATION OF LOGIC

One of the first non-mathematical applications of symbolic logic came in the 1937 master's thesis of Claude Shannon. Shannon, now a professor of mathematics at MIT, showed how logic could be used as an aid in designing electrical circuits. His work was immediately taken up by the designers of computers. These computers, then in the developmental stage, could be simplified and built for less money, using the ideas of Shannon.

To see how Shannon's ideas work, look at the electrical switch shown in Figure 2.9. We assume that current will flow through this switch when it is closed, and not when it is open.

Figure 2.9

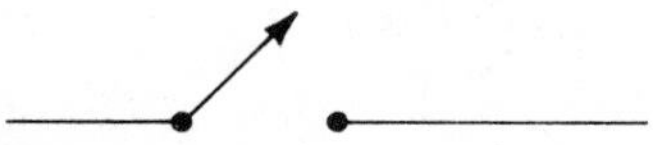

Figure 2.10
Series circuit

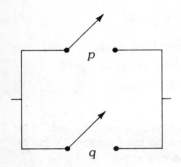

Figure 2.10 shows two switches connected in *series;* in such a circuit, current will flow only when both switches are closed. Note how closely a series circuit corresponds to the conjunction $p \wedge q$. We know that $p \wedge q$ is true only when both p and q are true.

We can find a circuit corresponding to the disjunction $p \vee q$ by drawing a *parallel* circuit, as in Figure 2.11. Here, current flows if either p or q is closed, or if both p and q are closed.

The way that logic is used to simplify an electrical circuit, depends on the idea of equivalent statements, which we defined in the exercise set for Section 2.4. Two statements are *equivalent* if they have exactly the same truth table. We use the symbol $\equiv$ to show that the two statements are equivalent. Some of the equivalent statements that we shall need are:

Figure 2.11
Parallel circuit

$$p \vee (q \wedge r) \equiv (p \vee q) \wedge (p \vee r) \qquad p \vee p \equiv p$$
$$p \wedge (q \vee r) \equiv (p \wedge q) \vee (p \wedge r) \qquad p \wedge p \equiv p$$
$$p \rightarrow q \equiv {\sim}q \rightarrow {\sim}p \qquad {\sim}p \vee {\sim}q \equiv {\sim}(p \wedge q)$$
$$p \rightarrow q \equiv {\sim}p \vee q \qquad {\sim}p \wedge {\sim}q \equiv {\sim}(p \vee q)$$

If T represents any true statement and F represents any false statement, then

$$p \vee T \equiv T \qquad p \vee {\sim}p \equiv T$$
$$p \wedge F \equiv F \qquad p \wedge {\sim}p \equiv F$$

We can use circuits as models of compound statements by letting a closed switch correspond to T, while an open switch corresponds to F. The method for simplifying circuits is explained in the following example.

Example 1 Simplify the circuit of Figure 2.12.

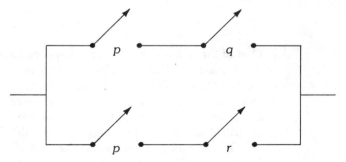

Figure 2.12

Solution. The circuit of Figure 2.12 can be written as

$$(p \wedge q) \vee (p \wedge r).$$

(Think of the two switches labeled *"p"* as being controlled by the same handle.) By one of the pairs of equivalent statements above,

$$(p \land q) \lor (p \land r) \equiv p \land (q \lor r),$$

which has the circuit of Figure 2.13. This new circuit is logically equivalent to the one above, and yet contains only three switches instead of four—which might well lead to a large savings in manufacturing costs.

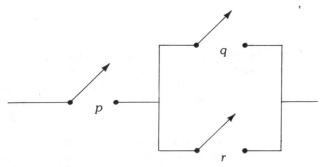

Figure 2.13   The circuit is logically equivalent to Figure 2.12

Example 2   Draw a circuit for $p \rightarrow (q \land \sim p)$.

Solution. From the list of equivalent statements above, we know that $p \rightarrow q$ is equivalent to $\sim p \lor q$. Thus, $p \rightarrow (q \land \sim p) \equiv \sim p \lor (q \land \sim p)$, which has the circuit diagram in Figure 2.14.

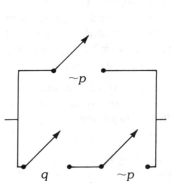

Figure 2.14

2.7 EXERCISES

Draw circuits representing each of the following statements.

1. $p \wedge (q \vee \sim p)$
2. $(\sim p \wedge \sim q) \wedge \sim p$
3. $(p \vee q) \wedge (\sim p \wedge \sim q)$
4. $(\sim q \wedge \sim p) \vee (\sim p \vee q)$
5. $[(p \vee q) \wedge r] \wedge \sim p$
6. $[(\sim p \wedge \sim r) \vee \sim q] \wedge (\sim p \wedge r)$
7. $\sim p \rightarrow (\sim p \vee \sim q)$
8. $\sim q \rightarrow (\sim p \rightarrow q)$

Write logical statements representing each of the following circuits. Simplify each circuit when possible.

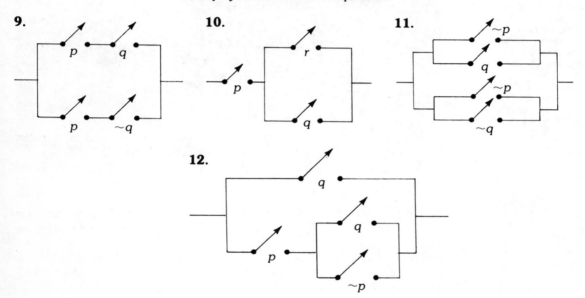

9.   10.   11.

12.

APPENDIX: EVERYDAY LOGIC

In everyday discussions, logical errors are very common. The conclusion reached in a discussion may be true, or it may be false, but the logic used to get the conclusion may be invalid. In this appendix we shall discuss some of the common types of fallacies; as you read through these fallacies, see if you can think of examples where such logical fallacies have been used.

Fallacy of Ambiguity This fallacy occurs when a word or phrase is used with one meaning in one premise, and with another meaning in another premise, or in the conclusion. Find the fallacy in the argument

> People should do what is right.
> People have the right to disregard good advice.
> People should disregard good advice.

The word *right* in the first premise means *correct*, while *right* in the second premise means *permitted*.

Fallacy of Composition   Here we assume that some whole object has a property because each of its parts has the property. For example,

> Each part of my new machine cost less than $1.
> My new machine cost less than $1.

Another version of the fallacy of composition occurs when we give a property to each part, based on a property of the whole. For example:

> Senator Wind gave a long speech.
> Each sentence in the speech is long.

It is certainly possible that a long-winded speech is made up of *short* sentences (but many of them).

Fallacy of Emotion   We commit a fallacy of emotion if we appeal to popular passions, pity, brute force, snob appeal, vanity, or some other emotion. The following argument shows the fallacy of emotion:

> I have a sick spouse and four sick children at home.
> You will be very satisfied with this new car, since I
> must sell five a week to make ends meet.

Fallacy of Experts   Here we quote an expert from one field on a matter outside of the person's field of knowledge. For example,

> Donna Shields is a master Chevrolet mechanic.
> She recommends Vitamin C for colds.

Fallacy of the Complex Question   Here we ask a question, not in a neutral way to discover facts, but in a loaded way, trying to lead the answer along certain lines. Some examples are:

> Have you stopped beating your dog?
> How does mind reading work?
> Are the people in your town still rude to visitors?
> You will really enjoy that blouse; will it be cash or charge?

Fallacy of Non Sequitur   (Literally: "it doesn't follow") We commit this fallacy when the conclusion is completely unrelated to the premises.

> South Carolina is on the Atlantic Ocean.
> The capital of South Carolina is Columbia.
> Many South Carolinians are Baptists.

Fallacy of False Cause   Here we imply that because two events are related in time, somehow one *caused* the other. For example,

> I walked under a ladder.
> I was hit by a car.
> Walking under a ladder brings bad luck.

This implies that walking under a ladder caused the bad luck of being hit by a car. Actually, it is only a coincidence that the two events occurred close together in time.

EXERCISES ON FALLACIES

Identify the type of fallacy in Exercises 1–20. More than one answer may be possible.

1. After Thompson insulted him, Smith was mad.
    Mad people should be put in a hospital.
    Smith should be put in a hospital.

2. Jane is an impossible student.
    Impossible things cannot exist.
    There is no Jane.

3. People in the hospital are ill.
    They should never have gone there.

4. It would be a mistake for our committee to do this job.
    We had better give the job to a subcommittee.

5. All people will eventually die.
    The species itself will therefore die out.

6. You are fond of working alone.
    I am fond of working alone.
    I am sure that we shall enjoy working together.

7. No designing persons are deserving of trust.
    Engravers are designers by profession.
    Engravers are to be distrusted.

8. We were told to be quiet when everybody is asleep.
    Everybody is awake now.
    Don't worry about the noise.

9. If everyone became an astronomer, civilization would disappear.
    You had better not become an astronomer.

10. Every person on that jury is a wise and sober person.
    The jury will reach the proper decision.

11. My wife is in the hospital.
    My car needs a new transmission.
    We have a baby.
    My work is deserving of a raise.

12. Senator, my organization can swing a lot of votes in your state.
    The organization can make substantial campaign contributions.
    You must see the merit in the opposition to that bill.

13. Dr. Lynn is against the farm policies of the administration.
    She is an expert chemist.
    The farm policies must be wrong.

14. Chevrolet is the most popular car in America today.
    It must be the best car.

15. Jewels are hard.
    She is a jewel.
    She is hard.

16. John had a heart attack.
    If he stays in bed, he'll get better.
    If I stay in bed, I'll feel even better.

17. All members of the Atlanta Falcons exercise daily.
    If you exercise daily, you can be on the Atlanta Falcons.

18. I just saw *King Kong Eats Toledo*, and it was bad.
    I just saw *Beach Ball Bongo,* and it was lousy.
    All movies are bad.

19. The new people on the block have a lovely living room set.
    They also have a beautiful car.
    They must be nice people.

20. Henry Fonda is a good actor.
    He recommends a certain brand of floor tile.
    That floor tile is excellent.

The following arguments (Exercises 21–28) are written in paragraph form. Decide why each is a fallacy. More than one answer may be possible.

21. The hats made by Warman and Company must be excellent since all movie stars wear Warman hats.

22. Rome has a lousy climate. I was there for two days last year, and it rained cats and dogs the whole time.

23. I will never see a doctor again. All my friends who were ill over the past winter went to doctors.

24. The number of automobiles in the city has greatly increased during the past five years, and the number of suicides has gone up steadily for the past four. It is clear that the automobile is making life miserable.

25. *People* is a very fine magazine. Shall I put you down for a one-year subscription, or would you rather have the three-year bargain rate?

26. Unemployment is bad, and whatever leads to a demand for work is good. We should therefore welcome floods, earthquakes, and other disasters, for all these create a good deal of work.

27. The two members of the test group were unable to work well after taking the new medicine. We may therefore conclude that the effects of this medicine on the general population will be bad.

28. The two top students in the class are able to keep up the pace set by the instructor. Therefore, the pace is suitable for the class as a whole.

*29. Find three examples of fallacies in the daily newspaper (not in the ads).

*30. Find three examples of fallacies in advertising.

APPENDIX: THE LOGIC OF A RENTAL CONTRACT

One place where logic can be used is in analyzing a rental contract. The many clauses, whereas's, and furthermore's make such contracts a fertile place for symbolic logic.

In the following exercises, we give a few of the clauses of a typical rental contract, and then write the clauses with symbols. In this work, we need the following definition, discussed in Exercise Set 2.4 of this chapter:

$$p \text{ unless } q \qquad \text{means} \qquad {\sim}q \rightarrow p$$

We also need De Morgan's Laws, from the exercises for Section 2.4:

$${\sim}p \vee {\sim}q \qquad \text{is the same as} \qquad {\sim}(p \wedge q)$$
$${\sim}p \wedge {\sim}q \qquad \text{is the same as} \qquad {\sim}(p \vee q).$$

Also, recall that the contrapositive of $p \rightarrow q$ is ${\sim}q \rightarrow {\sim}p$, and that the direct statement and the contrapositive have the same truth value.

In each of the following exercises, fill in the blanks with the correct letters and/or symbols.

31. *This contract is noncancelable by tenants for 90 days.*

Let p be   "You are a tenant."
$\quad\quad q$ be   "It is within 90 days."
$\quad\quad r$ be   "The contract may be canceled."

The contract sentence can be written

$$(p \wedge q) \rightarrow \underline{\quad\quad}.$$

32. Write the contrapositive of the symbolic statement from Exercise 31.

$$\underline{\quad\quad} \rightarrow {\sim}(\underline{\quad\quad})$$

33. By De Morgan's Laws, the result of Exercise 32 becomes

$$r \rightarrow (\underline{\quad\quad}).$$

34. Write the statement from Exercise 33 in words.

35. *The tenant agrees to lease the premises for 9 months, beginning on October 1 and at a monthly rental of $175.*

Let p be   "You are a tenant."
$\quad\quad s$ be   "You will lease the premises for 9 months."
$\quad\quad t$ be   "You will begin on October 1."
$\quad\quad u$ be   "You will pay $175 per month."

The contract sentence becomes

$$p \rightarrow [(\underline{\quad\quad} \wedge t) \wedge \underline{\quad\quad}].$$

36. *Utilities, except water and garbage, are paid by the tenant.*

Let   *p*   be   "You are a tenant."
       v   be   "You pay water."
      w   be   "You pay garbage."
      x   be   "You pay the other utilities."

The contract sentence becomes

$$p \rightarrow [\_\_\_\_ \wedge (\_\_\_\_ \vee \_\_\_\_)].$$

37. *Rents are prorated to the first day of the month except for the first month.*

Let   *p*   be   "You are the tenant."
      a   be   "Rents are prorated to the first."
      b   be   "It is the first month."

The contract sentence becomes

$$p \rightarrow (a \leftrightarrow \_\_\_\_).$$

38. *A late charge of $10 is added to the rent payment after the 5th of each month with the exception of the first month.*

Let   *p*   be   "You are the tenant."
      c   be   "A late charge of $10 is added."
      d   be   "It is after the 5th of the month."
      b   be   "It is the first month."

The contract sentence becomes

$$p \rightarrow [\sim b \_\_\_\_ (d \rightarrow \_\_\_\_)].$$

Rewrite the contract sentences in Exercises 39 and 40 in symbols.

*39. *No alterations, redecorating, tacks or nails shall be made in the building, unless written permission is obtained.*

Let   *p*   be   "You are the tenant."
      e   be   "Alterations to buildings are made."
      f   be   "Redecorating is done."
      g   be   "Tacks are used."
      h   be   "Nails are used."
      j   be   "Written permission is obtained."

*40. *The tenant shall not let or sublet the whole or any part of the premises to anyone for any purpose whatsoever, unless written permission is obtained.*

Let   *p*   be   "You are the tenant."
      k   be   "You let the whole premises."
     m   be   "You sublet the whole premises."
      n   be   "You let part of the premises."
      q   be   "You sublet part of the premises."
      j   be   "Written permission is obtained."

CHAPTER 2 SUMMARY

Keywords		
	Inductive reasoning (Induction)	Symbolic logic
	Hypothesis	Truth value
	Counterexample	Statement
	Deductive reasoning (Deduction)	Connective
	Proof	Compound statement
		Conjunction
	Argument	Disjunction
	Premise(s)	Exclusive disjunction
	Conclusion	Negation
	Euler diagram	Conditional (Implication)
	Valid	Biconditional
	Modus ponens	Only if
	Modus tollens	If and only if
	Invalid (Fallacy)	Direct statement
	Fallacy of the converse	Converse
	Fallacy of the inverse	Inverse
	Necessary	Contrapositive
	Sufficient	Equivalent statements
	Consistent	
	Contrary	Series circuit
	Tautology	Parallel circuit
	De Morgan's laws	

Symbols

Connectives	Symbols	Types of statement
and	$\wedge$	Conjunction
or	$\vee$	Disjunction
not	$\sim$	Negation
if . . . then	$\rightarrow$	Conditional
if and only if	$\leftrightarrow$	Biconditional

Truth tables

p	q	$p \wedge q$	$p \vee q$	$p \rightarrow q$	$p \leftrightarrow q$	$p \underline{\vee} q$	p unless q
T	T	T	T	T	T	F	T
T	F	F	T	F	F	T	T
F	T	F	T	T	F	T	T
F	F	F	F	T	T	F	F

Arguments

$p \rightarrow q$	$p \rightarrow q$	$p \rightarrow q$	$p \rightarrow q$	$p \rightarrow q$
p	$\sim q$	q	$\sim p$	$q \rightarrow r$
q	$\sim p$	p	$\sim q$	$p \rightarrow r$
Modus ponens	Modus tollens	FALLACY of the converse	FALLACY of the inverse	
Valid	**Valid**			**Valid**

CHAPTER 2 TEST

Are the following arguments valid?

1. If I become a doctor, then I can put a "Malpractice Reform" bumper sticker on my Mercedes.

I put a "Malpractice Reform" bumper sticker on my Mercedes.

I became a doctor.

2. If I can add fractions, then I can subtract fractions.

I can't subtract fractions.

I can't add fractions.

Which of the following are statements?

3. All cats have nine lives.   **5.** $42 + 5 = 50$

4. The Fonz is the greatest.   *(open sentence)*

Let *p* represent "She passes algebra" and let *q* represent "She passes history." Write each of the following in words.

6. $p \wedge \sim q$ **7.** $\sim p \to q$

Write a negation for each of the following.

8. All cluckers are chickens.   **9.** Some spiders eat their mates.

In each of the following, assume that *p* and *q* are false, with *r* true. Find the truth value of each statement.

10. $\sim p \vee \sim q$ **11.** $r \wedge (\sim p \vee q)$ **12.** $(\sim p \wedge q) \to (\sim r \vee q)$

Write a truth table for each of the following.

13. $\sim p \to q$ **14.** $p \vee (\sim p \vee q)$ **15.** $(\sim p \wedge q) \to (\sim p \vee \sim q)$

Write each of the following in the form *if . . . then*.

16. He would pass mathematics if he would work.

17. Snoopy flies a kite only if Lucy is not around.

18. Having yesterday's newspaper is necessary for owning a bird.

19. Being unemployed is sufficient for getting food stamps.

Write the converse, inverse, or contrapositive, as indicated.

20. If I buy a sandwich, I'll need mustard.   (*Converse*)

21. I never drink coffee if it's after 7 pm.   (*Inverse*)

22. If Sandra's not going, then Tom's not going.   (*Contrapositive*)

23. $\sim p \to (q \vee r)$ (*Converse and Contrapositive*)

Use truth tables to test the validity of each of the following arguments.

24. $\sim p \to \sim q$ **25.** $\sim p \vee \sim q$

$q \to p$ $\sim q \wedge p$

$p \vee q$ p

26. *Which of the following best describes your experience with this chapter?*
(a) *It stretched your mind*
(b) *It blew your mind*
(c) *It stretched your neck*
(d) *All of the above*
(e) *None of the above*

3 *Numeration Systems*

Today, numbers and the way we write numbers are among the most basic ideas of mathematics. Children can often count a fair number of items by about age four. Historically, however, it has been a different story. The idea of number developed fairly late in human history. Early cultures (and some existing even today) had words for only *one* and *two*. Any number past *two* would be named by some indefinite word suggesting "many."

A practical method of keeping accounts by matching may have developed as humans established permanent settlements and began to grow crops and raise livestock. People might have kept track of the number of sheep in a flock by matching pebbles, for example, with the sheep. The pebbles could then be kept as a record of the number of sheep.

An even simpler method is to keep a **tally stick.** With a tally stick, one notch or **tally** is made on a stick for each sheep. Tally sticks and tally marks have been found that appear to be at least 20,000 years old. We still use tally marks today: we commonly tally nine items by writing ⋉ ||||. The Lamba people of Zambia in southern Africa invented a number of tallying methods. Strokes were scratched on the ground, in groups of ten for larger numbers. Herdskeepers tallied cattle with sticks, a long one for ten, a short one for five, and remembered the number below five. Travellers recorded the days of journey on the "lever stick," a support for the baggage pole carried on one shoulder. Notches at one end of the lever stick gave the number of days enroute; the other end was notched for the journey back.

Number	Lamba word
1	chimo
2	fibili
3	fitatu
4	fine
5	fisano
6	fisano nacimo
7	fisano nafibili
8	fisano nafitatu
9	fisano nafine
10	ikumi
11	ikumi nacimo
12	ikumi nafibili
20	amakumi yabili
100	umwanda
1000	umukama

Tally sticks and groups of pebbles were an important advance. By these methods, the idea of number began to develop. Early people began to see that a group of three chickens and a group of three dogs had something in common: the idea of *three*. Gradually, people began to think of numbers separately from the things they represent. Words and symbols were developed for various numbers.

The Lamba people had a very sophisticated system of number words. They used 5 and 10 in a counting system that went up to 1000, as shown in the table on page 82. The prefix *fi* indicates "more than 1," while *na* and *ya* mean respectively "plus" and "so many times." Thus, the number 21 would be *amakumi yabili nacimo*.

As early cultures became more complex, more and more numbers came to be used. Cultures began developing *symbols* to represent numbers. A group of symbols that can be used to express a number is called a **numeration system.** The symbols of a numeration system are called **numerals.**

Numerals have all the advantages of other symbols, such as the letters of an alphabet. Numerals are concise, are few enough to be quickly learned, and can be combined easily for addition or multiplication. Numeration systems, like alphabet systems, are a medium of communication, making ideas and information available to people at great distances away and in other times. The numerical records of ancient people give us some idea of their daily lives and create a picture of them as producers and consumers. For example, Mary and Joseph went to Bethlehem to be counted in a census — a numerical record.

This chapter examines three types of numeration systems and then surveys numeration methods of the Egyptians, Romans, Chinese and Japanese, Mayan, and Babylonian civilizations. The important thing in this chapter is the **number base,** the size of the groupings in a numeration system. Base 10 (the one we use) is so common throughout human culture that it is probable that most people learned to count by using their two hands. However, some cultures have used base 5, as well as base 20 and base 60. The second half of the chapter studies bases other than our own base 10, including base 2, which is used in computers.

Peruvian quippu *This is part of a numerical record of the Incas. Knots in the strings represent numbers. The larger knots are multiples of the smaller.*

3.1 THREE METHODS OF NUMERATION

The tally method described above is not too hard to learn, but it is far too cumbersome to be practical today. (Just think how many tally marks you would need to record the monthly rent.) Because of the impractical nature of the tally method, most cultures eventually developed more convenient methods of recording numbers.

Throughout history, three fundamental types of numeration systems have evolved. We shall look at these three types of systems in this section. In addition to adopting some type of numeration system, a culture must also standardize the size of the basic group used to write numbers. Our culture takes *ten* as this basic group; other cultures have taken other numbers.

The size of the basic group is called the **number base,** or **base,** of the numeration system.

We shall explain the three types of numeration systems using *base five*. By using a base other than our own base ten, we may avoid any preconceived ideas about what to expect.

Simple grouping   In the tally method, we use one stroke for each item in a collection of items. Thus, we would write the number of items in a collection of nine as | | | | | | | | |. To simplify this cumbersome way of writing numerals, we can introduce a new symbol Λ to replace five strokes. (We let Λ represent *five* since we have chosen a base five system.)

Using this new symbol, we can rewrite nine strokes as Λ| | | |, or one 5 and four 1s. In the same way, 17 is three 5s and two 1s, or

$$ΛΛΛ| | \quad \text{instead of} \quad | | | | | | | | | | | | | | | | |$$

while 20 is four 5s, or ΛΛΛΛ.

By using this new symbol, we will never need more than four of the stroke symbols—anytime we have five strokes, we can replace them with Λ. By using just Λ and |, we can write numbers more readily than with a tally method. However, even fairly small numbers still need a lot of symbols. For example, to write 67 we would need thirteen Λs and two |s, or a total of fifteen symbols:

$$ΛΛΛΛΛΛΛΛΛΛΛΛΛ| |$$

This numeral is still too long to be useful. To shorten it, we let a new symbol Ν represent five Λs, or 5 × 5 = 25. By using Ν, we shall never need more than four Λs.

Example 1   Write Ν Ν Ν ΛΛ| | | | in our numeration system.

Solution.   Each Ν represents 25. There are three of these symbols, for a subtotal of 3 × 25 = 75. Each Λ represents 5. There are two of the Λ symbols, for a subtotal of 2 × 5 = 10. Finally, there are four | symbols, or 4 × 1 = 4. The total is 75 + 10 + 4 = 89, so Ν Ν Ν ΛΛ| | | | represents 89 in our system.

The system we are developing groups the correct number of symbols to represent numbers. This system is called a **simple grouping system.** Since we are using groups of size five, the system is a base five system.

Example 2   Write 107 using the base five simple grouping system.

Solution.   To write 107, we need four 25s, since 4 × 25 = 100. We also need one 5 and two 1s. Thus, 107 is Ν Ν Ν Ν Λ| |.

We can avoid using more than four Ν symbols if we introduce a symbol that represents five 25s, or 5 × 25 = 125. For 125, we introduce the symbol Μ. With this symbol, 125 is written Μ, 126 is Μ|, 130 is ΜΛ, 150 is ΜΝ, 159 is ΜΝΛ| | | |, and so on, up to (and including) 624.

Bride-price   *Below is a portion of a traditional marriage tablet from the Kai Islands, Indonesia. It records the "bride-price," which the groom pays to the bride's father in goods. The actual number of bracelets, axes, and combs are pictured rather than tallied. "Bride-price" is a misleading term; payment is not for the woman herself as property, but is compensation for loss of her labor in the extended family she is leaving.*

Number	Symbol
1	I
5	∧
25	N
125	M
625	∧∧
3125	MM
15,625	M∧

Powers of 5

$5^0 = 1$
$5^1 = 5$
$5^2 = 5 \times 5 = 25$
$5^3 = 5 \times 5 \times 5 = 5 \times 25$
$\qquad = 125$
$5^4 = 5 \times 5 \times 5 \times 5 = 625$
$5^5 = 5 \times 5 \times 5 \times 5 \times 5$
$\qquad = 3125$
$5^6 = 15,625$

The tiny numerals are called either **powers** or **exponents**. Here the base is 5, and expressions like 5^2 and 5^3 are called **exponentials**.

The next symbol, ∧∧, represents five M symbols, or $5 \times 125 = 625$. For example, 629 is ∧∧ I I I I, and 1272 is ∧∧ ∧∧ ∧∧∧∧ I I .

We could keep introducing symbols in this way, but the ones we already have are sufficient for our needs.

Example 3 Write the following in our system:

(a) M M N N N ∧ I I I I   (b) ∧∧ ∧∧ M N N N N ∧∧∧ I I

(a) Here M is used twice, giving $2 \times 125 = 250$. The three N s give $3 \times 25 = 75$; the one ∧ gives $1 \times 5 = 5$, and the four Is give $4 \times 1 = 4$. Adding these subtotals gives $250 + 75 + 5 + 4 = 334$. Thus, M M N N N ∧ I I I I represents 334. This work could also be done as follows.

M M	:	$2 \times 125 = 250$
N N N	:	$3 \times 25 = 75$
∧	:	$1 \times 5 = 5$
I I I I	:	$4 \times 1 = 4$
		334

(b)			
∧∧ ∧∧	:	$2 \times 625 = 1250$
M	:	$1 \times 125 = 125$
N N N N	:	$4 \times 25 = 100$
∧∧∧	:	$3 \times 5 = 15$
I I	:	$2 \times 1 = 2$
		1492

Multiplicative grouping   To see how multiplicative grouping works, take the number 1492 from Example 3(b) above. As shown in that example, 1492 is written ∧∧ ∧∧ M N N N N ∧∧∧ I I in the base five simple grouping system. The two ∧∧ symbols take up a lot of space. To save space, let us use a *multiplier,* along with only one ∧∧ symbol to show how many ∧∧s. Let us use ☉ to represent two of any symbol. We can now write ∧∧ ∧∧ as ☉∧∧ . In the same way, M becomes ☉M , while N N N N becomes ☺N . Also, ∧∧∧ is ☉∧ , and I I is ☉I. Using these symbols, ∧∧ ∧∧ M N N N N ∧∧∧ I I is shortened to

$$☉∧∧ ☉M ☺N ☉∧ ☉I$$

Here we *pair* a multiplier with a symbol for a number. For example, ☺N ☉∧ ☉I represents $(4 \times 25) + (3 \times 5) + (2 \times 1) = 100 + 15 + 2 = 117$.

Example 4   Write ☉M ☉∧ ☉I in our system.

Solution.   Take the symbols in pairs. First we have ☉M, which means three 125s, or $3 \times 125 = 375$. Next we have ☉∧, which means $1 \times 5 = 5$. Last, ☉I means $4 \times 1 = 4$. Adding these subtotals gives $375 + 5 + 4 = 384$. If you like, you can arrange the work as follows:

☉M	$3 \times 125 =$	375
☉∧	$1 \times 5 =$	5
☉I	$4 \times 1 =$	4
		384

Thus, ☉M ☉∧ ☉I is 384 in our numeration system.

Our new system still makes use of groups of 1s, 5s, 25s, and so on, but how many of each is now shown by a multiplier rather than by repetitions of a given symbol. For this reason, we call the new method a **multiplicative grouping method.** We are still using five for the base, but remember that the size is arbitrary. Any counting number except 1 could be used as a number base without changing the numeration method. Examples of numerals written in this system are shown in Figure 3.1.

Positional method You may have discovered by now that there is a rather natural way to change the multiplicative grouping method and save even more space. Note the numerals in Figure 3.1. The sequence of symbols, going from right to left, is as follows: a pair of symbols showing the number of 1s, then a pair showing the number of 5s, then a pair for 25s, and so on, as needed. Since this sequence always occurs in that order, it is possible to use the multipliers alone. For example, we can write 1416 as ②①①③①, using only multipliers. The idea of position then allows us to work this numeral out in our system:

② ① ① ③ ①

$$(2 \times 625) + (1 \times 125) + (1 \times 25) + (3 \times 5) + (1 \times 1)$$
$$= 1250 + 125 + 25 + 15 + 1 = 1416$$

With this **positional method,** the numerals from Figure 3.1 can be written with fewer symbols, as shown in Figure 3.2.

Figure 3.1 Multiplicative grouping, base five

Number	Breakdown into grouping units	Numerals
3	3×1	③|
19	$(3 \times 5) + (4 \times 1)$	③Λ④|
88	$(3 \times 25) + (2 \times 5) + (3 \times 1)$	③N②Λ③|
342	$(2 \times 125) + (3 \times 25) + (3 \times 5) + (2 \times 1)$	②M③N③Λ②|
1416	$(2 \times 625) + (1 \times 125) + (1 \times 25) + (3 \times 5) + (1 \times 1)$	②W①M①N③Λ①|

Figure 3.2

Number	Positional
3	③
19	③④
88	③②③
342	②③③②
1416	②①①③①

You may have already found a problem with this method of positional grouping. How, for example, could we write the number 27? To write 27, we need one 25, no 5s, and two 1s. In multiplicative grouping, we would write ①N②|. To express this same result in the positional system, we could write ① ②, leaving a space in the Λs position, but this could easily be mistaken for ①②, or $(1 \times 5) + (2 \times 1) = 7$. The number 50 presents an even greater problem, for we must leave blank spaces in both the Λs and |s positions. The resulting number, ②   , is hardly different from the one for 10, ② , or the one for 2, ②.

During the historical development of numeration methods, the solution of this problem took various peoples a good many years, in fact, many centuries. The solution of the problem was to develop the idea of a **placeholder,** for which we use 0 (zero) in our system. If we include the symbol ○

Figure 3.3

Number	Numeral
1	⊙
2	⊙
3	⊙
4	⊙
5	⊙○
6	⊙○
7	⊙○
10	⊙○
11	⊙○
15	⊙○
24	⊙⊙
25	⊙○○
26	⊙○⊙
50	⊙○○

in our list of multipliers, we then have a workable positional system. We can then write 27 as ⊙○⊙ , while 50 is ⊙○○ . The ○ is a placeholder, which shows that certain values are missing. Further examples of base five positional numerals are shown in Figure 3.3.

3.1 EXERCISES

Identify each base five numeral in Exercises 1–24 as simple grouping, multiplicative grouping, or positional. Write each numeral in our system.

1. N Λ Λ |
2. ⊙○
3. ⊙⊙
4. ⊙Λ⊙|
5. N ||
6. N Λ Λ Λ | | |
7. ⊙Λ⊙|
8. ⊙○⊙

9. N N N Λ |
10. ⊙○○
11. ⊙N⊙Λ
12. ⊙M⊙Λ
13. N N N N
14. ⊙○○○
15. ⊙○○○
16. ⊙N⊙|

17. N N M
18. N M N | | | |
19. ⊙N⊙M
20. N Λ Λ | | |
21. ⊙○○⊙
22. N N N N N N Λ
*23. ○○⊙⊙⊙○
*24. ⊙M⊙N⊙Λ⊙|

For each of the following, give a base five simple grouping numeral, a multiplicative grouping numeral, and a positional numeral.

25. 8   **29.** 37   **33.** 76   **37.** 320
26. 9   **30.** 41   **34.** 85   **38.** 372
27. 14   **31.** 47   **35.** 125   **39.** 1250
28. 22   **32.** 52   **36.** 250   **40.** 625

The more things change . . .
A sketch of the "Ishango bone" (near right) shows tallies on one side of a tool-handle found at the Ishango fishing site in Zaire, Africa. The tallies are believed to mark phases of the moon. Over 8000 years separate the bone from the cartoon. We are still scratching along. . . .

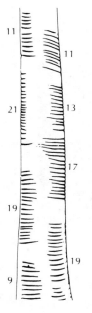

"I've prepared this simple chart to give you a clear picture of our financial situation."

3.2 EGYPTIAN AND ROMAN NUMERATION

Like other inventions, numeration systems were invented as needed. As long ago as 5000 years, the keeping of government and business records by Egyptians and Sumerians required the use of large numbers. Ancient documents tell us about the methods used by these peoples, as well as those of the Greeks, Romans, Chinese, Japanese, Central American Indians, and the Hindus. We shall look at just a few of these numeration systems in this section and the next.

Egyptian numeration   At least 5000 years ago, the Egyptians were using a simple grouping system. The Egyptians used ten as the base, with the set of symbols shown in Figure 3.4.

Tutankhamun, *pharaoh of Egypt from about 1334 to 1325 BC, in the Eighteenth Dynasty of the New Kingdom.*

Figure 3.4   Early Egyptian Symbols

Number	Symbol	Description
1	I	Stroke
10	∩	Heel bone
100	ϑ	Scroll
1000	ℒ	Lotus flower
10,000	⌐	Pointing finger
100,000	⌐⌐	Burbot fish
1,000,000	☥	Astonished person

These symbols are sufficient for writing numbers up to 9,999,999.

Hieroglyphics   *Picture symbols in Egyptian writing represented numbers (Figure 3.4) as well as sounds, syllables, or words. The lid of the box at left, from the tomb of Tutankhamun, gives his name and title. For example, the syllable "ankh" in his name is given by the symbol ♀, also the word for "life." A royal name was enclosed in an oval (a rope loop with the tied ends sticking out); that accounts for the shape of the box. Cartouches (as the ovals are called) were keys to deciphering hieroglyphics when the bilingual Rosetta Stone (Egyptian and Greek) was found: the names in cartouches were the same sounds in both languages.*

Example 1   Calculate the number expressed by

$$\text{↷↷} \mathcal{f}\mathcal{f}\mathcal{f}\mathcal{f}\mathcal{f}\,9999\,\frac{\text{∩∩∩∩∩III}}{\text{∩∩∩∩IIII}}$$

Solution.   Refer to Figure 3.4 for the value of the Egyptian symbols. Each ↷ represents 100,000. Therefore, two ↷ represent $2 \times 100{,}000$, or 200,000. Proceed as follows:

two ↷:	$2 \times 100{,}000 =$	200,000
five $\mathcal{f}$:	$5 \times 1000 =$	5,000
four 9 :	$4 \times 100 =$	400
nine ∩ :	$9 \times 10 =$	90
seven I :	$7 \times 1 =$	7
		205,497

The number is 205,497.

Example 2   Write 376,248 in Egyptian symbols.

Solution.   We need three ↷, seven ℓ, six $\mathcal{f}$, two 9, four ∩, and eight Is. The number 376,248 is written

$$\text{↷↷}\ \ell\ell\ell\ \mathcal{f}\mathcal{f}\mathcal{f}\,99\frac{\text{∩∩IIII}}{\text{↷}\ \ell\ell\ell\ell\ \mathcal{f}\mathcal{f}\mathcal{f}\,99\,\text{∩∩IIII}}$$

Notice that the position or order of the symbols really makes no difference in a simple grouping system. The numbers 99∩∩∩IIII, IIII∩∩∩99, and II∩∩99∩II would each be interpreted as 234. The most common order, however, is that shown in Examples 1 and 2, where like symbols are grouped.

A simple grouping system, such as the early Egyptian, is particularly well suited to arithmetic, and especially to addition and subtraction. For example, to add $\mathcal{f}\mathcal{f}$99 ∩∩∩ II and $\mathcal{f}$ 999∩IIIIIII, work as shown. Two Is plus six Is equal eight Is, and so on.

$$
\begin{array}{r}
\mathcal{f}\mathcal{f}\quad 99\ \text{∩∩∩}\ \text{II} \\
+\quad \mathcal{f}\quad 999\ \text{∩}\ \text{IIIIII} \\
\hline
\end{array}
$$

Sum: $\quad \mathcal{f}\mathcal{f}\mathcal{f}\dfrac{999\ \text{∩∩}\ \text{IIII}}{99\ \text{∩∩}\ \text{IIII}}$

Sometimes, we must regroup, or "carry," as in the example below, where the answer contains more than nine heelbones. To regroup, we get rid of ten heelbones from the tens group. To compensate for this, we place an extra scroll in the hundreds group.

$$
\begin{array}{l}
\quad \ell\ \ \mathcal{f}\mathcal{f}\ 99\ \dfrac{\text{∩∩∩∩}}{\text{∩∩∩}}\ \text{II} \\
+\ \ell\ell\quad\quad 999\ \dfrac{\text{∩∩∩}}{\text{∩∩}}\ \begin{array}{l}\text{II}\\ \text{III}\end{array} \\
\hline
\end{array}
$$

Regrouped

Sum: $\quad \ell\ell\ell\ell\mathcal{f}\mathcal{f}\dfrac{999\text{∩∩∩∩∩∩∩IIII}}{99\ \text{∩∩∩∩∩∩}\ \text{III}} \rightarrow \ell\ell\ell\ell\mathcal{f}\mathcal{f}\dfrac{999}{999}\text{∩∩}\begin{array}{l}\text{III}\\ \text{IIII}\end{array}$

Subtraction is done in much the same way, as shown in the next example.

Example 3   Subtract in each of the following.

(a)   999 ∩∩ ||||
      99 ∩∩ |||
 – 999 ∩ ||||

(b)
    99∩∩∩∩ ||
 – 9   ∩∩ ||||

Difference: 99 ∩∩∩ |||

In (b), we can't subtract four |s from two |s. We need to "borrow" one heel-bone, which is equivalent to ten |s. After writing ten |s on the right, we can finish the problem.

Regrouped:
                              ||||||
                      99∩∩∩ ||||||

               – 9 ∩∩ ||||

Difference: 9 ∩ |||||||||

Procedures such as those described above are called **algorithms.** An algorithm is a rule or procedure for working a problem. The Egyptians used an interesting algorithm for multiplication, which requires only an ability to add and to double numbers. Example 4 illustrates the way that the Egyptians multiplied, but we shall use our symbols rather than theirs for convenience.

Example 4   A stone used in building a pyramid has a rectangular base measuring 5 by 18 cubits. Find the area of the base.

Solution.   To find the area of a rectangle, we multiply the length by the width; in this problem, 5 times 18. To begin, build two columns of numbers, as shown below. Start the first column with 1, and the second column with 18. Each column is built downward by doubling the number above. Keep going until the first column contains numbers that can be added to make 5. Here we have 1 and 4. To find 5×18, add only those numbers from the second column that correspond to 1 and 4. Here we add 18 and 72 to get the answer 90. The area of the base of the stone was 90 square cubits.

$$\begin{array}{ll} \to 1 & 18 \leftarrow \text{corresponds to 1} \\ 2 & 36 \\ \to 4 & 72 \leftarrow \text{corresponds to 4} \end{array}$$

Add: $1 + 4 = 5$ $18 + 72 = 90$ Thus $5 \times 18 = 90$

Example 5   Use the Egyptian multiplication algorithm to find 19×70.

Solution.   Form two columns, headed by 1 and by 70. Keep doubling until you get numbers in the first column that add to 19. (Here: $1 + 2 + 16 = 19$.) Then add corresponding numbers from the second column: $70 + 140 + 1120 = 1330$. Thus $19 \times 70 = 1330$.

$$\begin{array}{ll} \to 1 & 70 \leftarrow \\ \to 2 & 140 \leftarrow \\ 4 & 280 \\ 8 & 560 \\ \to 16 & 1120 \leftarrow \end{array}$$

Roman numeration   You are probably familiar with the Roman numeration system, which is still used today, mainly for decorative purposes, on clock faces, for heading numbers in outlines, for chapter numbers in books, and so on. What are the characteristics of the Roman method? Apparently, base 10 appealed to them also, but they introduced "extra" symbols to reduce the number of repetitions within a group. The Roman symbol for 1 was similar to that of the Egyptians, with repetitions of this symbol used for 2, and 3. Instead of continuing in this way up to 9, though, they introduced a distinct symbol for 5. Likewise, their symbol for 10 was repeated for 20 and 30, but a distinct symbol denoted 50. Other symbols denoted 500, 5000, 50,000, and 500,000. Roman numeration is like a base 10 simple grouping system with a secondary base 5 grouping. More symbols had to be used than in a pure simple grouping system, but the necessity of repetition was reduced.

The Romans also saved space by using subtraction. A symbol to the left of another symbol of larger value indicated that the smaller value was to be subtracted from the larger value on the right. For example, 9 was written IX (10 minus 1) rather than VIIII (5 plus 4). The same method was often used to write 4 as IV rather than IIII. Figure 3.5 gives the most common Roman symbols, illustrated by the numerals in Figure 3.6. Multiples of 1000 were sometimes indicated by placing a bar over the entire numeral, and two such bars meant to multiply by 1,000,000.

Constantine I, *the Great, first Christian emperor of Rome, about 280–337 A.D.*

Adding and subtracting with Roman numerals is very similar to the Egyptian method, except that the subtractive device of the Roman system sometimes makes the processes more involved. With Roman numerals we cannot add IV and VII to get the sum VVIII by simply combining like symbols. (Even XIII would be incorrect.) The safest method is to rewrite IV as IIII, then add IIII and VII, getting VIIIIII; convert this to VVI, and then to XI by regrouping. Subtraction, which is similar, is shown in the following example.

Example 6   A Roman official has 26 servants. If, on a given Saturday, he has excused 14 of them to attend a Rolling Marbles concert at the Forum, how many are still at home to serve the banquet?

Solution.   To find the answer, we subtract XIV from XXVI. Set up the problem in terms of simple grouping numerals (that is, XIV is rewritten as XIIII):

```
Problem    XXVI       Problem restated in simple     XXVI
         − XIV        grouping form without        −XIIII
                      subtractive notation
```

```
         Regrouped form     XXIIIIII
                          −  XIIII
                             XII   Answer
```

Since four **I**s cannot be subtracted from one **I**, we have "borrowed" in the top numeral, writing XXVI as XXIIIIII. The subtraction can then be carried out.

Figure 3.5

Number	Symbol
1	I
5	V
10	X
50	L
100	C
500	D
1000	M

Figure 3.6

Number	Numerals
6	VI
12	XII
19	XIX
30	XXX
49	XLIX
85	LXXXV
35,000	$\overline{\text{XXXV}}$
7,000,000	$\overline{\overline{\text{VII}}}$

3.2 EXERCISES

Write the Egyptian numerals in our system.

1.   ϑϑ∩∩∩|

2.   ϑϑϑ∩||

3.   ℓℓℓℓℓϑϑ|

4.   ℓℓℓℓϑϑϑϑϑ∩∩∩∩|||||

5.   ☒☒☒⌒⌒⌒⌒∩∩|||||

6.   ⌒ℓℓℓℓ∩∩∩ |||
                    |||

7.   ☒☒ℓℓℓℓℓℓ∩∩∩∩||||
                     ∩∩∩ |||

8.   ⌒⌒⌒ϑϑϑ∩∩∩ |
      ⌒⌒ ϑϑϑ∩∩∩

9.   ℓℓℓℓ ℓℓℓϑϑϑ∩∩∩ |||
      ℓℓℓℓℓℓ ϑϑϑ ∩∩

10. ⌒⌒ℓℓℓℓℓℓℓϑϑϑ∩∩|||
      ⌒⌒ ℓℓℓ ℓℓℓ ϑϑ ∩∩ ||

11. ☒☒☒☒ ℓℓℓℓϑϑϑ∩∩∩||||
      ☒☒☒ ⌒ ℓℓℓ ϑϑϑ∩∩∩||||

12. ☒☒☒☒ℓℓℓℓ ϑϑ ∩∩∩|||||||
                    ϑϑ

Exercises 13–20 are made up of verses from the book of I Kings in the Bible. The verses describe the building of King Solomon's Temple. Suppose you were an Egyptian spy sending reports back to Pharaoh. Write each of the **boldface** numbers with Egyptian symbols.

13. It was the **four hundred and eightieth** year after the Israelites came out of Egypt,

14. In the **fourth** year of Solomon's reign over Israel . . . he began to build the house of the Lord.

15. . . . and Solomon supplied Hiram with **twenty thousand** kor of wheat for his household,

16. . . . and **twenty** kor of oil of pounded olives . . .

17. He uttered **three thousand** proverbs, and his songs numbered a **thousand and five.**

18. Solomon had **forty thousand** chariot horses in his stables and **twelve thousand** cavalry horses.

19. King Solomon raised a forced levy from the whole of Israel amounting to **thirty thousand** men. He sent them to Lebanon in monthly relays of **ten thousand,** so that the men spent **one** month in Lebanon and **two** at home.

20. Solomon also had **seventy thousand** hauliers and **eighty thousand** quarrymen, apart from the **three thousand three hundred** in charge of the work. . . .

Write each number in Egyptian symbols.

21.	2348	**23.**	7060	**25.**	5,824,000
22.	1000	**24.**	9004	**26.**	4,251,000

Write each Roman numeral in our system.

27.	XXIII	**31.**	MMCDL	**35.**	$\overline{\text{XXI}}$
28.	XLII	**32.**	CMLXXVI	**36.**	$\overline{\overline{\text{XXIII}}}$
29.	LXXXVI	**33.**	MMI	**37.**	$\overline{\text{XIV}}$
30.	LXVIII	**34.**	MMMCDLXXIX	**38.**	$\overline{\overline{\text{XXII}}}$

Write each of the following, using Roman symbols.

39.	12	**43.**	474	**47.**	1728
40.	37	**44.**	288	**48.**	3209
41.	47	**45.**	759	**49.**	12,000,000
42.	58	**46.**	983	**50.**	38,000,000

Work each of the addition or subtraction problems in Exercises 51–68, giving answers in the system used.

51.   ∩∩ IIII
     + ∩∩∩ I

52.   ∩∩∩III
        ∩∩ III
     + ∩∩ III

53.   ϱ ∩∩ III
                  II
     + ϱϱ∩∩∩IIIII
        ∩∩∩ IIII

54.   999 ⋂⋂ ⎮⎮
              ⋂⋂

+   99 ⋂⋂⋂⋂⎮⎮⎮
     99 ⋂⋂⋂
_____

55.   𝄉𝄉𝄉𝄉 99   ⋂⋂⋂ ⎮⎮⎮⎮
          𝄉𝄉𝄉   ⋂⋂ ⎮⎮⎮⎮

+   𝄉𝄉𝄉 99999⋂⋂⋂⋂ ⎮⎮
          9999 ⋂⋂⋂ ⎮⎮
_____

56.   𝄉𝄉𝄉 99⋂⋂⋂
                  ⋂⋂ ⎮⎮

+   𝄉𝄉𝄉 9 ⋂⋂⋂⎮⎮⎮⎮
     𝄉𝄉   ⋂⋂⋂⎮⎮⎮⎮
_____

57.   ⋂⋂⋂ ⎮⎮
           ⋂⋂ ⎮⎮

−   ⋂⋂⋂⎮⎮⎮
_____

58.   ⋂⋂⎮⎮⎮⎮
              ⎮⎮⎮⎮

−   ⋂ ⎮⎮⎮
          ⎮⎮
_____

59.   ⋂⋂⋂⋂ ⎮⎮
           ⋂⋂⋂

−   ⋂⋂⋂ ⎮⎮⎮
              ⎮⎮
_____

60.   ⋂⋂ ⎮⎮⎮
           ⋂⋂ ⎮⎮⎮

−   ⋂ ⎮⎮⎮⎮
          ⎮⎮⎮⎮
_____

61.   9999⋂⋂⋂ ⎮⎮

−   ⋂⋂⋂⎮⎮⎮⎮
     9   ⋂⋂ ⎮⎮⎮
_____

62.   999 ⋂ ⎮⎮

−   ⋂⋂⋂⋂⎮⎮⎮
     9   ⋂⋂⋂ ⎮⎮⎮
_____

63.	XXIX + XXI	**66.**	LIX +XLVII	**69.**	XXXV −XXVII
64.	XIX +XXI	**67.**	CCCLX + CDIV	**70.**	CCLXXIV − CXCII
65.	DCCLIX + LXXVII	**68.**	XCII −XLIII		

Use the Egyptian algorithm to find each product.

71. 4×17 **73.** 7×51 **75.** 19×38

72. 8×27 **74.** 11×86 **76.** 23×47

Convert all numbers in the following problems to Egyptian numerals. Multiply using the Egyptian algorithm, and add using the Egyptian symbols. Give the solution in our system.

*77. King Solomon told the King of Tyre (now Lebanon) that Solomon needed the best cedar for his temple, and that he would "pay you for your men whatever sum you fix." Find the total bill to Solomon if the King of Tyre used the following numbers of men: 5500 tree cutters at two shekels per week each, for a total of seven weeks; 4600 sawers of wood at three shekels per week each, for a total of 32 weeks; and 900 sailors at one shekel per week each, for a total of 16 weeks.

*78. The book of Ezra in the Bible describes the return of the exiles to Jerusalem. When they rebuilt the temple, the King of Persia gave them the following items: thirty golden basins, a thousand silver basins, four hundred ten silver bowls, and thirty golden bowls. Find the total value of this treasure, if each gold basin is worth 3000 shekels, each silver basin is worth 500 shekels, each silver bowl is worth 50 shekels, and each golden bowl is worth 400 shekels.

What about the Greeks?
Classical Greeks used letters of their alphabet as numerical symbols. The base of the system was 10, and numbers 1 through 9 were symbolized by the first nine letters of the alphabet. Rather than using repetition or multiplication, they assigned nine more letters to multiples of 10 (through 90), and more letters to multiples of 100 (through 900). This is called a ciphered system, and sufficed for small numbers. For example, 57 would be $\nu\zeta$; 573 would be $\phi o \gamma$; and 803 would be $\omega\gamma$. A small stroke was used with a units symbol for multiples of 1000 (up to 9000); thus 1000 would be ,α or α'. Often M would indicate tens of thousands (M for myriad = 10,000) with the multiples written above M.

Greek numerals

1	α	10	ι	100	ρ
2	β	20	κ	200	σ
3	γ	30	λ	300	τ
4	δ	40	μ	400	υ
5	ϵ	50	ν	500	ϕ
6	ς	60	ξ	600	χ
7	ζ	70	o	700	ψ
8	η	80	π	800	ω
9	θ	90	ϕ	900	λ

Bodhisattva, *the Enlightened Being, the Compassionate; one of many depictions in Chinese art*

3.3 CHINESE, BABYLONIAN, MAYAN NUMERATION

In this section we shall look at three more numeration systems. The Chinese-Japanese system is multiplicative, and the Babylonian and the Mayan systems are both positional.

Chinese-Japanese system One of the few multiplicative systems actually used was developed many years ago in China, and later adopted by the Japanese. Recall from Section 1 of this chapter that a multiplicative grouping system needs special symbols for all counting numbers less than the base, to be used as the multipliers. In addition, symbols are needed for powers of the base. Some of the traditional Chinese-Japanese symbols are shown in Figure 3.7. The system uses base 10.

We should mention three features of this system. First, numbers are usually written vertically, as shown in Example 1 below. Second, if only one of a certain symbol is needed, the *multiplier* 1 is omitted from a pair. Finally, the *symbol for* 1 is also omitted from a pair of symbols giving how many 1s. (This omission is perhaps a hint of transition toward a positional system.) An example will clarify these features.

Example 1 Identify the Chinese-Japanese numeral below.

Solution. The pair at the top represents 3×1000. Then comes a single symbol for 100 (meaning 1×100, with the multiplier 1 omitted); then the pair for 6×10; and finally 4 (meaning 4×1). The number is 3164.

Example 2 Write the Chinese-Japanese numeral for 7503.

We need seven 1000s, five 100s, and a three. The numeral would be written

Figure 3.7

Number	Symbol
1	―
2	⁼
3	≡
4	◎
5	五
6	亣
7	七
8	八
9	九
10	十
100	百
1000	千

Sargon, *king of Akkad, about 2800 BC, in Mesopotamia before the rise of Babylonia.*

Babylonian system   The Babylonians used a base of 60 in their system. Because of this, in theory they would then need distinct symbols for numbers from 1 through 59 (just as we have symbols for 1 through 9). However, the Babylonians didn't do it this way. In fact, their method of writing on clay with wedge-shaped sticks gave rise to only *two* symbols: ⟨ represented ten, and ❚ represented one. Thus, 47 would be written

$$⟨⟨⟨⟨❚❚❚❚❚❚❚ \qquad \text{or to save space} \qquad ⟨⟨❚❚❚❚\!/\!⟨⟨\,❚❚❚$$

Thus, the digits 1 through 59 in the Babylonian system were not distinct symbols, but combinations of ⟨ and ❚.

Since the Babylonian system had base 60, the "digit" on the right in a multi-digit number represented the number of 1s, with the second "digit" from the right giving the number of 60s. The third digit would give the number of 3600s (60 × 60 = 3600), and so on.

Example 3   Convert ⟨⟨⟨⟨⟨❚❚❚ to our system.

Solution.   Here we have five 10s and three 1s, or

$$5 \times 10 = 50$$
$$3 \times 1 = \underline{3}$$
$$53 \quad \text{Answer}$$

Example 4   Convert ⟨⟨⟨⟨❚❚❚❚ / ⟨⟨ ❚❚❚❚ ⟨⟨❚❚ to our system.

This "two-digit" Babylonian number represents twenty-two 1s and fifty eight 60s, or

$$22 \times 1 = 22$$
$$58 \times 60 = \underline{3480}$$
$$3502 \quad \text{Answer}$$

Example 5   Convert ⟨⟨❚❚❚❚❚⟨❚⟨⟨⟨⟨❚❚❚❚❚❚ to our system.

Solution.   Here we have a three-digit number.

$$36 \times 1 = 36$$
$$11 \times 60 = 660$$
$$25 \times 3600 = \underline{90,000}$$
$$90,696 \quad \text{Answer}$$

Example 6   Convert each number to Babylonian: (a) 733   (b) 75,904   (c) 43,233.

(a)   To write 733 in Babylonian, we will need some 60s and some 1s. Divide 60 into 733; the answer is 12, with a remainder of 13. Thus we need twelve 60s and thirteen 1s, or

$$⟨❚❚⟨❚❚❚$$

From tally to tablet *The clay tablet you see, despite damage, shows how durable is the mud of the Babylonians. Thousands of years later we can work out Babylonian algebra problems from the original writings.*

Out of this clay, but in the 20th century, herdsmen still make little clay "tokens" as tallies of animals. Similar tokens have been unearthed by archaeologists in the land that was Babylonia—some tokens are 10,000 years old. Shaped like balls, disks, cones, and other regular forms, they rarely exceed 5 cm in diameter. Were they toys? or game pieces? No; they may rather be tallies that were the first step in a technological process that ended with the system of cuneiform writing on clay tablets.

Denise Schmandt-Besserat, professor, University of Texas, puts forth such a theory, and she suggests a 4-stage process:

I Merchants, for example, made figures standing for bread, oil jugs, animals. They kept records by direct tallies.

II They sealed tokens in clay balls (called bullae) to go with shipments so that the receiver could check on the number of items actually sent.

III They pressed tokens on the outside of the bullae so that the check could be made without breaking them open.

IV They transferred the action of pressing on clay bullae to other clay surfaces—the **tablets.** *Impressions of the original tokens came to be symbols, further stylized by the scribes in working with a sharp-tipped stylus.*

For example, if the token for a sheep was a disk with an X, then eventually the symbol ⊕ would stand for "sheep" and later, perhaps, for the syllable "shee" in any word. The wedges for "shee" would eventually look completely different from the original picture-symbol.

(b) For 75,904 here we need some 3600s, as well as some 60s and some 1s. Divide 75,904 by 3600. The answer is 21 with a remainder of 304. Divide 304 by 60; the answer is 5, with a remainder of 4. Therefore, 75,904 is written

$$\text{⟨⟨𝌆 𝌆𝌆𝌆𝌆𝌆 𝌆𝌆𝌆𝌆}$$

(c) Divide 43,233 by 3600; the answer is 12 with a remainder of 33. We need no 60s here; in a system such as ours we would use a 0 to show that no 60s are needed. The early Babylonians had no such symbol; they merely left a space. They would have written 43,233 as

$$\text{⟨𝌆𝌆   ⟨⟨⟨𝌆𝌆𝌆}$$

You can see the problem presented by the lack of a symbol for zero. In our system we know that 202 is not the same as 2002 or 20,002. Without the zero symbol, we would have to try to distinguish 2 2 from 2 2 or 2   2. It would get very difficult. The lack of a zero symbol was a major difficulty with the very early Babylonian system. A symbol for zero was introduced about 300 B.C.

Xipe Tótec, *the red god of maize, Red Tezcatlipoca, the god of the East, rejuvenated in Springtime.*

Figure 3.8

Number	Symbol
0	☺
1	·
2	··
3	···
4	····
5	—
6	·⎯
7	··⎯
8	···⎯
9	····⎯
10	═
11	·═
12	··═
13	···═
14	····═
15	≡
16	·≡
17	··≡
18	···≡
19	····≡

Mayan system The Mayan Indians of Central America and Mexico used what is basically a positional system. Like the Babylonians, the Mayans did not use base 10—they used base 20, with a twist. In a true base 20 system, the digits would represent 1s, 20s, $20 \times 20 = 400$s, $20 \times 400 = 8000$s, and so on. For some reason, however, the Mayans used 1s, 20s, $18 \times 20 = 360$s, $20 \times 360 = 7200$s, and so on. It is possible that they multiplied 20 by 18 (instead of 20) since 18×20 is close to the number of days in a year, convenient for astronomy. The symbols of the Mayan system are shown in Figure 3.8.

The Mayans were one of the first civilizations to invent a placeholder. They had a zero symbol many hundreds of years before it reached western Europe. Mayan numerals are written from top to bottom, just as in the traditional Chinese-Japanese system.

Example 7 Convert each Mayan numeral to our system.

(a)

··═
····⎯

(b)

···
☺
══

(a) The top group of symbols represents twelve 20s, while the bottom group represents nine 1s. Thus, the numeral represents

$$
\begin{aligned}
12 \times 20 &= 240 \\
9 \times 1 &= \underline{9} \\
&\ 249 \quad \text{Answer}
\end{aligned}
$$

(b)
$$
\begin{aligned}
8 \times 360 &= 2880 \\
0 \times 20 &= 0 \\
15 \times 1 &= \underline{15} \\
&\ 2895 \quad \text{Answer}
\end{aligned}
$$

Example 8 Convert to Mayan: (a) 277   (b) 1238.

(a) The number 277 requires thirteen 20s (divide 277 by 20) and seventeen 1s, and would be written

···═

··≡

(b) Divide 1238 by 360. The result is 3, with 158 left over. Divide 158 by 20; the answer is 7, with 18 left over. Thus, we need three 360s, seven 20s, and eighteen 1s:

···

··⎯

···≡

3.3 EXERCISES

Identify each numeral in Exercises 1–24 as Babylonian, Mayan, or Chinese-Japanese. Give the equivalent in our system.

1. ⠒

2.  ⟨⟨𐎠

3.  ⟨⟨⟨⟨𐎠

4. ≣

5. 〇 千 三 十 九

6. 百 三 十 五

7. ⠒ ⠄

8. ⠘ ⠘

9.  ⟨⟨𐎠⟨⟨𐎠

10.  ⟨𐎠𐎠𐎠⟨⟨⟨𐎠

11.  ⟨⟨𐎠 ⟨⟨𐎠⟨⟨𐎠

12.  ⟨⟨⟨⟨𐎠𐎠𐎠 ⟨⟨ 𐎠𐎠⟨⟨⟨𐎠

13. 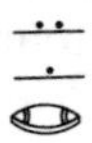 (Mayan)

14. 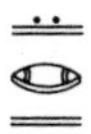 (Mayan)

15. 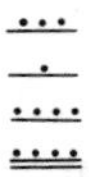 (Mayan)

16.  (Mayan)

17.  ⟨⟨𐎠𐎠⟨⟨𐎠⟨𐎠𐎠𐎠𐎠

18. 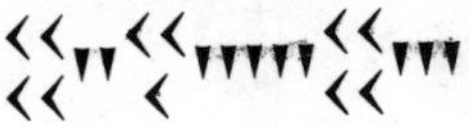 ⟨⟨𐎠𐎠⟨⟨𐎠𐎠𐎠𐎠𐎠⟨⟨𐎠𐎠𐎠
⟨⟨ ⟨ ⟨⟨

19. 七 百 七 十 二 九

20. 九 千 七 百 七 十 九

Write each number as a Chinese-Japanese numeral.

21. 39	**24.** 106	**27.** 2769
22. 51	**25.** 412	**28.** 9814
23. 92	**26.** 381	

Write each number as a Babylonian numeral.

29. 21	**33.** 1514	***37.** 43,205
30. 32	**34.** 3280	***38.** 90,180
31. 293	**35.** 5190	
32. 412	**36.** 7842	

Write each number as a Mayan numeral.

39. 12	**42.** 208	***45.** 64,712
40. 32	**43.** 4694	***46.** 61,598
41. 151	**44.** 4328	

3.4 THE HINDU-ARABIC SYSTEM

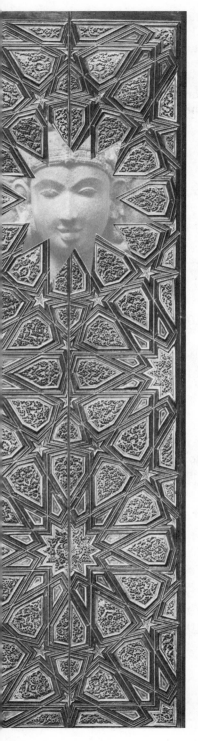

The positional system most familiar to us is our own numeration system, called the **Hindu-Arabic system.** The symbols of this system can be traced back to the Hindus of 200 B.C. The symbols were picked up from the Hindus by the Arabs, and eventually transmitted to Spain, slowly replacing Roman numerals throughout Europe.

The advantage of the Hindu-Arabic system over the Roman system is that it is positional. For example, each 4 in the number 46,424 represents a different value. The 4 on the right represents four 1s (the place value is 1). The 4 in the middle represents four 100s, while the 4 on the left represents four 10,000s.

The **place values** of the system, from the right, are 1s, 10s, 100s, 1000s, 10,000s, 100,000s, and so on.

As we work with Hindu-Arabic numerals, it would be easier if we didn't have to write out long numbers such as 10,000. To avoid writing all the zeros, *exponents* are used. For example, 10,000 is written as 10^4, where the 4 is called an **exponent** and the 10 is called the **base.** By definition,

$$10^4 = \underbrace{10 \times 10 \times 10 \times 10}_{\text{4 tens}} = 10,000.$$

In the same way, $10^2 = 10 \times 10 = 100$, $10^6 = 10 \times 10 \times 10 \times 10 \times 10 \times 10 = 1,000,000$, and so on. The base of an exponential number does not have to be 10; for example,

$$4^3 = 4 \times 4 \times 4 = 64, \qquad 2^2 = 2 \times 2 = 4,$$
$$3^5 = 3 \times 3 \times 3 \times 3 \times 3 = 243,$$

and so on.

In general, if m is a counting number, then

$$a^m = a \times a \times a \times \cdots \times a, \quad \text{where there are } m \text{ of the } a\text{'s.}$$

Example 1

(a) $10^3 = 10 \times 10 \times 10 = 1000$
(10^3 is read "10 cubed," or "10 to the third power.")

(b) $7^2 = 7 \times 7 = 49$
(7^2 is read "7 squared," or "7 to the second power.")

(c) $5^4 = 5 \times 5 \times 5 \times 5 = 625$
(5^4 is read "5 to the fourth power.")

Arabesque and Brahma *The Moslems were not allowed by their religion to portray animate forms in art, and they developed an abstract and intricate style called* arabesque. *You see the carved wood of a door, inlaid with ivory. Through it we have shown Brahma, creator god in the Hindu pantheon.*

To simplify work with exponents, it is generally agreed that $a^0 = 1$, for any non-zero number a. Thus, $7^0 = 1$, $10^0 = 1$, and $52^0 = 1$, and so on. At the same time, $a^1 = a$, for any number a. Thus $8^1 = 8$, $25^1 = 25$, and so on. The exponent 1 is usually omitted.

By using exponents, we can write numbers in **expanded form,** in which the value of the number in each position is made clear. For example, to write 924 in expanded form, we first note that 924 means nine 100s plus two 10s plus four 1s, or

$$924 = 900 + 20 + 4$$
$$924 = (9 \times 100) + (2 \times 10) + (4 \times 1).$$

By the definition of exponents, we know that $100 = 10^2$, $10 = 10^1$, and $1 = 10^0$. Thus, we can rewrite:

$$924 = (9 \times 10^2) + (2 \times 10^1) + (4 \times 10^0).$$

Example 2 The following are written in expanded form.

(a) $1906 = (1 \times 10^3) + (9 \times 10^2) + (0 \times 10^1) + (6 \times 10^0)$. Since $0 \times 10^1 = 0$, we could omit this term if we wished, but the form is clearer with it.

(b) $46{,}424 = (4 \times 10^4) + (6 \times 10^3) + (4 \times 10^2) + (2 \times 10^1) + (4 \times 10^0)$.

Example 3 Each of the following expansions is simplified.

(a) $(3 \times 10^5) + (2 \times 10^4) + (6 \times 10^3) + (8 \times 10^2) + (7 \times 10^1) + (9 \times 10^0) = 326{,}879$.

(b) $(2 \times 10^1) + (8 \times 10^0) = 28$.

New, improved system
Why is this person smiling? He has tried out the new system of numeration using Hindu-Arabic numerals and quill pens. He likes it! It's the 15th century and time for a change.

One reason that we look at expanded notation is to see why the standard algorithms for addition and subtraction really work. The key idea behind these algorithms is based on the **distributive property,** which can be written in one form as

$$(b \times a) + (c \times a) = (b + c) \times a.$$

For example,
$$(3 \times 10^4) + (2 \times 10^4) = (3 + 2) \times 10^4$$
$$= 5 \times 10^4.$$

Example 4 Use expanded notation to add 23 and 64.

Solution. We have
$$23 = (2 \times 10^1) + (3 \times 10^0)$$
$$+ 64 = \underline{(6 \times 10^1) + (4 \times 10^0)}$$

By the distributive property,
$$(2 \times 10^1) + (6 \times 10^1) = (2 + 6) \times 10^1 = 8 \times 10^1$$
and $\quad (3 \times 10^0) + (4 \times 10^0) = (3 + 4) \times 10^0 = 7 \times 10^0$.

Substitute: $\quad 23 + 64 = (8 \times 10^1) + (7 \times 10^0) = 87$.

Old Roman blues *Tired of the old counting board? Fed up with pushing counters from line to line? Those Roman numerals aren't very convenient anyhow. Well, don't just sit there and mope—switch over!*

Subtraction works in much the same way.

Example 5   Find $695 - 254$.

Solution.

$$695 = (6 \times 10^2) + (9 \times 10^1) + (5 \times 10^0)$$
$$-254 = (2 \times 10^2) + (5 \times 10^1) + (4 \times 10^0)$$
$$\overline{(4 \times 10^2) + (4 \times 10^1) + (1 \times 10^0) = 441}\quad \text{Answer}$$

Since our numeration system is based on powers of ten, as we have seen, it is often called the **decimal system,** from the Latin word *decem,* meaning "ten."*

3.4 EXERCISES

Write each number in expanded form.

1. 59
2. 74
3. 426
4. 546
5. 1984
6. 3712
7. 29,846
8. 30,897
9. 2,508,901
10. 7,603,490

11. four thousand seven hundred twelve

12. nine thousand eight hundred seventeen

*13. seven million, two thousand, seven

*14. twelve million, five hundred seven thousand, eighty

Simplify each of the following expansions.

15. $(9 \times 10^1) + (8 \times 10^0)$

16. $(7 \times 10^1) + (2 \times 10^0)$

17. $(6 \times 10^2) + (0 \times 10^1) + (0 \times 10^0)$

18. $(4 \times 10^2) + (8 \times 10^1) + (9 \times 10^0)$

19. $(7 \times 10^3) + (8 \times 10^2) + (0 \times 10^1) + (5 \times 10^0)$

20. $(2 \times 10^3) + (1 \times 10^2) + (3 \times 10^1) + (0 \times 10^0)$

21. $(5 \times 10^4) + (7 \times 10^3) + (8 \times 10^2) + (4 \times 10^1) + (3 \times 10^0)$

22. $(2 \times 10^4) + (0 \times 10^3) + (1 \times 10^2) + (5 \times 10^1) + (0 \times 10^0)$

*23. $(8 \times 10^6) + (2 \times 10^4) + (3 \times 10^2) + (5 \times 10^0)$

*24. $(9 \times 10^7) + (3 \times 10^5) + (1 \times 10^2) + (2 \times 10^1)$

December was the tenth month in an old form of the calendar. It is interesting to note that *decem* became *dix* in the French language; a ten-dollar bill, called "dixie," was in use in New Orleans before the Civil War. And "Dixie Land" was a nickname for that city before Dixie came to refer to the Southern states, as in Daniel D. Emmett's song, written in 1859.

Sliding beads *The wedge-shaped beads tell us that the abacus above is Japanese (soroban). The Chinese abacus (suan-pan) has round beads. The Russian s'choty, the Armenian choreb, and the Turkish coulba all have horizontal rods. Compare the Roman counting board opposite.*

In each of the following, add in expanded notation.

25. $42 + 35$ **27.** $859 + 120$

26. $27 + 51$ **28.** $763 + 220$

In each of the following, subtract in expanded notation.

29. $76 - 21$ **31.** $328 - 106$

30. $58 - 46$ **32.** $559 - 237$

We can also use expanded notation when "carrying" in addition.

Example: Add 67 and 19.

$$\begin{aligned} \text{We have} \quad 67 &= (6 \times 10^1) + (7 \times 10^0) \\ 19 &= \underline{(1 \times 10^1) + (9 \times 10^0)} \\ \text{Add:} \quad & (7 \times 10^1) + (16 \times 10^0) \end{aligned}$$

To simplify 16×10^0, proceed as follows:

$$16 \times 10^0 = 16 \times 1 = 16 = (1 \times 10^1) + (6 \times 10^0).$$

Replace 16×10^0 by this result:

$$\begin{aligned} 67 + 19 &= (7 \times 10^1) + (16 \times 10^0) \\ &= (7 \times 10^1) + (1 \times 10^1) + (6 \times 10^0) \\ &= (8 \times 10^1) + (6 \times 10^0) \\ 67 + 19 &= 86 \end{aligned}$$

Find each of the following, using expanded notation as in the example.

33. $74 + 92$ **35.** $427 + 398$ **37.** $2515 + 4395$

34. $63 + 55$ **36.** $242 + 689$ **38.** $7655 + 3928$

Use a method similar to that given for "carrying" to work each of the following subtractions in expanded notation. *Hint:* $(5 \times 10^3) = (4 \times 10^3) + (1 \times 10^3) = (4 \times 10^3) + (10 \times 10^2)$.

*39. $72 - 68$ ***41.** $126 - 19$ ***43.** $382 - 274$

*40. $83 - 55$ ***42.** $251 - 34$ ***44.** $507 - 286$

One device that uses a positional numeration system is the **abacus.** An abacus consists of a series of rods with sliding beads, and a dividing bar. The rods (right to left) have values of 1, 10, 100, 1000, and so on. The bead above the bar has five times the value of those below. Beads moved *toward* the *bar* are in the "active" position; those toward the *frame* are "neutral." The number shown on the sketch is found as follows:

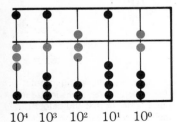

$10^4 \quad 10^3 \quad 10^2 \quad 10^1 \quad 10^0$

$$[(3 \times 10{,}000)] + [(1 \times 1000)] + [(1 \times 500) + (2 \times 100)] + 0 \\ + [(1 \times 5) + (1 \times 1)]$$

$$= 30{,}000 + 1000 + 500 + 200 + 0 + 5 + 1$$

$$= 31{,}706.$$

Identify the numbers represented in each of the following sketches.

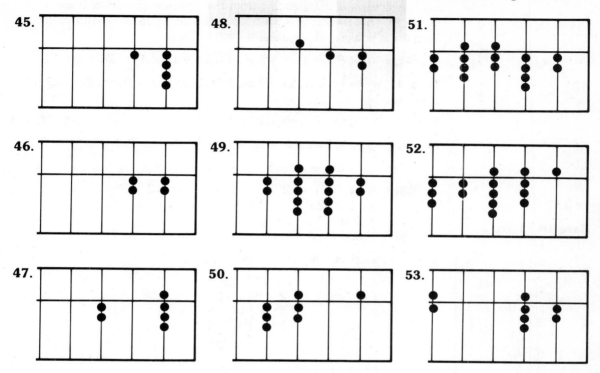

Sketch an abacus to show each of the following numbers.

Counting board *The Romans called it* abacus, *and we borrowed the word for similar devices. The sketch shows the numerical value of the lines (and spaces). The counters were usually pebbles—the Latin word for pebble is* calculus. *The Japanese also calculated on a board, ruled into rectangles, with sticks of bone or bamboo (*sanji*).*

54.	72	**57.**	3507	**59.**	23,190
55.	98	**58.**	4726	**60.**	64,720
56.	842				

Decide how an abacus could be used to add the following.

*61.	42 and 36	*63.	5824 and 7691
*62.	107 and 321	*64.	2143 and 9769

934 + 286

3.5 CONVERTING BETWEEN NUMBER BASES

We have studied number bases other than base 10, including a base 20 system (the Mayan) and a base 60 system (the Babylonian). In this section we study other number bases in more detail. Rather than use different symbols for the digits in other bases, we shall just use our familiar Hindu-Arabic symbols. To show the base that a number is expressed in, we use subscript numbers like $_5$ and $_7$. Thus 41_5 indicates that 41 is to be considered a base 5 numeral. Also, 41_7 would indicate a base 7 numeral.

In base 5, the digits indicate powers of 5, just as the digits in base 10 indicate powers of 10. Reading from the right, the powers of 5 are $5^0 = 1$, $5^1 = 5$, $5^2 = 5 \times 5 = 25$, $5^3 = 5 \times 5 \times 5 = 125$, $5^4 = 625$, $5^5 = 3125$, and so on. With these powers, 41_5 represents

$$41_5 = (4 \times 5^1) + (1 \times 5^0) = (4 \times 5) + (1 \times 1) = 20 + 1 = 21.$$

Thus, 41_5 is the same as 21 in base 10. (If no base is indicated, base 10 is understood.)

Example 1   Convert each to base 10:   (a) 34_5 (b) 4213_5 (c) 32144_5.

(a) We have three 5s and four 1s, or

$$34_5 = (3 \times 5) + (4 \times 1) = 15 + 4 = 19.$$

(b) $4213_5 = (4 \times 5^3) + (2 \times 5^2) + (1 \times 5^1) + (3 \times 5^0)$
$\qquad = (4 \times 125) + (2 \times 25) + (1 \times 5) + (3 \times 1)$
$\qquad = 500 + 50 + 5 + 3$
$4213_5 = 558.$

This same work can be done more rapidly as follows:

$$
\begin{aligned}
4 \times 125 &= 500 \\
2 \times 25 &= 50 \\
1 \times 5 &= 5 \;\cdot \\
3 \times 1 &= \underline{3} \\
& 558 \quad \text{Answer}
\end{aligned}
$$

(c) Using powers of 5, 32144_5 becomes

$$
\begin{aligned}
3 \times 625 &= 1875 \\
2 \times 125 &= 250 \\
1 \times 25 &= 25 \\
4 \times 5 &= 20 \\
4 \times 1 &= \underline{4} \\
& 2174 \quad \text{Answer}
\end{aligned}
$$

For reference, the first thirty counting numbers have been converted to base 5. See Figure 3.9.

As we saw above, the digits in base 5 represent, from the right, 1s, 5s, 25s, 125s, 625s, 3125s, and so on. We can use these values to convert a number from base 10 to base 5, as shown in the following example.

Example 2   Convert each base 10 number to base 5: (a) 89   (b) 2853.

(a) The number 89 is more than 25, but less than 125. Thus, we need some 25s, but no 125s. Divide 89 by 25; the answer is 3 with a remainder of 14. We need three 25s. Now, divide 14 by 5. The answer is 2, with a remainder of 4. Thus, we also need two 5s and four 1s. Finally,

$$89 = 324_5.$$

Figure 3.9

Base 10	Base 5
1	1
2	2
3	3
4	4
5	10
6	11
7	12
8	13
9	14
10	20
11	21
12	22
13	23
14	24
15	30
16	31
17	32
18	33
19	34
20	40
21	41
22	42
23	43
24	44
25	100
26	101
27	102
28	103
29	104
30	110

(b) The number 2853 is more than 625, but less than 3125. If we divide 2853 by 625, we get 4 with a remainder of 353. Divide this remainder by 125, and continue as shown below.

$$
\begin{array}{rr}
 & 2853 \\
4 \times 625 = 2500 & -\ 2500 \\
\hline
 & 353 \\
2 \times 125 = 250 & -\ \ \ 250 \\
\hline
 & 103 \\
4 \times 25 = 100 & -\ \ \ 100 \\
\hline
3 \times 1 = 3 & 3
\end{array}
$$

Therefore, $2853 = 42403_5.$

There is a shortcut that we can use when converting from base 10 to base 5. This shortcut consists of doing the necessary divisions as shown below. For example, we have just seen that $2853 = 42403_5$. To get this result with the shortcut, start with 2853 and divide by 5 again and again, as shown, until no remainder results. Keep track of the remainder at each step. The base 5 numeral is found by reading the remainders from bottom to top and forming the numeral 42403. Thus, again, $2853 = 42403_5$.

	Remainder
5 ⌐2853	3
5 ⌐570	0
5 ⌐114	4
5 ⌐22	2
5 ⌐4	4
0	

The methods we have used for converting back and forth between base 10 and base 5 can be used for any base. For example, to convert 365_7 to base 10, we expand the numeral by powers of 7, as follows.

$$
\begin{aligned}
365_7 &= (3 \times 7^2) + (6 \times 7^1) + (5 \times 7^0) \\
&= (3 \times 49) + (6 \times 7) + (5 \times 1) \\
&= 147 + 42 + 5 \\
365_7 &= 194.
\end{aligned}
$$

Thus, 365_7 is the same as 194 in base 10.

Example 3   Convert each number to base 10:   (a) 4239_{12} (b) 2132_4.

(a) Expand using powers of 12:

$$
\begin{aligned}
4239_{12} &= (4 \times 12^3) + (2 \times 12^2) + (3 \times 12^1) + (9 \times 12^0) \\
&= (4 \times 1728) + (2 \times 144) + (3 \times 12) + (9 \times 1) \\
&= 6912 + 288 + 36 + 9 \\
4239_{12} &= 7245
\end{aligned}
$$

(b)
$$
\begin{aligned}
2132_4 &= (2 \times 4^3) + (1 \times 4^2) + (3 \times 4^1) + (2 \times 4^0) \\
&= (2 \times 64) + (1 \times 16) + (3 \times 4) + (2 \times 1) \\
&= 128 + 16 + 12 + 2 \\
2132_4 &= 158.
\end{aligned}
$$

Selected powers of some of the most common numbers used as bases are shown in Figure 3.10.

Figure 3.10 Selected powers of some number bases

	Fourth power	Third power	Second power	First power	Zero power
Base 2	16	8	4	2	1
Base 3	81	27	9	3	1
Base 4	256	64	16	4	1
Base 5	625	125	25	5	1
Base 7	2401	343	49	7	1
Base 8	4096	512	64	8	1
Base 11	14,641	1331	121	11	1
Base 12	20,736	1728	144	12	1

To convert a numeral from base 10 to another base, divide the numeral repeatedly by the other base, as shown in the following examples.

Example 4 Convert 298 to base 7.

Solution. Divide 298 by 7 until no remainder results.

$$\begin{array}{ll} & \text{Remainder} \\ 7\underline{|298} & 4 \\ 7\underline{|42} & 0 \\ 7\underline{|6} & 6 \\ 0 & \end{array}$$

Read the remainders from bottom to top: $298 = 604_7$.

Example 5 Convert 43,825 to base 8.

Solution. Divide by 8.

$$\begin{array}{ll} & \text{Remainders} \\ 8\underline{|43,825} & 1 \\ 8\underline{|5478} & 6 \\ 8\underline{|684} & 4 \\ 8\underline{|85} & 5 \\ 8\underline{|10} & 2 \\ 8\underline{|1} & 1 \\ 0 & \end{array}$$

Thus, $43,825 = 125,461_8$.

We run into a problem when converting a base 10 numeral to base 12. A base 12 numeration system would require 12 symbols, and we have only ten available. To get around this problem, we can use the symbol T to represent the digit *ten*, while E represents *eleven*. The use of these symbols is shown in the following examples.

Example 6 Convert each base 10 number to base 12: (a) 138 (b) 1582.

(a) Divide 138 by 12:

	Remainder	Base 12 notation
12⌊138	6	6
12⌊11	11	E
0		

Using E to represent 11, we have $138 = E6_{12}$.

(b) Divide 1582 by 12.

	Remainder	Base 12 notation
12⌊1582	10	T
12⌊131	11	E
12⌊10	10	T
0		

Thus, $1582 = TET_{12}$.

There is a group of people who feel that 12 is a better base than 10, since more numbers divide evenly into 12 than into 10. This makes fractions easier than in base 10. These people have given a name to T and E: they call T *dek,* again from the Latin *decem* (or "ten"), while E is called *el* from "eleven." The symbol 10 would be read *do* from "dozen."

3.5 EXERCISES

Convert each number to base 10.

1. 43_5
2. 32_5
3. 62_7
4. 54_7
5. 37_{12}
6. 59_{12}
7. $T4_{12}$
8. $3E_{12}$
9. 423_5
10. 243_5
11. 306_7
12. 162_7
13. 734_8
14. 974_{12}
15. 1424_5
16. 6542_7

Convert each to the indicated base.

17. 59 to base 5
18. 77 to base 7
19. 342 to base 7
20. 274 to base 5
21. 193 to base 4
22. 587 to base 8
23. 1250 to base 7
24. 398 to base 12
25. 4372 to base 12
26. 1197 to base 11
27. 2587 to base 11
28. 5946 to base 8
29. 3884 to base 12
30. 6893 to base 12
31. 6645 to base 11
32. 7962 to base 11
33. 5000 to base 12
34. 5000 to base 8
35. 5000 to base 7
36. 5000 to base 4

To convert a number (not in base 10) from one base to another, convert the number first to base 10. Then convert the resulting base 10 number to the required base.

Example   Convert 5736_8 to base 3.

Solution.   First convert to base 10:

$$5736_8 = (5 \times 8^3) + (7 \times 8^2) + (3 \times 8^1) + (6 \times 8^0)$$
$$= (5 \times 512) + (7 \times 64) + (3 \times 8) + (6 \times 1)$$
$$= 2560 + 448 + 24 + 6$$
$$= 3038$$

Then convert 3038 to base 3, by dividing by 3. Thus,

$$5736_8 = 3038 = 11011112_3.$$

Convert each of the following as indicated.

37. 26_7 to base 4   **42.** 641_{11} to base 5

38. 53_7 to base 12   **43.** 814_{11} to base 8

39. 414_5 to base 7   **44.** 5134_8 to base 4

40. 2231_5 to base 7   **45.** 3132_4 to base 12

41. 729_{12} to base 8   **46.** 212110_3 to base 8

3.6 ARITHMETIC IN OTHER BASES

Now that we have seen how to convert numbers back and forth between base 10 and other bases, we can look at methods for doing arithmetic in these other bases. For example, to add in base 5 we might first make a table showing the various possible sums. To make such a *base 5 addition table,* we place 0, 1, 2, 3, and 4 (the symbols in a base 5 numeration system) across the top and down the side, as in Figure 3.11. To fill in the rows and columns we must add combinations of 0, 1, 2, 3, and 4. For example,

$$2 + 1 = 3, \quad 1 + 3 = 4, \quad 2 + 2 = 4, \quad \text{and} \quad 3 + 0 = 3.$$

When we try to add 2 and 3, we might be tempted to write "5." However, we are working in base 5, and "5" is not a symbol in this system. In base 5, the number 5 is written with *one* 5 and *zero* 1s, or

$$2_5 + 3_5 = 10_5.$$

In the same way, $3 + 4 = 7 = 12_5$, $4 + 4 = 8 = 13_5$, and so on. By working out sums such as these, we get numerals to complete the addition table of Figure 3.11.

We can use this table to work longer computations in base 5. For example, to find the sum $3 + 4$, find 3 at the side and 4 across the top. The sum, 12, is given where the row containing 3 and the column containing 4 meet. See Figure 3.12.

Figure 3.11

Base 5 addition table

+	0	1	2	3	4
0	0	1	2	3	4
1	1	2	3	4	10
2	2	3	4	10	11
3	3	4	10	11	12
4	4	10	11	12	13

Figure 3.12
In base five, $3 + 4 = 12$

+	0	1	2	3	4
0	0	1	2	3	4
1	1	2	3	4	10
2	2	3	4	10	11
3	3	4	10	11	**12**
4	4	10	11	12	13

Example 1 Add 422_5 and 243_5.

Solution. First, arrange the numbers just as you would for ordinary addition in base 10. Then add columns from the right, finding sums in the addition table above.

To begin, add 2 and 3: $2 + 3 = 10$. Write 0 and carry 1.

$$\begin{array}{r} \overset{1}{4}22 \\ +243 \\ \hline 0 \end{array}$$

Now add 1, 2, and 4: $1 + 2 + 4 = 12$. Write 2 and carry 1.

$$\begin{array}{r} \overset{11}{4}22 \\ +243 \\ \hline 20 \end{array}$$

Finally, $1 + 4 + 2 = 12$.

$$\begin{array}{r} \overset{11}{4}22 \\ +243 \\ \hline 1220 \end{array}$$

To check, convert all numbers to base 10 and work the problem in that base. In this example, we have $422_5 + 243_5 = 112 + 73 = 185 = 1220_5$.

Example 2 Find the sum of the base 5 numbers 1342, 224, and 2143.

Solution. Write the numbers in a column.

Step 1:	Step 2:	Step 3:	Step 4:
$1\overset{1}{3}42$	$1\overset{21}{3}42$	$\overset{121}{1}342$	$\overset{121}{1}342$
224	224	224	224
+2143	+2143	+2143	+2143
4	14	314	4314 Answer

$$\begin{array}{llll} 2+4+3 & 1+4+2+4 & 2+3+2+1 & 1+1+2=4 \\ = 14 & = 21 & = 13 & \end{array}$$

The base 5 addition table of Figure 3.11 can also be used for subtraction. For example, to find $12_5 - 3_5$, locate the number that would be added to 3_5 to get 12_5. This number is 4_5. Thus, $12_5 - 3_5 = 4_5$. In general, work subtraction problems as in base 10. It may be necessary to borrow. This process is explained in the next example.

Example 3 Work each base 5 subtraction problem.

Solution.

(a)
$$\begin{array}{r} 423 \\ -121 \\ \hline 302 \end{array}$$

(b)
$$\begin{array}{r} \overset{2\,121}{3}32 \\ -\ 43 \\ \hline 234 \end{array}$$

(c)
$$\begin{array}{r} \overset{3\,11\,21}{4}2030 \\ -24301 \\ \hline 12224 \end{array}$$

Figure 3.13
Base 5 multiplication table

×	0	1	2	3	4
0	0	0	0	0	0
1	0	1	2	3	4
2	0	2	4	11	13
3	0	3	11	14	22
4	0	4	13	22	31

To work problems in multiplication, we need a *base 5 multiplication table,* as in Figure 3.13. The numbers in the table were found as above. For example, $4 \times 4 = 16$ (base 10) which is written in base 5 with *three* 5s and *one* 1. Thus, $4_5 \times 4_5 = 31_5$. This number is found in the lower right corner of the table.

Example 4   Find each of the following products in base 5.

(a)
$$
\begin{array}{r}
34 \\
\times\,23 \\
\hline
212 \\
123 \\
\hline
1442
\end{array}
\quad \text{Answer}
$$

$34_5 \times 23_5 = 1442_5$

(b)
$$
\begin{array}{r}
342 \\
\times\,\,21 \\
\hline
342 \\
1234 \\
\hline
13232
\end{array}
\quad \text{Answer}
$$

$342_5 \times 21_5 = 13232_5$

3.6 EXERCISES

Work the following problems in base 5.

1. $\begin{array}{r} 42 \\ +31 \\ \hline \end{array}$

2. $\begin{array}{r} 23 \\ +41 \\ \hline \end{array}$

3. $\begin{array}{r} 242 \\ +314 \\ \hline \end{array}$

4. $\begin{array}{r} 422 \\ +314 \\ \hline \end{array}$

5. $\begin{array}{r} 213 \\ 422 \\ +\,\,34 \\ \hline \end{array}$

6. $\begin{array}{r} 242 \\ 30 \\ +214 \\ \hline \end{array}$

7. $\begin{array}{r} 321 \\ -\,\,30 \\ \hline \end{array}$

8. $\begin{array}{r} 214 \\ -\,\,32 \\ \hline \end{array}$

9. $\begin{array}{r} 1042 \\ -\,\,324 \\ \hline \end{array}$

10. $\begin{array}{r} 4121 \\ -3024 \\ \hline \end{array}$

11. $\begin{array}{r} 214 \\ \times\,\,\,3 \\ \hline \end{array}$

12. $\begin{array}{r} 132 \\ \times\,\,\,2 \\ \hline \end{array}$

13. $\begin{array}{r} 21 \\ \times 43 \\ \hline \end{array}$

14. $\begin{array}{r} 32 \\ \times 22 \\ \hline \end{array}$

15. $\begin{array}{r} 212 \\ \times\,\,43 \\ \hline \end{array}$

16. $\begin{array}{r} 322 \\ \times\,\,42 \\ \hline \end{array}$

17. 321^2

18. 224^2

*19. $2\overline{)13}$

*20. $3\overline{)41}$

*21. $4\overline{)31013}$

*22. $4\overline{)23124}$

*23. $12\overline{)144}$

*24. $22\overline{)121}$

25. Complete the base 7 addition table.

+	0	1	2	3	4	5	6
0	0	1	2	3	4	5	6
1	1	2	3	4	5	6	10
2	2	3	4	5	6		
3	3	4	5	6		11	
4	4	5	6		11		13
5	5	6		11		13	14
6							

26. Complete the base 7 multiplication table.

×	0	1	2	3	4	5	6
0	0	0	0	0	0	0	0
1	0	1	2	3	4	5	6
2	0	2	4	6	11	13	15
3	0	3	6			21	
4	0	4	11		22		33
5	0	5		21		34	42
6	0	6					51

Use the tables in Exercises 25 and 26 to work the following base 7 problems.

27.   36
           +41

28.   44
           +62

29.   365
           242
          +216

30.   245
           345
          +466

31.   253
          − 46

32.   412
          −105

33.   4651
         − 266

34.   3654
         − 666

35.   43
           × 6

36.   25
           × 6

37.   325
          × 42

38.   416
          × 55

39. Write a base 12 addition table.

*40. Write a base 12 multiplication table.

Use the tables from Exercises 39 and 40 to work the following base 12 problems.

41.   27
            46
           +5E

42.   9T
            21
           +35

43.   2E7
           4T8
          +290

44. 213ET
         + 5072

*45.   25
           ×37

*46.   98
           ×22

*47.   2E5
          × 42

*48.   T74
          × 29

*49. TE²

*50. ET²

3.7 THE BINARY SYSTEM (BASE 2) AND ITS APPLICATIONS

In this section, we shall look in detail at one particular base, *base 2*, often called the **binary system.** The base 2 number system has perhaps the greatest number of applications of any system other than base 10. These applications depend on the fact that only two symbols are used in base 2, namely, 0 and 1. Thus, numbers can be represented in calculating machines by switches where "on" represents 0 and "off" represents 1.

First, let us review methods of converting from base 10 to base 2, and back again.

Example 1   Convert 1011_2 to base 10.

Solution.   In the binary system, the place values from right to left represent powers of 2: 2^0 (or 1), 2^1 (or 2), 2^2 (or 4), 2^3 (or 8), 2^4 (or 16), 2^5 (or 32), and so on. Using these powers, 1011_2 expands to

$$1011_2 = (1 \times 2^3) + (0 \times 2^2) + (1 \times 2^1) + (1 \times 2^0)$$
$$= (1 \times 8) + (0 \times 4) + (1 \times 2) + (1 \times 1)$$
$$= 8 + 0 + 2 + 1$$
$$1011_2 = 11.$$

Figure 3.14

Base 10	Base 2
1	1
2	10
3	11
4	100
5	101
6	110
7	111
8	1000
9	1001
10	1010
11	1011
12	1100
13	1101
14	1110
15	1111
16	10000
17	10001
18	10010
19	10011
20	10100

Example 2   Convert 11100110_2 to base 10.

Solution.   We have

$$11100110_2 = (1 \times 2^7) + (1 \times 2^6) + (1 \times 2^5) + (0 \times 2^4) + (0 \times 2^3)$$
$$+ (1 \times 2^2) + (1 \times 2^1) + (0 \times 2^0)$$
$$= 128 + 64 + 32 + 0 + 0 + 4 + 2 + 0$$
$$11100110_2 = 230.$$

Example 3   Convert each number to base 2:   (a) 29   (b) 99.

(a) Divide 29 by 2, writing the remainders at the side.

	Remainder
2⌋29	1
2⌋14	0
2⌋7	1
2⌋3	1
2⌋1	1
0	

Write the remainders from bottom to top: $29 = 11101_2$.

(b) Divide 99 by 2:

	Remainder
2⌋99	1
2⌋49	1
2⌋24	0
2⌋12	0
2⌋6	0
2⌋3	1
2⌋1	1
0	

Read from bottom to top: $99 = 1100011_2$.

The binary system is widely used in the design of computers and pocket calculators. If you punch the "6" key on a calculator, the machine converts the "6" into 110_2 for its own use. All calculations are done in base 2, with the answer converted back to base 10 before it is displayed.

Several games and tricks are based on the binary number system. For example, use the table of Figure 3.15 to find the age of a person 31 years old or younger. The person need only tell you the columns that contain his or her age. For example, suppose Francisco says that his age appears in columns B, D, and E only. To find his age, add the numbers from the top row of these three columns:

Francisco is $2 + 8 + 16 = 26$ years old.

Figure 3.15

A	B	C	D	E
1	2	4	8	16
3	3	5	9	17
5	6	6	10	18
7	7	7	11	19
9	10	12	12	20
11	11	13	13	21
13	14	14	14	22
15	15	15	15	23
17	18	20	24	24
19	19	21	25	25
21	22	22	26	26
23	23	23	27	27
25	26	28	28	28
27	27	29	29	29
29	30	30	30	30
31	31	31	31	31

3.7 EXERCISES

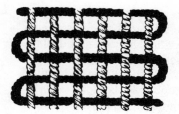

Woven fabric is a binary system *of threads going lengthwise (warp threads; white in the diagram above) and threads going crosswise (weft, or woof). At any point in a fabric, either warp or weft is on top, and the variation creates the pattern. Weaving is done on a loom, and there must be some way to lift the warp threads wherever the pattern dictates. Nineteenth century looms operated using punched cards, "programmed" for pattern. The looms were set up with hooked needles, the hooks holding the warp. Where there were holes in cards, the needles moved, the warp lifted, the weft passed under. Where no holes were, the warp did not lift, and the weft was on top. The system parallels the on-off system in calculators and computers. In fact, the looms described here were models in the development of modern calculating machinery.*

Joseph Marie Jacquard (1752–1823) is credited with improving the mechanical loom so that mass production of fabric was feasible. Jacquard invented a better card-conveyor about 1804. The result was that more cards could run in faster sequence, and the loom could be quickly "reprogrammed" for a desired pattern.

Convert each binary number to base 10.

1. 11   **4.** 1001   **7.** 1110011   **10.** 1111111

2. 10   **5.** 11100   **8.** 1011111

3. 111   **6.** 101101   **9.** 111111

Convert each base 10 number to base 2.

11. 4   **15.** 17   **19.** 47

12. 5   **16.** 21   **20.** 92

13. 9   **17.** 30   **21.** 286

14. 11   **18.** 38   **22.** 345

+	0	1
0		
1		10

23. Complete the base 2 addition table above.

Find each of the following sums in base 2.

24.
$$10$$
$$+\underline{10}$$

25.
$$1010$$
$$+\underline{1101}$$

26.
$$100111$$
$$+\ \underline{11101}$$

27.
$$1001011$$
$$+\underline{1011010}$$

28.
$$10$$
$$1$$
$$+\underline{11}$$

29.
$$11$$
$$10$$
$$+\underline{11}$$

30.
$$100$$
$$111$$
$$+\underline{101}$$

31.
$$1001$$
$$101$$
$$+\underline{1101}$$

32.
$$1101$$
$$101$$
$$+\underline{1100}$$

×	0	1
0		
1		1

33. Complete the base 2 multiplication table above.

Find each of the following products in base 2.

34.
$$10$$
$$\times\underline{10}$$

35.
$$10$$
$$\times\underline{11}$$

36.
$$101$$
$$\times\ \underline{10}$$

37.
$$111$$
$$\times\ \underline{10}$$

38.
$$10001$$
$$\times\ \underline{101}$$

39.
$$1110$$
$$\times\ \underline{110}$$

*40.
$$110110$$
$$\times\ \underline{1011}$$

*41.
$$110111$$
$$\times\ \underline{111}$$

*42.
$$1111$$
$$\times\ \underline{101}$$

*43. Explain how the columns in Figure 3.15 were determined; that is, what do the numbers in a given column have in common? (*Hint:* Rewrite the numbers in base 2.)

*44. How does the "trick" in connection with Figure 3.15 work?

*45. Extend the table in Figure 3.15 so that it will accommodate any age up to 63 years.

*46. How many rows would be needed in such a table in order to include ages up to 127?

Photographs from space are often sent back to earth by the following method. The spacecraft first divides the picture up into thousands of little squares. A light meter then looks at each square individually and assigns it a number from 0 through 63. Here 0 represents pure white, and 63 represents pure black. Numbers between 0 and 63 stand for shades of gray. Each number is converted to a binary number of six digits (0 is 000000 and 63 is 111111). These numbers can then be transmitted to earth as a string of "on's" and "off's."

*47. Suppose a photograph is divided into 1000 squares horizontally and 500 vertically. How many squares is this altogether?

*48. Each square needs six binary digits for transmission to earth. How many digits would be needed altogether?

CHAPTER 3 SUMMARY

Keywords

Number	Algorithm
Numeral	Abacus
Number base (Base)	Addition table
Numeration system	Multiplication table
Simple grouping	Matching
Multiplicative grouping	Tally
Positional method	Tally stick
Placeholder	Hindu-Arabic system
Place value	Decimal system
Exponent	Binary system (base 2)
Base of an exponential	Distributive property:
Expanded form	$(b \times a) + (c \times a) = (b + c) \times a$

Numeration systems

Egyptian		Roman		Chinese-Japanese		Mayan			
1	I	1	I	1	ー	0	⬭	10	=
10	∩	5	V	2	ニ	1	.	11	≐
100	၅	10	X	3	三	2	..	12	⸬
1000	₰	50	L	4	囗	3	...	13	⸬.
10,000	၇	100	C	5	五	4	....	14	⸬..
100,000	☜	500	D	6	六	5	—	15	≡
1,000,000	🜨	1000	M	7	七	6	≐	16	≡.
				8	八	7	⸬	17	≡..
Babylonian				9	九	8	⸬.	18	≡...
				10	十				
				100	百	7	⸬.	17	≡..
1	▼			100	百	8	⸬..	18	≡...
10	◄			1000	千	9	....	19	≡....

CHAPTER 3 TEST

Identify each of the following base five numerals as simple grouping, multiplicative grouping, or positional. Convert to our system.

1. ∧∧∧|| **2.** ☺∧☺∧☺| **3.** ☺☺○☺○

Identify the numeration system for each of the following. Then convert to our system.

4. ⌎⌎⌎ ⁹⁹⁹∩∩∩ ⁹⁹ ∩∩ ||||| **7.** (cuneiform-style dots)   **9.** LXVII

5. ⟨⟨⟩

6. ⟨⟩⟨⟩⟩⟨⟩⟩⟩   **8.** (Mayan-style symbol)

Write each of the following in expanded notation.

10. 28   **11.** 4690

12. Simplify: $(6 \times 10^3) + (5 \times 10^2) + (7 \times 10^1) + (4 \times 10^0)$.

Convert each of the following to our system.

13. 432_5 **14.** 216_7 **15.** $34T_{12}$ **16.** 110111_2

Convert as indicated.

17. 98 to base 4   **19.** 2481 to base 12   **21.** 242 to base 2
18. 2914 to base 5   **20.** 683 to base 8

Perform the following calculations.

22. 432
  + 24 (base 5)

24. 22
  ×31 (base 5)

23. 3124
  −2432 (base 5)

25. 1101
  + 101 (base 2)

Digits *This Iranian stamp should remind us that counting on fingers is an age-old practice—fingers and toes, too. In fact, our word digit, referring to the numerals 0–9, comes from a Latin word for "finger" (or "toe"). It seems reasonable to connect so natural a counting method with the fact that number bases of 5, 10, or 20 are the most frequent in human cultures. Aristotle first noted the relationship between fingers and base 10 in Greek numeration. Anthropologists go along with the notion. Some cultures, however, have used 2, 3, or 4 as number bases, for example, counting on the joints of the fingers or the spaces between them. All in all, mathematics is in your own hands!*

Chapter 4 *The Rational Numbers*

The word "rational" comes from *ratio,* an idea that goes back to the ancient Greeks over 500 years before Christ. Pythagoras (about 580 BC – 501 BC) assumed that all quantities in nature could be expressed as ratios of counting numbers. When this proved to be false (as we see in the next chapter), the Greek value system was shaken.

In this chapter, we first look at *prime numbers,* which are counting numbers that can only be divided by themselves and 1. We see how useful they are in working with fractions. We find that not all problems in mathematics can be solved with counting numbers, so we expand the set to the *whole numbers.*

Everything is number. *So said Pythagoras, as tradition has it. You see here Raphael's image of him from* The School of Athens *mural—painted with reverence.*

Pythagoras was born, according to the legend, on the island of Samos, which is close to Asia Minor. He left Samos because of the tyrant Polycrates, and may have visited Egypt, Babylonia, and India. (The mysticism of his teachings suggests influence from the East—he lived at the time of Buddha and Lao-Tse.)

Pythagoras turned up in Italy, at Crotona, a southeastern port. There he gathered disciples from among the town's well-off young men and founded a brotherhood. *(Cont. p. 119.)*

We look at the *properties* of the whole numbers — the characteristics that distinguish them from other types of numbers. We find that not *every* problem can be solved with the whole numbers either, so we study the *integers:* the whole numbers together with their negatives. Then, finally, we study numbers which are quotients of integers, the *rational numbers.* The chapter ends with some applications of rational numbers to sequences.

4.1 PRIME NUMBERS

We begin our study of number sets by looking at a special subset of the set of counting numbers, the prime numbers, which Pythagoras and other Greek mathematicians studied at length. In fact, prime numbers have been one of the key ideas studied throughout the history of mathematics. The definition of a prime number is bound up with the idea of divisibility — how one counting number divides into another.

There are many pairs of counting numbers where the quotient is also a counting number. For example, $8 \div 4 = 2$, $30 \div 5 = 6$, and so on. In general, the counting number a is **divisible** (without a remainder) by the counting number b if the quotient $a \div b$ is a counting number. Thus, 8 is divisible by 4 since the quotient $8 \div 4$ is a counting number, 2. Also, 30 is divisible by 5, 72 is divisible by 36, and 121 is divisible by 11. The counting number 60 is not divisible by 7, since the quotient $60 \div 7$ is not a counting number.

If the counting number a is divisible (without remainder) by the counting number b, then b is a **factor** of a, and a is a **multiple** of b. Thus, 5 is a factor of 30, and 30 is a multiple of 5. Also, 6 is a factor of 30, and 30 is a multiple of 6. The number 30 equals $6 \cdot 5$; this product $6 \cdot 5$ is called a **factorization** of 30. (Here we use a dot for multiplication, instead of $\times$.) Other factorizations of 30 include $3 \cdot 10$, $2 \cdot 15$, and $1 \cdot 30$.

Disciples tried out as "hearers" before they became "mathematicians."

Pythagoras preached that number underlies the properties of the material world. He valued arithmetic, music, geometry, and astronomy as steps toward wisdom. His teachings spread throughout Greece when the brotherhood broke up because of political intervention. Pythagoras himself was exiled.

Through the centuries the myth of Pythagoras has inspired number theory and numerology both, astronomy and astrology both, science and magic both. Perhaps this century will emerge with a Pythagorean unity out of apparent differences.

Example 1   Find all the factors of each number: (a) 36   (b) 50   (c) 11.

(a) To find the factors of 36, start by trying to divide 36 by 1, 2, 3, 4, 5, 6, and so on. By doing this, we can find the counting number factors of 36 to be

$$1, \quad 2, \quad 3, \quad 4, \quad 6, \quad 9, \quad 12, \quad 18, \quad \text{and} \quad 36.$$

(b) The factors of 50 are 1, 2, 5, 10, 25, and 50.

(c) The only counting number factors of 11 are 11 and 1.

In Example 1 (c), we saw that the number 11 has only two counting number factors, itself and 1. A counting number that has only itself and 1 as factors is called a **prime number.** A number that is not prime is called **composite.** To simplify certain formulas and rules, mathematicians generally agreed that the counting number 1 is neither prime nor composite. The Pythagoreans did not consider 1 to be prime because they believed that unity is the generating principle of all numbers.

Example 2   Decide whether each number is prime or composite: (a) 37 (b) 59,872   (c) 697.

(a) The only factors of 37 are 37 and 1. Thus, 37 is prime.

(b) The number 59,872 is even, and thus divisible by 2. It is composite.

Note that there is only one even prime, the number 2 itself.

(c) For 697 to be composite, we must be able to find a number other than 697 and 1 that divides into it without remainder. Start by trying 2, and then 3. Neither works. There is no need to try 4. (If 4 divides without remainder into a number, then 2 will also.) Try 5. There is no need to try 6 or any succeeding even number. (Why?) Try 7. Try 11. (Why not try 9?) Try 13. We keep trying numbers until we find one that works, or until we try one whose square exceeds the given number. Try 17:

$$697 \div 17 = 41.$$

The number 697 is composite.

This method of testing individual numbers can be very time consuming. A faster way to identify primes has come down to us from the Greek geographer, poet, and astronomer Eratosthenes. (He lived from about 276 to 192 BC.) His method is as follows. List all counting numbers from 2 through some given counting number, such as 100. The number 2 is prime, but all multiples of 2 (4, 6, 8, 10, and so on) are composite. Thus, draw a circle around the prime 2, and cross out all the other multiples of 2. Since 3 is a prime, it should be circled, while all other multiples of 3 (6, 9, 12, 15, and so on) that are not already crossed out should be crossed out. The next prime is 5, which is circled, while all other multiples of 5 are crossed out. Circle 7, and cross out all the other multiples of 7.

Figure 4.1   The process that we used to obtain primes is called the **Sieve of Eratosthenes**

2	3	4	5	6	7	8	9	10	11	12	13	14
15	16	17	18	19	20	21	22	23	24	25	26	27	28
29	30	31	32	33	34	35	36	37	38	39	40	41	42
43	44	45	46	47	48	49	50	51	52	53	54	55	56
57	58	59	60	61	62	63	64	65	66	67	68	69	70
71	72	73	74	75	76	77	78	79	80	81	82	83	84
85	86	87	88	89	90	91	92	93	94	95	96	97	98
99	100												

This is far enough. Every number in the list that is circled or not crossed out is a prime. We can stop at 7 since $7 \cdot 7 = 49$, which is less than 100, while the next prime is 11, and $11 \cdot 11 = 121$ is greater than 100. From Figure 4.1, check that the primes less than 100 are

2,   3,   5,   7,   11,   13,   17,   19,   23,   29,   31,   37,   41,
43,   47,   53,   59,   61,   67,   71,   73,   79,   83,   89,   97

We can find all the prime factors of a given number, as shown by the following example.

Example 3 Find all the prime factors of each number: (a) 90 (b) 504.

(a) The first prime is 2, and 90 is divisible by 2.

$$90 = 2 \cdot 45$$

The number 45 is not divisible by 2, but it is divisible by 3.

$$90 = 2 \cdot 3 \cdot 15$$

We can divide by 3 again.

$$90 = 2 \cdot 3 \cdot 3 \cdot 5$$

Each of these factors is prime. We call $2 \cdot 3 \cdot 3 \cdot 5$ the **prime factorization** of 90.

(b) To find the prime factors of 504, first divide by 2 as many times as possible.

$$
\begin{aligned}
504 &= 2 \cdot 252 \\
&= 2 \cdot 2 \cdot 126 \\
&= 2 \cdot 2 \cdot 2 \cdot 63
\end{aligned}
$$

Now divide by 3 as many times as possible

$$
\begin{aligned}
504 &= 2 \cdot 2 \cdot 2 \cdot 3 \cdot 21 \\
&= 2 \cdot 2 \cdot 2 \cdot 3 \cdot 3 \cdot 7
\end{aligned}
$$

All the factors are prime, so we have found the prime factorization $504 = 2 \cdot 2 \cdot 2 \cdot 3 \cdot 3 \cdot 7$.

To simplify the writing of prime factorizations, we use exponents. Recall from Chapter 3 that $2 \cdot 2 \cdot 2$ can be written as 2^3, while $3 \cdot 3$ can be written as 3^2. With exponents,

$$504 = 2^3 \cdot 3^2 \cdot 7.$$

These same steps can be done in a more compact form:

$$
\begin{array}{r|l}
2 & 504 \\
2 & 252 \\
2 & 126 \\
3 & 63 \\
3 & 21 \\
\hline
& 7
\end{array}
$$

$504 = 2 \cdot 2 \cdot 2 \cdot 3 \cdot 3 \cdot 7$

$504 = 2^3 \cdot 3^2 \cdot 7$

Prospecting for primes
During the Gold Rushes sieves (pans) were used to trap particles of gold while debris washed away through perforations in the pans.

Keep going, using if possible the primes 2, 3, 5, 7, 11, and so on, until the last number is a prime.

Example 4 Find the number whose prime factorization is $2^2 \cdot 3^4 \cdot 5^2$.

Solution. Multiply:
$$
\begin{aligned}
2^2 \, 3^4 \, 5^2 &= 2 \cdot 2 \cdot 3 \cdot 3 \cdot 3 \cdot 3 \cdot 5 \cdot 5 \\
&= 8100.
\end{aligned}
$$

We have seen examples of how counting numbers can be broken down to their prime factorization. It is always possible to do this, and also, there is only one possible way to do it (disregarding the order of factors). These statements come from the Fundamental Theorem of Arithmetic, a form of which was known to the ancient Greeks.*

The Fundamental Theorem of Arithmetic
Every composite counting number can be expressed
in one and only one way as a product of primes
(if the order of the factors is disregarded).

This theorem is sometimes called the *Unique Factorization Theorem*, which reflects the idea that there is only one (unique) factorization possible for any given counting number.

4.1 EXERCISES

1. Continue the Sieve of Eratosthenes of Figure 4.1 from 101 to 200. List the primes between 100 and 200.

Use the results given in the text, together with the results from Exercise 1, to tell how many prime numbers there are between

2. 1 and 50   4. 100 and 150
3. 50 and 100   5. 150 and 200

6. As you go further out in the list of counting numbers, do the primes occur more often or less often?

7. List two primes which are consecutive counting numbers.

8. Can there possibly be any other pair of primes which are consecutive counting numbers?

A Conditional Goldbach Theorem *has been proved by use of a new approach where standard methods in number theory have not worked. H. A. Pogorzelski (University of Maine) believes that the Goldbach Conjecture is directly related to the consistency of arithmetic itself. He has been able to show that the consistency along with certain propositions imply that the Conjecture is true. Details concerning one proposition must still be verified — the reason for the label "conditional."*

A mathematical system is called consistent if its assumptions cannot lead to contradictory results. The proof of Cantor's Hypothesis (see page 33) also used an argument on consistency.

The mathematician Christian Goldbach (1690–1764) made a famous conjecture (guess) about prime numbers. He said that *every even number greater than 2 can be written as the sum of two prime numbers.* For example, $8 = 5 + 3$ and $10 = 5 + 5$. Mathematicians have tried for several centuries to prove Goldbach's Conjecture either true or false, but without success.

Write each even number as the sum of two primes.

9. 12   11. 24   13. 46
10. 18   12. 30   14. 52

---

*A *theorem* is a statement which can be proven true from other statements. For a proof of this theorem, see *What Is Mathematics?*, by Richard Courant and Herbert Robbins (Oxford University Press, 1941), page 23.

Srinivasa Ramanujan *developed many results in number theory. His friend and collaborator on occasion was G. H. Hardy, also a number theorist and professor at Cambridge University.*

Ramanujan introduced himself to Hardy by letter in 1913, in which he stated without proof a number of complicated formulas. Hardy at first thought it was a crank letter, but soon realized that no one could have made up the formulas—they had to be the work of a genius. Hardy arranged for Ramanujan and his wife to receive a modest stipend, enough to live fairly well at Cambridge.

*Ramanujan continued his work in number theory. One interest of his was the problem about determining how many primes are less than **n,** for an arbitrary number **n.** Ramanujan was elected Fellow of the Royal Society, the first East Indian to achieve that honor.*

Two primes (such as 3 and 5) whose difference is 2 are called **twin primes.** Use the Sieve of Erathosthenes to find a pair of twin primes between

15. 15 and 25	**17.** 50 and 75	**19.** 100 and 105
16. 30 and 50	**18.** 75 and 100	**20.** 105 and 110

Find all factors of each number.

21. 12	**24.** 28	**27.** 120	**30.** 972
22. 18	**25.** 52	**28.** 172	**31.** 154
23. 20	**26.** 63	**29.** 850	**32.** 179

Write the prime factorization of each number.

33. 15	**35.** 36	**37.** 240	**40.** 425
34. 21	**36.** 54	**38.** 300	**41.** 663
		39. 360	**42.** 885

Find the numbers having the following prime factorizations.

43. $2 \cdot 3^2 \cdot 5$	**46.** $2 \cdot 3 \cdot 5 \cdot 7 \cdot 11$	**49.** $3^2 \cdot 5 \cdot 7$
44. $2^3 \cdot 3 \cdot 5^2$	**47.** $2^3 \cdot 3^2 \cdot 7 \cdot 11$	**50.** $2^5 \cdot 3^2 \cdot 11$
45. $2 \cdot 3 \cdot 5^2 \cdot 7^2$	**48.** $3^2 \cdot 5^2 \cdot 13$	

Euclid, about 300 BC wrote a famous textbook on mathematics. This book is valued today mainly for the geometry it contains (see Chapter 8). However, the book also contains a lot of information on prime numbers. For example, Euclid proved that the following statement is true:

There is no largest prime.

He proved it by the method of *indirect proof.* That is, he assumed the opposite of the statement, and showed that this assumption leads to a false statement. Euclid thus began by assuming that there indeed *is* a largest prime. Call this prime *p.* Form a number by multiplying together all the primes from 2 through *p,* and add 1 to this result. Call this number *N.* That is,

$$N = (2 \cdot 3 \cdot 5 \cdot 7 \cdot 11 \cdot 13 \cdots p) + 1.$$

*51. Explain why *N* cannot be divided by 2.

*52. Or 3.   ***53.** Or 5.   ***54.** Or any prime up to *p.*

*55. Why can *N* not be composite?   ***56.** Why can *N* not be prime?

Since *N* can be neither prime nor composite, we have reached a contradiction. Thus, by the method of indirect proof we must reject the assumption that led to the contradiction. Therefore, it follows that the original statement is true: there is no largest prime; in other words, the set of primes is infinite.

4	9	2
3	5	7
8	1	6

Magic Squares have been popular throughout history. A magic square is an array of numbers with the special property that the sum of the numbers along either diagonal, along any row, or along any column, is the same. The sum is called the *magic number* for the magic square. For example, in the magic square at the side, the sum of the numbers in any row, column, or diagonal, is 15.

According to a fanciful story by Charles Trigg in *Mathematics Magazine* (September 1976, page 212), the Emperor Charlemagne (742–814) ordered a five-sided fort to be built at an important point in his kingdom. As good-luck charms, he had magic squares placed on all five sides of the fort. He had one restriction for these magic squares: all the numbers in them must be prime. These magic squares are listed below. Find the magic number for each square. You should get the same magic number for each of the five squares. This number gives the year that the fort was built.

57.

479	71	257
47	269	491
281	467	59

58.

389	191	227
107	269	431
311	347	149

59.

389	227	191
71	269	467
347	311	149

60.

401	227	179
47	269	491
359	311	137

61.

401	257	149
17	269	521
389	281	137

4.2 APPLICATIONS OF PRIME NUMBERS

An important application of prime numbers is in finding the *greatest common factor* and *least common multiple* of a group of counting numbers.

The **greatest common factor,** used in problems with fractions, is the largest counting number that divides into all the numbers in a given group. For example, 18 is the greatest common factor of 36 and 54, since 18 is the largest counting number dividing into both 36 and 54. Also, 1 is the greatest common factor of 7 and 18.

Greatest common factors can be found by using prime factorizations. To see how, let us find the greatest common factor of 36 and 54. First, write the prime factorization of each number:

$$36 = 2^2 \cdot 3^2 \quad \text{and} \quad 54 = 2^1 \cdot 3^3.$$

The greatest common factor can be found by finding the product of the primes common to the factorizations, with each prime raised to the *smallest* exponent that it has in any factorization. Here, the prime 2 has 1 as smallest exponent (in $54 = 2^1 \cdot 3^2$), while the prime 3 has 2 as smallest exponent (in $36 = 2^2 \cdot 3^2$). Thus, the greatest common factor of 36 and 54 is

$$2^1 \cdot 3^2 = 2 \cdot 9 = 18,$$

as we found above.

REPUBLIQUE FRANÇAISE

0.60   CHARLEMAGNE   POSTES

**Charlemagne
crowned emperor**
*Charles (or Carol) the Great
founded the Holy Roman Empire.
Like other monarchs earlier and
later he fostered learning. He
himself studied rhetoric, logic, and
astronomy, and also learned to
reckon. So says Einhard, an
eyewitness who wrote the first
biography of Charlemagne. The
emperor also planned for his
children's education. Girls and
boys alike were instructed in the
Seven Liberal Arts.*

*Alcuin, a Saxon theologian,
was the spirit behind the rebirth
of learning that historians call
the Carolingian renaissance. (This
was during the "dark ages" in
Europe.) Alcuin instituted a
system of elementary education
to fight illiteracy, and devised
the Seven Liberal Arts, which
became the basic curriculum in
Europe for centuries.*

*The first four liberal arts
(quadrivium) were straight from
Pythagoras: arithmetic, geometry,
astronomy, and music. The other
three (trivium) were grammar,
logic, and rhetoric. You can guess
whose authority prevailed—Euclid
in geometry, Pythagoras in music,
Aristotle in logic. Ptolemy's ideas
prevailed in astronomy (see
Chapter 8).*

Example 1   Find the greatest common factor of 360 and 2700.

Solution.   First, write the prime factorization of each number:

$$360 = 2^3 \cdot 3^2 \cdot 5 \quad \text{and} \quad 2700 = 2^2 \cdot 3^3 \cdot 5^2.$$

Then, form the product of primes common to each factorization, with each prime having as exponent the smallest exponent from either product:

$$\text{Greatest common factor} = 2^2 \cdot 3^2 \cdot 5 = 180.$$

Thus, 180 is the largest counting number that will divide into both 360 and 2700.

Example 2   Find the greatest common factor of the three numbers 720, 1000, and 1800.

Solution.   To begin, write the prime factorization for each number:

$$720 = 2^4 \cdot 3^2 \cdot 5, \quad 1000 = 2^3 \cdot 5^3, \quad \text{and} \quad 1800 = 2^3 \cdot 3^2 \cdot 5^2.$$

Use the smallest exponent on each prime common to the factorizations:

$$\text{Greatest common factor} = 2^3 \cdot 5 = 40.$$

(We did not use the prime 3 in the greatest common factor since the prime 3 does not appear in the prime factorization of 1000.)

Example 3   Find the greatest common factor of 80 and 63.

Solution.   We have

$$80 = 2^4 \cdot 5 \quad \text{and} \quad 63 = 3^2 \cdot 7.$$

There are no primes in common here, so

$$\text{Greatest common factor} = 1.$$

Thus, 1 is the largest number that will divide into both 80 and 63.

Two numbers, such as 80 and 63, that have a greatest common factor of 1 are called **relatively prime** numbers.

Applications of primes to fractions   Prime numbers are very helpful when working with fractions, as we shall see in the rest of this section. For example, we can reduce a fraction to lowest terms by using the idea of a greatest common factor. A fraction is in **lowest terms** when the numbers on the top and bottom of the fraction are relatively prime. Reducing fractions to lowest terms depends on the following property:

If a, b, and k are counting numbers, then

$$\frac{a \cdot k}{b \cdot k} = \frac{a}{b}.$$

This is called the *fundamental property of rational numbers.*

For example, we can reduce 6/15 to lowest terms by writing 6 as $2 \cdot 3$ and 15 as $5 \cdot 3$. Then by the theorem we have

$$\frac{6}{15} = \frac{2 \cdot 3}{5 \cdot 3} = \frac{2}{5}.$$

In general, a fraction can be reduced to lowest terms by finding the greatest common factor of the numbers on the top (the *numerator*) and bottom (the *denominator*) of the fraction. For example, to reduce 360/2700 to lowest terms we first find the greatest common factor of 360 and 2700. In Example 1 above, we found this factor to be 180. Thus,

$$\frac{360}{2700} = \frac{2 \cdot 180}{15 \cdot 180} = \frac{2}{15}.$$

Example 4   Reduce 36/54 to lowest terms.

Solution.   From the example we worked above, the greatest common factor of 36 and 54 is 18. Therefore,

$$\frac{36}{54} = \frac{2 \cdot 18}{3 \cdot 18} = \frac{2}{3}.$$

We need to use a **least common multiple** to add or subtract fractions. For example, to find the sum

$$\frac{2}{15} + \frac{1}{10},$$

we need to replace the fractions 2/15 and 1/10 with equivalent fractions having a common denominator. To find the lowest common denominator, we need to find the smallest counting number that can be divided by both 15 and 10. To find this number, we can list the counting number multiples of 15 and of 10.

Multiples of 15: $\{15, 30, 45, 60, 75, 90, 105, \ldots\}$
Multiples of 10: $\{10, 20, 30, 40, 50, 60, 70, \ldots\}.$

The set of counting numbers that are multiples of *both* 15 and 10 form the set of *common multiples*:

$$\{30, 60, 90, 120, \ldots\}.$$

The number 30, the smallest number in the set of common multiples, is called the **least common multiple** of the numbers 15 and 10.

The least common multiple of two counting numbers can be found using prime factorizations, in a manner very similar to that one used above, as shown in the next example.

Example 5   Find the least common multiple of 72 and 150.

Solution.   Write the prime factorization of each number:

$$72 = 2^3 \cdot 3^2 \quad \text{and} \quad 150 = 2 \cdot 3 \cdot 5^2.$$

Now use every different prime from both factorizations. Use as exponents the *highest* exponent from either factorization. Doing this gives

$$\text{Least common multiple} = 2^3 \cdot 3^2 \cdot 5^2 = 1800.$$

Example 6   Find the least common multiple of 135, 280, and 300.

Solution.   Write the prime factorizations:

$$135 = 3^3 \cdot 5, \qquad 280 = 2^3 \cdot 5 \cdot 7, \qquad \text{and} \qquad 300 = 2^2 \cdot 3 \cdot 5^2.$$

Form the product of all the primes that appear in any of the factorizations. As exponents, use the largest exponent from any product.

$$\text{Least common multiple} = 2^3 \cdot 3^3 \cdot 5^2 \cdot 7 = 37{,}800.$$

Thus, 37,800 is the smallest counting number divisible by 135, 280, and 300.

To add the fractions

$$\frac{2}{15} + \frac{1}{10}$$

we need the least common multiple of the two denominators. The least common multiple of 15 and 10 is 30. Thus, we must write both 2/15 and 1/10 as fractions with a denominator of 30. Proceed as follows:

$$\text{Since } 30 \div 15 = 2, \qquad \frac{2}{15} = \frac{2 \cdot 2}{15 \cdot 2} = \frac{4}{30}.$$

$$\text{Since } 30 \div 10 = 3, \qquad \frac{1}{10} = \frac{1 \cdot 3}{10 \cdot 3} = \frac{3}{30}.$$

Thus,

$$\frac{2}{15} + \frac{1}{10} = \frac{4}{30} + \frac{3}{30} = \frac{7}{30}.$$

Example 7   Simplify $\dfrac{173}{180} - \dfrac{69}{1200}$.

Solution.   To find the least common multiple of 180 and 1200, write the prime factorization of each number:

$$180 = 2^2 \cdot 3^2 \cdot 5 \qquad \text{and} \qquad 1200 = 2^4 \cdot 3 \cdot 5^2.$$

Thus, the least common multiple is $2^4 \cdot 3^2 \cdot 5^2 = 3600$. Convert both fractions into fractions having a denominator of 3600:

$$\text{Since } 3600 \div 180 = 20, \qquad \frac{173}{180} = \frac{173 \cdot 20}{180 \cdot 20} = \frac{3460}{3600}.$$

$$\text{Since } 3600 \div 1200 = 3, \qquad \frac{69}{1200} = \frac{69 \cdot 3}{1200 \cdot 3} = \frac{207}{3600}.$$

We can now subtract:

$$\frac{173}{180} - \frac{69}{1200} = \frac{3460}{3600} - \frac{207}{3600} = \frac{3460 - 207}{3600} = \frac{3253}{3600}.$$

Tests for divisibility   We end this section by listing tests telling when one counting number is divisible by another.

Divisible by	Test	Example
2	Number ends in 0, 2, 4, 6, or 8. (The number is even.)	9,489,994 ends in 4; is divisible by 2
3	Sum of the digits is divisible by 3.	897432 is divisible by 3, since $8 + 9 + 7 + 4 + 3 + 2 = 33$ is divisible by 3
4	Last two digits form a number divisible by 4.	7693432 is divisible by 4, since 32 is divisible by 4
5	Number ends in 0 or 5.	890 and 7635 are divisible by 5
6	Number is divisible by both 2 and 3	27,342 is divisible by 6 since it is divisible by 2 and 3
8	Last three digits form a number divisible by 8.	1,437,816 is divisible by 8, since 816 is
9	Sum of the digits is divisible by 9.	111,111,111 is divisible by 9 since sum of digits $= 9$, which is divisible by 9
10	Number ends in 0.	897,463,940 is divisible by 10
12	Number is divisible by 4 and 3 both.	376,984,032 is divisible by 12

4.2 EXERCISES

Find the greatest common factor of each set of numbers.

1. 20 and 30
2. 70 and 120
3. 180 and 300
4. 480 and 1800
5. 168 and 504

6. 99 and 275
7. 342 and 380
8. 130 and 455
9. 310 and 460
10. 234 and 470

11. 24, 48, 60
12. 28, 35, 56
13. 12, 18, 30
14. 252, 308, 504

Reduce each fraction to lowest terms.

15. $\dfrac{20}{30}$
16. $\dfrac{40}{80}$
17. $\dfrac{40}{75}$

18. $\dfrac{36}{42}$
19. $\dfrac{63}{70}$
20. $\dfrac{27}{45}$

21. $\dfrac{120}{150}$
22. $\dfrac{30}{66}$
23. $\dfrac{36}{63}$

24. $\dfrac{195}{390}$
25. $\dfrac{150}{273}$
26. $\dfrac{96}{132}$

27. $\dfrac{132}{144}$
28. $\dfrac{230}{455}$

Find the least common multiple for each set of numbers.

29. 12 and 32   **33.** 48 and 96   **37.** 24, 36, and 48

30. 24 and 30   **34.** 105 and 210   **38.** 30, 40, and 70

31. 28 and 70   **35.** 52 and 68   **39.** 24, 54, and 72

32. 56 and 96   **36.** 60 and 95   **40.** 15, 45, and 60

Simplify each of the following.

41. $\dfrac{1}{8} + \dfrac{1}{4}$ **44.** $\dfrac{2}{9} + \dfrac{1}{3}$ **47.** $\dfrac{7}{20} - \dfrac{1}{8}$ **50.** $\dfrac{5}{6} - \dfrac{3}{8}$

42. $\dfrac{1}{2} + \dfrac{1}{6}$ **45.** $\dfrac{5}{9} + \dfrac{1}{10}$ **48.** $\dfrac{5}{8} - \dfrac{3}{14}$

43. $\dfrac{5}{16} + \dfrac{7}{12}$ **46.** $\dfrac{3}{5} + \dfrac{1}{12}$ **49.** $\dfrac{8}{9} - \dfrac{2}{3}$

	12	18	40
2	6	9	20
2	3	9	10
2	3	9	5
3	1	3	5
3	1	1	5
5	1	1	1

A shortcut way of finding the least common multiple of the numbers 12, 18, and 40 is diagrammed at the side.

Answer: the least common multiple of 12, 18, and 40 is
$2 \cdot 2 \cdot 2 \cdot 3 \cdot 3 \cdot 5 = 360.$

51. Explain how the answer was obtained.

Use this method to find the least common multiples for each of the following lists of numbers.

52. 12, 24, 30   **53.** 20, 32, 50   **54.** 16, 22, 132

In Exercises 55–66, decide if the given number is divisible by the indicated factor.

55. 762; 3   **61.** 2,598,960; 8

56. 8217; 3   **62.** 11,259,936; 8

57. 65,924; 4   **63.** 8,243,976; 9

58. 83,976; 4   **64.** 42,974,872,213; 9

59. 21,495; 6   **65.** 24,946,382; 12

60. 837,694; 6   **66.** 98,721,482; 12

To test the number 8,493,969 for divisibility by 11, proceed as follows.

*67. Starting at the left, add together every other digit.

*68. Add together the remaining digits.

*69. Subtract the sum in Exercise 68 from the sum in Exercise 67.

*70. The number 8,493,969 is divisible by 11 if the difference in Exercise 69 is divisible by 11. Is 8,493,969 divisible by 11?

Test each of the following numbers for divisibility by 11.

71. 216,293   **72.** 847,667,931

Pythagoras, teacher of arithmetic *The fanciful picture of Pythagoras above is part of the title page of Filippo Calandri's arithmetic book of 1491. Arithmetic in those days included number theory, so it was fitting for Calandri to invoke the name of Pythagoras.*

The Pythagorean brotherhood studied numbers as geometric arrangements of points, such as the triangular numbers 1, 3, 6, 10, 15, 21, and so on. The first several triangular numbers are shown below. The dots in each new row are 1 more than in the preceding row: 1 + 2 = 3, 1 + 2 + 3 = 6, 1 + 2 + 3 + 4 = 10, and so on. Convince yourself by pictures that the sum of the first n counting numbers is a triangular number.

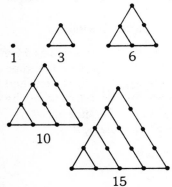

Answer *true* or *false* for each statement in Exercises 73–80. In these statements, a and b represent counting numbers and p represents a prime.

*73. The set of all common multiples of two given counting numbers is finite.

*74. Any two counting numbers have a greatest common factor.

*75. If a divides into b, then the least common multiple of a and b is b.

*76. Any two consecutive counting numbers are relatively prime.

*77. If a is not divisible by p, then a and p are relatively prime.

*78. The least common multiple of two different primes is their product.

*79. Two composite numbers can be relatively prime.

*80. The product of the greatest common factor of a and b and the least common multiple of a and b equals the product of a and b.

4.3 WHOLE NUMBERS

Thus far in this book, most of our work has been with the set of counting numbers,

Counting numbers = {1, 2, 3, 4, 5, . . .}.

In this section, we study the set of whole numbers:

Whole numbers = {0, 1, 2, 3, 4, 5, . . .}.

The set of counting numbers is a subset of the set of whole numbers. The whole numbers, in turn, are a more extensive set of numbers, even though only one new element, zero, has been included. The single number 0 took society a long time to develop, however.

We saw in Chapter 3 that the Mayan Indians had one of the earliest numeration systems that included a symbol for 0. The very early Babylonians had a positional system, but they placed only a space between "digits" to indicate a missing power. When the Greeks absorbed Babylonian astronomy, they quickly saw the need for a symbol to represent "no powers." They used o or ō to represent "no power," or "zero," as early as 150 AD. By the year 500, the Greeks were using the letter omicron, o, of their alphabet for "zero." The Greek number system gradually lost out to the Roman number system throughout Western Europe.

The Roman system was the one most commonly used in Europe from the time of Christ up until perhaps 1400 AD, when the Hindu-Arabic system began to take over. Zero was fundamental to the Hindu-Arabic system. The original Hindu word for zero was *sunya*, meaning "void." The Arabs adopted this word as *sifr*, or "vacant." There was a considerable battle over the new system in Europe, with most people sticking with the Roman system. Gradually, however, the advantages of the new Hindu-Arabic system became

Figurate numbers *is the general name for triangular numbers (as on the opposite page), square and pentagonal numbers (as shown below), and so on. You can use figurate number diagrams to demonstrate, for example, that every square number greater than 1 is the sum of two successive triangular numbers. Also, you can demonstrate that the fifth pentagonal number is equal to the sum of three triangular numbers plus 5. Does a similar result hold for the pentagonal numbers in general?*

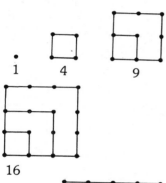

1   4   9

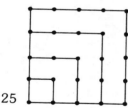

16

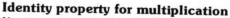

25

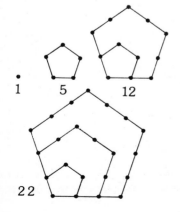

1   5   12

22

clear, and it replaced the cumbersome Roman system. The word *sifr* passed into Latin as *zephirum*, which over the years became *zevero, zepiro,* and finally, *zero.*

The number 0 has a special property that no other whole number has: the sum of any whole number and 0 is that original whole number. Thus, $8 + 0 = 8$, and $0 + 75 = 75$. Adding 0 to a whole number does not change the identity of the whole number. This special property of 0 is called the identity property, which we can write in symbols as follows:

Identity property for addition
If a is any whole number, then
$$0 + a = a \quad \text{and} \quad a + 0 = a.$$

The number 0 is called the **identity element** *for addition* of whole numbers. There is no identity element for addition of counting numbers, since 0 is not a counting number. The identity property for addition is one property that the set of whole numbers has and the set of counting numbers does not.

Example 1 Is there an identity property for subtraction of whole numbers?

Solution. If we subtract 0 from a whole number, we get the original whole number as the answer, for example, $12 - 0 = 12$. However, if we subtract 12 from 0, we do not get 12 since $0 - 12 \neq 12$. Therefore, there is no identity property for subtraction.

While there is no identity property for subtraction, there is one for multiplication. If we choose any whole number and multiply it by 1, we get the original whole number as the answer. Thus, $7 \cdot 1 = 7$ and $1 \cdot 7 = 7$. Also $18 \cdot 1 = 18$ and $1 \cdot 18 = 18$. We call 1 the **identity element** *for multiplication* of whole numbers. The identity property for multiplication can be stated as follows:

Identity property for multiplication
If a is any whole number, then
$$a \cdot 1 = a \quad \text{and} \quad 1 \cdot a = a.$$

Is there an identity property for division?

If we earn \$15 today and \$26 tomorrow, the result is the same as if we had earned \$26 today and \$15 tomorrow. That is, $15 + 26$ gives the same result as $26 + 15$. In general, given two numbers, we can add the first and second, or we can add the second and first. The answer is the same, by the commutative property of addition:

Commutative property of addition
If a and b are any whole numbers, then
$$a + b = b + a.$$

Ancient Chinese number lore
Male-female reciprocity is seen in the yang-yin symbol above, and appears in the magic square below (the lo-shu). Odd numbers (white) were yang, even numbers (black) were yin. The lo-shu is contained in the ancient book of divination, I Ching (Book of Changes). Its fundamental eight trigrams (see opposite at bottom) incorporate unbroken lines— (yang) and broken lines - - (yin). Yang lines have values of 7 or 9, yin lines have values of 6 or 8, where 7 and 8 are static, and 6 and 9 changing phases. Note that the sums 7 + 8 and 6 + 9 equal 15, the number of the Tao (the mystic Way) and also the magic number of the lo-shu. (Combinatorial aspects of I Ching are discussed in Chapter 9.)

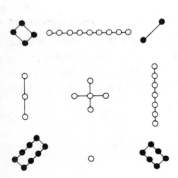

What about subtraction? Is it always true that $a - b = b - a$ for any whole numbers a and b?

We can find the product $6 \cdot 7$ and get the same answer as if we had found the product $7 \cdot 6$. In general, $a \cdot b = b \cdot a$ for any whole numbers a and b. This is the commutative property of multiplication:

Commutative property of multiplication
If a and b are any whole numbers, then
$a \cdot b = b \cdot a$.

The operations of addition, subtraction, multiplication, and division are called **binary operations,** since the operations apply to only *two* numbers at a time. For example, we can add 8 and 9 as a single problem, getting 17. However, to work the problem $8 + 9 + 11$ we must add two of the numbers, and then add the third number to this answer. There are two ways we can do this. We could first add 8 and 9, and then add 11 to the answer. To show that we are working the problem in this order, we use parentheses:

$$(8 + 9) + 11 = 17 + 11$$
$$= 28.$$

Or, we could add 9 and 11, and then add 8 to the result:

$$8 + (9 + 11) = 8 + 20$$
$$= 28.$$

Since both answers are the same, we have

$$(8 + 9) + 11 = 8 + (9 + 11).$$

It turns out that we always get the same answer, no matter which two numbers in a group of three we add first. This fact about the whole numbers is the associative property:

Associative property for addition
If a, b, and c are any whole numbers, then
$(a + b) + c = a + (b + c)$.

Multiplication works in the same way. We can find the product $6 \cdot 4 \cdot 2$ by first multiplying 6 and 4:

$$(6 \cdot 4) \cdot 2 = 24 \cdot 2 = 48,$$

or by first multiplying 4 and 2:

$$6 \cdot (4 \cdot 2) = 6 \cdot 8 = 48.$$

Thus, $(6 \cdot 4) \cdot 2 = 6 \cdot (4 \cdot 2)$; and, in general:

Associative property for multiplication
If a, b, and c are any whole numbers, then
$a \cdot (b \cdot c) = (a \cdot b) \cdot c$.

Rhythm of Life *That is the book by Wilhelm Fliess, published in 1906, containing his theories of male and female cycles. He associated the number 28 with female cycles, and 23 with male. Fliess was a doctor in Berlin, with wide scientific interests. He exerted a rather unwholesome influence on Sigmund Freud in the period 1887–1902 (as Ernest Jones implies in his Life and Works of Freud). Freud said that Fliess was expert at mathematics and could arrive at any desired number by manipulating 23 and 28 arithmetically.*

Cycles help to describe the nature of biorhythm, business fluctuation, musical tone, light propagation, and many other phenomena. Cyclic curves such as those above can be drawn with the aid of trigonometry, a rich area of mathematics, but beyond the scope of this book.

The 17-year cycle of locusts was first noted by Benjamin Banneker, a mathematician from Maryland (see Chapter 5).

What about the remaining properties for subtraction?

Example 2   Is there an associative property for subtraction?

Solution.   Suppose we have the subtraction problem $30 - 12 - 5$. Do we get the same answer no matter how we perform the subtraction? Let's check and find out.

First subtract the *first* two numbers:

$$(30 - 12) - 5 = 18 - 5$$
$$= 13$$

First subtract the *last* two numbers:

$$30 - (12 - 5) = 30 - 7$$
$$= 23$$

Since the answers are different, $(30 - 12) - 5 \neq 30 - (12 - 5)$, and there is no associative property for subtraction.

Example 3   Identify each statement as an example of the commutative or the associative property.

(a) $5 + 9 = 9 + 5$

Solution.   The order of the numbers changes from one side of the equals sign to the other. Thus, this statement is an example of the commutative property of addition.

(b) $9 + (6 + 12) = (9 + 6) + 12$

Solution.   The order of the numbers is the same on both sides of the equals sign. Only the order in which the additions are to be performed, indicated by parentheses, has been changed. This statement is an example of the associative property of addition.

(c) $(5 \cdot 6) \cdot 2 = (6 \cdot 5) \cdot 2$

Solution.   The order of the numbers changes here, so that the statement is an example of the commutative property of multiplication.

Addition of whole numbers has one last property that we should discuss: namely, the fact that the sum of two whole numbers is always a whole number. For example, if we add 9 and 6, we get 15, which is a whole number. This property is the closure property for addition:

> **Closure property for addition**
> If a and b are any whole numbers, then
> $a + b$ is a whole number.

There is no closure property for *subtraction* of whole numbers. If we subtract the whole numbers a and b, we are not at all sure that the difference, $a - b$, is a whole number. For example, if we subtract the whole numbers 8 and 17, the difference, $8 - 17$, is not a whole number. This one counterexample is enough to show that subtraction of whole numbers does not have a closure property.

Leo Decimus *(Pope Leo the Tenth) was "converted" via numerology by Michael Stifel, mathematician of the early 16th century. Stifel, in training to take holy orders, fell under the sway of Martin Luther, and even got himself in jail briefly. His anti-clerical numerology went as follows:*

Leo Decimus is the Pope's name in Latin, and can be expressed Leo DeCIMVs. The Roman numerals from the name can be arranged as MDCLVI. The M (for Mystery) can be removed; an X for Leo X can be added. The result is DCLXVI, or 666 — the number of the Beast in the Book of Revelations. Thus, argued Stifel, Leo the Tenth is the Beast!

However, if we multiply two whole numbers, their product is always a whole number:

Closure property for multiplication
If a and b are whole numbers, then
$a \cdot b$ is a whole number.

There is no closure property for division of whole numbers. For example, the quotient of 6 and 8 is not a whole number.

Example 4   Do the following sets have the closure property for the indicated operations?

(a) $\{1, 3, 5, 7, 9, \ldots\}$ Multiplication

Solution.   The set $\{1, 3, 5, 7, 9, \ldots\}$ includes all the odd whole numbers. Is the product of two odd numbers always an odd number? Try some examples to verify that it is. Since the product of two odd numbers is always an odd number, this set is closed for multiplication.

(b) $\{1\}$ Addition

Solution.   If we select any two numbers in the set $\{1\}$, and add them, is the answer always in the set? We can choose the numbers 1 and 1 (the numbers we choose need not be distinct) and add them: $1 + 1 = 2$. The answer is not in the set, so that the set is not closed for addition.

All the properties discussed so far in this section involve either addition or multiplication, but not both. There is one property, perhaps the most important of all, which does involve both operations. To see how this final property works, let us look at an example.

Suppose a student has taken 7 tests in the first half of the semester, and 5 tests in the second half. Suppose also that the student earned a score of 82 points on each test (a very consistent student). We could find the total score that the student has earned in the course in either of two ways:

(1)   The student got 82 on each of $7 + 5 = 12$ tests, for a total point score of

$$82 \cdot (7 + 5) = 82 \cdot 12 = 984.$$

(2)   The student got 82 on the first 7 tests, and 82 on the last 5 tests, for a total point score of

$$82 \cdot 7 + 82 \cdot 5 = 574 + 410 = 984.$$

Both methods lead to the same answer, so that

$$82 \cdot (7 + 5) = 82 \cdot 7 + 82 \cdot 5.$$

The 82 on the left was *distributed* over the 7 and the 5. Here's a further example:

$$9 \cdot (11 + 15) = 9 \cdot 26 \qquad 9 \cdot 11 + 9 \cdot 15 = 99 + 135$$
$$= 234 \qquad\qquad\qquad = 234$$

The 9 was distributed over the 11 and the 15.

This property of both addition and multiplication is the distributive property of multiplication over addition:

Distributive property of multiplication over addition
If a, b, and c are whole numbers, then
$a \cdot (b + c) = a \cdot b + a \cdot c$.

We can now summarize the properties of the whole numbers that we have developed. If a, b, and c are any whole numbers, then:

	Addition	**Multiplication**
Identity	$a + 0 = a$ and $0 + a = a$	$a \cdot 1 = a$ and $1 \cdot a = a$
Commutative	$a + b = b + a$	$a \cdot b = b \cdot a$
Associative	$a + (b + c) = (a + b) + c$	$a \cdot (b \cdot c) = (a \cdot b) \cdot c$
Closure	$a + b$ is a whole number	$a \cdot b$ is a whole number
Distributive		$a \cdot (b + c) = a \cdot b + a \cdot c$

4.3 EXERCISES

Identify the property illustrated by each of the following.

1. $6 + 9 = 9 + 6$
2. $8 \cdot 4 = 4 \cdot 8$
3. $7 + (2 + 5) = (7 + 2) + 5$
4. $(3 \cdot 5) \cdot 4 = 4 \cdot (3 \cdot 5)$
5. $9 + 6$ is a whole number
6. $12 + 0 = 12$
7. $9 \cdot 1 = 9$
8. $1 \cdot 4 = 4$
9. $0 + 283 = 283$
10. $6 \cdot (4 \cdot 2) = (6 \cdot 4) \cdot 2$
11. $2 \cdot (4 + 3) = 2 \cdot 4 + 2 \cdot 3$
12. $9 \cdot 6 + 9 \cdot 8 = 9 \cdot (6 + 8)$

13. Complete the following chart for whole numbers.

	Operation			
Property	Addition	Subtraction	Multiplication	Division
Identity	yes	no	yes	
Commutative	yes		yes	
Associative	yes		yes	no
Closure	yes		yes	

14. What changes would have to be made in the chart in Exercise 13 if only counting numbers could be used? (*Hint:* There is only one change.)

Decide whether or not each set is closed for the given operation. If the set is not closed for the given operation, give an example of an arithmetic problem whose answer is not in the given set.

15. $\{0, 2, 4, 6, 8, \ldots\}$, $+$
16. $\{0, 2, 4, 6, 8, \ldots\}$, $\cdot$
17. $\{0, 2, 4, 6, 8, \ldots\}$, $-$
18. $\{0, 2, 4, 6, 8, \ldots\}$, $\div$
19. $\{1, 3, 5, 7, \ldots\}$, $+$
20. $\{1, 3, 5, 7, \ldots\}$, $-$
21. $\{0\}$, $+$
22. $\{0\}$, $\cdot$
23. $\{0, 1\}$, $\cdot$
24. $\{-1, 0, 1\}$, $\cdot$
25. the set of counting numbers, $\cdot$
26. the set of counting numbers, $\div$

Supply the missing properties in the following verifications of problems from arithmetic.

27.
$$\begin{aligned}
39 + 48 &= (30 + 9) + (40 + 8) && \text{positional number system} \\
&= 30 + [9 + (40 + 8)] && \underline{\hspace{4cm}} \\
&= 30 + [(40 + 8) + 9] && \underline{\hspace{4cm}} \\
&= 30 + [40 + (8 + 9)] && \underline{\hspace{4cm}} \\
&= (30 + 40) + (8 + 9) && \underline{\hspace{4cm}} \\
&= (3 \cdot 10) + (4 \cdot 10) + 17 && \text{number facts} \\
&= [(3 + 4) \cdot 10] + 17 && \underline{\hspace{4cm}} \\
&= (7 \cdot 10) + (10 + 7) && \text{number facts} \\
&= [7 \cdot 10 + 1 \cdot 10] + 7 && \text{(2 reasons)} \\
&= [(7 + 1) \cdot 10] + 7 && \underline{\hspace{4cm}} \\
&= 8 \cdot 10 + 7 && \text{number facts} \\
&= 87. && \text{positional number system}
\end{aligned}$$

29.

4	9	2
3	5	7
8	1	6

30.

16	3	2	13
5	10	11	8
9	6	7	12
4	15	14	1

28.
$$\begin{aligned}
25 \cdot 3 &= (20 + 5) \cdot 3 && \text{number facts} \\
&= (20 \cdot 3) + (5 \cdot 3) && \underline{\hspace{4cm}} \\
&= 60 + 15 && \text{number facts} \\
&= 60 + (10 + 5) && \text{number facts} \\
&= (60 + 10) + 5 && \underline{\hspace{4cm}} \\
&= (6 \cdot 10 + 1 \cdot 10) + 5 && \text{number facts; identity} \\
&= [(6 + 1) \cdot 10] + 5 && \underline{\hspace{4cm}} \\
&= (7 \cdot 10) + 5 && \text{number facts} \\
&= 75. && \text{positional number system}
\end{aligned}$$

One application of numbers that has been popular throughout history is the *magic square*. As we mentioned earlier, a magic square is an array of numbers with the special property that the sum of all the numbers in a row, or column, or along a diagonal, is always the same. In Exercises 29–31, show that each array is a magic square by adding the numbers in each row, column, and diagonal.

31.

11	24	7	20	3
4	12	25	8	16
17	5	13	21	9
10	18	1	14	22
23	6	19	2	15

The magic square at the top of the next page was developed by Benjamin Franklin. Find the sums described in Exercises 32–36.

Benjamin Franklin *admitted that he would amuse himself while in the Pennsylvania Assembly with magic squares or circles "or any thing to avoid Weariness." He wrote about the usefulness of mathematics in the Gazette in 1735, saying that no employment can be managed without arithmetic, no mechanical invention without geometry. He also thought that mathematical demonstrations are better than academic logic for training the mind to reason with exactness and distinguish truth from falsity, even outside of mathematics.*

200	217	232	249	8	25	40	57	72	89	104	121	136	153	168	185
58	39	26	7	250	231	218	199	186	167	154	135	122	103	90	71
198	219	230	251	6	27	38	59	70	91	102	123	134	155	166	187
60	37	28	5	252	229	220	197	188	165	156	133	124	101	92	69
201	216	233	248	9	24	41	56	73	88	105	120	137	152	169	184
55	42	23	10	247	234	215	202	183	170	151	138	119	106	87	74
203	214	235	246	11	22	43	54	75	86	107	118	139	150	171	182
53	44	21	12	245	236	213	204	181	172	149	140	117	108	85	76
205	212	237	244	13	20	45	52	77	84	109	116	141	148	173	180
51	46	19	14	243	238	211	206	179	174	147	142	115	110	83	78
207	210	239	242	15	18	47	50	79	82	111	114	143	146	175	178
49	48	17	16	241	240	209	208	177	176	145	144	113	112	81	80
196	221	228	253	4	29	36	61	68	93	100	125	132	157	164	189
62	35	30	3	254	227	222	195	190	163	158	131	126	99	94	67
194	223	226	255	2	31	34	63	66	95	98	127	130	159	162	191
64	33	32	1	256	225	224	193	192	161	160	129	128	97	96	65

B. Franklin inv. I. Ferguson delin.   *J. Mynde sc.*

32. The numbers in each row. (The columns do not always add up to the same number here.)

33. The first four numbers in rows 2, 4, 6, and so on.

34. The eight numbers on the left half of each of the nine "bent diagonals."

35. The four corners and the four middle numbers.

36. Any four numbers that are opposite each other and at equal distances from the center.

Find the numbers **(a)**–**(h)** in each magic square:

37.

13	24	5	6	12
10	**(a)**	17	23	4
(b)	3	9	15	16
14	20	**(c)**	2	8
(d)	**(e)**	**(f)**	**(g)**	**(h)**

38.

15	18	**(a)**	4	7
(b)	2	10	13	16
8	11	19	**(c)**	5
17	25	3	6	**(d)**
(e)	9	**(f)**	**(g)**	**(h)**

In the remaining exercises for this section, we shall see how to use sets to define addition of whole numbers. To do this, we use the symbol $n(A)$ from Chapter 1. Recall that $n(A)$ represents the number of elements in set A. For example, $n(\{1, 3, 4, 7, 9\}) = 5$, $n(\{4, 6, 8\}) = 3$, and $n(\varnothing) = 0$. To define what we mean by the sum $4 + 3$, we choose a set A having $n(A) = 4$. For example, we might let $A = \{m, n, p, q\}$. Then choose a set B (where A and B have no elements in common) such that $n(B) = 3$. For example, B might be the set $\{c, d, e\}$.

*39.   If $A = \{m, n, p, q\}$, and $B = \{c, d, e\}$, find $A \cup B$.

*40.   Find $n(A \cup B)$.   ***41.**   Does $4 + 3 = n(A \cup B)$?

*42.   Complete this statement: Let a and b be whole numbers. Choose sets A and B, with $n(A) = $ _____ and $n(B) = $ _____, where A and B have no elements in common. Then $a + b = n($_____$)$.

*43.   Why do we need to use sets A and B that have no elements in common?

Multiplication of whole numbers can be defined in terms of addition as follows. We say that $3 \cdot 4$ means that 4 is to be added 3 times, or that $3 \cdot 4$ is the repeated sum

$$3 \cdot 4 = 4 + 4 + 4.$$

Write the following as repeated sums.

*44. $6 \cdot 5$ ***45.** $3 \cdot 2$ ***46.** $8 \cdot 4$

4.4 INTEGERS

So far in this chapter, we have looked at prime numbers, the set of counting numbers, $\{1, 2, 3, 4, \ldots\}$, and the set of whole numbers, $\{0, 1, 2, 3, 4, 5, \ldots\}$. We have seen several applications of these numbers. However, useful as these numbers are, they cannot be used for all the number problems we see in daily life. For example, we cannot use the whole numbers to write the temperatures on January days in Anchorage—such temperatures are often less than 0. Also, we cannot use whole numbers to express the profits of a company that loses money. Businessmen in the Renaissance wrote losses in red ink, leading to the expression "in the red." Profits were written in black ink.

Instead of using red ink to show numbers less than 0, we use a negative sign, to distinguish the numbers in the set of *negative integers*:

$$\{\ldots, -4, -3, -2, -1\}.$$

By joining the set of negative integers with the set of whole numbers, we get the set of integers:

Integers = {. . . , −4, −3, −2, −1, 0, 1, 2, 3, . . .}

Thus, the whole numbers are a subset of a larger set of numbers, the integers.

A **number line** provides a good picture of the set of integers. Begin with a horizontal line, and locate 0 on it. We can place, by equal spacing, the numbers 1, 2, 3, 4, and so on, to the right of 0. (See Figure 4.2.) The portion of the number line to the left of 0 seems a perfect place to put numbers smaller than 0, as we have done in Figure 4.3. (It's like laying a thermometer on its side, with "below zero" becoming "to the left of zero.")

Figure 4.2

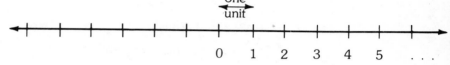

The whole numbers can be marked off on a number line by repeating a convenient unit distance

Figure 4.3

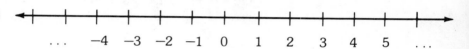

Negative numbers can be traced back to the Chinese between 200 BC and 220 AD. Mathematicians at first found negative numbers ugly and unpleasant, even though they kept cropping up in the solutions of problems. For example, an Indian text of about 1150 AD gives the solution of an equation as −5, and then makes fun of anything so useless. Even a mathematician as excellent as Descartes (see Chapter 7) found negative numbers "false." As time passed, however, the usefulness of negative numbers in algebra became clear, so that since the seventeenth century they have become a standard set of numbers in mathematics.

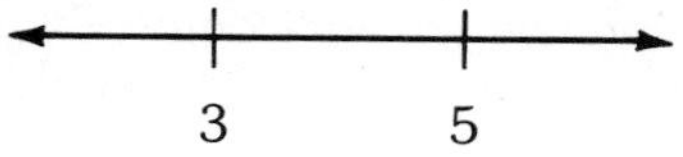

We can use a number line to good advantage as we study integers. For example, we can use a number line to tell which of two integers is smaller. The number line of Figure 4.4 shows the points for 3 and 5. We know that 3 is smaller than 5. On the number line, 3 is to the left of 5. In general,

An integer a is smaller than an integer b
if a is to the left of b on the number line.

The symbol $<$ replaces the words "is smaller than" or "is less than." Thus, "3 is less than 5" is written

$$3 < 5.$$

Since 3 is less than 5, we know that 5 is greater than 3, which is written

$$5 > 3.$$

Figure 4.4

3   5

Example 1 Use the number line of Figure 4.5 to decide whether each of the following statements is true or false.

Figure 4.5

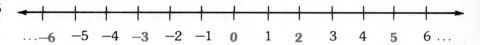

...−6 −5 −4 −3 −2 −1 0 1 2 3 4 5 6 ...

(a) −3 < −6.

Solution. The smaller number is always to the left of the larger number on the number line. On the number line of Figure 4.5, −3 is to the right of −6. Thus, −3 < −6 is false.

(b) −3 < 2. True.

(c) 0 > −5. True, since −5 < 0.

We can also use the number line when adding integers. For example, to add −6 and 4, start at 0 on the line and draw an arrow to −6, as shown in Figure 4.6. From the end of this arrow, draw an arrow 4 units to the right (representing the addition of +4). This second arrow ends at −2, so that

$$-6 + 4 = -2.$$

Figure 4.6

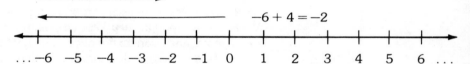

...−6 −5 −4 −3 −2 −1 0 1 2 3 4 5 6 ...

Example 2 Find each sum: (a) 5 + (−3) (b) −3 + (−4) = −7.

(a) Draw an arrow from 0 to 5. (See Figure 4.7.) From the end of this arrow, draw an arrow of 3 units to the left, to represent the addition of −3. This arrow ends at 2, so that 5 + (−3) = 2.

Figure 4.7

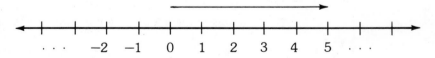

. . . −2 −1 0 1 2 3 4 5 . . .

(b) See Figure 4.8.

Figure 4.8

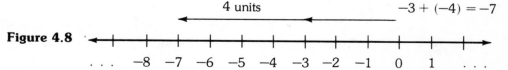

. . . −8 −7 −6 −5 −4 −3 −2 −1 0 1 . . .

The numbers 12 and −12 are not the same; having $12 is not the same as owing $12, a temperature of 12° above zero is not the same as a temperature 12° below zero, and so on. Nevertheless, 12 and −12 *do* share one common feature: they are the same distance from 0 on the number line. Two distinct numbers that are the same distance from 0 are called **opposites** of each other. Thus, 12 and −12 are opposites, as are −6 and 6. The number 0 is its own opposite.

Numbers which are opposites of each other share another property: their sum is 0.

$$12 + (-12) = 0, \qquad -6 + 6 = 0, \qquad \text{and} \qquad 0 + 0 = 0.$$

In general, we have the additive inverse property of integers:

Additive inverse property

If a is any integer, then there exists an integer $-a$ such that
$$a + (-a) = 0 \qquad \text{and} \qquad -a + a = 0.$$

All the properties of the whole numbers that we discussed in the previous section are also valid for the integers.

To **subtract** 3 from −2, written −2 − 3, start at 0 on the number line and draw an arrow to −2. (See Figure 4.9.) From the end of this arrow, draw another arrow 3 units to the *left*, to represent subtraction. The second arrow ends at −5, so that

$$-2 - 3 = -5.$$

Figure 4.9

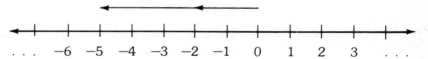

3 units

. . .   −6   −5   −4   −3   −2   −1   0   1   2   3   . . .

To subtract 3 from −2, move three units to the left of −2. Thus −2 − 3 = −5.

As shown in Figure 4.9, −2 − 3 = −5. The same result would have been found had we *added* −2 and −3: −2 + (−3) = −5. It is always true that *subtracting b from a* leads to the same answer as *adding a and the opposite of b*, namely, −b. We use this fact to define **subtraction of integers:**

$$a - b = a + (-b), \qquad \text{for any integers } a \text{ and } b.$$

Example 3 Convert each subtraction problem to addition, and find the answers.

Solution.

Subtraction	Addition	Answer
4 − 7	4 + (−7)	−3
−6 − 14	−6 + (−14)	−20
−3 − (−9)	−3 + (+9)	6
8 − (−4)	8 + (+4)	12
12 − 12	12 + (−12)	0

What about multiplication? Any rules we work out for multiplication of integers ought to be consistent with the usual multiplication of positive numbers. For example, we would want the product of 0 and *any* integer (positive or negative) to be 0, or

$$n \cdot 0 = 0$$

for any integer *n*.

How can we now define the product of a positive and a negative integer so that the result is consistent with past work? Look at the pattern of products.

$$
\begin{aligned}
4 \cdot 5 &= 20 \\
4 \cdot 4 &= 16 \\
4 \cdot 3 &= 12 \\
4 \cdot 2 &= 8 \\
4 \cdot 1 &= 4 \\
4 \cdot 0 &= 0 \\
4 \cdot (-1) &= ?
\end{aligned}
$$

What number should we assign as the product $4 \cdot (-1)$ so that the pattern is maintained? The numbers just to the left of the equals signs decrease by 1 each time, and the products to the right decrease by 4 each time. To maintain the pattern, the number to the right in the bottom row must be 4 less than 0, which is −4. Therefore, we must have

$$4 \cdot (-1) = -4.$$

The pattern continues with

$$
\begin{aligned}
4 \cdot (-2) &= -8 \\
4 \cdot (-3) &= -12 \\
4 \cdot (-4) &= -16
\end{aligned}
$$

and so on. In the same way, we could find that

$$
\begin{aligned}
(-4) \cdot 2 &= -8 \\
(-4) \cdot 3 &= -12 \\
(-4) \cdot 4 &= -16,
\end{aligned}
$$

and so on. In general,

The product of a positive integer and a negative integer
is a negative integer.

Example 4 (a) $8 \cdot (-7) = -56$ (c) $-6 \cdot 8 = -48$
(b) $-9 \cdot 3 = -27$

The product of two positive integers is positive, and we have seen that the product of a positive and a negative integer is negative. What about the product of two negative integers? Look at another pattern.

$$
\begin{aligned}
(-5) \cdot 4 &= -20 \\
(-5) \cdot 3 &= -15 \\
(-5) \cdot 2 &= -10 \\
(-5) \cdot 1 &= -5 \\
(-5) \cdot 0 &= 0 \\
(-5) \cdot (-1) &= ?
\end{aligned}
$$

The numbers just to the left of the equals signs decrease by 1 each time. The products on the right increase by 5 each time. To maintain the pattern, we will have to agree that $(-5) \cdot (-1)$ is 5 more than $(-5) \cdot (0)$. Therefore, we must have

$$(-5) \cdot (-1) = 5.$$

Continuing this pattern gives

$$
\begin{aligned}
(-5) \cdot (-2) &= 10 \\
(-5) \cdot (-3) &= 15 \\
(-5) \cdot (-4) &= 20,
\end{aligned}
$$

and so on.

In general,

The product of two negative integers is positive.

Example 5   (a) $(-9) \cdot (-7) = 63$ (c) $(-8) \cdot (-4) = 32$
(b) $(-14) \cdot (-5) = 70$

4.4 EXERCISES

Identify each statement as *true* or *false*.

1. -12 is an integer.
2. 0 is an integer.
3. Every counting number is an integer.
4. Every integer is a counting number.
5. The opposite of -9 is -9.
6. The difference of two integers is always an integer.
7. Every whole number is an integer.
8. Every integer is either positive or negative.
9. The product of two integers is an integer.
10. The sum of two integers is an integer.

Graph each set of numbers on a number line.

11. $\{-4, 2, 3, 5\}$
12. $\{-6, -3, 0, 2, 1\}$
13. $\{-2, -1, 0, 1, 2\}$
14. $\{-5, -4, -3, -2, -1, 0\}$
15. {the integers between -4 and 1}*
16. {the integers between 3 and 7}
17. {the opposites of 1, 2, 3, and 4}
18. {the opposites of 3, 4, 5, and 6}

Write either $<$ or $>$ in each blank.

19.	-6 ____ 4	**22.**	-10 ____ 0	**25.**	0 ____ -8
20.	-8 ____ 3	**23.**	-9 ____ -12	**26.**	0 ____ -15
21.	-7 ____ 0	**24.**	-8 ____ -16		

Find each sum in Exercises 27–36:

27.	$8 + (-3)$	**29.**	$-3 + 5$
28.	$9 + (-6)$	**30.**	$-9 + 15$

---

*-4 is *not* between -4 and 1.

ADAM RIESE

DEUTSCHE BUNDESPOST

10

Practical arithmetic   *From the time of Egyptian and Babylonian merchants, practical aspects of arithmetic complemented mystical (or "Pythagorean") tendencies. This was certainly true in the time of Adam Riese (1489–1559), a "reckon master" influential when commerce was growing in Northern Europe. Riese's likeness on the stamp above comes from the title page of one of his popular books on Rechnung (or "reckoning"). He championed new methods of reckoning using Hindu-Arabic numerals and quill pens. (The Roman methods in use moved counters on a ruled board.) Riese thus fulfilled Fibonacci's efforts 300 years earlier to supplant Roman numerals and methods (see page 157). Examples of the new reckoning can be seen opposite, as they were brought over to the New World.*

It is interesting to note that Martin Luther encouraged Riese and other reckon masters, and advocated in 1525 that children should learn not only language and history but music and "all of mathematics."

31. $-8 + (-6)$ **34.** $-6 + (-5) + (-2)$

32. $-12 + (-15)$ **35.** $-8 + 12 + (-3)$

33. $-9 + (-8) + (-6)$ **36.** $-15 + (-10) + (-8)$
       (Add from left to right.)

Perform the following subtractions:

37. $-9 - 2$ **41.** $8 - (-2)$ **45.** $-9 - (-15)$
38. $-6 - 5$ **42.** $6 - (-9)$ **46.** $-6 - (-12)$
39. $-12 - 14$ **43.** $4 - (-15)$ **47.** $-15 - (-2)$
40. $-15 - 32$ **44.** $17 - (-21)$ **48.** $-21 - (-5)$

Perform the following multiplications:

49. $6 \cdot (-2)$ **52.** $-7 \cdot (-6)$ **55.** $-6 \cdot (-25)$

50. $-3 \cdot (5)$ **53.** $-12 \cdot (3)$ **56.** $-15 \cdot (-32)$

51. $-9 \cdot (-2)$ **54.** $-9 \cdot (12)$

Although we shall not prove it, the *quotient* of two integers follows the same rules as multiplication regarding positive or negative answers.

Examples: $\dfrac{-9}{-3} = 3; \quad -15 \div (5) = -3; \quad -36 \div (-9) = 4.$

Perform the following divisions:

57. $\dfrac{-25}{5}$ **59.** $\dfrac{-6}{-3}$ **61.** $-18 \div (-6)$

                                                            62. $-30 \div (-5)$

58. $\dfrac{-120}{12}$ **60.** $\dfrac{-20}{-4}$ **63.** $50 \div (-25)$

                                                            64. $100 \div (-10)$

When a problem involves several operations, for example,

$$(-6) \cdot (3) \div 9 + 2 \cdot (-5),$$

use the following **order of operations:**

(1)   Do any work inside parentheses.
(2)   Start at the left and do multiplications or divisions as you come to them.
(3)   Start again at the left and do additions or subtractions as you come to them.

Example: $(-6) \cdot (3) \div 9 + 2 \cdot (-5) = -18 \div 9 + 2 \cdot (-5)$
$$= -2 + 2 \cdot (-5)$$
$$= -2 + (-10)$$
$$= -12.$$

Use these "order of operations" rules to work Exercises 65–78.

65. $9 \cdot (6 - 10)$ **67.** $-6 \cdot (3 - 5)$
66. $8 \cdot (12 - 19)$ **68.** $-12(5 - 10)$

69. $(-4 - 3) \cdot (-2) + 4$

70. $(-5 - 10) \cdot (-2) + 6$

71. $(-9 - 2) \cdot (-3) - (-7)$

72. $(-8 - 5) \cdot (-2) - (-9)$

73. $-8 \cdot (-2) \div 4$

74. $-12 \cdot (-3) \div (-12)$

75. $-2 \cdot (6) + 3 \div 3$

76. $-5 \cdot (2) + 3 \cdot (-2) \div (-6)$

77. $\dfrac{-9 \cdot (-2) - (-4) \cdot (-2)}{-2(3) - (-2) \cdot (2)}$

78. $\dfrac{4 \cdot (-2) - 5 \cdot (-3)}{2 \cdot (-1 - 3) - (-7)}$

Certain general statements about integers can be proved by the method illustrated as follows:

Example: Show that the product of two even integers is even.

PROOF   If an integer is even, it can be written in the form $2k$, where k is an integer. Let $2k$ and $2m$ represent two even integers. Their product is

$$2k(2m) = 2(2km).$$

The integer on the right is a multiple of 2, and thus is even. This is what we wanted to show.

Use a similar method to show that each statement in Exercises 79–82 is true.

*79. The sum of two even integers is even.

*80. The sum of two odd integers is even.
(*Hint:* Let $2k + 1$ and $2m + 1$ represent the two odd integers.)

*81. The product of an odd and an even integer is even.

*82. The product of two odd integers is odd.

People who work in precision manufacturing often give measurements such as 175 ± 2 feet. This means that 175 feet is the best possible single measure of the length in question, but that there is a possibility of a 2 foot error either way. Thus, the *actual* length is somewhere between $175 - 2 = 173$ and $175 + 2 = 177$ feet. Give the highest and lowest possible values for each of the following.

83. 28 ± 2 inches

84. 31 ± 1 miles

85. Smith has $47\% \pm 3\%$ in the poll.

86. The public favors that bill by $61\% \pm 4\%$.

```
 875\978
 726 460
  56 95
  63 54
   4 5
   4 3
 _____
 855 750
```

```
      00,
      12
    0 300
 11.4400|4400
   26666
    222
```

New (World) math   *The two computations (at left) are from the* Sumario Compendioso *by Brother Juan Diez, the first mathematics text published in the New World (Mexico City, 1556). It contained 18 pages on arithmetic, 6 on algebra. The top computation is 978 × 875, using the algorithm called "per copa" (meaning "by the cup," since it resembled the stemmed cup of the times, without handle). The bottom computation is 114,400 ÷ 26, using the "galley" algorithm (resembling a ship).*

4.5 RATIONAL NUMBERS

The set of integers is not closed for division. That is, the quotient of two integers need not be an integer. For example, none of the quotients

$$\frac{15}{7}, \quad \frac{-8}{3}, \quad \text{or} \quad \frac{7}{25}$$

is an integer. Quotients of integers are called **rational numbers.** Think of the rational numbers as being made up of all the fractions (quotients or ratios of integers) and all the integers. Any integer can be written as the quotient of two integers. For example, the integer 9 can be written as the quotient 9/1, or 18/2, or 27/3, and so on. Also, −5 can be expressed as the quotient of integers −5/1 or −10/2, and so on. (How can the integer 0 be written as a quotient of integers?) Since both fractions and integers can be written as quotients of integers, we use this idea to define the set of rational numbers:

Rational numbers
{all numbers which can be expressed as the quotient of two integers, with denominator not zero}.

We can restate this definition using set-builder notation:

$$\left\{ \frac{p}{q} \;\middle|\; p \text{ and } q \text{ are integers, with } q \neq 0 \right\}$$

In both of these definitions we were careful to say that the denominator of the quotient cannot equal 0. This is because division by 0 is not meaningful. To see why, first note that 0 pies can be divided equally among 7 people—each person gets 0 pies. On the other hand, the idea of trying to divide 7 pies equally among 0 people is meaningless.

When we discussed the integers, we found that the integers have an inverse property for addition. That is, if a represents any integer, then there exists an integer $-a$ such that

$$a + (-a) = 0.$$

We called $-a$ the *opposite* or *additive inverse* of a. We can now establish a similar property for multiplication of rational numbers. We know that 1 is the identity element for multiplication. (That is, $a \cdot 1 = a$ for any number a.) To have an inverse property for multiplication, we must be able to start with some rational number, say a/b, and find a rational number, say x, such that

$$\frac{a}{b} \cdot x = 1.$$

A rational number that works is b/a:

$$\frac{a}{b} \cdot \frac{b}{a} = 1.$$

Pythagoras and music *The lyra player above is from a Greek vase painting. On such an instrument Pythagoras may have discovered that musical tones are related to the length of stretched strings by ratios of counting numbers. You can test this on a cello, for example (pictured opposite). Stop any string midway, so that the ratio of the whole string to the part is 2/1. If you pick the free half of the string, you get the octave above the fundamental tone of the whole string. The ratio 3/2 gives you the fifth above the octave, and so on. Pythagoras noted that simple ratios of 1, 2, 3, 4 give the most harmonious musical intervals. He claimed that the intervals between planets must also be ratios of counting numbers. The idea came to be called "music of the spheres." (The planets were believed to orbit around the Earth on crystal spheres.)*

The numbers a/b and b/a are called the **multiplicative inverses,** or **reciprocals,** of each other. For example, 4/3 and 3/4 are reciprocals of each other, since

$$\frac{4}{3} \cdot \frac{3}{4} = 1.$$

Not every rational number has a reciprocal: there is no reciprocal for 0. That is, we cannot find a rational number x such that $0 \cdot x = 1$. Every rational number except 0 has a reciprocal.

In summary, we have the multiplicative inverse property:

Multiplicative inverse property

If a/b is a rational number (except 0),
then there exists a rational number b/a such that

$$\frac{a}{b} \cdot \frac{b}{a} = 1 \qquad \text{and} \qquad \frac{b}{a} \cdot \frac{a}{b} = 1.$$

Example 1 The following list shows several numbers and their reciprocals.

Number	Reciprocal
2	$\frac{1}{2}$
$\frac{3}{8}$	$\frac{8}{3}$
$\frac{-7}{3}$	$\frac{-3}{7}$
0	none

The idea of reciprocal is used in division of rational numbers. First, we define the **product of two rational numbers** a/b and c/d as

$$\frac{a}{b} \cdot \frac{c}{d} = \frac{ac}{bd}$$

Example 2 (a) $\dfrac{3}{4} \cdot \dfrac{7}{10} = \dfrac{3 \cdot 7}{4 \cdot 10} = \dfrac{21}{40}$ (b) $\dfrac{5}{18} \cdot \dfrac{3}{10} = \dfrac{5 \cdot 3}{18 \cdot 10} = \dfrac{15}{180} = \dfrac{1}{12}.$

In part (b) the product, 15/180, was reduced to lowest terms.

To find the **quotient** $a/b \div c/d$, we first write

$$\frac{a}{b} \div \frac{c}{d} = \frac{\dfrac{a}{b}}{\dfrac{c}{d}}$$

Then multiply top and bottom of this fraction by d/c, to make the denominator equal 1.

$$\frac{a}{b} \div \frac{c}{d} = \frac{\dfrac{a}{b}}{\dfrac{c}{d}} = \frac{\dfrac{a}{b} \cdot \dfrac{d}{c}}{\dfrac{c}{d} \cdot \dfrac{d}{c}} = \frac{\dfrac{a}{b} \cdot \dfrac{d}{c}}{1} = \frac{a}{b} \cdot \frac{d}{c}$$

Thus,

$$\frac{a}{b} \div \frac{c}{d} = \frac{a}{b} \cdot \frac{d}{c} = \frac{a \cdot d}{b \cdot c} = \frac{ad}{bc}$$

(This shows the origin of the rule "To divide fractions, invert the second and multiply.")

Example 3 (a) $\dfrac{3}{5} \div \dfrac{7}{15} = \dfrac{3}{5} \cdot \dfrac{15}{7} = \dfrac{45}{35} = \dfrac{9}{7}$ (b) $\dfrac{4}{7} \div \dfrac{3}{14} = \dfrac{4}{7} \cdot \dfrac{14}{3} = \dfrac{8}{3}$

(c) $\dfrac{2}{9} \div 4 = \dfrac{2}{9} \div \dfrac{4}{1} = \dfrac{2}{9} \cdot \dfrac{1}{4} = \dfrac{1}{18}$ (reduced to lowest terms)

There is no integer between the integers 3 and 4. However, we can always find a rational number between any two given rational numbers. One rational number that is between two given rationals is found by taking the average of the two numbers. (Add the numbers and divide by 2.) This procedure is shown in the next example.

Example 4 Find a rational number between $\dfrac{2}{3}$ and $\dfrac{5}{6}$.

Solution. Add the numbers:

$$\frac{2}{3} + \frac{5}{6} = \frac{4}{6} + \frac{5}{6} = \frac{9}{6} = \frac{3}{2}$$

Take half this sum:

$$\frac{1}{2} \cdot \frac{3}{2} = \frac{3}{4}$$

The number $\dfrac{3}{4}$ is between $\dfrac{2}{3}$ and $\dfrac{5}{6}$.

Which of the rational numbers 3/4 and 11/15 is larger? We could find out by converting each to fractions having the same denominator. A common denominator for 4 and 15 is 60:

$$\frac{3}{4} = \frac{45}{60} \quad \text{and} \quad \frac{11}{15} = \frac{44}{60}.$$

Thus, 45/60, or 3/4 is larger than 44/60, or 11/15.

4.5 EXERCISES

Perform the following operations.

1. $\dfrac{3}{4} \cdot \dfrac{9}{5}$

2. $\dfrac{3}{8} \cdot \dfrac{2}{7}$

3. $\dfrac{1}{10} \cdot \dfrac{6}{3}$

4. $2 \cdot \dfrac{1}{3}$

5. $\dfrac{9}{4} \cdot \dfrac{8}{6}$

6. $\dfrac{3}{5} \cdot \dfrac{5}{3}$

7. $\dfrac{3}{8} \div \dfrac{5}{4}$

8. $\dfrac{9}{16} \div \dfrac{3}{8}$

9. $\dfrac{5}{12} \div \dfrac{15}{4}$

10. $\dfrac{15}{16} \div \dfrac{30}{8}$

11. $\dfrac{15}{32} \cdot \dfrac{8}{25}$

12. $\dfrac{24}{25} \cdot \dfrac{50}{3}$

13. $\dfrac{-2}{3} \cdot \dfrac{-5}{8}$

14. $\dfrac{-2}{4} \cdot \dfrac{3}{9}$

Simplify each of the following. Write all answers in lowest terms.

15. $\left(\dfrac{3}{4} \cdot \dfrac{2}{9}\right) + \dfrac{1}{3}$

16. $\left(\dfrac{1}{3} \div \dfrac{1}{2}\right) + \dfrac{5}{6}$

17. $\left(\dfrac{2}{5} - \dfrac{3}{8}\right) + \left(\dfrac{1}{5} \cdot \dfrac{3}{4}\right)$

18. $\left(\dfrac{3}{4} \div \dfrac{3}{16}\right) - \left(\dfrac{2}{3} \cdot \dfrac{3}{4}\right)$

19. $\left(\dfrac{5}{6} - \dfrac{3}{8} + \dfrac{1}{2}\right) \div \dfrac{5}{2}$

20. $\left(\dfrac{3}{4} \div \dfrac{2}{3}\right) \cdot \dfrac{5}{8}$

Place either $<$ or $>$ in each blank.

21. $\dfrac{1}{2}$ ____ $\dfrac{3}{7}$

22. $\dfrac{3}{4}$ ____ $\dfrac{5}{7}$

23. $\dfrac{5}{12}$ ____ $\dfrac{7}{20}$

24. $\dfrac{8}{9}$ ____ $\dfrac{7}{8}$

25. $\dfrac{11}{15}$ ____ $\dfrac{5}{6}$

26. $\dfrac{3}{17}$ ____ $\dfrac{4}{13}$

27. $\dfrac{7}{9}$ ____ $\dfrac{11}{15}$

28. $\dfrac{3}{7}$ ____ $\dfrac{6}{11}$

29. $\dfrac{9}{21}$ ____ $\dfrac{7}{23}$

30. $\dfrac{11}{40}$ ____ $\dfrac{3}{20}$

Write the fractions in each list in order, from smallest to largest.

31. $\dfrac{12}{25}, \dfrac{13}{30}$

32. $\dfrac{23}{24}, \dfrac{14}{15}$

33. $\dfrac{4}{5}, \dfrac{6}{7}, \dfrac{7}{9}$

34. $\dfrac{5}{8}, \dfrac{6}{11}, \dfrac{3}{5}$

35. $\dfrac{3}{8}, \dfrac{1}{3}, \dfrac{3}{7}, \dfrac{3}{10}$

36. $\dfrac{9}{10}, \dfrac{6}{7}, \dfrac{5}{6}, \dfrac{10}{11}$

37. $\dfrac{4}{7}, \dfrac{1}{2}, \dfrac{5}{12}, \dfrac{8}{15}$

38. $\dfrac{3}{4}, \dfrac{2}{3}, \dfrac{11}{15}, \dfrac{12}{17}$

Find a rational number between the numbers in each pair.

39. $\dfrac{1}{2}, \dfrac{3}{4}$　　**44.** $\dfrac{5}{6}, \dfrac{9}{10}$　　**49.** $\dfrac{-3}{4}, \dfrac{-1}{2}$

40. $\dfrac{1}{3}, \dfrac{5}{12}$　　**45.** $\dfrac{3}{7}, \dfrac{1}{14}$　　**50.** $\dfrac{-2}{3}, \dfrac{-5}{6}$

41. $\dfrac{3}{5}, \dfrac{2}{3}$　　**46.** $\dfrac{2}{9}, \dfrac{1}{12}$　　**51.** $\dfrac{-3}{2}, 2$

42. $\dfrac{7}{12}, \dfrac{5}{8}$　　***47.** $\dfrac{5}{9}, 4$　　**52.** $\dfrac{-4}{3}, -1$

43. $\dfrac{3}{20}, \dfrac{1}{4}$　　***48.** $\dfrac{2}{3}, 2$

Each of the following recipes serves four people. On Monday night Tom is cooking just for his wife and himself, while on Thursday, he is cooking for a group of eight. Take half of each ingredient in each recipe for Monday, and double each recipe for Thursday.

53. **Zucchini Bread**
2 cups sugar
3 cups flour
$\frac{1}{2}$ teaspoon salt
1 teaspoon soda
$\frac{1}{2}$ teaspoon baking powder
3 teaspoons cinnamon
(with)
2 cups grated zucchini
3 eggs beaten foamy
2 sticks melted butter
3 teaspoons vanilla
$\frac{1}{4}$ cup or more of nuts

*54. **Cheese Cake**
$1\frac{1}{3}$ cup graham cracker crumbs
$\frac{1}{3}$ cup brown sugar
$\frac{1}{2}$ teaspoon cinnamon
$\frac{1}{3}$ cup melted butter
12 ounces cream cheese
2 eggs
$\frac{1}{3}$ cup granulated sugar
$\frac{1}{2}$ teaspoon vanilla
1 cup sour cream

*4.6 NUMBER SEQUENCES

In defining the counting numbers, we write them as the infinite set $\{1, 2, 3, 4, 5, \ldots\}$. This actual listing is sequential, in counting order. A list of numbers having a first number, a second number, a third number, and so on, is called a **sequence.** The numbers in a sequence are called the **terms** of the sequence.

Example 1 Examples of sequences:

(a) Even positive integers: 2, 4, 6, 8, 10, . . .

(b) Positive multiples of 5: 5, 10, 15, 20, 25, . . .

(c) Powers of 2:   2, 4, 8, 16, 32, 64, . . .

(d) Divisions on a ruler: $1, \dfrac{1}{2}, \dfrac{1}{4}, \dfrac{1}{8}, \dfrac{1}{16}, \dfrac{1}{32}$

(e) Periodic withdrawals of $10 from a bank account of $500 generate the sequence

$$500, 490, 480, 470, 460, \ldots, 20, 10, 0.$$

This sequence would stop when $0 is reached.

The sequences in parts (a), (b), and (e) are examples of **arithmetic sequences.** In an arithmetic sequence, each term after the first is found by adding the same number to the preceding term. For example,

$$9, 17, 25, 33, 41, \ldots$$

is an arithmetic sequence. Each term after the first is found by adding 8 to the preceding term. The number 8 is called the **common difference.** To find the common difference for an arithmetic sequence, choose any term except the first and subtract the preceding term. (Or, choose any two adjacent terms and subtract the first from the second.)

Example 2   Find the common difference in each arithmetic sequence.

(a) 11, 14, 17, 20, 23, 26, 29, . . .
Choose any two adjacent terms and subtract the first from the second. If we choose 14 and 17, we have

$$\text{Common difference} = 17 - 14 = 3.$$

(b) 13, 7, 1, −5, −11, −17, . . .   Let us choose 7 and 1.

$$\text{Common difference} = 1 - 7 = -6.$$

If we know the first term and common difference of an arithmetic sequence, we can write as many terms of the sequence as we might like. For example, suppose the first term of the sequence is 3. The first term of a sequence is often named a_1 (read "a sub one"). Thus,

$$a_1 = 3.$$

If the common difference is 5, then the second term of the sequence, a_2, is

$$a_2 = a_1 + 5 = 3 + 5 = 8.$$

We can find further terms:

$$a_3 = a_2 + 5 = (a_1 + 5) + 5 = a_1 + 2 \cdot 5 = 13$$
$$a_4 = a_3 + 5 = (a_2 + 5) + 5 = (a_1 + 5) + 5 + 5 = a_1 + 3 \cdot 5 = 18$$
$$a_5 = a_4 + 5 = a_1 + 4 \cdot 5 = 23,$$

and so on.

$a_2 = a_1 + 1 \cdot 5$
$a_3 = a_1 + 2 \cdot 5$
$a_4 = a_1 + 3 \cdot 5$
$a_5 = a_1 + 4 \cdot 5,$
and so on.

In summary, given a_1, we have the pattern shown at the side. Note that the numbers in color are 1 less than the corresponding "sub" number. Following this pattern, we would have, for example,

$$a_{12} = a_1 + 11 \cdot 5 \quad \text{(Note: } 11 = 12 - 1)$$
$$a_{20} = a_1 + 19 \cdot 5 \quad \text{(Note: } 19 = 20 - 1).$$

We can generalize this result: if a_1 is the first term of an arithmetic sequence, and if d represents the common difference for the sequence, then

$$a_n = a_1 + (n - 1)d,$$

where a_n represents the **general term**, or *n*-th *term*, of the sequence.

Example 3   Find the indicated term for each arithmetic sequence.

(a) 7, 12, 17, 22, 27, ...   Find the fifteenth term.

Solution.   Here the first term is $a_1 = 7$, and the common difference is $d = 5$. (We found 5 by subtracting two adjacent terms.) To find the fifteenth term, a_{15}, replace n with 15 in the formula for a_n:

$$\begin{aligned} a_n &= a_1 + (n - 1)d && \text{Formula} \\ a_{15} &= 7 + (15 - 1)5 && \text{Let } n = 15, d = 5, a_1 = 7 \\ &= 7 + (14)5 \\ &= 7 + 70 \\ a_{15} &= 77. && \text{Fifteenth term} \end{aligned}$$

(b) 23, 16, 9, 2, −5, −12, ...   Find a_{19}.

Solution.   For this sequence, $a_1 = 23$ and $d = -7$. Thus,

$$\begin{aligned} a_{19} &= a_1 + (19 - 1)d && \text{Let } n = 19 \\ &= 23 + (18)(-7) \\ &= 23 + (-126) \\ a_{19} &= -103. && \text{Nineteenth term} \end{aligned}$$

We can also find a formula for the **sum of the first *n* terms** of an arithmetic sequence. This formula can be developed by a method used by the mathematician C. F. Gauss (1777–1855) when he was a small child.

According to an old story, one of Gauss's teachers was a petty tyrant. One day he asked the students to find the sum of the first hundred counting numbers, or the sum

$$1 + 2 + 3 + 4 + 5 + 6 + \cdots + 100.$$

The other students in the class began by adding two numbers at a time, but Gauss found a quicker way. He first wrote the sum twice:

$$1 + 2 + 3 + 4 + \cdots + 99 + 100$$
$$100 + 99 + 98 + 97 + \cdots + 2 + 1.$$

He then noticed that by adding vertically, each pair of numbers adds up to 101.

$$\begin{array}{r} 1+ \quad 2+ \quad 3+ \quad 4+\cdots+ \quad 99+100 \\ \underline{100+ \quad 99+ \quad 98+ \quad 97+\cdots+ \quad 2+ \quad 1} \\ 101+101+101+101+\cdots+101+101 \end{array}$$

There are 100 of these sums of 101, for a total of $100 \cdot 101$, or 10,100. However, in finding the total, the numbers 1 through 100 were added twice. So the final result should actually be half of 10,100, or

$$1+2+3+4+\cdots+99+100 = \frac{10,100}{2} = 5050.$$

Using a method very similar to this, we could find a formula for the sum of the first n terms of an arithmetic sequence having first term a_1, n-th term a_n, and common difference d. If S_n represents this sum, then

$$S_n = \frac{n}{2}(a_1 + a_n).$$

Complete the Gauss example:

$$S_{100} = \frac{100}{2}(1 + 100) = 50 \cdot 101 = 5050.$$

Example 4   Find the sum of the indicated number of terms in each of the following arithmetic sequences.

(a) Find S_9 in 12, 19, 26, 33, 40, 47, 54, 61, 68. . . . .

Solution.   We need the sum of the first 9 terms of this arithmetic sequence. By the formula $a_n = a_1 + (n-1)d$, or by counting in the sequence above, verify that $a_9 = 68$. Also $a_1 = 12$, and $n = 9$. Thus,

$$S_n = \frac{n}{2}(a_1 + a_n)$$

$$S_9 = \frac{9}{2}(12 + 68)$$

$$= \frac{9}{2}(80)$$

$$S_9 = 360.$$

(b) Find S_{20} in 5, 13, 21, 29, 37, ...

Solution.   To find S_{20}, we need to find a_{20}. We know that $a_1 = 5$ and $d = 8$. Thus, $a_{20} = 5 + (20 - 1)8 = 5 + 152 = 157$. Thus,

$$S_{20} = \frac{20}{2}(5 + 157) = 10(162) = 1620.$$

Doubling "cubes"   *Doubling in backgammon is a vital aspect of the modern game. A player who is ahead offers to double the stakes; the opponent must agree or concede the game immediately. In return, only the player who accepted the double may redouble later in the game.*

The photo shows "supercubes" from the Chicago Chess and Backgammon Shop. Two of them were shot together to show you what either of them looks like if you go all around the sides. (The shape is technically an 8-sided prism.) Numbers on the sides are in geometric sequence, beginning with 2, and doubling. (It is also a sequence of powers of 2 from 2^1 to 2^8.)

In an arithmetic sequence, each term after the first is found by *adding* the same number to the previous term. In a **geometric sequence,** each term after the first is found by *multiplying* the previous term by a fixed number. For example,

$$1, 2, 4, 8, 16, 32, 64, \ldots$$

is a geometric sequence. Here each term after the first is found by multiplying the previous term by 2. The number 2 is called the **common ratio.** The common ratio can be found in a geometric sequence by choosing any term except the first, and then dividing it by the preceding term. In the sequence above, we could have found the common ratio by choosing, say, the term 16. We then divide it by the preceding term: $16/8 = 2$.

Example 5   Identify any of the following sequences that are geometric. If a sequence is geometric, find the common ratio.

(a) 6, 24, 96, 388, . . .

Solution.   Here each term after the first is found by multiplying the preceding term by 4. Thus, the common ratio is 4. The common ratio could also have been found by dividing: $96/24 = 4$, for example.

(b) 8, −16, 32, −64, 128, . . .

Solution.   Each term is found by multiplying the preceding term by −2. The sequence is geometric and the common ratio is −2.

(c) 100, 90, 81, 73, 66, 60, . . .

Solution.   This sequence is not geometric. (Is it arithmetic?)

Cotanto monta loſcacchue diſcacchi átado pianto ciaſcnna chuſa

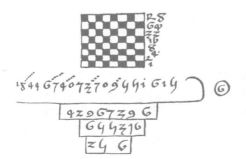

Chessboard problem   *The drawing here is from the early 1400s, but the problem is much older. One medieval Arabic version says that after Sissah ibn Dahir invented chess, his king was happy enough to give him any reward he wanted. Sissah asked for 1 wheat grain on the first square of the board, 2 grains on the second, 4 grains on the third, and so on. Again we get a geometric (and doubling) sequence. The grains total*

$$2^{64} - 1 =$$

18,446,744, 073,709,551,615.

By a mathematical twist we come back to primes (see opposite).

4.6 EXERCISES

In each of Exercises 1–16, a formula for the general term of a sequence is given. Use the given formula to find the first five terms of the sequence. Do this by replacing *n*, in turn, with 1, 2, 3, 4, and 5. Then identify each sequence as *arithmetic, geometric,* or *neither.*

1. $a_n = 6n + 5$
2. $a_n = 12n - 3$
3. $a_n = 3n - 7$
4. $a_n = 5n - 12$
5. $a_n = -6n + 4$
6. $a_n = -11n + 10$

7. $a_n = 2^n$
8. $a_n = 3^n$
9. $a_n = (-2)^n$
10. $a_n = (-3)^n$
11. $a_n = 3 \cdot 2^n$
12. $a_n = -4 \cdot 2^n$

13. $a_n = \dfrac{n+1}{n+2}$
14. $a_n = \dfrac{2n}{n+1}$
15. $a_n = \dfrac{1}{n+1}$
16. $a_n = \dfrac{1}{n+8}$

Identify each sequence as *arithmetic, geometric,* or *neither.* For an arithmetic sequence, give the common difference. For a geometric sequence, give the common ratio.

17. 6, 14, 22, 30, 38, 46, . . .
18. 40, 46, 52, 58, 64, . . .
19. 5, 8, 11, 14, 17, 20, 23, . . .
20. 23, 34, 45, 56, 67, 78, . . .
21. 4, 12, 36, 108, . . .
22. 7, 14, 28, 56, 112, . . .
23. 2, 5, 9, 14, 20, 27, . . .

24. 1, 4, 9, 16, 25, 36, . . .
25. 12, 9, 6, 3, 0, −3, −6, . . .
26. 37, 31, 25, 19, 13, 7, . . .
27. −18, −15, −12, −9, −6, . . .
28. −21, −17, −13, −9, −5, . . .
29. 3, −6, 12, −24, 48, −96, . . .
30. −5, 10, −20, 40, −80, . . .

31. −5, 6, −7, 8, −9, 10, −11, . . .
32. −12, 9, −6, 3, . . .

Find the indicated term for each arithmetic sequence.

33. $a_1 = 10$, $d = 5$; find a_{13}
34. $a_1 = 6$, $d = 9$; find a_8
35. $a_1 = 8$, $d = 3$; find a_{20}
36. $a_1 = 13$, $d = 7$; find a_{11}

37. $a_1 = 28$, $d = -3$; find a_8
38. $a_1 = 47$, $d = -5$; find a_7
39. $a_1 = 12$, $d = -7$; find a_{12}
40. $a_1 = 18$, $d = -9$; find a_{13}

41. 6, 9, 12, 15, 18, . . . ; find a_{25}
42. 14, 17, 20, 23, 26, 29, . . . ; find a_{13}
*43. −9, −13, −17, −21, −25, . . . ; find a_{15}
*44. −4, −11, −18, −25, −32, . . . ; find a_{18}

Find the sum of the first six terms in each arithmetic sequence.

45.   3, 6, 9, 12, 15, 18, . . .   **51.** $a_1 = 8, d = 9$
46.   11, 13, 15, 17, 19, 21, . . .   **52.** $a_1 = 12, d = 6$
47.   9, 17, 25, 33, 41, 49, . . .   **53.** $a_1 = 7, d = -2$
48.   28, 32, 36, 40, 44, 48, . . .   **54.** $a_1 = 13, d = -5$
49.   88, 98, 108, 118, 128, 138, . . .   **55.** $a_1 = -8, d = -7$
50.   92, 95, 98, 101, 104, 107, . . .   **56.** $a_1 = -15, d = -2$

*57.   Find the sum of the first 1000 counting numbers.

*58.   Find the sum of the first 10,000 counting numbers.

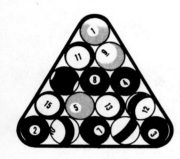

Fifteen balls are used in the game of pool. The balls are numbered from 1 through 15. In the game of "rotation," when a player "sinks" a ball in a pocket of the table, the player receives as many points as the number on the ball. Answer the following questions about this game. For some of them you can use the formula for the sum of the terms of an arithmetic sequence.

59.   How many points would a player score if he or she sunk all fifteen balls?

60.   Suppose player A sunk balls 1 through 10 in the pockets, and player B sunk balls 11 through 15. Find the total score for player A.

61.   Find the total score for player B.

62.   Who sunk more balls into the pockets? Who won the game?

63.   Suppose now that player A sunk all the even numbered balls, and player B sunk all the odd numbered balls. What was each player's score? Who won the game?

Suppose you save the amounts of money during each week as shown in the table.

Week	1st	2d	3d	4th	5th
Money	28¢	45¢	62¢	79¢	96¢

64.   Find the value of a_1.   **65.**   Find d.

66.   Find the amount that you would save in the 52nd week.

*67.   Find the total amount that you would save in 52 weeks.

A woman is offered a $10,000 starting salary, with an annual raise of $800.

68.   List her annual salaries for the first five years. Start with $10,000.

69.   Find d.   **70.**   Find her salary in year 10.

*4.7 THE FIBONACCI SEQUENCE

A useful sequence which is neither arithmetic nor geometric shows up in nature in a variety of ways, for example, in the family tree of a male bee. Male bees hatch from eggs which have not been fertilized, while female bees hatch from fertilized eggs. Because of this, a male bee has only one parent, his mother. On the other hand, female bees have both mothers and fathers.

The chart in Figure 4.10 shows the family tree of a single male bee. A male bee has 1 parent, 2 grandparents, 3 great grandparents, 5 great great grandparents, 8 great great great grandparents, and so on. Extend the chart of Figure 4.10 for one more generation—you should find 13 bees in this generation.

The numbers from the chart,

$$1, \quad 1, \quad 2, \quad 3, \quad 5, \quad 8, \quad 13,$$

are the same as the first few terms of the **Fibonacci sequence.**

You may have noticed that each term in the sequence is found by adding the two preceding terms. The next term in the sequence would be found by adding 8 and 13, or 21.

The next term would be $13 + 21 = 34$, then $21 + 34 = 55$, then $34 + 55 = 89$, and so on. The first fifteen terms of the Fibonacci sequence are as follows:

$$1, \quad 1, \quad 2, \quad 3, \quad 5, \quad 8, \quad 13, \quad 21,$$
$$34, \quad 55, \quad 89, \quad 144, \quad 233, \quad 377, \quad 610.$$

Fibonacci *discovered the sequence named after him in a problem on rabbits (see page 160). More than that, he is considered one of the greatest medieval mathematicians. Fibonacci (son of Bonaccio) is one of several names for Leonardo of Pisa, where he was born (1170). His father managed a warehouse in present-day Bougie (or Bejaia), in Algeria. Thus it was that Leonardo Pisano studied with a Moorish teacher and learned the "Indian" numbers which the Moors and other Moslems brought with them in their westward drive.*

Leonardo wandered around the Mediterranean, observing various systems of arithmetic in use among merchants, and he talked to mathematicians. Out of this came his Liber Abaci in 1202 ("abacus" here refers to computation). In it, Fibonacci presented the Hindu-Arabic numerals as the preferred numeration system. He covered the four operations with integers; fractions; prices of goods and other business math; problem solving; and some geometry and algebra. The Liber Abaci was an important means by which the new system was introduced into Europe. As it was, the conversion from Roman methods took two centuries more in Italy alone.

Fibonacci also wrote books on algebra, geometry, trigonometry. These contain Arabian mathematics as well as his own original work. He died about 1250, around the birthdate of Marco Polo.

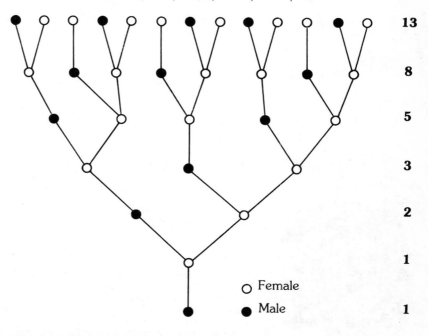

Figure 4.10 Family tree of a male bee

4.7 EXERCISES

The first fifteen Fibonacci numbers are given in the text. Use these numbers to find the following terms in the Fibonacci sequence.

1.	sixteenth	**3.**	eighteenth	**5.**	twentieth
2.	seventeenth	**4.**	nineteenth	**6.**	twenty-first

Various numbers from the Fibonacci sequence are given below. Use the given numbers to find the indicated terms. Here F_n denotes the n-th term of the Fibonacci sequence.

7. $F_{23} = 28,657$, $F_{24} = 46,368$; find F_{25}

8. $F_{27} = 196,418$, $F_{28} = 317,811$; find F_{29}

9. $F_{36} = 14,930,352$, $F_{38} = 39,088,169$; find F_{37}

10. $F_{40} = 102,334,155$, $F_{42} = 267,914,296$; find F_{41}

A plant is started in a greenhouse at the beginning of January. The plant grows for two months, and then adds a new branch. Each new branch grows for two months, and then adds another branch. After the second month, each branch adds a new branch every month.

January	February	March	April	May	June

11. Use the diagram to find the number of branches on the plant in each month. Complete the following table.

Month	January	February	March	April	May	June
Branches	____	____	____	____	____	____

How many branches would be on the plant in each month?

12. July   **13.** August   **14.** September   **15.** October

16. Are the answers to Exercises 11–15 terms of the Fibonacci sequence?

Fibonacci numbers in nature *Seeds in a sunflower are arranged in two sets of spirals, clockwise from center and counterclockwise. (Look closely at the top photo.) A typical sunflower head has 34 spirals going one way, 55 the other way. The ratio may vary with the size of the flower, but the ratio is always of two Fibonacci numbers. The pine cone (left) was photographed from the bottom to show its spiral patterns, again ratios of F-numbers. Other examples of F-numbers are in the number of petals in some species of daisies and in arrangements of leaves in some plants.*

The Parthenon, *Acropolis, Athens, is pleasing despite its ruined condition because of the ratio of height to width. First imagine that the peak above the columns (the triangular pediment) is complete. From its tip to just below the first three steps is the height that you can compare with the width of the roof. Is the ratio of height to width a golden ratio? (See the last paragraph on this page.)*

To the ancient Greeks, the rectangle with corners at A, B, C, and D, or the one with corners at C, B, E, and F, had the most pleasing shape. (See the figure.) Rectangles of other shapes, such as the one with corners at A, D, F, and E, were considered unattractive.

17. The rectangle whose corners are at points A, B, C, and D (called "rectangle ABCD") has the shape that the Greeks considered pleasing. Count the squares and find the length and width of this rectangle.

18. Do the numbers of Exercise 17 appear in the Fibonacci sequence?

19. The ratio of the length of rectangle ABCD to its width is 21/13. Divide 13 into 21 and write the ratio 21/13 as a decimal to the nearest tenth.

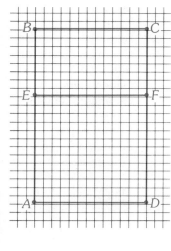

20. Line EF was drawn in the figure so that AEFD is a square. Locate rectangle BCFE, and find its length and width.

21. Are the numbers from Exercise 20 in the Fibonacci sequence?

22. Find the ratio of the length to the width for rectangle BCFE as a decimal to the nearest tenth.

23. Draw line GH so that EGHB is a square. Find the length and width of rectangle FGHC.

24. Find the ratio of the length to the width for rectangle FGHC as a decimal to the nearest tenth.

In Exercises 19, 22, and 24, you should have computed the answer 1.6. This number is called the *Golden Ratio*. A rectangle having sides in this ratio was considered the most pleasing by the ancient Greeks. (See also page 199.)

Fibonacci's original problem
The "rabbit problem"—found in Liber Abaci—concerns the number of rabbits alive after a given number of generations of rabbit matings. Assume that a pair of rabbits produces a new pair of rabbits in their second month, and then one pair in every month thereafter. All new rabbits reproduce in the same way. The number of pairs of rabbits that will be present is shown at the right. The rabbit population is given by numbers that we now recognize as "Fibonacci."

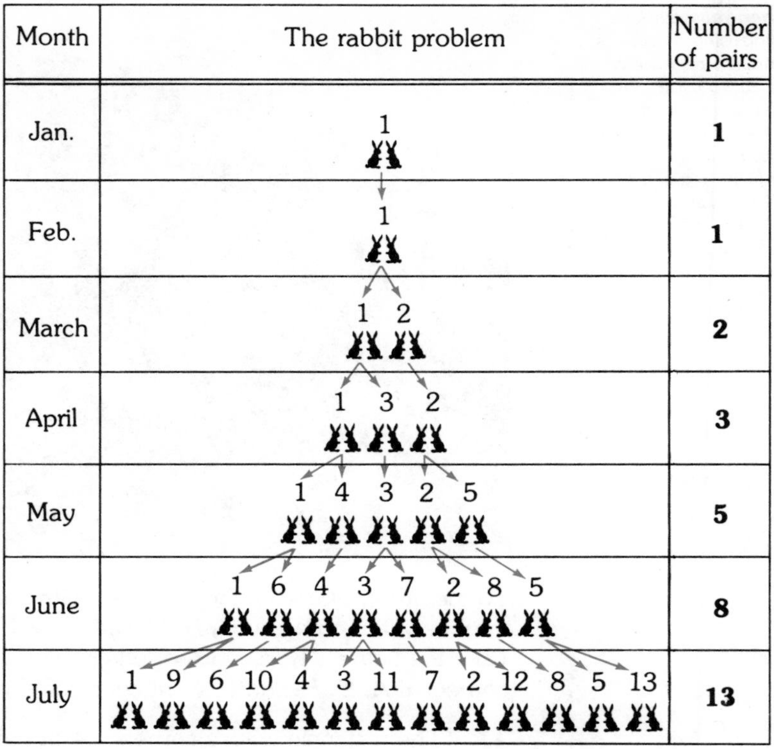

Month	The rabbit problem	Number of pairs
Jan.	1	1
Feb.	1	1
March	1 2	2
April	1 3 2	3
May	1 4 3 2 5	5
June	1 6 4 3 7 2 8 5	8
July	1 9 6 10 4 3 11 7 2 12 8 5 13	13

CHAPTER 4 SUMMARY

Keywords

Multiple	Number line	Theorem
Divisible	Whole numbers	Indirect proof
Factor	Integers	
Factorization	Rational numbers	Sequence
Prime	Numerator	Term
Composite	Denominator	Arithmetic sequence
Sieve of Eratosthenes	Lowest terms	Common difference
Prime factorization	Quotient	n-th term
Fundamental Theorem		General term
of Arithmetic	Closure	Geometric sequence
	Commutative	
Unique Factorization	Associative	Common ratio
Theorem	Identity	Fibonacci sequence
Relatively prime	Identity element	Golden Ratio
Twin primes	Additive inverse	
Greatest common factor	(opposite)	Magic square
Least common multiple	Multiplicative inverse	
Tests for divisibility	(reciprocal)	

Sets of numbers

Counting Numbers   {1, 2, 3, 4, 5, 6, . . .}

Whole Numbers   {0, 1, 2, 3, 4, 5, 6, . . .}

Integers   {. . . , −4, −3, −2, −1, 0, 1, 2, 3, 4, . . .}

Rational Numbers $\left\{\dfrac{p}{q}\middle| p \text{ and } q \text{ are integers, with } q \neq 0\right\}$

Properties

If a, b, and c represent rational numbers, then the following properties are valid.

	Addition	*Multiplication*
Identity	$a + 0 = a$ and $0 + a = a$	$1 \cdot a = a$ and $a \cdot 1 = a$
Commutative	$a + b = b + a$	$a \cdot b = b \cdot a$
Associative	$a + (b + c) = (a + b) + c$	$a \cdot (b \cdot c) = (a \cdot b) \cdot c$
Closure	$a + b$ is a rational number	$a \cdot b$ is a rational number
Distributive		$a \cdot (b + c) = a \cdot b + a \cdot c$

Formulas

If a_1 is the first term of an arithmetic sequence having common difference d, and if a_n represents the n-th term, with S_n representing the sum of the first n terms, then

$$a_n = a_1 + (n - 1) \cdot d \qquad \text{and} \qquad S_n = \frac{n}{2}(a_1 + a_n).$$

CHAPTER 4 TEST

Identify the property illustrated by each statement.

1. $8 + 21 = 21 + 8$ **3.** $6 + 0 = 6$

2. $9 \cdot (6 \cdot 8) = (9 \cdot 6) \cdot 8$ **4.** $9 \cdot (6 + 12) = 9 \cdot 6 + 9 \cdot 12$

Are the following sets closed for the given operation?

5. {. . . , −3, −1, 1, 3, 5, 7, . . .},   Multiplication

6. {. . . , −6, −4, −2, 0, 2, 4, 6, . . .},   Division

Write *prime* or *not a prime* for each of the following.

7. 33   **8.** 926,482

Write out the prime factorization of each number. Use exponents when possible.

9. 60   **10.** 325

Decide if the given number is divisible by the given factor.

11. 6,482,   4   **12.** 1,986,492,   3   **13.** 896,425,805,   5

Find the greatest common factor for each set of numbers.

14. 50 and 85   **15.** 560 and 720   **16.** 36, 48, and 60

Find the least common multiple for each set of numbers.

17. 48 and 36   **18.** 12, 18, and 30

Simplify each of the following.

19. $\dfrac{3}{16} + \dfrac{1}{2}$ **20.** $\dfrac{9}{20} - \dfrac{3}{32}$ **21.** $\dfrac{3}{8} \cdot \dfrac{16}{15}$ **22.** $\dfrac{7}{9} \div \dfrac{14}{27}$

Write either < or > in each blank.

23. -9 ____ 6 **24.** 0 ____ -14 **25.** $\dfrac{5}{12}$ ____ $\dfrac{6}{13}$

Perform each of the following operations so that a single number results.

26. $6 + (-2)$

27. $-12 + (-6) + 15$

28. $-8 - 2$

29. $-14 - (-3)$

30. $\dfrac{-36}{6}$

31. $9(-5 - 8)$

32. $(-6 - 5)(-4) - 8$

33. $-14(-2) \div 7 + 8$

Write the first five terms for each sequence.

34. $a_n = 6n - 2$ **35.** $a_n = n(n + 2)$

36. Arithmetic sequence, first term 6, common difference -2.

37. Geometric sequence, first term 3, common ratio 4.

38. Geometric sequence, $a_1 = 6$, $r = -2$.

39. Find the sum of the first six terms for the arithmetic sequence having first term 9 and common difference 5.

40. Find the sum of the first 500 counting numbers.

Dichotomy   *This is another of Zeno's paradoxes (his paradox Achilles and the tortoise is in Chapter 1). Suppose I want to go all the way (1 unit of measure). To do so, I must go half the way. But first I must go half of that, and so on infinitely—I cannot even begin to move.*

Some say that Zeno directed his paradoxes against Pythagoras, whose followers thought of space as the sum of points. In any case, you have not gone "all the way" until you go through the next chapter, since the rational numbers are not all the numbers we use!

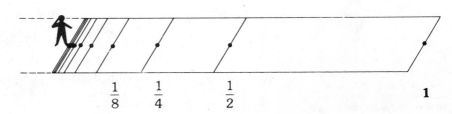

$\dfrac{1}{8}$ $\dfrac{1}{4}$ $\dfrac{1}{2}$ **1**

Chapter 5 The Real Numbers
With Applications

Applications *of mathematics are key ideas in this chapter, and Simon Stevin is the right person to lead off with. He started as a bookkeeper in Bruges (in present-day Belgium) and became an engineer in the Netherlands army. His work on decimals was an attempt to place whole numbers and fractions on common ground, so to speak. A unified system of numbers for measurement was of great benefit to surveyors, for example, at a time when many civilian and military works were under construction. Stevin also invented a method of finding the force of water on dams or canal walls, which anticipates some methods in calculus almost a century before Newton.*

In the previous chapter we looked at counting numbers, whole numbers, integers, and quotients of integers—all of which are rational numbers. There are useful numbers, such as $\sqrt{2}$ and π, which cannot be represented as quotients of integers, although they may be approximated that way. The set of all such numbers, a very numerous set, is called the set of *irrational numbers.* The rational numbers and the irrational numbers together make up the set of *real numbers.*

All real numbers can be represented as *decimals,* a fact that suits our base 10 Hindu-Arabic system just fine. Instead of a variety of ratios such as 1/2, 2/3, 5/16, and the like, we have tenths, hundredths, thousandths, and so on, to the right of the decimal point. Thus, $\sqrt{2} = 1.414$ (to three decimal places), and $\pi = 3.14159$ (to five decimal places).

Since the metric system of weights and measures works with multiples of ten, and since pocket calculators work with decimal numbers, it is natural to consider these topics in this chapter, as well as common applications of percent. The final section of the chapter discusses a set of numbers that includes the real numbers, the *complex numbers.*

Credit for the invention of decimals is usually given to Simon Stevin (1548–1620). His pamphlet of about 1600 explained the essentials of decimal numbers, but in such a muddy way that nothing came of it for many years. Stevin's notation was clumsy; he did not use a decimal point. In fact, historians do not agree on who invented it. There is, however, agreement that the decimal point or comma did not come into common use until the seventeenth century.

A comma, instead of a decimal point, is still in use in many European countries. In Britain, a point is used, but in an elevated position, as 23 · 298. To get around this lack of uniformity of symbols, several scientific groups have recently agreed to write decimals with a space left between each three digits. For example, 23,768.432567 would be written in this system as 23 768 · 432 567.

5.1 REAL NUMBERS AS DECIMALS

We have defined a rational number as a number that can be expressed as the quotient of two integers, with denominator not zero. We can also express rational numbers as decimals. To see how, let us write 3/8 as a decimal. The fraction bar in 3/8 indicates the division of 3 by 8. We convert 3/8 to a decimal by actually carrying out this division.

$$
\begin{array}{r}
.375 \\
8)\overline{3.000} \\
\underline{2\,4} \\
60 \\
\underline{56} \\
40 \\
\underline{40} \\
0
\end{array}
$$

(Move the decimal point straight up. Add on as many 0's as necessary.)

Thus, $3/8 = .375$. A decimal such as .375, which stops, is called a **terminating decimal.** Other examples of terminating decimals are

$$\frac{1}{4} = .25, \qquad \frac{7}{10} = .7, \qquad \frac{89}{1000} = .089.$$

$$
\begin{array}{r}
.3636\ldots \\
11)\overline{4.0000}\ldots \\
\underline{3\,3} \\
70 \\
\underline{66} \\
40 \\
\underline{33} \\
70 \\
\underline{66} \\
40
\end{array}
$$

Not all rational numbers lead to terminating decimals. For example, we can convert 4/11 into a decimal by dividing 11 into 4, as shown at the side. By continuing this division, we can obtain as many decimal digits as we might desire. Note the pattern that begins to appear: the digits 36 repeat indefinitely, so that we have 4/11 written as the **repeating decimal** .363636 . . . , or

$$\frac{4}{11} = .363636\ldots.$$

To simplify the symbols, we use a bar, and write

$$\frac{4}{11} = .\overline{36}$$

The bar shows that the two digits, 36, repeat indefinitely.

Example 1 By dividing, we can find the following decimal values for rational numbers.

(a) $\dfrac{5}{11} = .454545\ldots = .\overline{45}$

(b) $\dfrac{1}{3} = .33333\ldots = .\overline{3}$

(c) $\dfrac{5}{6} = .833333\ldots = .8\overline{3}$

(d) $\dfrac{10}{13} = .769230769230769230\ldots$

$\phantom{\textbf{(d)} \dfrac{10}{13}} = .\overline{769230}$

Example 2 Is the rational number 3/7 represented by a terminating decimal, a repeating decimal, or neither of these?

$$
\begin{array}{r}
.428571 \\
7)\overline{3.0000000} \\
2\,8 \\
\hline
20 \\
14 \\
\hline
60 \\
56 \\
\hline
40 \\
35 \\
\hline
50 \\
49 \\
\hline
10 \\
7 \\
\hline
30
\end{array}
$$

Solution. To find out, we divide 7 into 3, as shown at the side. The next step would be to divide 30 by 7, which is exactly where we began. Thus, the digits of the quotient will now start repeating, and

$$\frac{3}{7} = .428571428571\ldots = .\overline{428571}.$$

By appeal to Example 2, we can see that any rational number can be represented by a terminating or repeating decimal. At any point in the division, if we find a remainder of 0, then the division terminates, and we have obtained a terminating decimal.

What happens if we never get a remainder of zero? Each step of the division process must produce a remainder which is less than the divisor (the divisor is 7 in Example 2). Since the number of different possible remainders is less than the divisor, the remainders must eventually begin to repeat. This makes the digits of the quotient repeat, producing a repeating decimal. In summary,

> Any rational number can be expressed
> as a terminating decimal or a repeating decimal.

This is illustrated in Figure 5.1, which shows the reciprocals of the first twenty counting numbers. Recall that the reciprocal of a number n is $1/n$. Reciprocals of counting numbers are rational numbers.

Figure 5.1

Number	Reciprocal	Number	Reciprocal
1	1.0	11	$.\overline{09}$
2	.5	12	$.08\overline{3}$
3	$.\overline{3}$	13	$.\overline{076923}$
4	.25	14	$.0\overline{714285}$
5	.2	15	$.0\overline{6}$
6	$.1\overline{6}$	16	.0625
7	$.\overline{142857}$	17	$.\overline{0588235294117647}$
8	.125	18	$.05\overline{5}$
9	$.\overline{1}$	19	$.\overline{052631578947368421}$
10	.1	20	.05

What about the reverse process? That is, if we are given a terminating decimal or a repeating decimal, must it represent a rational number? The answer is *yes*. For example, the terminating decimal .6 represents a rational number:

$$.6 = \frac{6}{10} = \frac{3}{5}.$$

Example 3   Write each terminating decimal as a rational number.

(a) $.437 = \dfrac{437}{1000}$ **(b)** $8.2 = 8\dfrac{2}{10} = \dfrac{82}{10} = \dfrac{41}{5}$

We can't convert repeating decimals into rational numbers quite so quickly. To see how this is done, follow the steps in the next example.

Example 4   Find the rational number equal to $.\overline{85}$.

Step 1   Let $x = .\overline{85}$, or $x = .858585\ldots$.

Step 2   Multiply both sides of the equation $x = .858585\ldots$ by 100. (We use 100 since there are two decimal positions in the part that repeats.)

$$x = .858585\ldots$$
$$100x = 100(.858585\ldots)$$
$$100x = 85.858585\ldots$$

Step 3   Subtract the statement in Step 1 from the final statement in Step 2:

$$100x = 85.858585\ldots \qquad \text{(From algebra: } x = 1x \text{ and}$$
$$\underline{x = .858585\ldots} \qquad 100x - x = 99x.)$$
$$99x = 85$$

Step 4   Solve the equation $99x = 85$ by dividing both sides by 99:

$$99x = 85$$
$$\frac{99x}{99} = \frac{85}{99}$$
$$x = \frac{85}{99}.$$

Since we defined x to equal $.\overline{85}$, we have

$$.\overline{85} = \frac{85}{99}.$$

To check, divide 99 into 85.

Example 5   Express $.3\overline{2}$ as the quotient of two integers.

Solution.   Follow the steps given above.

Step 1 $x = .3\overline{2} = .322222\ldots$

Step 2 $10x = 3.2222\ldots$ (Why did we use 10 and not 100?)

Step 3 $10x = 3.22222\ldots$
$$\underline{x = .32222\ldots}$$
$$9x = 2.9$$

Step 4   Since $9x = 2.9$, we have $x = \dfrac{2.9}{9}$.

Since 2.9 is not an integer as required, we multiply numerator and denominator by 10:

$$x = \frac{2.9}{9} = \frac{2.9 \cdot 10}{9 \cdot 10} = \frac{29}{90}.$$

Thus, $.3\overline{2} = \frac{29}{90}.$

We have seen that any rational number can be expressed as a repeating or terminating decimal. Also, we have seen that every repeating or terminating decimal represents a rational number. Some decimals, however, neither repeat nor terminate. For example, the decimal

$$.102001000200001000002 \ldots$$

does not terminate and does not repeat. (It is true that there is a pattern in this decimal, but no single block of digits repeats indefinitely.)

A number represented by a non-repeating, non-terminating decimal is called an **irrational number.** The decimal above is an irrational number. Common irrational numbers include $\sqrt{2}$, $\sqrt{3}$, $\sqrt{5}$, and so on. Also, π, the ratio of the circumference to the diameter of a circle, is an irrational number. We discuss irrational numbers in more detail in Section 5.6.

As we have seen, rational numbers can be expressed as repeating or terminating decimals, while irrational numbers have non-repeating decimal expansions. The union of the set of rational numbers and the set of irrational numbers forms the set of *all* decimals, terminating, repeating, or non-terminating, non-repeating. This larger set is called the set of **real numbers:**

Real numbers = {all decimals}.

Thus, all counting numbers are real numbers, all integers are real numbers, all rational numbers are real numbers, and all irrational numbers are real numbers. The relationships between the various sets of real numbers are shown in Figure 5.2.

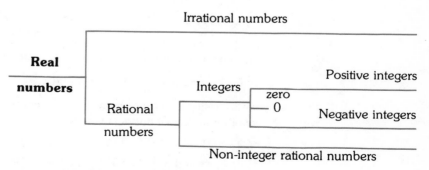

Figure 5.2   Real number tree

5.1 EXERCISES

Write each repeating decimal, using a bar.

1. .4444444 . . .	**4.** .65656565 . . .	**7.** .913455555 . . .
2. .585858 . . .	**5.** .469469469 . . .	**8.** .21766666 . . .
3. .929292 . . .	**6.** .810810810 . . .	**9.** .083252525 . . .

Convert each rational number into either a repeating or a terminating decimal.

10. $\dfrac{3}{4}$ **13.** $\dfrac{9}{32}$ **16.** $\dfrac{3}{11}$ **19.** $\dfrac{11}{15}$

11. $\dfrac{7}{8}$ **14.** $\dfrac{5}{12}$ **17.** $\dfrac{9}{11}$ **20.** $\dfrac{2}{15}$

12. $\dfrac{3}{16}$ **15.** $\dfrac{11}{12}$ **18.** $\dfrac{2}{7}$

Convert each terminating decimal into a rational number. Reduce each to lowest terms.

21. .4   **23.** .85   **25.** .105

22. .9   **24.** .36   **26.** .934

Convert each repeating decimal into a rational number. Reduce each to lowest terms.

27. $.\overline{8}$ **30.** $.\overline{36}$ **33.** $4.\overline{92}$ **36.** $.2\overline{6}$

28. $.\overline{1}$ **31.** $.\overline{123}$ **34.** $6.\overline{13}$ ***37.** $.\overline{9}$

29. $.\overline{54}$ **32.** $.\overline{516}$ **35.** $.4\overline{3}$ ***38.** $.4\overline{9}$

Decide whether each decimal represents a rational number or an irrational number.

39. .81818181 . . .   **43.** 8.4355555 . . .

40. .035035035 . . .   **44.** 15.1432189 . . .

41. .21359876 . . .   **45.** 23.25819724

42. .59148273109 . . .   **46.** 51.9276325

Find an irrational number between each pair of decimals. (*Hint:* Find a nonrepeating decimal.) An infinite number of answers is possible.

***47.** .5 and .6   ***48.** 1.8 and 1.9

Find a rational number between each pair of rational numbers. An infinite number of answers is possible.

***49.** .91 and .92   ***51.** $.\overline{16}$ and $.\overline{17}$

***50.** .03 and .04   ***52.** $.\overline{798}$ and $.\overline{799}$

It can be shown that a rational number a/b in lowest terms results in a terminating decimal if the only primes that divide into the denominator b are 2 or 5. If a prime other than 2 or 5 divides into b, then a/b will give a repeating decimal. Use this idea to decide whether the following rational numbers would yield a repeating or a terminating decimal. (Reduce to lowest terms before trying to decide.)

53. $\dfrac{8}{15}$ **55.** $\dfrac{13}{125}$ **57.** $\dfrac{22}{55}$

54. $\dfrac{8}{35}$ **56.** $\dfrac{3}{24}$ **58.** $\dfrac{24}{75}$

Answer *true* or *false* for each statement.

59. Every integer is a rational number.
60. All real numbers are rational numbers.
61. 0 is a counting number. F
62. 0 is an integer.
63. $\sqrt{2}$ is a real number. T
64. $\sqrt{2}$ is a rational number.
65. Every real number is a repeating or terminating decimal. F
66. Every repeating decimal is a real number.
67. .01001000100001 . . . is a repeating decimal. F
68. All irrational numbers are real numbers.

Complete the following chart.

	Number	Natural	Whole	Integer	Rational	Irrational	Real
69.	5	yes					
70.	−9				yes		
71.	0					no	
72.	3/4		no				yes
73.	π					yes	
74.	$\sqrt{2}$	no					
75.	1.444 . . .	no	no	no			
76.	3.62	no					
77.	.819342 . . .				no		yes
78.	9.345218 . . .	no					
***79.**	9/0	no	no				
***80.**	−4/0				no	no	

5.2 DECIMALS AND PERCENTS

Decimals and the related subject of percents have many practical applications in daily life. We shall see some of these applications in this section and the next two. We begin by developing some of the rules for working with decimals.

The decimal .35 represents the rational number 35/100. Also, the decimal .58 represents 58/100. We can use these fractional forms to find the sum of .35 and .58:

$$.35 + .58 = \frac{35}{100} + \frac{58}{100} = \frac{35 + 58}{100} = \frac{93}{100} = .93.$$

We could have obtained the same answer by placing .35 and .58 in a vertical column, with the decimal points lined up:

$$\begin{array}{r} .35 \\ .58 \\ \hline .93 \end{array}$$

Example 1   Add or subtract in each of the following.

(a) .46 + 3.9 + 12.58

Solution.   Write the numbers vertically, with decimal points lined up.

$$\begin{array}{r} .46 \\ 3.9 \\ 12.58 \\ \hline 16.94 \end{array}$$

(b) 9.43 − 6.8

Solution.

$$\begin{array}{r} 9.43 \\ -6.8 \\ \hline 2.63 \end{array}$$

We could write 6.8 as 6.80, if desired:

$$\begin{array}{r} 9.43 \\ -6.80 \\ \hline 2.63 \end{array}$$

(c) 12.1 − 8.723

Solution.   Write 12.1 as 12.100:

$$\begin{array}{r} 12.100 \\ -8.723 \\ \hline 3.377 \end{array}$$

Decimals are multiplied with the following rule:

Multiply as for integers. The product has as many decimal places as the *sum* of the number of decimal places in the two factors.

Example 2   Multiply: (a) 2.9 × 5.8   (b) 4.613 × 2.52.

(a) 2.9 × 5.8

$$\begin{array}{r} 5.8 \longleftarrow \text{1 place} \\ 2.9 \longleftarrow \text{1 place} \\ \hline 522 \\ 116 \\ \hline 16.82 \longleftarrow 1 + 1 = \text{2 places} \end{array}$$

(b) 4.613 × 2.52

$$\begin{array}{r} 4.613 \longleftarrow \text{3 places} \\ 2.52 \longleftarrow \text{2 places} \\ \hline 9226 \\ 23065 \\ 9226 \\ \hline 11.62476 \longleftarrow 3 + 2 = \text{5 places} \end{array}$$

To divide 11.5 by 2.3, write the quotient as

$$\frac{11.5}{2.3}$$

This problem would be easier if the denominator were a whole number. We can convert the denominator to a whole number by multiplying both numerator and denominator by 10:

$$\frac{11.5}{2.3} = \frac{11.5 \times 10}{2.3 \times 10} = \frac{115}{23}.$$

Instead of dividing 11.5 by 2.3, we divide 115 by 23:

$$\begin{array}{r} 5 \\ 23{\overline{\smash{\big)}\,115}} \\ \underline{115} \\ 0 \end{array}$$
   Thus, $\dfrac{11.5}{2.3} = 5.$

Example 3   Divide in each of the following.

(a) $\dfrac{25.97}{4.9}$.   Multiply numerator and denominator by 10:

$$\frac{25.97}{4.9} = \frac{25.97 \times 10}{4.9 \times 10} = \frac{259.7}{49}$$

Now divide 259.7 by 49:

$$\begin{array}{r} 5.3 \\ 49{\overline{\smash{\big)}\,259.7}} \\ \underline{245} \\ 14\,7 \\ \underline{14\,7} \\ 0 \end{array}$$
   (The decimal point moves straight up.)

Here $\dfrac{25.97}{4.9} = 5.3.$

$$\begin{array}{r} 7.9 \\ 825{\overline{\smash{\big)}\,6517.5}} \\ \underline{5775} \\ 742\,5 \\ \underline{742\,5} \\ 0 \end{array}$$

(b) $\dfrac{65.175}{8.25} = \dfrac{65.175 \times 100}{8.25 \times 100} = \dfrac{6517.5}{825} = 7.9.$

Breakfast sale   *According to a literal reading of this sign at the local Montgomery Ward store, can you get breakfast for less than a penny?*

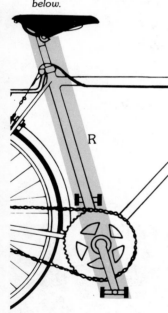

One of the main applications of decimals is in **percents.** Percents are widely used. In consumer mathematics, interest rates and discounts are often given as percents, as we shall see in the next two sections.

The word *percent* means "per hundred." This idea is used in the basic definition of percents: we say that

$$1\% = \frac{1}{100} = .01.$$

Thus, $45\% = 45(1\%) = 45(.01) = .45$.

Example 4   Convert each percent to a decimal.

(a) $98\% = 98(1\%) = 98(.01) = .98$

(b) $3.4\% = .034$

(c) $.2\% = .002$

We can convert a decimal to a percent in much the same way.

Example 5   Convert each decimal to a percent.

(a) $.13 = 13(.01) = 13(1\%) = 13\%$ **(c)** $.4 = 40\%$
(b) $.532 = 53.2(.01) = 53.2(1\%) = 53.2\%$ **(d)** $2.3 = 230\%$

Example 6   Convert each fraction to a percent.

(a) $\frac{3}{5}$ First write 3/5 as a decimal. Dividing 5 into 3 gives $\frac{3}{5} = .6 = 60\%$.

(b) $\frac{14}{25} = .56 = 56\%$

The following examples show how to work the various types of problems involving percents.

Example 7   What is 18% of 250?

Solution.   The key word here is "of." The word "of" translates as "times." Thus, 18% of 250 is given by

$$(18\%)(250) = (.18)(250) = 45.$$

Example 8   What percent of 500 is 75?

Solution.   First answer the question, what fraction of 500 is 75? The answer is 75/500. Convert this fraction to a decimal, and finally to a percent.

$$\frac{75}{500} = .15 = 15\%$$

5.2 EXERCISES

Work each of the following problems involving decimals.

1.	8.6 + 4.9	**4.**	58.3 − 2.9	**9.**	28 × 3.92
				10.	74 × 6.53
2.	12.3 + 8.5	**5.**	6.5 × 3.8	**11.**	$\dfrac{87.3}{3}$
3.	47.4 − 8.5	**6.**	7.4 × 5.3	**12.**	$\dfrac{103.2}{4}$

11. $\dfrac{87.3}{3}$ **14.** $\dfrac{114.66}{4.9}$

12. $\dfrac{103.2}{4}$ **15.** $\dfrac{6.44}{.23}$

7. $2.5 + 8.97 + 6.3 + 19.481$

8. $7.42 + 12.38 + 5.9 + 6.243$

13. $\dfrac{60.5}{2.5}$ **16.** $\dfrac{35.25}{.75}$

Solve each of the following problems.

17. During a recent month, a Honda dealer sold $36,948.21 worth of motorcycles, $12,408.20 worth of parts, and $41,793.59 worth of service. What was the total income for the month from these sources?

18. At the beginning of school, Joann paid $46.28 for textbooks (a bargain, we think), $6.49 for supplies, and $12.56 for some Rolling Stones albums. How much did she spend altogether?

19. In a busy week at the local McDonald's, Tom earned $104.28 at the regular rate, $39.83 at the overtime rate, and $29.38 at the holiday rate. How much did he earn that week?

The bank balance of Tammy's Hobby Shop was $1856.12 on March 1. During March, Tammy deposited $1742.18 received from the sale of goods, $4271.94 paid by customers on their accounts, and a $28.37 tax refund. She paid out $1496.14 for merchandise, $268.37 in salaries, and $793.46 in other expenses.

20. How much did Tammy deposit during March?

21. How much did she pay out?

22. What was her bank balance at the end of March?

23. For the lodge picnic, Kathleen bought 120 hamburgers at $.47 each, 140 bags of french fries at $.36 each, and 21 sixpacks at $1.93 each. How much did she spend?

24. George went to Sears and bought 4 tires at $29.38 each, 12.9 gallons of gas at $.67 per gallon, 2.3 cans of air conditioner propellant at $1.70 per can (yes, one can buy a partial can of this), and one shovel for $8.43. How much did he pay all together?

Wheel in the wind *See this at Oklahoma State University, Stillwater, where scientists are testing a "bicycle windmill" for the Energy Research and Development Administration (ERDA). This 15-foot, 5 kilowatt model can provide electricity for ordinary uses in a rural home or farm. (See Chapter 7 for more about energy from windmills.)*

25. Samantha drove 350.45 miles on 16.3 gallons of gas. How many miles per gallon did she get?

26. In a recent year, the average price paid by United Airlines for jet fuel increased by 5¢ per gallon. This caused the company's total fuel bill to increase by $80,000,000. How many gallons of fuel did United buy that year?

Sometimes we must *round off* a decimal. For example, to round 17.639 to the nearest tenth, look at the tenths digit, which is 6 in this case. Then look at the first digit to the right of the tenths digit, 3 in this case. If this digit is 0, 1, 2, 3, or 4, leave the tenths digit unchanged. If this digit is 5, 6, 7, 8, or 9, increase the tenths digit by 1. Thus, 17.639 rounded to the nearest tenth is 17.6. Also, 17.639 rounded to the nearest hundredth is 17.64.

Round each of the following to: (a) the nearest tenth; (b) the nearest hundredth.

27.	78.414	**29.**	.0837	**31.**	12.68925
28.	3689.537	**30.**	.0658	**32.**	43.897142

Convert each decimal to a percent.

33.	.42	**35.**	.365	**37.**	.008	**39.**	2.1
34.	.87	**36.**	.792	**38.**	.0093	**40.**	8.9

Convert each percent to a decimal.

41.	46%	**43.**	8%	**45.**	159%	**47.**	.5%
42.	92%	**44.**	3%	**46.**	274%	**48.**	.08%

Work each of the following problems involving percents.

49. Find 28% of 596.

50. Find 74% of 382.

51. What number is 118% of 36?

52. What number is 212% of 50?

*53. 50.7 is what percent of 78?

*54. 39.42 is what percent of 146?

55. Lodi charges a property tax equal to 3.1% of the market value of the property that is taxed. Mary's home has a value of $52,000. Find her property tax.

56. According to an article in *The Wall Street Journal*, senior airline pilots make as much as $99,500 per year. Suppose such a pilot received a cost of living raise of 6%. Find the amount by which the salary would increase.

57. The same article says that 10% of all Mercedes cars are sold to pilots. If Mercedes sells 165 cars to pilots in a certain area, how many cars are sold there all together?

58. A family earns a total of $1653 per month. The family spends 89% of its income and saves the rest. How much does the family save in one month?

59. This term, there are 1746 married students on campus, which is 34% of the total enrollment. Find the total enrollment. (Round to the nearest whole number.)

60. Sid's Pharmacy has a total monthly payroll of $6896, of which $1724 goes for employee fringe benefits. What percent of the total payroll goes for fringe benefits?

*61. In some states, the sales tax is 5%. Suppose a merchant has $250 in her cash box at the end of a day, including money paid for merchandise, plus the tax. How much tax goes to the state? (*Hint:* Do *not* multiply .05 times 250.)

*62. Mr. and Mrs. Carver want to receive $42,000 from the sale of their house, after the real estate agent's commission of 6% of the selling price is deducted. Find the selling price that the Carvers must get.

The local Montgomery Ward store often has "15% off sales," where most everything in the store is for sale at 15% off. Find the sale price for each of the following items.

63. color television; regular price $300

64. drapes; regular price $80

65. carpet; regular price $1100

66. dress; regular price $42

Many stores find **percent of markup** by the formula

$$\frac{\text{selling price} - \text{cost}}{\text{selling price}}$$

with this decimal converted to a percent. Find the percent of markup on the following items. (We found these selling prices and costs at a local store not known for bargains. We rounded the amounts slightly to make the calculation easier.)

67. Scotch tape; cost 30¢; selling price, 50¢

68. One-A-Day-Vitamins; cost $2.05; selling price, $3.00

69. 1/2 gallon Jim Beam; cost $9.35; selling price, $12.50

70. Contac Cold Capsules; cost (20 capsules) $1.60; selling price, $2.00

*5.3 CALCULATORS

We have looked at decimals in this chapter, and now we can discuss one of the most important applications of decimals, the pocket calculator. Pocket calculators as we know them would be almost impossible without decimals. (Try to imagine a calculator operating with the Egyptian numerals of Chapter 3.)

Calculators today come in a huge array of types, sizes, and prices. It can be difficult to know which machine to buy for your needs. To help you decide on the best machine, we list some of the available features.

Floating decimal   A machine with a floating decimal will locate the decimal point exactly where it should be in the final answer. For example, to buy 55.99 square yards of Armstrong Solarian at $13.99 per square yard, proceed as follows.

$$\boxed{5}\ \boxed{5}\ \boxed{\cdot}\ \boxed{9}\ \boxed{9}\ \boxed{\times}\ \boxed{1}\ \boxed{3}\ \boxed{\cdot}\ \boxed{9}\ \boxed{9}\ \boxed{=}\quad 783.3001$$

The final number displayed by the machine, 783.3001, can be rounded to the nearest cent, $783.30.

As this example shows, a decimal point is entered with the decimal point key, as needed. For example, $47 has no decimal point, and would be entered just as

$$\boxed{4}\ \boxed{7}$$

with no decimal point needed, while 95¢ (or $.95) would be entered as

$$\boxed{\cdot}\ \boxed{9}\ \boxed{5}$$

On a machine without a floating decimal, we might have to enter $47 as

$$\boxed{4}\ \boxed{7}\ \boxed{0}\ \boxed{0}$$

One problem that you might have with floating decimals is shown by the following example. If we add $21.38 and $1.22, we get $22.60. However, the calculator would leave off the final 0 here and display the answer as

$$22.6$$

We must remember that we are using dollars and cents, and write the answer as $22.60.

Operations   All machines feature the four operations of addition, subtraction, multiplication, and division. Machines which do only these four operations are called *four function calculators*. More advanced machines do more things, as we shall see.

How do you account for this?   *At McEwen's Department Store in Melbourne, Don Pressor (on calculator) challenged Johnson Lowe (on abacus). Guess who won? Lowe, of course. Whenever the pocket calculator and the abacus are put in competition (with a very skilled user of the abacus), the abacus almost always wins on ordinary arithmetic calculations.*

Logic Most machines use *algebraic logic.* These can be recognized by $\boxed{+}$ and $\boxed{-}$ keys (instead of $\boxed{+=}$ and $\boxed{-=}$ keys.) On these machines, the problem $9 + 8$ would be worked as

$$\boxed{9}\ \boxed{+}\ \boxed{8}\ \boxed{=}\quad 17$$

with 17 displayed as the answer. We would work $29 - 6$ as

$$\boxed{2}\ \boxed{9}\ \boxed{-}\ \boxed{6}\ \boxed{=}\quad 23$$

with 23 appearing in the display. On a machine without algebraic logic, this problem would be worked as

$$\boxed{2}\ \boxed{9}\ \boxed{+=}\ \boxed{6}\ \boxed{-=}\quad 23$$

Constant key This key is used when repeated multiplications or divisions are needed. For example, if you work in a store in a state with a 6% sales tax, you would find the sales tax on an item selling for \$169 as follows.

$$\boxed{1}\ \boxed{6}\ \boxed{9}\ \boxed{\times}\ \boxed{\cdot}\ \boxed{0}\ \boxed{6}\ \boxed{=}\quad 10.14$$

The tax on an item costing \$259.16 is

$$\boxed{2}\ \boxed{5}\ \boxed{9}\ \boxed{\cdot}\ \boxed{1}\ \boxed{6}\ \boxed{\times}\ \boxed{\cdot}\ \boxed{0}\ \boxed{6}\ \boxed{=}\quad 15.55 \text{ (rounded)}$$

By continuing in this way, we could find the tax on any of the items in the store. However, we can get the same results more quickly by using the $\boxed{\text{CONST}}$ key. Read your calculator instructions and find the way to activate this key. Then enter .06. The sales tax can now be found as follows.

$$\boxed{1}\ \boxed{6}\ \boxed{9}\ \boxed{=}\quad 10.14$$
$$\boxed{2}\ \boxed{5}\ \boxed{9}\ \boxed{\cdot}\ \boxed{1}\ \boxed{6}\ \boxed{=}\quad 15.55$$

By this process, the tax on each item in the store can be found quickly.

Memory key This key saves the trouble of writing down intermediate steps. For example, suppose 6 bottles of Pepsi cost 43¢ each, and 5 bags of potato chips cost 89¢ each. The total cost of the Pepsi is $6 \times .43 = \$2.58$. We could find this number using a calculator, and then write it down before finding the cost of the potato chips. With a memory key $\boxed{\text{M}}$ just store 2.58 in memory. Then, find the cost of the potato chips and, finally, the total.

Clearing keys All machines have a $\boxed{\text{C}}$ key. By pressing this key, you cause everything in the machine to be erased, or "cleared," and you can start over. Some machines also have a $\boxed{\text{CE}}$ key. By pressing this key, you can "clear entry." Only the number displayed will be erased. This helps in correcting errors without having to start over. For example, if we want to add 17 and 36, we should proceed as follows.

$$\boxed{1}\ \boxed{7}\ \boxed{+}\ \boxed{3}\ \boxed{6}\ \boxed{=}\quad 53$$

Suppose we enter 17 correctly, and then hit the $\boxed{+}$ key. We should then enter 36. If we make a mistake and enter 37 instead, we can correct the error with the $\boxed{\text{CE}}$ key. Proceed as follows.

$$\boxed{1}\ \boxed{7}\ \boxed{+}\ \boxed{3}\ \boxed{7}\ \boxed{\text{CE}}\ \boxed{3}\ \boxed{6}\ \boxed{=}\quad 53$$

Other keys $\boxed{\sqrt{}}\ \boxed{\sin}\ \boxed{\cos}\ \boxed{\log}\ \boxed{x^y}$ These keys appear on more advanced (and more expensive) machines. There is little if any need for such keys in everyday life. Buy a machine with these keys only if you expect to use them.

Few calculators do fractions directly. However, problems involving fractions can be worked on a calculator by converting each fraction to a decimal. For example, to add 3/4 and 5/6,

$$\frac{3}{4} + \frac{5}{6},$$

convert 3/4 to a decimal:

$$\boxed{3}\ \boxed{\div}\ \boxed{4}\ \boxed{=}\quad .75$$

Then convert 5/6 to a decimal:

$$\boxed{5}\ \boxed{\div}\ \boxed{6}\ \boxed{=}\quad .833 \text{ (rounded to the nearest thousandth)}$$

Add the two decimals:

$$\boxed{.}\ \boxed{7}\ \boxed{5}\ \boxed{+}\ \boxed{.}\ \boxed{8}\ \boxed{3}\ \boxed{3}\ \boxed{=}\quad 1.583$$

This work can be done faster with the following rule:

To add a fraction to what is already in the calculator, push the following keys:

$\boxed{\times}$ Denominator $\boxed{+}$ Numerator $\boxed{\div}$ Denominator

By this procedure, no intermediate steps need be written down or stored in memory. Thus,

$$\frac{3}{4} + \frac{5}{6}$$

is worked with this rule as follows:

$$\boxed{3}\ \boxed{\div}\ \boxed{4}\ \boxed{\times}\ \boxed{6}\ \boxed{+}\ \boxed{5}\ \boxed{\div}\ \boxed{6}\ \boxed{=}\quad 1.583 \text{ (rounded)}$$

To subtract, push $\boxed{-}$ instead of $\boxed{+}$. For example,

$$\frac{7}{8} - \frac{9}{17}$$

becomes

$$\boxed{7}\ \boxed{\div}\ \boxed{8}\ \boxed{\times}\ \boxed{1}\ \boxed{7}\ \boxed{-}\ \boxed{9}\ \boxed{\div}\ \boxed{1}\ \boxed{7}\ \boxed{=}\quad .346 \text{ (rounded)}$$

Note: On some calculators, you must push $\boxed{=}$ after each step.

The Centennial Problem:
Arrange the ten digits and four dots (see the blackboard) to form numbers that add up to 100. You may use the dots as decimal points or instead of bars above numbers to indicate repeating decimals.

Sam Loyd (1841–1911) invented this puzzle in 1876, as well as many other brainteasers and games. His interest in chess led him to mathematical problem solving, and his name became a household word in the United States. The Brooklyn Daily Eagle and Woman's Home Companion brought such pastimes to an avid public. (This was long before TV.) Martin Gardner edited a 2-volume collection, Mathematical Puzzles of Sam Loyd *(Dover, 1959).*

5.3 EXERCISES

Work each of the following problems on a calculator. Round all answers to the nearest hundredth.

1. 28.96
   34.25
   19.78
   +21.59

2. 358.42
   76.98
   +217.91

3. 8974.2
   15892
   +   38.55

4. 25198.3
   7168.9
   +215897.38

5. 251.9876
   347.489
   +1258.3

6. 589.41
   376.974
   +258.441

7. 21.5
   −13.82

8. 769.2
   − 35.75

9. $12 - 10.798$

10. $44 - 36.405$

11. 89.7
    × .63

12. 47.82
    × .13

13. $\dfrac{9225}{25}$

14. $\dfrac{4594.2}{78}$

15. $\dfrac{3676.5}{142.5}$

16. $\dfrac{9963.52}{389.2}$

17. $\dfrac{17}{9}$

18. $\dfrac{36}{42}$

The **reciprocal** of a number n is $1/n$. Use a calculator to find the reciprocal (in decimal form) of each of the following. Round to the nearest thousandth, if necessary.

19. 2   20. 5   21. 3   22. 7   23. 11   24. 15

Tricentennial solution *Union City, Michigan, is all set up financially for celebrating the Tricentennial year, thanks to Eli Hooker, a leading citizen. He chaired the Bicentennial committee and wanted to prevent the 2076 committee from running out of money as his committee did. He raised money ($25 from each of 42 local citizens) and deposited it in a Union City bank at 7% compound interest. Yield: $1,000,000 in 2076!*

(Similar matters of interest are coming up in Section 5.4.)

Work each of the following fraction problems. Round all answers to the nearest thousandth.

25. $\dfrac{1}{2}+\dfrac{1}{3}$ **28.** $\dfrac{5}{9}+\dfrac{7}{12}$ **31.** $\dfrac{7}{5}+\dfrac{19}{3}$ **34.** $\dfrac{9}{17}-\dfrac{3}{8}$

26. $\dfrac{1}{8}+\dfrac{1}{9}$ **29.** $\dfrac{9}{4}+\dfrac{8}{3}$ **32.** $\dfrac{8}{7}+\dfrac{5}{4}$ **35.** $\dfrac{7}{3}-\dfrac{5}{9}$

27. $\dfrac{3}{5}+\dfrac{2}{3}$ **30.** $\dfrac{5}{2}+\dfrac{7}{5}$ **33.** $\dfrac{2}{3}-\dfrac{1}{7}$ **36.** $\dfrac{2}{5}-\dfrac{4}{37}$

Some problems involve both addition and multiplication. In these problems, we must be careful with the order of operations. A common example of these mixed calculations comes in figuring federal income tax. Some of the tax rates for married couples filing a joint return are as follows.

If taxable income is over	but not over	the tax is	of excess over
$4000	$8000	$620 + 19%	$4000
$8000	$12,000	$1380 + 22%	$8000
$12,000	$16,000	$2260 + 25%	$12,000
$16,000	$20,000	$3260 + 28%	$16,000
$20,000	$24,000	$4380 + 32%	$20,000

"Taxable income" is the amount left over after all deductions and exemptions have been subtracted from total income.

Example: Suppose your taxable income is $14,896. Your tax would then be figured as follows.

Step 1 $14,896 is between $12,000 and $16,000.
Step 2 The tax is $2260 plus 25% of the excess over $12,000.
Step 3 The excess over $12,000 is $14,896 − $12,000 = $2896.
Step 4 25% of the excess is .25 × 2896 = $724.
Step 5 The total tax is $2260 + $724 = $2984.

On a calculator:

| 1 | 4 | 8 | 9 | 6 | − | 1 | 2 | 0 | 0 | 0 | = |

| × | · | 2 | 5 | + | 2 | 2 | 6 | 0 | = | 2984

Find the income tax for each of the following taxable incomes.

37. $5240   **39.** $9712   **41.** $13,920   **43.** $19,846

38. $7643   **40.** $10,845   **42.** $17,825   **44.** $22,490

To find the (arithmetic) *mean* of a collection of numbers, add the numbers and divide by the number of numbers. Find the mean for each of the following collections of numbers. Round to the nearest thousandth, if necessary.

45.   396, 512, 243, 159, 728

46.   2580, 3290, 7258, 976, 12,512

47.   3.72, 5.91, 2.86, 3.41, 5.90

48.   2.1435, 3.7298, 42.147, 38.710

49.   .065, .0423, .0173, .02981

50.   .0042, .00398, .001, .00213, .00276

Calculators can also be used for games. For example, one game is played in the following way. Player 1 chooses any positive integer (such as 97 or 386) and enters it into the keyboard. Player 2 then pushes the subtraction key, and then any digit key except 0. The game continues with each player in turn subtracting any digit except 0. A player loses if a negative number appears on his or her turn. There is one condition: on each subtraction after the first, a player must choose a key adjacent (sides or corners) to the key just pushed.

*51.   Suppose a player has just pushed "5". What keys are then available for the next player to push?

*52.   Which keys are available if the last player has pushed "2"?

*53.   Which keys are available if the last player has pushed "6"?

*54.   There is a strategy by which the second player can always win. Look it up on page 126 of the July 1976 issue of *Scientific American* magazine.

5.4 CONSUMER APPLICATIONS

Many of the consumer applications of mathematics involve percents, because many such applications involve the calculation of *interest,* which is usually expressed as a percent per year or per month.

    Interest is a charge for borrowing money. Clay tablets from Babylonia (2000 years before Christ) describe interest (see the note at the side). Interest rates in Babylonia ran to 33% per year. In Rome, interest could go to 48% per year, with 60% rates in twelfth-century India.

    Today, the interest rate charged the most creditworthy large borrowers is called the **prime rate.** This is the rate that appears from time to time in the news as it fluctuates up and down. Consumers can sometimes borrow money for about 9% per year, but a range of 15–20% is much more common. In this section, we look at various ways of calculating consumer interest. We then look at ways to find the dealer's price on a new car and the house payment necessary to pay off a mortgage.

King Hammurabi *tried to hold interest rates at 20 percent for both silver and gold, but moneylenders ignored his decrees.*

Heating a house is another cost that may get you involved with banks and interest rates after you finally get a roof over your head. The roofs you see above and opposite do more than keep the rain off. They hold glass tube solar collectors, part of the solar heating systems in the houses. The photos show two of four townhouses in Vernon Hills, Illinois, set up with such solar systems in a 2-year experimental program to see whether solar heating is feasible in and around Chicago, where winter sunshine is not as frequent as in the Southwest, say.

Investigations into solar energy have increased since 1977, when President Carter projected that 2.5 million homes would have solar units by 1985.

Add-on interest Many companies that sell on the installment plan calculate interest by the *add-on method*. The add-on method is especially common with car dealers and furniture stores.

Suppose you buy a car for $4200. If the down payment (either cash or a trade-in) is $700, then the amount that you owe is $4200 − $700 = $3500. Suppose the dealer says that the add-on interest rate is 7%, and you expect to pay for the car over 36 months (3 years). The total amount of interest is calculated by the formula for simple interest, interest = principal × rate × time, or

$$\text{Interest} = \text{principal} \cdot \text{rate} \cdot \text{time}$$
$$= 3500 \cdot (.07) \cdot (3)$$
$$\text{Interest} = \$735.$$

This interest is then added to the amount owed, $3500, to find the total amount that must be repaid.

$$\$3500 + \$735 = \$4235.$$

The monthly payment is found by dividing the total amount to be repaid, $4235, by the number of payments, 36.

$$\text{Monthly payment} = \frac{4235}{36} = \$118$$

(rounded to the nearest dollar). We are paying this loan back in 36 monthly installments. Because of this, we have not had use of the full loan of $3500 for three years. (In the exercises below, we see that a 7% add-on rate is actually greater than a 7% true annual interest rate.)

Example 1 A family buys $2500 worth of furniture, with no down payment. They agree to pay off the furniture in 24 months at a 10% add-on interest rate. Find the amount of interest they must pay, and the monthly payment.

Solution. The interest is found as follows.

$$\text{Interest} = \text{principal} \cdot \text{rate} \cdot \text{time}$$
$$= 2500 \cdot (.10) \cdot (2)$$
$$\text{Interest} = 500.$$

Thus, the interest is $500. The total amount that must be repaid is $2500 + $500 = $3000. This amount will be repaid in 24 months, so that the monthly payment is

$$\text{Monthly payment} = \frac{3000}{24} = \$125.$$

Compound interest When you deposit money in a bank, the bank pays you interest on the money. After a period of time, the bank pays you interest again. This time the bank pays you interest on the money originally

deposited, and also on the interest paid earlier. This type of interest is called *compound interest*. There are formulas for compound interest, but it is more common to use special compound interest tables, such as the table in the Appendix. The use of this table is explained in the next example.

Example 2 Find the total amount on deposit after 21 years if $3500 is deposited in an account paying 5% interest compounded annually.

Solution. Look in the table. Find the column with 5% at the top and the row with 21 along the side. This row and column lead to the number 2.78596. To find the total amount on deposit, multiply this number by the amount deposited, or $3500. This gives

$$\text{Amount on deposit} = \text{principal} \cdot \text{number from table}$$
$$= (3500) \cdot (2.78596)$$
$$\text{Amount on deposit} = 9750.85.$$

Thus the amount on deposit is $9750.85. (How much of this amount represents interest?)

Buying a house There are three things that most people need to know about the cost of a home — the purchase price, the down payment, and the monthly payment. The purchase price depends on many factors. The down payment is determined by general conditions in the money market. If money is plentiful, and you have good credit, you may need to make only a 5% down payment. You will then have to pay a higher interest rate and a higher monthly payment. Normally, the down payment is around 20% of the purchase price.

The monthly payment depends on the amount financed (the balance that you owe), the term (or length) and interest rate of your loan (the mortgage), and any property taxes or insurance that you must pay monthly. The amount that you must pay for principal and interest is found from a table, such as the one in the Appendix.

Example 3 Suppose you buy a house for $40,000 with a 30-year mortgage at 8½% interest. The down payment is 20%. Find the monthly payment for principal and interest.

Solution. The down payment is 20% of the purchase price of $40,000, or

$$.20 \cdot 40,000 = \$8000.$$

This down payment leaves a balance of $40,000 − $8000 = $32,000 to be financed. From the table, the payment for $32,000 is found from the 8½% column of the table, under "30 years." The payment is $246.06. This amount pays the principal and interest on the loan. You may also need to pay a sum for property taxes and insurance, as explained in the next example.

Before *The supporting structures are made of recycled materials in the Garbage House Project at Rensselaer Polytechnic Institute, Troy, New York. Note the wall module made of ten tin cans in a triangular arrangement.*

Example 4 The house in Example 3 requires a monthly payment of $246.06 for principal and interest. In addition, the lender requires the payment of property taxes and fire insurance monthly. Annual taxes are $840 and the annual fire insurance premium is $240. Find the total monthly payment for principal, interest, taxes, and insurance.

Solution. Taxes amount to $840 per year; per month, this is $840 ÷ 12, or $70. Fire insurance is $240 ÷ 12 = $20 per month. The total monthly payment is thus

$$\$246.06 + \$70 + \$20 = \$336.06.$$

The $246.06 part of the total will not change, but the taxes and insurance premiums may well go up every year.

Buying a car One of the difficulties in buying a new car is in deciding on a fair price to pay for the car. The following table gives a rough estimate that may help you decide.

Type of car	Ford, Chrysler, General Motors	American Motors
Subcompact	.87	.88
Compact	.85	.86
Intermediate	.81	.82
Full size	.77	——

This table is used to estimate the dealer's price for the car. To figure this estimate, find the sticker price for the car, and then subtract the freight charge. Next, multiply the result by the proper number from the table, and then add the freight back on. The result is the *approximate* price paid by the dealer for the car. Dealers sometimes pay less than this figure for the car, but seldom pay more.

You must allow the dealer a certain overhead (operating expenses) and profit. In our experience, dealers are usually happy to sell a car for about 10% over their cost. A smart buyer, willing to spend days or weeks haggling with several dealers might get as low a price as 5% over cost. However, on cars that are very popular and in short supply, you may have to pay the full sticker price.

Example 5 Find the dealer's cost on a Ford compact. The sticker price is $4250, which includes a freight charge of $130.

Solution. First, subtract the freight charge.

$$\$4250 - \$130 = \$4120.$$

Multiply this result by the correct number from the table, namely, .85:

$$.85 \cdot 4120 = \$3502.$$

Add the freight back on: $3502 + 130 = $3632.

The dealer probably paid around $3632 for the car.

After *The tin cans embedded in cement give support, insulation, and also an unusual design. So maybe instead of buying a house, you can build one (and borrow money from a bank at a given rate of interest).*

True annual interest   Different lenders compute interest in different ways. This makes it difficult for the consumer to compare the cost of borrowing money. For example, how does $1\frac{1}{2}$% per month at Sears compare with 8% add-on at a furniture store? To help consumers with the problem of knowing and comparing true interest rates, Congress passed the Truth in Lending Act in 1969. This law requires that all sellers must reveal the *true annual interest rate* they charge, as well as the total finance charge. The consumer is then in a better position to compare. The formulas for calculating true annual interest rates are very involved. The easiest way to find these rates is to use the table at the bottom of the page.

To find the true annual interest rate, first use the formula

$$\frac{\text{amount of interest}}{\text{amount financed}} \times 100.$$

Then look up this number in the row of the table that corresponds to the correct number of payments.

Example 6   You use a charge account to buy $1600 worth of furniture. Your monthly payment will be $77.20 for 24 months. The total amount that you must repay is

$$\$77.20 \cdot 24 = \$1852.80.$$

Subtracting the cost of the furniture (the amount financed) from the total amount to be repaid gives the interest charge:

$$\$1852.80 - \$1600 = \$252.80.$$

Now use the formula from above.

$$\frac{\text{amount of interest}}{\text{amount financed}} \times 100 = \frac{\$252.80}{\$1600} \cdot 100 = .1580 \cdot 100 = \$15.80.$$

Look at the table for the row "24 payments"; then read across the table until you find the number closest to the amount of $15.80. It turns out that $15.80 is in the table. Read up the column to find the true annual interest rate of $14\frac{1}{2}$%.

True annual interest rate

Number of payments	14%	$14\frac{1}{2}$%	15%	$15\frac{1}{2}$%	16%	$16\frac{1}{2}$%	17%
	(Finance charge per $100 of amount financed)						
6	$ 4.12	4.27	4.42	4.57	4.72	4.87	5.02
12	7.74	8.03	8.31	8.59	8.88	9.16	9.45
18	11.45	11.87	12.29	12.72	13.14	13.57	13.99
24	15.23	15.80	16.37	16.94	17.51	18.09	18.66
30	19.10	19.81	20.54	21.26	21.99	22.72	23.45
36	23.04	23.92	24.80	25.68	26.57	27.46	28.35
42	27.06	28.10	29.15	30.19	31.25	32.31	33.37
48	31.17	32.37	33.59	34.81	36.03	37.27	38.50

5.4 EXERCISES

Find the total interest and the monthly payment for each of the following installment loans. Interest is calculated by the add-on method. Round to the nearest dollar.

	Amount purchased	Down payment	Interest rate	Number of payments
1.	$3800	$300	7%	36
2.	$4500	$500	8%	36
3.	$2800	0	10%	24
4.	$4750	0	9%	36
5.	$8720	$1000	10%	48
6.	$1540	$320	8%	24

John wants to buy a new Datsun "Z." The purchase price is $9700; John gets a $1400 trade-in on his old Ford. The dealer charges 10% add-on interest.

7. Find the interest charge if John pays off the car in 36 payments.
8. What would the monthly payment be? (Round to the nearest dollar.)
9. Find the interest charge if John takes 48 months to pay off the car.
10. What would the monthly payment be? (Round to the nearest dollar.)

Department stores, such as Sears, Wards, and Penneys, and bank credit cards often charge $1\frac{1}{2}$% per month interest on any unpaid balance. In Exercises 11–16 find the interest charge on the unpaid balances. (*Hint:* $1\frac{1}{2}$% = .015.) Round to the nearest cent.

11.	$500	13.	$287	15.	$482.51
12.	$300	14.	$96	16.	$397.29

Part of a typical statement from a Mastercharge bank is shown on the opposite page. Use this bill to answer the questions in Exercises 17–22.

17. What was the previous balance owed?
18. What payment was made during the month?
19. How much was charged to the account during the month?
20. Check that the interest charge is correct.
21. Find the new balance. This is the amount that must be paid to avoid further interest charges.
22. What is the credit limit assigned to this customer? How much credit is left? (Banks thoughtfully provide the "remaining credit" amount so that the customer will think he or she has some sort of "free money" lying around.)

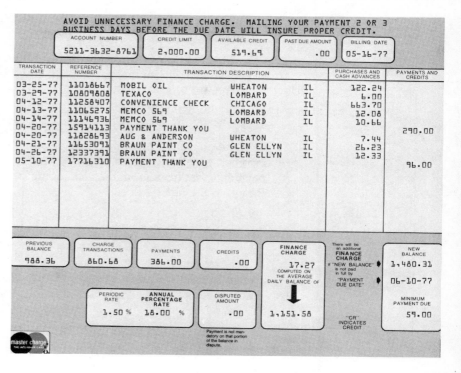

Find the monthly payment for principal and interest for each of the following 30-year home loans.

23. $40,000 at $8\frac{1}{2}\%$ **24.** $30,000 at 9%

25. $28,000 at 9% (*Hint:* Add the payments for $25,000 and $3000.)

26. $43,000 at $8\frac{1}{2}\%$

In Example 3 on page 183, we decided that a monthly payment of $246.06 would be necessary for principal and interest to pay off the loan of $32,000 in 30 years.

27. Find the total amount of money that will be repaid.
(*Hint:* 30 years contain how many months?)

28. Which costs more, the interest or the house?

For Exercises 29–32, find the monthly payment including taxes and insurance. Assume a 30-year loan.

	Amount of loan	Interest rate	Annual taxes	Annual insurance
29.	$30,000	$8\frac{1}{2}\%$	$600	$180
30.	$40,000	9%	$1080	$252
31.	$43,000	9%	$780	$155
32.	$39,000	$8\frac{1}{2}\%$	$580	$130

Many lenders use the following method to decide on how much a person can afford to pay for a house payment.

(1) Find your gross monthly income, before any deductions.
(2) Subtract all monthly payments that will not be paid off within 6 months.
(3) Divide the result by 4. Your monthly house payment should not exceed this amount.

Find the maximum monthly house payment for each of the following incomes. Monthly gross income is given first, and current monthly payments that have more than six months to go are given second. Round to the nearest dollar.

33. $975; $30   **35.** $1562; $174

34. $1200; $79   **36.** $1836; $275

In one state, the seller of a house must pay the following fees: a 6% commission for the salesperson, a 1% title insurance charge, and $398 of miscellaneous charges. Find the net amount received by the sellers of the following houses after these deductions.

37. selling price $42,000   **39.** selling price $75,200

38. selling price $34,800   **40.** selling price $52,000

Solar powered vehicles *One of two electric cars used during the Bicentennial celebration in Washington, D.C. This car is fitted with a vacuum cleaner and used for refuse pickup. The other car transported workers. The car batteries are charged by photovoltaic cells, converting sunlight directly into electrical energy. Such cells were first used to power space satellites.*

Universal Product Code (UPC)
The ten-digit number above is what the bars and spaces stand for—the first five digits code the manufacturer; the other five code the given product. The bars are "read" by an optical scanner, part of a supermarket checkout system that may be time-saving for consumers. The checker is exposing the UPC symbol to the dark window (the scanner), which reads it electronically and relays it to a minicomputer that decodes the symbol and retrieves the price. Consumers see the information on a display unit and get a detailed printout on register tape.

Find the dealer cost for each of the following new cars. Sticker price is given first, with the freight charge second. Round to the nearest dollar.

41. General Motors intermediate; $6450; $150

42. Ford full size; $7283; $174

43. Chrysler subcompact; $4673; $128

44. American Motors compact; $5063; $158

45. Ford compact; $5893; $205

46. General Motors full size; $6978; $194

Find the true annual interest rate for each of the following. Use the table on page 185.

	Amount financed	Interest charge	Number of payments
*47.	$1200	$96	12
*48.	$4800	$576	18
*49.	$5000	$850	24
*50.	$4900	$975	30

*51. Cammie bought a television set for $222. He paid nothing down, and paid for the set in 12 monthly payments of $20 each. Find the interest that he paid. What was his true annual interest rate?

*52. Wei-Jen Luan bought a compact car for $3750. She paid $1000 down, and then made payments of $96 per month for 36 months. Find the total interest that she paid. Find the true annual interest rate that she paid.

For Exercises 53–57, find the final amount in a savings account if interest is compounded annually.

53. $1000 at 5% for 11 years

54. $10,000 at 4% for 8 years

55. $2500 at 6% for 15 years

56. $5500 at 5% for 6 years

57. $8750 at 4% for 24 years

Find the amount of *interest* earned by each deposit. Assume interest is compounded annually. (Subtract the original deposit from the final amount.)

58. $1500 at 5% for 7 years

59. $2000 at 6% for 9 years

60. $3700 at 4% for 20 years

61. $5400 at 5% for 18 years

62. $12,500 at 6% for 21 years

Joseph Louis Lagrange
(1736 – 1813) was considered in his day to be the greatest living mathematician. He was born in Turin, Italy, and became a professor at age 19. In 1776 he came to Berlin at the request of Frederick the Great to take the position Euler left. A decade later Lagrange settled permanently in Paris. Napoleon was among many who admired and honored him.

Lagrange's greatest work was in the theory and application of calculus. He carried forward Euler's work of putting calculus on firm algebraic ground in his theory of functions. Lagrange's notation for the derivative (the function f') is in use today. His Analytic Mechanics (1788) applied calculus to the motion of objects. Lagrange's contributions to algebra had great influence on Galois and hence the theory of groups (see Section 6.6). He also wrote on number theory, and proved, for example, that every integer is the sum of four squares (or less than four). His study of the Moon led to methods for finding longitude.

5.5 THE METRIC SYSTEM OF MEASUREMENT

The metric system of weights and measures is based on decimals and powers of ten. Because of this, calculations and changes of units are much easier than in the English system that we now use. For example, in our system a foot is used to measure short distances, while a mile is used for long distances. The measures "foot" and "mile" are not conveniently related to one another. In the metric system, short distances are measured with a meter. For a longer distance, we attach the prefix "kilo" and get one thousand meters, or one kilometer. For a shorter distance, we attach the prefix "centi," and get one centimeter, or one-hundredth of a meter. The prefixes of the metric system are shown in Figure 5.3.

The metric system was developed by a committee of the French Academy of Sciences just after the French Revolution of 1789. The president of the committee was the mathematician Joseph Louis Lagrange. He urged that some natural measure be found for length, with weight and volume measures found from the measure of length. It was decided that one **meter** would be the basic unit of length, with a meter defined as one ten-millionth of the distance from the equator to the North Pole. (A slight mistake was made in calculating this distance, so that today's meter is a little shorter than it should be.)

To obtain measures longer than a meter, Greek prefixes were added. For measures smaller than a meter, Latin prefixes were used. Thus, a centimeter is one-hundredth of a meter, and a millimeter is one-thousandth of a meter.

The basic unit of weight is the **gram** (a common paperclip weighs about one gram.) A gram is the weight of water that fits into a cube one centimeter on a side. Volume is measured by **liters,** the space enclosed by a cube one-tenth of a meter on a side.

The metric system was unpopular at first. Finally, in 1843 the French government levied stiff fines for use of the old systems. This hastened the acceptance of the metric system. Germany adopted the metric system when it became a nation in 1871. More and more countries switched over, until today, only the United States and four small nations do not use the metric system.

In the United States, many industries are going to the metric system. The changeover is being led by export industries, such as the aircraft and computer industries, but the rest of industry is not far behind. The metric system is being taught extensively in elementary schools today, and in perhaps twenty years should take over from our current system to a large degree.

As we said, the basic unit of length in the metric system is the *meter*. A meter is a little longer than a yard. Shorter lengths are measured in **centimeters** and **millimeters**. The prefix "centi" means hundredth, so that one centimeter is one-hundredth of a meter. Thus,

$$100 \text{ centimeters} = 1 \text{ meter}.$$

Figure 5.3
A list of prefixes in the metric system. The most common prefixes are in **bold** type.

Prefix	Multiple	Prefix	Multiple
exa	1,000,000,000,000,000,000	deci	0.1
peta	1,000,000,000,000,000	**centi**	0.01
tera	1,000,000,000,000	**milli**	0.001
giga	1,000,000,000	micro	0.000001
mega	1,000,000	nano	0.000000001
kilo	1,000	pico	0.000000000001
hecto	100	femto	0.000000000000001
deka	10	atto	0.000000000000000001

The prefix "milli" means thousandth, so one millimeter means one-thousandth of a meter, and

$$1000 \text{ millimeters} = 1 \text{ meter.}$$

The word "meter" is abbreviated m; "centimeter" is cm; and "millimeter" is mm. We can convert from centimeters to millimeters to meters by multiplying or dividing, as the following example shows.

Example 1   Convert each of the following measurements.

(a) 6.4 m to cm

Solution.   A centimeter is only 1/100 of a meter. Thus, we need 100 centimeters to make one meter. For this reason, *multiply* by 100:

$$6.4 \text{ m} = (6.4)(100) = 640 \text{ cm.}$$

(b) .98 m to mm

Solution.   Multiply by 1000:

$$.98 \text{ m} = .98(1000) = 980 \text{ mm.}$$

(c) 34 cm to m

Solution.   A meter is a larger unit of measure, so we need fewer than 34 of them to equal the same length as 34 cm. Thus, *divide* by 100:

$$34 \text{ cm} = \frac{34}{100} = .34 \text{ m.}$$

Long distances are measured in **kilometers.** The prefix "kilo" means one thousand, so that one kilometer (km) means one thousand meters, or

$$1 \text{ kilometer} = 1000 \text{ meters.}$$

Since 1 meter is about a yard, 1000 meters is about 1000 yards, or 3000 feet. Therefore, 1 km is about 3000 feet. One mile is 5280 feet, so that 1 km is about 3000/5280 of a mile. By dividing 5280 into 3000, we find that a kilometer is about .6 mile.

Thomas Jefferson *was a strong advocate of the decimal system as the basis for measurement. He proposed a system of weights and measures, for example: "Let the foot be divided into 10 inches; the inch into 10 lines; the line into 10 points." His proposal was not accepted by Congress, although it adopted the dollar system he devised.*

Jefferson was influential in encouraging American students to study the works of European mathematicians, and in allowing mathematics more importance in the curriculum at the University of Virginia than it had at other U.S. colleges.

The basic unit of volume in the metric system is the **liter** (abbreviated l). A liter is a little more than a quart. We can again use the prefixes "milli" and "centi."

$$1 \text{ liter} = 100 \text{ centiliters}\quad(\text{cl})$$
$$1 \text{ liter} = 1000 \text{ milliliters}\quad(\text{ml})$$

Milliliters and centiliters are such small volumes that they find their main uses in science. In particular, drug dosages are often expressed in milliliters.

Weight is measured in **grams** (g). A nickel weighs almost 5 grams. Milligrams (one-thousandth of a gram) and centigrams (one-hundredth of a gram) are so small that they, too, are used mainly in science. A more common measure is the **kilogram** (kg), or 1000 grams. A kilogram weighs about 2.2 pounds. Kilogram is often shortened to **kilo.** Thus

$$1000 \text{ g} = 1 \text{ kg}.$$

Example 2   Convert each of the following measures.

(a) 650 g to kg

Solution.   A gram is a small unit, and a kg is a large unit. This, we *divide* by 1000:

$$650 \text{ g} = \frac{650}{1000} = .65 \text{ kg}.$$

(b) 9.4 l to cl

Solution.   Multiply by 100:
$$9.4 \text{ l} = 9.4(100) = 940 \text{ cl}.$$

(c) 4350 mg to g

Solution.   Divide by 1000:

$$4350 \text{ mg} = \frac{4350}{1000} = 4.35 \text{ g}.$$

Eventually, everyone will think in the metric system as easily as we now think in the *English system* of feet, quarts, pounds, and so on. During the period of changeover from the English system, it will be necessary for many people to convert from one system to the other. *Approximate* conversions can be made with the aid of the tables on the next page.

Example 3   Convert each of the following measurements as indicated.

(a) 15 m to yards

Solution.   Look in the "Metric to English" table for "meters–yards." Read across and find the number 1.09. Multiply 15 times 1.09.

$$15 \text{ m} = 15(1.09) = 16.35 \text{ yards}.$$

METRIC to ENGLISH			ENGLISH to METRIC		
FROM	*TO*	Multiply by	FROM	*TO*	Multiply by
meters	yards	1.09	yards	meters	.914
meters	feet	3.281	feet	meters	.3048
meters	inches	39.37	inches	meters	.0254
kilometers	miles	.62	miles	kilometers	1.609
grams	pounds	.0022	pounds	grams	454
kilograms	pounds	2.20	pounds	kilograms	.454
liter	quart	1.06	quart	liter	.946
liter	gallon	.264	gallon	liter	3.785

(b) 39 yards to meters

Solution. Read the yards–meters row of the "English to Metric" table. The number .914 appears. Multiply 39 times .914.

$$39 \text{ yards} = 39(.914) = 35.646 \text{ m}.$$

(c) 47 m = 47(39.37) = 1850.39 inches.

(d) 87 km = 87(.62) = 53.94 miles.

(e) 598 miles = 598(1.609) = 962.182 km.

(f) 12 quarts = 12(.946) = 11.352 liters.

5.5 EXERCISES

In Exercises 1–20, estimate the answer directly in the metric system. Do not convert from the current system.

1. Find your height in centimeters.

2. Find your height in meters.

3. Find the length of your longest finger in centimeters.

4. What are the dimensions of the cover of this book in millimeters?

5. Give your levi size, waist and length in centimeters.

6. What is your weight in kilograms?

7. What is the distance in kilometers between the two largest cities in your state?

8. How many liters in a six-pack of 12-ounce cans of beer?

9. One nickel weighs 5 g. How many nickels weigh a total of 1 kg?

10. Sea water contains about 3.5 grams of salt per 1000 milliliters of water. How many grams of salt would five liters of sea water contain?

11. Helium weighs about .0002 grams per milliliter. A balloon contains three liters of helium. How much does the helium weigh?

12. About 1500 grams of sugar can be dissolved in a liter of warm water. How much sugar can be dissolved in one milliliter of warm water?

13. Mihaly Mezaros is $32\frac{5}{8}$ inches tall. Find his height in centimeters.

14. Tom Thumb weighed 52 pounds. What was his weight in kilograms?

15. It is 108 miles from Phoenix to Tucson. Find this distance in kilometers.

16. A dollar bill is approximately 2.5 by 6.2 inches. Convert these dimensions to centimeters.

17. Tom Thumb was $26\frac{1}{2}$ inches tall. How many centimeters is this?

18. William J. Cobb weighs 802 pounds. Find his weight in kilograms.

19. How big is a 23-inch television screen in centimeters?

20. The photograph shows a liter bottle of 7-up. Is this more or less than a quart?

Temperature in the metric system is measured in **degrees Celsius.** In the Celsius scale, water freezes at 0° and boils at 100°. This is much more sensible than the Fahrenheit scale that we use now, where a mixture of salt and water freezes at 0°, and 100° represents the temperature inside Gabriel Fahrenheit's mouth. To convert a reading from Fahrenheit to Celsius, use the formula

$$C = \frac{5(F - 32)}{9}.$$

Example: Convert 68°F to Celsius.

Solution.   Substitute 68 for F in the formula above.

$$C = \frac{5(68 - 32)}{9}$$

$$= \frac{5(36)}{9}$$

$$= \frac{180}{9}$$

$$C = 20.$$

Thus, 68°F is the same as 20°C.

Convert each Fahrenheit temperature to Celsius. Round to the nearest degree.

21.	104°F	23.	536°F	25.	98°F	*27.	−40°F
22.	86°F	24.	464°F	26.	114°F	*28.	−112°F

To convert from Celsius to Fahrenheit, use the formula

$$F = \frac{9}{5}C + 32.$$

Convert each Celsius temperature to Fahrenheit.

29. 35°C   **31.** 10°C   *__33.__ −15°C

30. 100°C   **32.** 25°C   *__34.__ −40°C

In Exercises 35–40, round to the nearest degree.

35. The highest temperature ever recorded in Death Valley is 134°F. Convert this measurement to Celsius.

36. The coldest temperature ever recorded in New York state is −52°F. Convert this to Celsius.

37. To help in the changeover to the metric system, some television stations now give the daily temperatures in both Fahrenheit and Celsius. On the photograph of a television set shown here, the Fahrenheit temperature is not shown. Use the Celsius temperature shown to find the missing measurement.

38. A recipe calls for a temperature of 200°C. Convert this to Fahrenheit.

39. What is the approximate highest temperature reached in your town yesterday, in degrees Celsius?

40. What is the coldest temperature your town typically records on the coldest winter night, in degrees Celsius?

The metric system is used in medicine almost exclusively. In Exercises 41–44, a doctor's prescription is listed. Some of these are reasonable and some are wrong. Decide which you think must not be correct.

Benjamin Bannaker *(or Banneker, 1731–1806) spent the first half of his life tending a farm in Maryland. He gained a reputation locally for his mechanical skills and abilities in mathematical problem-solving. In 1772 he acquired astronomy books from a concerned neighbor, and devoted himself to learning astronomy, observing the skies, and making calculations. In 1789 Bannaker joined the team that surveyed what is now the District of Columbia.*

Bannaker published almanacs yearly from 1792 to 1802. (The above likeness is from the cover of one of them.) His almanacs contained the usual astronomical data and information about the weather and seasonal planting. He also wrote social commentary and made proposals for a peace office to be in the President's Cabinet, for a department of the interior, and for a league of nations. He sent a copy of his first almanac to Thomas Jefferson along with an impassioned letter against slavery. Jefferson had much praise for Bannaker, and sent the almanac to the French Academy to show what an American could achieve.

*41. 1940 grams of Kaopectate after each meal

*42. 76.8 centiliters of cough syrup every two hours

*43. 94.3 milliliters of antibiotic every six hours

*44. 1.3 kilograms of vitamins every three hours

Convert each measurement as indicated.

45. 68 cm to m
46. 934 mm to m
47. 4.7 m to mm
48. 7.43 m to cm
49. 8.9 kg to g
50. 4.32 kg to g

51. 39 cl to l
52. 469 cl to l
53. 46,000 g to kg
54. 35,800 g to kg
55. .976 kg to g
56. .137 kg to g

Convert each measurement as indicated. Round to the nearest tenth.

57. 36 m to yards
58. 76.2 m to yards
59. 55 yards to m
60. 893 yards to m
61. 4.7 m to feet
62. 1.92 m to feet
63. 3.6 feet to m
64. 12.8 feet to m
65. 496 km to miles

66. 138 km to miles
67. 768 miles to km
68. 1042 miles to km
69. 683 g to pounds
70. 1792 g to pounds
71. 4.1 pounds to g
72. 12.9 pounds to g
73. 38.9 kg to pounds
74. 40.3 kg to pounds

5.6 IRRATIONAL NUMBERS

Most of the numbers we deal with in daily life are rational. However, a surprising number of non-rational, or **irrational numbers,** come up. The first number that was known to be irrational was $\sqrt{2}$. The Pythagoreans, of 500 BC, discovered that this number could not equal any rational number, and thus was irrational.

We can draw a line segment having length equal to $\sqrt{2}$ units by starting with a square, one unit on a side (such as a floor tile one foot on a side). A diagonal of this square cuts the square into two right triangles. By the Pythagorean theorem, the length of the diagonal (the *hypotenuse* of one of the right triangles) is given by

$$c^2 = 1^2 + 1^2$$
$$= 1 + 1$$
$$c^2 = 2, \qquad \text{or} \qquad c = \sqrt{2}.$$

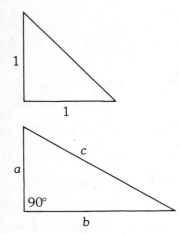

In a right triangle
$a^2 + b^2 = c^2$

Thus, the diagonal of a square one unit on a side is given by an irrational number. The point representing $\sqrt{2}$ can be located on a number line, as shown in Figure 5.4.

There is no general method for finding the decimal values of irrational numbers. Most of the methods in common use require procedures of the calculus. (One exception: square roots can be found on a calculator, as shown in the exercises below.) We can also use a *square root table*, such as the one in the Appendix.

This table gives decimal values of square roots of all integers $1-100$ to the nearest thousandth. Integers are listed in the first column, headed n. Square roots are given to 3 decimal places in the column headed $\sqrt{n}$. For example,

$$\sqrt{2} = 1.414 \qquad \text{and} \qquad \sqrt{5} = 2.236.$$

The square root table gives only approximations of many square roots. For example, 1.732 is listed for $\sqrt{3}$. In fact, $(\sqrt{3})^2 = 3$, while $(1.732)^2 = 2.999824$. However, to the nearest thousandth, 1.732 is the best possible approximation to $\sqrt{3}$. Some pocket calculators give more decimal places of accuracy, for example,

$$\sqrt{3} = 1.7320508,$$

To show that 1.732 is only an approximation of $\sqrt{3}$, we use the symbol $\approx$ (read "is approximately equal to") and write

$$\sqrt{3} \approx 1.732 \qquad \text{or} \qquad \sqrt{3} \approx 1.7320508.$$

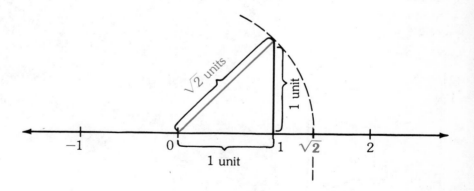

Figure 5.4 To find $\sqrt{2}$ on the number line, first construct a right triangle with each short side equal to 1 unit. The hypotenuse must then be $\sqrt{2}$ units. Draw a circle with the point 0 as center and the hypotenuse of the triangle as radius. The dashed line in the figure indicates a portion of the circle. The circle crosses the number line at the point $\sqrt{2}$. Why? (What is the line segment from 0 to $\sqrt{2}$?)

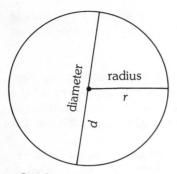

Circle

Circumference $= \pi d$

$\qquad\qquad\qquad = 2\pi r$

Area $= \pi r^2$

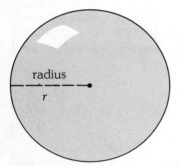

Sphere

Volume $= \frac{4}{3}\pi r^3$

Surface area $= 4\pi r^2$

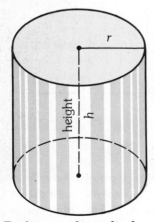

Right circular cylinder

Volume $= \pi r^2 h$

Surface

$\qquad$ area $= 2\pi rh + 2\pi r^2$

One of the single most useful irrational numbers is π, the ratio of the circumference to the diameter of a circle. Some of the most common formulas using π are shown at the side. For some four thousand years, mathematicians have been finding better and better approximation for π. The ancient Egyptians used a method for finding the area of a circle that is equivalent to a value of 3.1605 for π. The Babylonians used numbers that give $3\frac{1}{8}$ for π. In the Bible (I Kings 7:23) we find a verse describing a circular pool at King Solomon's temple, about 1000 BC. The pool is said to be ten cubits across, "and a line of 30 cubits did compass it round about." This implies a value of 3 for π.

Through the centuries mathematicians have calculated π in various ways. For example, in 1579 François Viète (see Chapter 7) used polygons having 393,216 sides to find π to nine decimal places. A computer was programmed by Shanks and Wrench in 1961 to compute 100,265 decimal places of π. Here are the first seven hundred:

PI = 3.+

1415926535	8979323846	2643383279	5028841971	6939937510	5820974944	5923078164
0628620899	8628034825	3421170679	8214808651	3282306647	0938446095	5058223172
5359408128	4811174502	8410270193	8521105559	6446229489	5493038196	4428810975
6659334461	2847564823	3786783165	2712019091	4564856692	3460348610	4543266482
1339360726	0249141273	7245870066	0631558817	4881520920	9628292540	9171536436
7892590360	0113305305	4882046652	1384146951	9415116094	3305727036	5759591953
0921861173	8193261179	3105118548	0744623799	6274956735	1885752724	8912279381
8301194912	9833673362	4406566430	8602139494	6395224737	1907021798	6094370277
0539217176	2931767523	8467481846	7669405132	0005681271	4526358082	7785771342
7577896091	7363717872	1468440901	2249534301	4654958537	1050792279	6892589235
4201995611	2129021960	8640344181	5981362977	4771309960	5187072113	4999999837
2978049951	0597317328	1609631859	5024459455	3469083026	4252230825	3344685035
2619311881	7101000313	7838752886	5875332083	8142061717	7669147303	5982534904

Yes, you can! *Why not measure the value of π yourself? All you need is a can—a large one works best—and a measuring tape. Measure the distance around the can; measure the distance across; then divide:*

$$\pi = \frac{\text{distance around}}{\text{distance across}}.$$

The Golden Ratio *Index cards usually measure 3×5 or 5×8. In both cases the ratio of the sides is close to 1.6. Greeks of antiquity thought rectangles having sides in the ratio 1.6 to 1 were aesthetically most pleasing (see exercises 17–24 in Section 4.7).*

Actually, 1.6 has been used as an approximation of the true value,

$$\frac{1 + \sqrt{5}}{2},$$

called the "Golden Ratio." This number is irrational. Its first few digits are 1.61803.

The Golden Ratio recurs in diverse places throughout mathematics. In the Fibonacci sequence 1, 1, 2, 3, 5, 8, 13, 21, . . . , the ratio of consecutive terms tends toward the Golden Ratio as we go out further in the sequence. Unexpected occurrences of the Golden Ratio include Le Corbusier's Modulor, the Parthenon, bee honeycombs, and sunflower florets.

Modulor *The foreground sculpture is modeled after diagrams of a universal human figure worked out by architect Le Corbusier in 1948. He sought some basis for adjusting architectural scale to human dimensions. Thus he devised the Modulor, a figure divided at the navel (see white dot on the sculpture) so that the two sections are in the ratio 5 to 8. This is roughly the Golden Ratio. (Photograph is by Howard Kaplan.)*

The number e *is a fundamental number in our universe. For this reason,* **e,** *like* π*, is called a universal constant. If there are intelligent beings elsewhere, they too will have to use* **e** *to do higher mathematics.*

The letter e is used to honor Leonhard Euler, who published extensive results on the number in 1748. The numerical value of **e** *is 2.718. . . . Since it is an irrational number, its decimal expansion never repeats.*

One way of defining **e** *involves the idea of an infinite series of terms to be added. Thus* **e** *is equal to the series*

$$1 + \frac{1}{1} + \frac{1}{1 \cdot 2} + \frac{1}{1 \cdot 2 \cdot 3}$$

$$+ \frac{1}{1 \cdot 2 \cdot 3 \cdot 4} + \ldots .$$

We can approximate **e** *as closely as desired by adding more and more terms of the series.*

Logarithms are occasionally taken to the base 10, but much more often **e** *is the base. The properties of* **e** *are used in calculus and higher mathematics extensively. It has been proved that*

$$e^{i\pi} = -1.$$

This formula relates four of the most important constants in mathematics, and has been given a mystical significance by some mathematicians.

5.6 EXERCISES

Use the square root table to find a decimal approximation for each of the following.

1. $\sqrt{13}$ **3.** $\sqrt{31}$ **5.** $\sqrt{78}$

2. $\sqrt{42}$ **4.** $\sqrt{54}$ **6.** $\sqrt{92}$

Approximations for square roots can be found using a pocket calculator, as the following example shows.

Example: Approximate $\sqrt{33}$ to the nearest thousandth.

Step 1   Make a guess. Let us guess 5 here.

Step 2   Divide 33 by 5.

   6.6

Step 3   Find the average of 5 and 6.6.

   5.8

Step 4   Divide 33 by 5.8.

$\boxed{3}\ \boxed{3}\ \boxed{\div}\ \boxed{5}\ \boxed{\cdot}\ \boxed{8}\ \boxed{=}$ 5.6896552

Step 5   Find the average of 5.8 and 5.6896552. You should get 5.7448276.

Step 6   Divide 33 by 5.7448276. You get 5.7442977.

Step 7   Find the average of 5.7448276 and 5.7442977. You get 5.74456265.

Step 8   Divide 33 by 5.74456265. You get 5.744562643.

Since the answer here and the answer in Step 5 are the same to three decimals, we have (to the nearest thousandth)

$$\sqrt{33} = 5.745$$

Use the method of the above example to find each square root to the nearest thousandth:

7. $\sqrt{3}$ **9.** $\sqrt{12}$ **11.** $\sqrt{50}$ **13.** $\sqrt{61}$

8. $\sqrt{8}$ **10.** $\sqrt{15}$ **12.** $\sqrt{70}$ **14.** $\sqrt{85}$

Exercises 15–25 give an algebraic proof that $\sqrt{2}$ is irrational. Give a reason for each numbered step.

There are two possibilities:

(i)   There is a rational number a/b such that $\left(\dfrac{a}{b}\right)^2 = 2.$

(ii)   There is no such rational number.

We work with assumption (i). If it leads to a contradiction, or false statement, then we know (ii) must be correct. We assume there exists a

Tsu Ch'ung-chih (about 500 AD), the Chinese mathematician honored in the above stamp, calculated π as 3.1415929 . . . , which is quite accurate. **Aryabhata,** his Indian contemporary, gave 3.1416 as the value.

rational number a/b such that $\left(\dfrac{a}{b}\right)^2 = 2$. Let us also assume that a/b is in lowest terms.

*15. Since $\left(\dfrac{a}{b}\right)^2 = 2$, we also have $\dfrac{a^2}{b^2} = 2$, or $a^2 = 2b^2$.

*16. $2b^2$ is an even number.

*17. Therefore, a^2 must be an even number, and a must be an even number. (*Hint: can* the square of an odd number be even?)

*18. Since a is an even number, it must be a multiple of 2. That is, we can find a counting number c such that $a = 2c$. Hence, $a^2 = 2b^2$ becomes $(2c)^2 = 2b^2$.

*19. Therefore, $4c^2 = 2b^2$ . . .

*20. or $2c^2 = b^2$.

*21. $2c^2$ is an even number.

*22. Therefore, b^2 is even . . .

*23. and b is even.

*24. We have reached a contradiction. (*Hint:* We have shown that a and b are both even, while a/b is written in lowest terms.)

*25. We must accept assumption (ii) from above: $\sqrt{2}$ is not rational.

Use the formulas given in the text to answer Exercises 26–31. Use 3.14 as an approximation for π. Round to the nearest tenth.

26. The area of a circle having a radius of 12.4 centimeters.

27. The area of a circle having a diameter of 4.9 meters.

28. The volume of a sphere having a radius of 2 meters.

29. The surface area of a sphere having a radius of 6 meters.

30. The surface area of a right circular cylinder having a radius of 14 centimeters and a height of 8 centimeters.

31. The volume of a right circular cylinder having a radius of 4 centimeters and a height of 10 centimeters.

*32. Some mathematics books say "let $\pi = 22/7$," which is a rational number; and yet, π is irrational. Where does 22/7 come from?

*5.7 COMPLEX NUMBERS

When working problems, mathematicians would come up with a solution such as $-2 + \sqrt{-16}$. If negative numbers and square roots were bad enough, what sense could be made from the square root of a negative number? These numbers were called *imaginary* by the early mathematicians, who would not permit these numbers to be used as solutions to problems. Gradually, however, applications were found that required the use of these numbers.

It thus became necessary to enlarge the set of real numbers to form the set of **complex numbers.** By doing this, we have reached an end; the set of

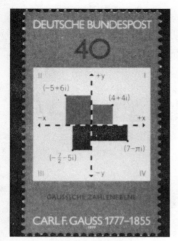

Gauss and the complex numbers *The above stamp honors the many contributions made by Gauss to our understanding of complex numbers. In about 1831 he was able to show that numbers of the form a + bi can be represented as points on the plane (as the stamp diagrams) just as real numbers are. He shares this contribution with Robert Argand, a bookkeeper in Paris, who wrote an essay on the geometry of the complex numbers in 1806. This went unnoticed at the time.*

Gauss also found a practical application of complex number theory, one of the more advanced fields of mathematics. Gauss was making a geodetic survey, and he wondered whether it is possible to make a "conformal" map of a curved surface. ("Conformal" here means that angles on a flat map conform to (or replicate) the actual angles on a portion of the Earth's surface.) Gauss showed that a conformal map can be made for a small portion; having one such map, all other such maps can be obtained by using complex function theory.

complex numbers provides a solution for just about any equation that we can write. Complex numbers depend on the number *i*, defined by

$$i = \sqrt{-1}.$$

The number *i* is not a real number, since there is no real number whose square is -1 (the square of a real number cannot be negative). The number *i*, and any nonzero multiple of *i*, is called an **imaginary number.** Examples of imaginary numbers include

$$i, \quad -4i, \quad \frac{3}{2}i, \quad \sqrt{2}i, \quad -\pi i,$$

and so on. The sum of a real number and any real multiple of *i*, is called a *complex number.* Examples of complex numbers include

$$2 + 5i, \quad -3 - 2i, \quad 4 + \sqrt{2}i,$$
$$0 + 8i \quad (= 8i), \quad 9 + 0i \quad (= 9).$$

As we have seen, a complex number is the sum of a real number and a real multiple of *i*. There is no restriction on the real multiple of *i*, so that it might well be 0*i*. Thus, a real number such as 6 can be written as a complex number by writing $6 = 6 + 0i$. In this way, *every* real number is a complex number. Now we can show the complete relationship between the various sets of numbers, as in Figure 5.5.

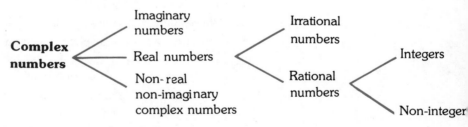

Figure 5.5 Complex number tree

5.7 EXERCISES

Identify each number as *real, imaginary,* or *complex.* (There may well be more than one answer.)

1. $6 + 4i$	**5.** $5i$	**9.** $\sqrt{6}i$
2. $8 - 3i$	**6.** $-12i$	**10.** $-\sqrt{3}i$
3. 12	**7.** 0	
4. $7/4$	**8.** $3 + 0i$	

CHAPTER 5   SUMMARY

Keywords

Real number
Terminating decimal
Repeating decimal
Irrational number
Imaginary number
Complex number
Real axis
Imaginary axis
Vector

Metric system
Meter
Centimeter
Millimeter
Gram
Kilogram (Kilo)
Liter
Celsius
Fahrenheit
English system

Percent
Prime rate
Compound interest
Add-on interest
True annual interest

Floating decimal

CHAPTER 5   TEST

Convert each rational number into a repeating or a terminating decimal.

1. $\dfrac{7}{16}$ **2.** $\dfrac{9}{20}$ **3.** $\dfrac{7}{3}$ **4.** $\dfrac{5}{12}$

Convert each decimal into a rational number.

5. $.72$ **6.** $.\overline{4}$ **7.** $.\overline{58}$ **8.** $.92\overline{3}$

Decide whether each of the following represents a rational or an irrational number.

9. $.8$ **10.** $.9121214121214\ldots$ **11.** $6.435125985415\ldots$

Find the answer to each decimal problem.

12. $4.6 + 9.21 + .87$ **14.** $86(.45)$ **15.** $\dfrac{236.439}{9.73}$
13. $12 - 3.725$

Work each of the following problems involving percent.

16. Convert .746 to percent.   **18.** Find 28% of 1478.
17. Convert 38% to a decimal.   **19.** What number is 145% of 70?
20. Airline pilots actually fly about 80 hours per month. A flight from New York to Tokyo requires 14 hours of flying. One such flight uses up what percent of a pilot's monthly workload?

For problems 21 and 22, find the total interest and the monthly payment for each installment loan. Interest is calculated by the add-on method. Round to the nearest dollar.

	Amount purchased	Down payment	Interest rate	Number of payments
21.	$5000	$500	9%	36
22.	$3750	$250	10%	24

23. A department store charges an interest rate of $1\frac{1}{2}\%$ on the unpaid balance. Find the interest charge on an unpaid balance of $279.00.

24. Find the monthly payment for principal and interest for a home loan of $44,000 at an interest rate of 9%. Assume a 30-year mortgage.

25. Find the total monthly payment for the house in Exercise 24 if annual taxes are $720 and annual fire insurance premium is $240.

Convert each measurement as indicated. Round to the nearest tenth.

26. 4 m to cm   **30.** 29 miles to km

27. .9 km to m   **31.** 6 yards to meters

28. 7.8 l to cl   **32.** 12.4 meters to feet

29. 297 g to kg   **33.** 9.8 pounds to kg

CUMULATIVE REVIEW FOR CHAPTERS 4 AND 5

Sets of numbers

Counting numbers $\{1, 2, 3, 4, 5, \ldots\}$

Whole numbers $\{0, 1, 2, 3, 4, \ldots\}$

Integers $\{\ldots, -3, -2, -1, 0, 1, 2, 3, 4, \ldots\}$

Rational numbers $\left\{\dfrac{p}{q} \middle| p \text{ and } q \text{ are integers, } q \neq 0\right\}$,

or   {all repeating or terminating decimals}

Irrational numbers   {all non-repeating, non-terminating decimals}

Real numbers   {all decimals}

Complex numbers $\{a + bi | i = \sqrt{-1}, a \text{ and } b \text{ real numbers}\}$

For any real numbers a, b, and c:

Properties	**Addition**	**Multiplication**
Commutative	$a + b = b + a$	$ab = ba$
Associative	$(a + b) + c = a + (b + c)$	$(ab)c = a(bc)$
Closure	$a + b$ is a real number	ab is a real number
Identity	$a + 0 = a$ and $0 + a = a$	$a \cdot 1 = a$ and $1 \cdot a = a$
Inverse	There exists a real number $-a$ such that $a + (-a) = 0$ and $(-a) + a = 0$	If $a \neq 0$, there exists a real number $\dfrac{1}{a}$ such that $a \cdot \dfrac{1}{a} = 1$ and $\dfrac{1}{a} \cdot a = 1$.
Distributive		$a(b + c) = ab + ac$

REVIEW EXERCISES

Complete the following table.

	Number	Whole	Integer	Rational	Irrational	Real	Complex
1.	12	yes			no		yes
2.	−9			yes			
3.	3/8	no					
4.	−5/9		no				yes
5.	2.4				no	yes	
6.	4.8					yes	
7.	$2.\overline{91}$		no				
8.	$4.\overline{8}$	no					
9.	$\sqrt{6}$	no	no	no			yes
10.	$-\sqrt{8}$		no				
***11.**	$\sqrt{9}$				no	yes	yes
***12.**	$\sqrt{25}$				no	yes	yes
***13.**	π			no		yes	yes
***14.**	$-\pi$			no			
***15.**	$3 + 2i$	no				no	
***16.**	$4 - i$	no				no	

Give the name of the property which shows why each statement is true.

17. $6 + (-6) = 0$

18. $8 \cdot \dfrac{1}{8} = 1$

19. $-3 + 3 = 0$

20. $\dfrac{5}{7} \cdot \dfrac{7}{5} = 1$

21. $6 + \sqrt{2} = \sqrt{2} + 6$

22. $\pi \cdot 4 = 4 \cdot \pi$

23. $8 + (9 + \sqrt{3}) = (8 + 9) + \sqrt{3}$

24. $6 \cdot \sqrt{2}$ is a real number.

25. $8(\pi + 2) = 8 \cdot \pi + 8 \cdot 2$

26. $9\sqrt{2} + 3\sqrt{2} = (9 + 3)\sqrt{2}$

Give the additive inverse of each number.

27. −4   **29.** 8   **31.** 0   ***33.** $\sqrt{5}$

28. −12   **30.** 30   **32.** −1/2   ***34.** $-\sqrt{6}$

Give the reciprocal of each number that has one.

35. $\dfrac{2}{3}$ **37.** $\dfrac{9}{11}$ **39.** 7   **42.** 16

40. −12   **43.** 0

36. $\dfrac{3}{4}$ **38.** $\dfrac{15}{23}$ **41.** −3   **44.** −0

Gram-o-phone *A preview of what may appear in the 1985 Sears catalog, when all products will have been converted to metric sizes. The gram-o-phone can be turned on while you are sleeping and will repeat information on the metric system. It helps you learn in a quick and easy way.*

Chapter 6 *Mathematical Systems*

In this chapter we look at some of the abstractions in mathematics. We begin with a common idea, a clock face, and use it to lead into some of the abstract systems of mathematics. A *mathematical system* is made up of two things:

(i)   A set of elements.

(ii)   One or more operations for combining the elements.

A common mathematical system is the set of whole numbers, {0, 1, 2, 3, 4, 5, . . .} and the operation of addition. Here the set of elements is a set of numbers. Elements from this set are combined using the operation of addition. For example, the elements 5 and 8 are combined to get 13. Also, 12 and 9 are combined to get 21.

The operation of the mathematical system must be applicable to the set of the system. It would be meaningless to speak of the system made up of a set of rabbits and the operation of multiplication, since the operation has not been given meaning for the set of rabbits. We do not know how to multiply two rabbits. (Two rabbits may know how to multiply, but we don't know how to multiply rabbits.) We can only count the results.

6.1 CLOCK ARITHMETIC

The mathematical systems we have seen so far have an unlimited number of elements. In this section we shall look at a different type of system, one with only a finite number of elements. By studying a system like this, one that is probably not familiar, we can get away from any preconceived ideas that we might have. This process lets us get at the fundamental structure that underlies our more familiar mathematical systems of integers and rational numbers under addition, multiplication, and so on.

Figure 6.1

The first non-traditional system that we shall look at is called the **12-hour clock system.** This system is based on an ordinary clock face, with the difference that 12 is replaced with 0, and the minute hand is left off. See Figure 6.1.

The clock face yields the finite set $\{0, 1, 2, 3, 4, 5, 6, 7, 8, 9, 10, 11\}$. As an operation for this clock system, we define addition as follows: add by moving the hour hand in a clockwise direction. For example, to add 5 and 2 on a clock, first move the hour hand to 5, as in Figure 6.2. Then, to add 2, move the hour hand 2 more hours in a clockwise direction. The hand ends at 7, so that

$$5 + 2 = 7.$$

This result agrees with our traditional addition. However, the sum of two numbers from the 12-hour clock system is not always what you might expect, as the next example shows.

Example 1 Find each sum in clock arithmetic: (a) $8 + 9$ (b) $11 + 3$.

(a) Move the hour hand to 8, as in Figure 6.3. Then advance the hand clockwise through 9 more hours. It ends at 5, so that

$$8 + 9 = 5.$$

Not many people would say that this result is traditional.

(b) Proceed as shown in Figure 6.4. Check that

$$11 + 3 = 2.$$

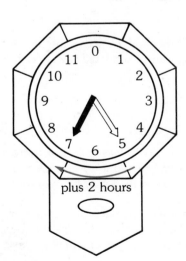

Figure 6.2 Clock addition: $5 + 2 = 7$

Figure 6.3 Clock addition: $8 + 9 = 5$

Figure 6.4 Clock addition: $11 + 3 = 2$

We could never hope to write a complete addition table for addition of whole numbers. For such a table, we would need to list every possible combination of two whole numbers. Since there are an infinite number of whole numbers, an addition table would have to contain an infinite number of rows and columns, which is impossible to construct.

On the other hand, our 12-hour clock system uses only the whole numbers 0, 1, 2, 3, 4, 5, 6, 7, 8, 9, 10, and 11. Thus, we can list every possible combination of two of these numbers in a table containing only 12 rows and 12 columns. We have done this with the 12-hour clock *addition table* in Figure 6.5.

Figure 6.5 Table for 12-hour clock addition

+	0	1	2	3	4	5	6	7	8	9	10	11
0	0	1	2	3	4	5	6	7	8	9	10	11
1	1	2	3	4	5	6	7	8	9	10	11	0
2	2	3	4	5	6	7	8	9	10	11	0	1
3	3	4	5	6	7	8	9	10	11	0	1	2
4	4	5	6	7	8	9	10	11	0	1	2	3
5	5	6	7	8	9	10	11	0	1	2	3	4
6	6	7	8	9	10	11	0	1	2	3	4	5
7	7	8	9	10	11	0	1	2	3	4	5	6
8	8	9	10	11	0	1	2	3	4	5	6	7
9	9	10	11	0	1	2	3	4	5	6	7	8
10	10	11	0	1	2	3	4	5	6	7	8	9
11	11	0	1	2	3	4	5	6	7	8	9	10

Example 2 Use the 12-hour clock addition table to find each sum.

(a) $7 + 11$

Solution. Find 7 on the left of the addition table and 11 across the top. At the intersection of the row headed 7 and the column headed 11 we find the number 6. Thus, $7 + 11 = 6$.

(b) Also from the table, $10 + 9 = 7$.

(c) $11 + 1 = 0$.

The mathematical system of the 12-hour clock uses only the twelve numbers 0, 1, 2, 3, 4, 5, 6, 7, 8, 9, 10, and 11. Since the system involves only a finite set of elements, it is called a **finite system.** Other finite systems will be discussed later in this chapter.

We don't speak of negative time, but we can define negative numbers in our system of clock arithmetic. For example, let us say that -5 is the number we get by going 5 hours counterclockwise from 0 to 7. See Figure 6.6. Thus,

$$-5 = 7.$$

In the same way, $-1 = 11, -2 = 10, -8 = 4$, and so on, with the negative numbers and their equivalent clock values shown in the following chart

Negative number	−1	−2	−3	−4	−5	−6	−7	−8	−9	−10	−11
Clock value	11	10	9	8	7	6	5	4	3	2	1

We can use the idea of a negative to find the difference of two numbers in clock arithmetic. In general, we define the *difference* of two numbers a and b, written $a − b$, as the *sum* of a and the negative of b. That is, $a − b = a + (−b)$. This is the same thing we did with integers in Chapter 4.

Example 3 Find each of the following differences.

(a) $8 − 5 = 8 + (−5)$ (Change subtraction to addition of the negative)

$\qquad\quad = 8 + 7$ $(−5 = 7,$ from the table above)

$8 − 5 = 3$ (From the addition table above)

This result agrees with traditional arithmetic.

(b) $6 − 11 = 6 + (−11)$ (c) $1 − 10 = 1 + (−10)$

$\qquad\quad = 6 + 1$ $\qquad\qquad = 1 + 2$

$6 − 11 = 7$ $1 − 10 = 3$

When we studied the mathematical system using the set of all integers and the operation of addition, we found that the system satisfied the closure, commutative, associative, identity, and inverse properties. Which of these properties are satisfied by our finite clock system? As we saw in the addition table above, the sum of two numbers on a clock face is always a number on the clock face. Because of this, clock arithmetic satisfies the *closure property* for addition. Check examples from the addition table above and convince yourself that the answer to an addition problem in clock arithmetic is the same, no matter whether you add clock numbers $a + b$ or $b + a$. For example, $6 + 8 = 8 + 6, 5 + 3 = 3 + 5, 9 + 11 = 11 + 9$, and so on. Since $a + b = b + a$, for any clock numbers a and b, clock addition is *commutative*.

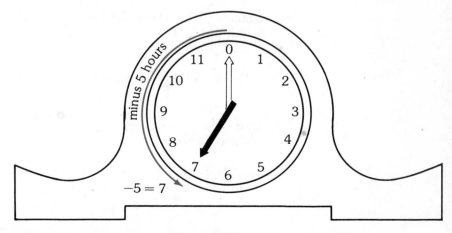

Figure 6.6 Negative numbers by counterclockwise motion

Example 4   Is 12-hour clock addition associative?

Solution.   To be associative, clock addition must satisfy

$$a + (b + c) = (a + b) + c$$

for any clock numbers *a, b,* and *c*. We really have no way to *prove* that this statement is always true (other than trying *every* possible combination of three clock face numbers). However, we can convince ourselves of the *plausibility* of the statement by trying a few examples. For example, if we try the clock numbers 3, 9, and 11, we get

$$
\begin{array}{ll}
3 + (9 + 11) & (3 + 9) + 11 \\
= 3 + 8 & = 0 + 11 \\
= 11 & = 11
\end{array}
$$

Thus, $3 + (9 + 11) = (3 + 9) + 11$. Let's try another example.

$$
\begin{array}{ll}
8 + (7 + 10) & (8 + 7) + 10 \\
= 8 + 5 & = 3 + 10 \\
= 1 & = 1
\end{array}
$$

Also, $8 + (7 + 11) = (8 + 7) + 11$. Any similar examples that we might try would also work. Clock addition *does* satisfy the associative property.

By the identity property for addition of whole numbers,

$$a + 0 = a \quad \text{and} \quad 0 + a = a$$

for any real number *a*. A similar property is valid in clock arithmetic:

$$a + 0 = a \quad \text{and} \quad 0 + a = a,$$

for any clock number *a*. To see this, use the addition table above. From the table, $6 + 0 = 6$, $1 + 0 = 1$, and so on.

Example 5   Does each clock number have an inverse for addition?

Solution.   The inverse for the clock number 5, for example, is a number *x* such that

$$5 + x = 0.$$

To get from 5 to 0 on the clock face, we need to go 7 more hours, so that

$$5 + 7 = 0.$$

Thus, 7 is the additive inverse of 5. Check that *every* clock number has an *additive inverse*.

We can also multiply clock numbers. For example, to find the product $5 \cdot 4$, add 4 a total of 5 times:

$$5 \cdot 4 = 4 + 4 + 4 + 4 + 4 = 8,$$

so that $5 \cdot 4 = 8$.

Example 6 Find each product, using clock arithmetic.

(a) $6 \cdot 9 = 9 + 9 + 9 + 9 + 9 + 9 = 6$

(b) $7 \cdot 3 = 3 + 3 + 3 + 3 + 3 + 3 + 3 = 9$

(c) $3 \cdot 4 = 4 + 4 + 4 = 0$

(d) $6 \cdot 0 = 0 + 0 + 0 + 0 + 0 + 0 = 0$

(e) $0 \cdot 8 = 0$

The system made up of the set of 12-hour clock numbers and the operation of multiplication satisfies the closure, commutative, associative, and identity properties. (Try examples to verify this.) The next example involves the inverse property for multiplication.

Example 7 Is there an inverse property for multiplication of clock numbers?

Solution. Can we find an inverse for the clock number 5? If y is an inverse for 5 for multiplication, then we must have

$$5 \cdot y = 1.$$

(That's the definition of an inverse for multiplication.) Try the numbers on the clock face: $5 \cdot 1$, $5 \cdot 2$, $5 \cdot 3$, and $5 \cdot 4$ do not equal 1. However,

$$5 \cdot 5 = 1,$$

so 5 is its own inverse for multiplication.

If there is an inverse for 2, then we must be able to find a clock number y such that

$$2 \cdot y = 1.$$

To multiply the various clock numbers times 2, we start with the clock hand at 2, and add 2 again and again. If we start at 2, and always add 2, we will get an *even* answer. Thus, we can never end at 1. There is no inverse for 2 for multiplication. Since we have found a non-zero number in the system that has no inverse, namely 2, the system of clock arithmetic for multiplication does not have the inverse property. (Can you find any other non-zero numbers on the 12-hour clock face that do not have multiplication inverses?)

Chess clock *The double clock you see below is used to time chess, backgammon, and scrabble games. Push one button, and that clock stops—the other begins simultaneously. In the photo, one player's flag has fallen at 6:00, meaning that the time allotted for the game has expired. If the player has not made the required number of moves, he or she will lose.*

Mathematics and chess are often thought to be closely related. Actually that is not so. Both arts demand logical thinking. Chess requires psychological acumen and knowledge of the opponent. No mathematical knowledge is needed in chess.

Emanuel Lasker was able to achieve mastery in both fields. He is best known as a World Chess Champion for 27 years, until 1921. Lasker also was famous in mathematical circles for his work concerning the theory of primary ideals, algebraic analogies of prime numbers. An important result, the Lasker-Noether Theorem, bears his name along with that of Emmy Noether (see pg. 237). Noether extended Lasker's work. Her father had been Lasker's PhD advisor.

6.1 EXERCISES

Find each product using a 12-hour clock.

1. $4 \cdot 2$	**4.** $6 \cdot 8$	**7.** $8 \cdot 2$	**10.** $3 \cdot 4$
2. $3 \cdot 3$	**5.** $9 \cdot 3$	**8.** $7 \cdot 7$	**11.** $10 \cdot 11$
3. $5 \cdot 4$	**6.** $2 \cdot 7$	**9.** $11 \cdot 11$	**12.** $8 \cdot 8$

13. Complete the following 12-hour clock multiplication table.

	0	1	2	3	4	5	6	7	8	9	10	11
0	0	0	0	0	0	0	0	0	0	0	0	0
1	0	1	2	3	4	5	6	7	8	9	10	11
2	0	2	4	6	8	10		2	4		8	
3	0	3	6	9	0	3	6			3	6	
4	0	4	8			8		4			4	8
5	0	5	10	3	8		6	11	4			
6	0	6	0		0	6	0	6		6		6
7	0	7	2	9				1			10	
8	0	8	4	0				8	4		8	4
9	0	9			0		6		0			
10	0	10	8			2						2
11	0	11										1

Find each of the following differences on the 12-hour clock.

14. $11 - 6$	**18.** $2 - 3$	**22.** $0 - 9$
15. $10 - 8$	**19.** $8 - 9$	**23.** $0 - 1$
16. $9 - 11$	**20.** $4 - 11$	
17. $7 - 10$	**21.** $3 - 10$	

The clock we used in the text had 12 hours on it. However, the same methods of clock arithmetic are just as valid if the clock face has only 7 hours, represented by the numbers 0, 1, 2, 3, 4, 5, and 6.

Use the 7-hour clock to find each of the following sums.

24. $6 + 4$	**26.** $4 + 4$	**28.** $1 + 6$
25. $3 + 5$	**27.** $2 + 5$	**29.** $6 + 6$

30. Complete the following 7-hour clock addition table.

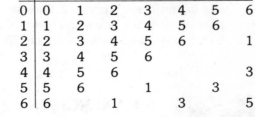

+	0	1	2	3	4	5	6
0	0	1	2	3	4	5	6
1	1	2	3	4	5	6	
2	2	3	4	5	6		1
3	3	4	5	6			
4	4	5	6				3
5	5	6		1		3	
6	6		1		3		5

7-hour clock system

Which properties are satisfied by the mathematical system made up of 7-hour clock numbers and addition?

31. closure	**33.** associative	**35.** inverse
32. cummutative	**34.** identity	

Multiply, using a 7-hour clock.

36. $6 \cdot 2$ **38.** $4 \cdot 6$ **40.** $5 \cdot 5$ **42.** $4 \cdot 2$

37. $3 \cdot 5$ **39.** $5 \cdot 2$ **41.** $6 \cdot 6$ **43.** $3 \cdot 4$

44. Construct a multiplication table for 7-hour clock arithmetic.

Which of the following properties are satisfied by the system of multiplication with 7-hour clock numbers?

45. closure   **47.** associative

46. commutative   **48.** identity

49. inverse   (*Hint:* The inverse property for multiplication is satisfied if each number except 0 has an inverse.)

List the inverse of each of the following numbers for multiplication on the 7-hour clock.

50. 1   **51.** 2   **52.** 3   **53.** 4   **54.** 5   **55.** 6

The military uses a 24-hour clock, to avoid the problems of "A.M." and "P.M." Thus, 1100 is 11 A.M., while 2100 is 9 P.M. (12 noon + 9 more hours). Find each of the following sums in the 24-hour clock system.

*56. $1000 + 500$ ***58.** $1500 + 1900$ ***60.** $2000 + 2000$

*57. $1300 + 600$ ***59.** $1700 + 2200$

The inverse property did not hold for multiplication on a 12-hour clock, but it did hold for multiplication on a 7-hour clock. Check whether or not the inverse property holds for multiplication in the following clock systems. (Don't forget: no inverse is needed for 0.)

*61. 3   ***62.** 4   ***63.** 5   ***64.** 9   ***65.** 11

*66. Complete this statement: The inverse property holds for multiplication only when the number of hours on a clock face is an _____ number.

6.2 MODULAR SYSTEMS

In the previous section we looked at the arithmetic of a clock face. For example, on an ordinary 12-hour clock, we found that $8 + 7 = 3$, $9 + 10 = 7$, $2 \cdot 8 = 4$, $5 \cdot 9 = 9$, and so on. In this section, we shall generalize the ideas of clock arithmetic.

If the hand on a 12-hour clock is at 0, and we move it 7 hours, the hand ends at 7. If we start at 0 and move the hand 31 hours, it again ends at 7. Thus, 7 and 31 lead to the same final position of the hand. Because of this, 7 and 31 are said to be *congruent modulo* 12, written

$$7 \equiv 31 \ (\text{mod } 12). \qquad \text{(The sign} \equiv \text{indicates } congruence.)$$

Also, 21 and 69 lead to the same final position of the hand, so that

$$21 \equiv 69 \pmod{12}.$$

The reason that 7 and 31 lead to the same final position of the hour hand is that the difference, $31 - 7 = 24$, is a multiple of 12, the number of hours on the clock. Also, $69 - 21 = 48$, a multiple of 12. In general, the integers a and b are **congruent modulo k** (where k is an integer greater than 1) if the difference $a - b$ is divisible by k.

The basic ideas of congruence were introduced by Karl F. Gauss in 1801, when he was 24 years old. For more information on the life of Gauss, see Chapter 10.

Example 1 Mark each statement as *true* or *false*.

(a) $16 \equiv 10 \pmod{2}$

Solution. The difference $16 - 10 = 6$ is divisible by 2. Thus, $16 \equiv 10 \pmod{2}$ is true.

(b) $49 \equiv 32 \pmod{5}$

False, since $49 - 32 = 17$, which is not divisible by 5.

(c) $30 \equiv 345 \pmod{7}$

True, since $30 - 345 = -315$ is divisible by 7. (It doesn't matter if we find $30 - 345$ or $345 - 30$.)

We can perform arithmetic in a modulo system just as we did with clock numbers in the previous section.

Example 2 Find each of the following sums.

(a) $9 + 14 \pmod{3}$

Solution. We know that $9 + 14 = 23$. We need to find the smallest non-negative integer which is congruent to 23 modulo 3. A practical way to find this integer is to divide 23 by 3. The remainder is 2, and we can check that $23 \equiv 2 \pmod{3}$, so

$$9 + 14 \equiv 2 \pmod{3}.$$

(b) $18 + 27 \pmod{6}$

Solution. $18 + 27 = 45$. Divide 45 by 6, obtaining 3 as a remainder. Thus,

$$18 + 27 \equiv 3 \pmod{6}.$$

(c) $50 + 34 \pmod{7}$

Solution. $50 + 34 = 84$. When 84 is divided by 7, a remainder of 0 is found. Thus,

$$50 + 34 \equiv 0 \pmod{7}.$$

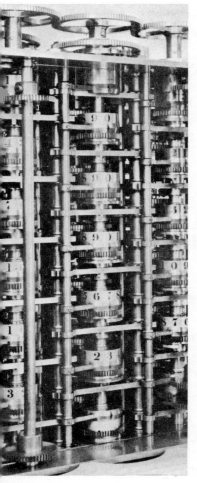

Example 3 Find each of the following products.

(a) $8 \cdot 9$ (mod 10)

Since $8 \cdot 9 = 72$, and 72 leaves a remainder of 2 when divided by 10, we have

$$8 \cdot 9 \equiv 2 \text{ (mod 10)}.$$

(b) $12 \cdot 10$ (mod 5); $\quad 12 \cdot 10 \equiv 120 \equiv 0$ (mod 5).

We can also solve equations in modulo systems. A *modulo equation* or just an *equation* is a statement such as $3 + x \equiv 5$ (mod 7) which is true for some replacements of the variable x and false for others. A method of solving these equations is given in the next example.

Example 4 Solve $3 + x \equiv 5$ (mod 7).

Solution. In a modulo 7 system, any integer will be congruent to one of the integers 0, 1, 2, 3, 4, 5, or 6. Thus, to solve the equation $3 + x \equiv 5$ (mod 7) we try, in turn, each of these integers in place of x.

$$\begin{array}{llll} \text{If } x = 0, & \text{is } 3 + 0 \equiv 5 \text{ (mod 7)}? & \text{No} \\ \text{If } x = 1, & \text{is } 3 + 1 \equiv 5 \text{ (mod 7)}? & \text{No} \\ \text{If } x = 2, & \text{is } 3 + 2 \equiv 5 \text{ (mod 7)}? & \text{Yes} \end{array}$$

Try $x = 3$, $x = 4$, $x = 5$, and $x = 6$ to see that none of them works. Thus, 2 is a solution of the equation $3 + x \equiv 5$ (mod 7).

Since 2 is a solution, we can find other solutions by repeatedly adding 7:

$$\begin{array}{l} 2 \\ 2 + 7 = 9 \\ 2 + 7 + 7 = 16 \\ 2 + 7 + 7 + 7 = 23, \end{array}$$

and so on. The set of all positive solutions of $3 + x \equiv 5$ (mod 7) is

$$\{2, 9, 16, 23, 30, 37, \ldots\}.$$

Example 5 Solve the equation $5x \equiv 4$ (mod 9).

Solution. Here we must try 0, 1, 2, 3, 4, 5, 6, 7, and 8:

$$\begin{array}{llll} \text{If } x = 0, & \text{is } 5 \cdot 0 \equiv 4 \text{ (mod 9)}? & \text{No} \\ \text{If } x = 1, & \text{is } 5 \cdot 1 \equiv 4 \text{ (mod 9)}? & \text{No} \end{array}$$

Continue trying numbers. You should find that none works except $x = 8$:

$$5 \cdot 8 \equiv 40 \equiv 4 \text{ (mod 9)}.$$

The set of all positive solutions to the equation $5x \equiv 4$ (mod 9) is thus

$$\{8, 8 + 9, 8 + 9 + 9, 8 + 9 + 9 + 9, \ldots\} \quad \text{or}$$
$$\{8, 17, 26, 35, 44, 53, \ldots\}.$$

Example 6   Solve the equation $6x \equiv 3 \pmod 8$.

Solution.   We must try the numbers 0, 1, 2, 3, 4, 5, 6, and 7. Try them; you should find that none works. Therefore, the equation $6x \equiv 3 \pmod 8$ has no solutions at all. We write the set of all solutions as the empty set, $\varnothing$.

This result is reasonable since $6x$ will always be even, no matter which whole number is used for x. Since $6x$ is even and 3 is odd, the difference $6x - 3$ will be odd, and therefore not divisible by 8.

Example 7   Solve $8x \equiv 8 \pmod 8$.

Solution.   By trying the integers 0, 1, 2, 3, 4, 5, 6, and 7, we see that *any* integer can be used as a solution. A statement like $8x \equiv 8 \pmod 8$ that is true for any values of the variable (x, y, and so on) is called an **identity.** Other identities include $2x + 3x = 5x$ and $y^2 \cdot y^3 = y^5$.

6.2 EXERCISES

Answer *true* or *false* for each of the following.

1. $6 \equiv 7 \pmod 2$ **6.** $179 \equiv 39 \pmod 8$

2. $15 \equiv 9 \pmod 2$ **7.** $48 \equiv 1{,}345{,}672{,}421 \pmod 2$

3. $10 \equiv 3 \pmod 7$ **8.** $46{,}791 \equiv 11{,}896 \pmod 5$

4. $18 \equiv 7 \pmod 4$ **9.** $0 \equiv 1551 \pmod 3$

5. $132 \equiv 13 \pmod 7$ **10.** $234 \equiv 102 \pmod 6$

Find each of the following sums or products.

11. $8 + 15 \pmod 9$ **21.** $8 \cdot 12 \pmod 4$

12. $21 + 36 \pmod 6$ **22.** $38 \cdot 5 \pmod 9$

13. $14 + 29 \pmod 7$ **23.** $21 \cdot 12 \pmod 6$

14. $38 + 146 \pmod{10}$ **24.** $38 \cdot 5 \pmod{12}$

15. $29 + 182 \pmod 9$ **25.** $59 \cdot 7 \pmod 3$

16. $385 + 121 \pmod{11}$ **26.** $8 \cdot 71 \pmod{11}$

17. $987 + 213 \pmod 3$ **27.** $3 \cdot (9 + 2) \pmod 5$

18. $217 + 499 \pmod 2$ **28.** $5 \cdot (8 + 17) \pmod 7$

19. $3 \cdot 8 \pmod 5$ **29.** $(12 + 9) \cdot (3 + 4) \pmod{10}$

20. $7 \cdot 6 \pmod 2$ **30.** $(16 + 21) \cdot (8 + 19) \pmod 6$

31.   Modulo 3

+	0	1	2
0	0	1	2
1	1		
2	2		

In each of Exercises 31–34:

(a)   Complete the given addition table.

(b)   Decide whether the closure, commutative, associative, identity, and inverse properties are satisfied.

(c)   If the inverse property is satisfied, give the inverse of each number.

32. Modulo 4

+	0	1	2	3
0	0	1	2	3
1	1			
2	2			
3	3			

33. Modulo 5

+	0	1	2	3	4
0	0	1	2	3	4
1	1	2	3	4	
2	2	3	4		
3	3				
4	4				

34. Modulo 7

+	0	1	2	3	4	5	6
0	0	1	2	3	4	5	6
1	1	2	3	4	5	6	
2	2	3	4	5	6		
3	3						
4	4						
5	5						
6	6	0	1	2	3	4	5

In each of Exercises 35–40:
(a) Complete the given multiplication table.
(b) Decide whether the closure, commutative, associative, identity, and inverse properties are satisfied.
(c) Give the inverse of each non-zero number that has an inverse.

35. Modulo 2

•	0	1
0	0	0
1	0	

36. Modulo 3

•	0	1	2
0	0	0	0
1	0	1	2
2	0	2	

37. Modulo 4

•	0	1	2	3
0	0	0	0	0
1	0	1	2	3
2	0	2		
3	0	3		

38. Modulo 5

•	0	1	2	3	4
0	0	0	0	0	0
1	0	1	2	3	4
2	0	2	4		
3	0	3		4	
4	0	4			

39. Modulo 7

•	0	1	2	3	4	5	6
0	0	0	0	0	0	0	0
1	0	1	2	3	4	5	6
2	0	2	4	6			
3	0	3	6		5		4
4	0	4					
5	0	5		6		2	
6	0	6		4		1	

40. Modulo 9

•	0	1	2	3	4	5	6	7	8
0	0	0	0	0	0	0	0	0	0
1	0	1	2	3	4	5	6	7	8
2	0	2	4	6	8		5		
3	0	3	6	0		6		3	6
4	0	4	8		7		6		5
5	0	5	1		2		3	8	
6	0	6	3	0	6	3	0	6	3
7	0	7	5			8			2
8	0	8	7		4	3			1

41. Complete this statement: a modulo system satisfies the inverse property for multiplication only if the modulo number is a _____ number.

In Exercises 42–60, find all positive solutions for each equation. Note any identities.

42. $x \equiv 4 \pmod 5$

43. $x \equiv 7 \pmod 9$

44. $x \equiv 2 \pmod 4$

45. $x \equiv 3 \pmod 7$

46. $x \equiv 0 \pmod 5$

47. $x + 2 \equiv 8 \pmod{10}$

48. $x + 6 \equiv 9 \pmod{11}$

49. $1 + x \equiv 3 \pmod 7$

50. $4 + x \equiv 5 \pmod 8$

51. $2x \equiv 3 \pmod 7$

52. $3x \equiv 5 \pmod 9$ **57.** $9x \equiv 3 \pmod 3$
53. $8x \equiv 1 \pmod 3$ **58.** $12x \equiv 8 \pmod 4$
54. $3x \equiv 2 \pmod 4$ **59.** $4x \equiv 1 \pmod 5$
55. $6x \equiv 4 \pmod 6$ **60.** $11x \equiv 5 \pmod{12}$
56. $8x \equiv 2 \pmod 4$

*61. The odometer (dial that records the total distance traveled) in an ordinary car can be considered as an example of a modulo system. Mileage totals are recorded modulo what number?

*62. Suppose that a car odometer shows a reading of 33,421. *In theory,* how many miles might the car have gone?

In modulo 15, we have $5 \cdot 3 \equiv 0 \pmod{15}$. Since the product of the non-zero numbers 5 and 3 is 0, we call 5 and 3 *zero divisors.* Find all possible zero divisors less than or equal to the modulo numbers in each of the following modulo systems.

*63. 4   ***65.** 8   ***67.** 10   ***69.** 7   ***71.** 13
*64. 6   ***66.** 9   ***68.** 12   ***70.** 11

*72. Complete this statement: there are zero divisors in a modulo system only if the modulo number is *not* a _____ number.

6.3 APPLICATIONS OF MODULO SYSTEMS

In this section we shall look at two applications of modulo systems. The first of these applications, *casting out nines,* is very useful in checking answers for problems in arithmetic. Casting out nines depends on the following property:

> Any whole number is congruent, modulo 9,
> to the sum of its digits.

For example, the sum of the digits in 5782 is $5 + 7 + 8 + 2 = 22$, and $5782 \equiv 22 \pmod 9$. To check, verify that 9 divides into the difference $5782 - 22 = 5760$. We can continue: $22 \equiv 2 + 2 \pmod 9 \equiv 4 \pmod 9$, so that

$$5782 \equiv 4 \pmod 9.$$

Example 1   **(a)** $46{,}720 \equiv (4 + 6 + 7 + 2 + 0) \pmod 9$
$\equiv 19 \pmod 9$
$\equiv (1 + 9) \pmod 9$
$46{,}720 \equiv 1 \pmod 9$

(b) $84{,}732 \equiv 24 \pmod 9 \equiv 6 \pmod 9$

The process of finding a number congruent, modulo 9, to a given number is called **casting out nines** since any combination of numbers that add to 9 can be "cast out." For example,

$$439,681 \equiv (4 + 3 + 9 + 6 + 8 + 1)\,(\text{mod } 9).$$

We can cross out the 9, the 3 and 6 (since $3 + 6 = 9$), and the 8 and 1. Thus,

$$439,681 \equiv (4 + \cancel{3} + \cancel{9} + \cancel{6} + \cancel{8} + \cancel{1})\,(\text{mod } 9)$$
$$439,681 \equiv 4\,(\text{mod } 9).$$

The method of casting out nines to check answers in arithmetic is shown in the next few examples.

Example 2   Check the addition problem at left by casting out nines.

Solution.   Proceed as follows:

$$
\begin{array}{rll}
89,174 \equiv 8 + 9 + 1 + 7 + 4 & \equiv 2\ (\text{mod } 9) \\
+\ \ 36,295 \equiv 3 + 6 + 2 + 9 + 5 & \equiv 7\ (\text{mod } 9) \\
\hline
125,469 \equiv 1 + 2 + 5 + 4 + 6 + 9\ (\text{mod } 9) & \equiv 0\ (\text{mod } 9)
\end{array}
$$

$$
\begin{array}{r}
89,174 \\
+\ 36,295 \\
\hline
125,469
\end{array}
$$

On the right, we have $(2 + 7)\ (\text{mod } 9) \equiv 9\ (\text{mod } 9) \equiv 0\ (\text{mod } 9)$. This is the same result we got for the answer, so that 125,469 is probably the correct answer. We say "probably" the correct answer, since an incorrect answer will still check as correct if it differs from the correct answer by a multiple of 9. However, most incorrect answers differ from the correct one only by one or two digits. Casting out nines catches most of the common types of errors.

Example 3   Check the subtraction problem by casting out nines.

$$
\begin{array}{r}
7438 \\
-\ 5716 \\
\hline
1722
\end{array}
$$

Solution.

$$
\begin{array}{rll}
7438 \equiv 7 + 4 + 3 + 8 & \equiv 4\ (\text{mod } 9) \\
-\ 5716 \equiv 5 + 7 + 1 + 6 & \equiv 1\ (\text{mod } 9) \\
\hline
1722 \equiv 1 + 7 + 2 + 2\ (\text{mod } 9) & \equiv 3\ (\text{mod } 9)
\end{array}
$$

On the right, $(4 - 1)\ (\text{mod } 9) \equiv 3\ (\text{mod } 9)$, so that the answer is probably correct.

Example 4   Check the subtraction problem by casting out nines.

Solution.   Proceed as above.

$$
\begin{array}{r}
39,712 \\
-\ 25,234 \\
\hline
14,478
\end{array}
$$

$$
\begin{array}{rll}
39,712 \equiv 3 + 9 + 7 + 1 + 2 & \equiv 4\ (\text{mod } 9) \\
-\ 25,234 \equiv 2 + 5 + 2 + 3 + 4 & \equiv 7\ (\text{mod } 9) \\
\hline
14,478 \equiv 1 + 4 + 4 + 7 + 8\ (\text{mod } 9) & \equiv 6\ (\text{mod } 9)
\end{array}
$$

On the right, we have $(4 - 7)\ (\text{mod } 9)$. It is not immediately clear that $(4 - 7)\ (\text{mod } 9) \equiv 6\ (\text{mod } 9)$, which is what we need. If we calculate the difference

$$(4 - 7) - 6 = -3 - 6 = -9,$$

we get -9. We can divide 9 evenly into -9, so that $(4 - 7)\ (\text{mod } 9) \equiv 6\ (\text{mod } 9)$, and the problem is probably correct.

Example 5   Check the multiplication problem by casting out nines.

$$896$$
$$\times\ 76$$
$$\overline{68{,}096}$$

Solution.

$$896 \equiv 5 \ (\text{mod } 9)$$
$$\times\ \ 76 \equiv 4 \ (\text{mod } 9)$$
$$\overline{68{,}096 \equiv 2 \ (\text{mod } 9)}$$

On the right, we have $(5 \times 4) \ (\text{mod } 9) \equiv 20 \ (\text{mod } 9) \equiv 2 \ (\text{mod } 9)$, so that the answer is probably correct.

The second type of application of modulo systems is based on the type of word problem in the next example.

Example 6   Shawn is a clerk in a shoe store. Customer 1 comes in and looks at all the shoes Shawn has in stock. As he shows the shoes, he stacks the boxes in stacks of 7 boxes each, with 2 boxes left over. After Shawn returns all the boxes to the shelf, Customer 2 enters to look at the shoes. This time, Shawn puts the boxes in stacks of 5 each, with 4 boxes left over. The last customer, Customer 3, looks at the shoes as Shawn places them in stacks of 3 each, with 2 left over. Shawn knows there are fewer than 120 boxes of shoes, and none of the customers bought any. How many boxes of shoes does he have?

Solution.   We are told nothing about the number of stacks at each showing of the shoes, but we do have information on the number of boxes in each stack, and the number left over. Let s stand for the total number of boxes of shoes. If the boxes are stacked 7 high, there are 2 left over, so that

$$s \equiv 2 \ (\text{mod } 7).$$

In the same way,

$$s \equiv 4 \ (\text{mod } 5) \qquad \text{and} \qquad s \equiv 2 \ (\text{mod } 3).$$

We must find a value of s, less than 120, that satisfies all three of these equations.

If we solve the first equation, $s \equiv 2 \ (\text{mod } 7)$, by the methods of the previous section, we get

$$\{2, 9, 16, 23, 30, 37, 44, 51, 58, 65, 72, 79, 86, 93, 100, 107, 114\}.$$

(Why did we stop at 114?) The answer must also satisfy the second equation, $s \equiv 4 \ (\text{mod } 5)$. Go through the set of solutions above and find all the numbers congruent to 4 modulo 5. You should find the numbers to be

$$\{9, 44, 79, 114\}.$$

Finally, identify the numbers from this second set that also satisfy $s \equiv 2$ $(\text{mod } 3)$, the last equation from above. The only number from the set $\{9, 44, 79, 114\}$ that satisfied $s \equiv 2 \ (\text{mod } 3)$ is 44, so that

$$s = 44.$$

Shawn has 44 boxes of shoes.

Biorhythm and modulo equations   *According to the biorhythm theory, everyone has three inner rhythms that start at birth: a 23-day physical cycle, a 28-day emotional cycle, and a 33-day intellectual cycle. Each cycle is made up of a high period, a low period, and a critical transition day when a person moves from one cycle to the next.*

To find your "highs" and "lows," first calculate the total number of days that you have been alive. (Don't forget leap years.) Then for the physical cycle, find the smallest whole number x such that

number of days $\equiv$ x (mod 23).

If x comes out 0, you are at the critical transition day. If x is around 6, you are at a physical "high"; if x is around 17 or 18, you are at a "low."

The corresponding equations for the emotional and intellectual cycles are respectively

number of days $\equiv$ x (mod 28),

number of days $\equiv$ x (mod 33).

Most medical researchers agree that we have definite biorhythms, but say that the cycles are not at all predictable in length. See Biological Rhythms in Human and Animal Physiology by Gay Gaer Luce (Dover, 1971).

6.3 EXERCISES

Use casting out nines to check the computations in Exercises 1 – 14.

1.
```
    6892
    4314
  + 5968
   17,174
```

2.
```
   3715
   4824
 + 3697
  12,236
```

3.
```
   9274
   8325
   4719
 + 3820
  26,148
```

4.
```
   7158
   3442
   7414
 + 8923
  26,837
```

5.
```
   21,983
 −  7,642
   14,341
```

6.
```
   89,425
 − 31,708
   57,717
```

7.
```
   84,362
 − 39,480
   44,882
```

8.
```
   79,643
 − 54,377
   25,266
```

9.
```
   27,342
 − 15,549
    1,793
```

10.
```
   32,594
 − 23,758
    8,856
```

11.
```
    891
  ×  43
  38,313
```

12.
```
    258
  ×  95
  24,510
```

13.
```
   1,762
  ×   53
  93,486
```

14.
```
   9,482
  ×    38
  359,316
```

Casting out nines can be used to check division problems. To see how, let us first divide 3096 by 43, as shown at the side.

```
      72
43) 3096
    301
     86
     86
      0
```

To check in the normal way, *multiply* 43 and 72: $43 \times 72 = 3096$. We use the same idea to check by casting out nines. For example, we can check the division problem below by first casting out nines:

$$
\begin{array}{r}
68 \\
35) \overline{2380}
\end{array}
\qquad
\begin{array}{l}
35 \equiv 8 \ (\text{mod } 9) \\
68 \equiv 5 \ (\text{mod } 9) \\
2380 \equiv 4 \ (\text{mod } 9)
\end{array}
$$

Now multiply: $8 \times 5 \equiv 40 \equiv 4 \ (\text{mod } 9)$. This matches the results for 2380, since $2380 \equiv 4 \ (\text{mod } 9)$. Thus, 68 is probably the correct answer. Check each of the following problems by casting out nines.

15.
```
      38
15) 570
```

16.
```
      45
21) 945
```

17.
```
       257
74) 19,092
```

18.
```
       345
85) 29,410
```

*19.
```
        372, remainder 4
98) 36,460
```

*20.
```
        514, remainder 14
85) 43,704
```

Find all the positive integers less than 50 that satisfy the equations in Exercises 21–28.

21. $s \equiv 5 \pmod 6$
$s \equiv 2 \pmod 7$

22. $x \equiv 1 \pmod 3$
$x \equiv 4 \pmod 5$

23. $k \equiv 5 \pmod 7$
$k \equiv 1 \pmod 5$

24. $z \equiv 2 \pmod 4$
$z \equiv 5 \pmod 7$

25. $r \equiv 1 \pmod 3$
$r \equiv 3 \pmod 5$
$r \equiv 1 \pmod 7$

26. $a \equiv 5 \pmod 6$
$a \equiv 1 \pmod 5$
$a \equiv 2 \pmod 3$

27. $m \equiv 1 \pmod 3$
$m \equiv 1 \pmod 5$
$m \equiv 5 \pmod 7$

28. $x \equiv 6 \pmod 7$
$x \equiv 5 \pmod 6$
$x \equiv 4 \pmod 5$

Find all whole numbers satisfying each of the following.

29. $x \equiv 1 \pmod{10}$
$x \equiv 2 \pmod 3$
$x \equiv 0 \pmod 7$
x between 100 and 200

30. $x \equiv 1 \pmod 2$
$x \equiv 2 \pmod 3$
$x \equiv 4 \pmod 5$
x between 100 and 200

Solve the word problems in Exercises 31–34.

31. Fran is a zoo-keeper. One day she noted that the orangutan stacked coconuts in stacks of 7 with 1 left over. The baboon then stacked the same nuts in stacks of 5 with 3 left over. The gorilla placed them in stacks of 4 with 3 left over. Fran has fewer than 50 nuts. How many does she have?

32. Anna, Kerry, and Shamus are little Chihuahuas. By standing on each other's backs, they were able to reach the table top and knock off a box of doggie treats. Anna arranged the treats in piles of 10, with 3 left over. Kerry arranged them in piles of 7 with 3 left over. Shamus arranged them in piles of 8, with one left over. There were fewer than 100 treats in the box. How many were there?

33. Lupe shows color slides at the Sisters United meetings. At the March meeting, the seating was arranged so that the Sisters saw the slides of Pennsylvania while sitting in rows of 4 with 2 left over. The April meeting had the same number of people for her slides of Arizona, this time in rows of 5 with 1 standing. In May, the Sisters saw the slides of South Carolina. This time the seating was arranged in rows of 11, with 9 in the last row. If fewer than 100 people saw the slides, how many were there?

34. Ronald is a miser who counts his $100 bills every night before putting them back in his mattress. He can put them in stacks of 7, 5, or 2 with none left over. If he has between 100 and 200 bills, how much wealth does he sleep on?

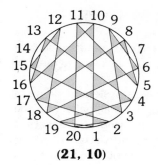

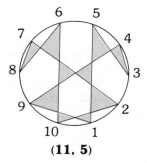

(21, 10)

Modulo numbers can be used to create **modulo designs.** For example, to construct the design (11, 3), proceed as follows.

35. Draw a circle and divide the circumference into 10 equal parts. Label the division points as 1, 2, 3, . . ., 10.

36. Since $1 \cdot 3 \equiv 3$ (mod 11), connect 1 and 3. (We use 3 as a multiplier since we are making an (11, 3) design.)

37. $2 \cdot 3 \equiv 6$ (mod 11)   Therefore, connect 2 and ____.

38. $3 \cdot 3 \equiv$ ____ (mod 11)   Connect 3 and ____.

39. $4 \cdot 3 \equiv$ ____ (mod 11)   Connect 4 and ____.

40. $5 \cdot 3 \equiv$ ____ (mod 11)   Connect 5 and ____.

41. $6 : 3 \equiv$ ____ (mod 11)   Connect 6 and ____.

42. $7 \cdot 3 \equiv$ ____ (mod 11)   Connect 7 and ____.

43. $8 \cdot 3 \equiv$ ____ (mod 11)   Connect 8 and ____.

44. $9 \cdot 3 \equiv$ ____ (mod 11)   Connect 9 and ____.

45. $10 \cdot 3 \equiv$ ____ (mod 11)   Connect 10 and ____.

46. Shade in the regions you have found. Try to make an interesting pattern. Other modulo designs are shown here. For more information, see "Residue Designs," by Phil Locke, *The Mathematics Teacher*, March 1972, pages 260–263.

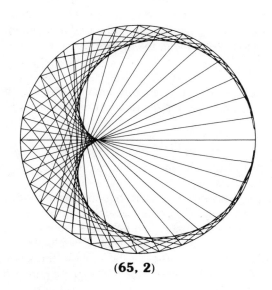

(11, 5)

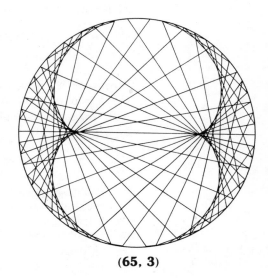

(65, 2)

(65, 3)

6.4 MORE FINITE MATHEMATICAL SYSTEMS

The systems discussed thus far in this chapter have sets of elements which are finite, and so are called *finite mathematical systems*. We shall look at more of these finite systems in this section.

☆	a	b	c	d
a	a	b	c	d
b	b	d	a	c
c	c	a	d	b
d	d	c	b	a

To begin, let us introduce a new finite mathematical system made up of the set of elements {a, b, c, d}, and an operation we shall write with the symbol ☆. We give meaning to operation ☆ by showing an **operation table,** which tells how operation ☆ is used to find the answer whenever we use any two elements from the set {a, b, c, d}. The operation table for ☆ is shown at the side.

To use the table to find, say, c ☆ d, we first locate c on the left, and d across the top (see Figure 6.7). This row and column give b, so that

$$c \; ☆ \; d = b.$$

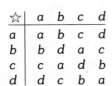

Figure 6.7

We shall mainly be interested in deciding on the properties possessed by the mathematical systems that we study. The properties that we investigate for finite systems, associative, closure, commutative, identity, and inverse, are the same ones we have looked at all along. The only difference now is that the operations are new, and must be looked at with a fresh eye — it is hard to have preconceived ideas about operation ☆ and the table above. Let us now decide which properties are satisfied by the system made up of {a, b, c, d} and operation ☆.

Closure property   For this system to be closed, the answer to any possible problem must be in the set {a, b, c, d}. A glance at the table above that defines operation ☆ shows that the answers in the body of the table are all elements of this set. Thus, the system is closed. If we had found an element other than a, b, c, or d in the body of the table, the system would not have been closed.

Commutative property   The system has the commutative property if Γ ☆ Δ = Δ ☆ Γ, where Γ and Δ stand for any elements from the set {a, b, c, d}. Try some examples:

$$c \; ☆ \; d = b \quad \text{and} \quad d \; ☆ \; c = b, \quad \text{so that} \quad c \; ☆ \; d = d \; ☆ \; c,$$
$$b \; ☆ \; c = a \quad \text{and} \quad c \; ☆ \; b = a, \quad \text{so that} \quad b \; ☆ \; c = c \; ☆ \; b.$$

☆	a	b	c	d
a	a	b	c	d
b	b	d	a	c
c	c	a	d	b
d	d	c	b	a

Figure 6.8

Examples such as these suggest that the system has the commutative property, but do not prove it completely. (There is always a chance that by further looking we could find an example that would fail.) In a table such as the one defining operation ☆ there is a test which can be used to tell for sure. Draw a diagonal from the upper left-hand corner of the table to the lower right-hand corner, as shown in Figure 6.8. If it is possible to fold the chart on this diagonal and have the corresponding elements match, then the system has the commutative property. We can fold the table for ☆ in this way, so that the system is commutative.

Associative property   The system is associative in the case $(\Gamma \star \Delta) \star \Upsilon$ $= \Gamma \star (\Delta \star \Upsilon)$, where Γ, Δ, and Υ represent any elements from the set $\{a, b, c, d\}$. There is no quick way to check a table for the associative property, as there was for the commutative property. All we can do here is try some examples. Using the table that defines operation $\star$, we have

$$(a \star d) \star b = d \star b = c, \quad \text{and} \quad a \star (d \star b) = a \star c = c,$$

so that

$$(a \star d) \star b = a \star (d \star b).$$

In the same way,

$$b \star (c \star d) = (b \star c) \star d.$$

In both these examples, changing the location of parentheses did not change the answers. Since these two examples worked, we suspect that the system is associative. We cannot be sure of this, however, unless we check it for *every* possible choice of three letters from the set.

Identity property   For the identity property to hold, we must be able to find an element Δ from the set of the system such that $\Delta \star X = X$ and $X \star \Delta = X$, where X represents any element from the set $\{a, b, c, d\}$. By inspecting the table that defines operation $\star$, we see that a is the element that we want, since

$$a \star a = a$$
$$a \star b = b \quad \text{and} \quad b \star a = b$$
$$a \star c = c \quad \text{and} \quad c \star a = c$$
$$a \star d = d \quad \text{and} \quad d \star a = d.$$

The element a is called the **identity element** for the system. We can locate the identity element in an operation table by looking for a column in the body of the table that is identical with the column at the left side of the table. If such a column is found, then the element at the top of that column is possibly the identity. If, when that element is located at the left, the corresponding row in the body of the table is identical with the row at the top, then that element is indeed the identity.

Inverse property   We found above that a is the identity element for the system using operation $\star$. Can we find an inverse in this system for, say, the element b? If Δ represents the inverse of b in this system, then we must have

$$b \star \Delta = a \quad \text{and} \quad \Delta \star b = a \quad \text{(since } a \text{ is the identity element)}.$$

By inspecting the table for operation $\star$, we find that Δ should be replaced with c:

$$b \star c = a \quad \text{and} \quad c \star b = a.$$

Thus, b and c are inverses of each other in this system. Since $d \star d = a$ and $a \star a = a$, the elements d and a are each their own inverses.

To find inverses when a table is given, look in each row of the body of the table for the identity element. If it does not appear in a particular row (or appears more than once) then the element at the left in that row has no inverse.

In summary, the mathematical system made up of the set $\{a, b, c, d\}$ and operation ☆ satisfies the closure, commutative, associative, identity, and inverse properties. The distributive property requires two operations, so that operation ☆ alone could not satisfy it. We shall discuss the distributive property in more detail in the next section.

Let us now list the basic properties that may (or may not) be satisfied by a mathematical system. Here a, b, and c represent elements from the set of the system, and ∘ represents the operation of the system.

CLOSURE   The system is closed if for any two elements a and b,

$$a \circ b$$

is in the set of the system.

COMMUTATIVE   The system has the commutative property if

$$a \circ b = b \circ a$$

for any two elements a and b from the system.

ASSOCIATIVE   The system has the associative property if

$$(a \circ b) \circ c = a \circ (b \circ c)$$

for every choice of three elements a, b, and c of the system.

IDENTITY   The system has an identity element e (where e is in the set of the system) if

$$a \circ e = a \quad \text{and} \quad e \circ a = a,$$

for every element a in the system.

INVERSE   The system satisfies the inverse property if, for every element a of the system, there is an element b in the system such that

$$a \circ b = e \quad \text{and} \quad b \circ a = e,$$

where e is the identity element of the system.

Example 1   The table in the margin is a multiplication table for the set $\{0, 1, 2, 3, 4, 5\}$ with the operation of multiplication modulo 6. Which of the properties above are satisfied by this system?

Solution.   All the numbers in the body of the table come from the set $\{0, 1, 2, 3, 4, 5\}$. Thus, the system is closed. If we draw a line from upper left to lower right, we could fold the table along this line and have the corresponding elements match; the system has the commutative property.

To check for the associative property, we must work out some examples.

$$2 \cdot (3 \cdot 5) = 2 \cdot 3 = 0 \quad \text{and} \quad (2 \cdot 3) \cdot 5 = 0 \cdot 5 = 0,$$

so that

$$2 \cdot (3 \cdot 5) = (2 \cdot 3) \cdot 5.$$

Also,

$$5 \cdot (4 \cdot 2) = (5 \cdot 4) \cdot 2.$$

●	0	1	2	3	4	5
0	0	0	0	0	0	0
1	0	1	2	3	4	5
2	0	2	4	0	2	4
3	0	3	0	3	0	3
4	0	4	2	0	4	2
5	0	5	4	3	2	1

Any other examples that we might try would also work. The system has the associative property.

The column at the left of the multiplication table is repeated under 1 in the body of the table. Thus, 1 is a candidate for the identity element in the system. To see that 1 is indeed the identity element here, check that the row corresponding to 1 at the left is identical with the row at the top of the table.

To find inverse elements, look for the identity element, 1, in the rows of the table. The identity element appears in the second row, $1 \cdot 1 = 1$; and in the bottom row, $5 \cdot 5 = 1$; so that 1 and 5 are each their own inverses. There is no identity element in rows 3, 4, and 5, so that none of these elements have inverses.

In summary, the system made up of the set {0, 1, 2, 3, 4, 5} and multiplication modulo 6 satisfies the closure, associative, commutative and identity properties, but not the inverse property.

Example 2 The table in the margin is a multiplication table for the set of numbers {1, 2, 3, 4, 5, 6} and the operation of multiplication modulo 7. Which of the properties are satisfied by this system?

•	1	2	3	4	5	6
1	1	2	3	4	5	6
2	2	4	6	1	3	5
3	3	6	2	5	1	4
4	4	1	5	2	6	3
5	5	3	1	6	4	2
6	6	5	4	3	2	1

Solution. Check that the system satisfies the closure, commutative, associative, and identity properties, with identity element 1. Let us now check for inverses. The element 1 is its own inverse, since $1 \cdot 1 = 1$. In row 2, the identity element 1 appears under the number 4, so that $2 \cdot 4 = 1$ (and $4 \cdot 2 = 1$), with 2 and 4 inverses of each other. Also, 3 and 5 are inverses of each other, and 6 is its own inverse. Since each number in the set of the system has an inverse, the system satisfies the inverse property.

6.4 EXERCISES

For each system in Exercises 1–12, decide which of the properties listed above are satisfied. If the inverse property is satisfied, give the inverse of each element.

1. {1, 2, 3, 4};
multiplication
modulo 5

•	1	2	3	4
1	1	2	3	4
2	2	4	1	3
3	3	1	4	2
4	4	3	2	1

2. {1, 2};
multiplication
modulo 3

•	1	2
1	1	2
2	2	1

3. {1, 2, 3, 4, 5};
multiplication
modulo 6

•	1	2	3	4	5
1	1	2	3	4	5
2	2	4	0	2	4
3	3	0	3	0	3
4	4	2	0	4	2
5	5	4	3	2	1

4. {1, 2, 3, 4, 5, 6, 7};
multiplication
modulo 8

•	1	2	3	4	5	6	7
1	1	2	3	4	5	6	7
2	2	4	6	0	2	4	6
3	3	6	1	4	7	2	5
4	4	0	4	0	4	0	4
5	5	2	7	4	1	6	3
6	6	4	2	0	6	4	2
7	7	6	5	4	3	2	1

5. {1, 3, 5, 7};
multiplication
modulo 8

•	1	3	5	7
1	1	3	5	7
3	3	1	7	5
5	5	7	1	3
7	7	5	3	1

6. {1, 3, 5, 7, 9};
multiplication
modulo 10

•	1	3	5	7	9
1	1	3	5	7	9
3	3	9	5	1	7
5	5	5	5	5	5
7	7	1	5	9	3
9	9	7	5	3	1

7. {m, n, p};
operation **J**

J	m	n	p
m	n	p	n
n	p	m	n
p	n	n	m

8. {A, B, F};
operation ☆

☆	A	B	F
A	B	F	A
B	F	A	B
F	A	B	F

9. {A, J, T, U};
operation #

#	A	J	T	U
A	A	J	T	U
J	J	T	U	A
T	T	U	A	J
U	U	A	J	T

10. {r, s, t, u};
operation **Z**

Z	r	s	t	u
r	u	t	r	s
s	t	u	s	r
t	r	s	t	u
u	s	r	u	t

11. {m, n, p, q, r, s}; operation ∘

∘	m	n	p	q	r	s
m	m	n	p	q	r	s
n	n	p	m	s	q	r
p	p	m	n	r	s	q
q	q	r	s	m	n	p
r	r	s	q	p	m	n
s	s	q	r	n	p	m

12. {p, q, r, s, t, u, v, w}; operation **A**

A	p	q	r	s	t	u	v	w
p	p	q	r	s	t	u	v	w
q	q	r	s	p	w	v	t	u
r	r	s	p	q	u	t	w	v
s	s	p	q	r	v	w	u	t
t	t	v	u	w	p	r	q	s
u	u	w	t	v	r	p	s	q
v	v	u	w	t	s	q	p	r
w	w	t	v	u	q	s	r	p

The tables in the finite mathematical systems that we developed in this section can be obtained in a variety of ways. For example, let us begin with a square, as shown in the figure. Let the symbols *a*, *b*, *c*, and *d* be defined as shown in the figure.

Let *a* represent zero rotation— leave the original square as is

Let *b* represent rotation of 90° clockwise from original position

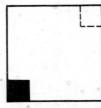

Let *c* represent rotation of 180° clockwise from original position

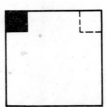

Let *d* represent rotation of 270° clockwise from original position

Start with a

We can define an operation ☆ for these letters as follows. To evaluate $b ☆ c$, for example, first perform b, by rotating the square 90°. (See the figure.) Then perform operation c, by rotating the square an additional 180°. The net result is the same as if we had performed d only. Thus,

$$b ☆ c = d.$$

Use this method to find each of the following.

Perform b

13. $b ☆ d$ **15.** $b ☆ b$ **17.** $d ☆ d$ **19.** $a ☆ a$

14. $c ☆ d$ **16.** $c ☆ c$ **18.** $d ☆ b$ **20.** $a ☆ b$

21. Complete the following table.

☆	a	b	c	d
a	a	b	c	d
b	b	c		a
c	c		a	
d	d	a		

Start with b, and perform c

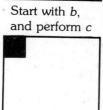

22. Which of the properties from this section are satisfied by this system?

*23. Define a universal set U as the set of counting numbers. Form a new set which contains all possible subsets of U. This new set of subsets together with the operation of set intersection (from Chapter 1) forms a mathematical system. Which of the properties listed in this section are satisfied by this system?

*24. Replace the word "intersection" with the word "union" in Exercise 23; then answer the same question.

*25. Complete the following table so that the result is *not* the same as operation ☆ of the text, but so that the five properties listed in this section hold.

	a	b	c	d
a				
b				
c				
d				

6.5 THE DISTRIBUTIVE PROPERTY

All the properties we have studied so far in this chapter, closure, commutative, associative, identity, and inverse, involve only one operation at a time. In this section, we shall study the distributive property, which involves *two* operations.

We have already seen one kind of example of the distributive property. For example, as we saw in Chapter 4,

$$\overbrace{3 \cdot (5 + 9)} = 3 \cdot 5 + 3 \cdot 9. \qquad *$$

[Check that $3 \cdot (5 + 9) = 3(14) = 42$, while $3 \cdot 5 + 3 \cdot 9 = 15 + 27 = 42$.] Here the 3 on the left is "distributed" over the 5 and the 9. In general, for real numbers a, b, and c,

$$a \cdot (b + c) = a \cdot b + a \cdot c.$$

Because of this property, we say that *multiplication is distributive over addition*.

Example 1   Is addition distributive over multiplication?

Solution.   To find out, we exchange $\cdot$ and $+$ in the equation labeled (*) above. We get

$$a + (b \cdot c) = (a + b) \cdot (a + c).$$

We need to find out if this statement is true for *every* choice of three numbers that we might make. Try an example. If $a = 3$, $b = 4$, and $c = 5$, we have

$$a + (b \cdot c) = 3 + (4 \cdot 5) = 3 + 20 = 23,$$

while $(a + b) \cdot (a + c) = (3 + 4) \cdot (3 + 5) = 7 \cdot 8 = 56.$

Since $23 \neq 56$, we have $3 + (4 \cdot 5) \neq (3 + 4) \cdot (3 + 5)$. This false result is a *counterexample* (an example showing that a general statement is false). Because of this counterexample, addition is *not* distributive over multiplication.

The distributive property of multiplication over addition is used to simplify some calculations in arithmetic, as shown by the next examples. (For simplicity, the distributive property of multiplication over addition will be called only "the distributive property.")

Example 2   Use the distributive property to find the product $8 \cdot 27$.

Solution.   Write 27 as $20 + 7$. We then have

$$
\begin{aligned}
8 \cdot 27 &= 8 \cdot (20 + 7) \\
&= 8 \cdot 20 + 8 \cdot 7 \qquad \text{(by the distributive property)} \\
&= 160 + 56 \\
8 \cdot 27 &= 216.
\end{aligned}
$$

Example 3   Find $12 \cdot 998$.

Solution.
$$
\begin{aligned}
12 \cdot 998 &= 12 \cdot (1000 - 2) \\
&= 12 \cdot 1000 - 12 \cdot 2 \\
&= 12{,}000 - 24 \\
12 \cdot 998 &= 11{,}976.
\end{aligned}
$$

'Rithmetic tables   *The tables you see throughout this chapter are very similar to the old pencil box multiplication tables. (Or did you lose your pencil box again?)*

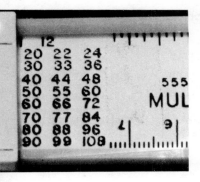

Example 4 Simplify the algebraic expression $8n + 10n$.

Solution. The symbol $8n$ represents the product of 8 and n. Here we use the distributive property "backwards" from the way that it was used above:

$$8n + 10n = 8 \cdot n + 10 \cdot n$$
$$= (8 + 10) \cdot n$$
$$8n + 10n = 18n.$$

The distributive property also applies to more than two products.

Example 5 $9z - 4z + 7z - 5z = (9 - 4 + 7 - 5)z = 7z.$

The distributive property also is used when *factoring,* or writing a sum as a product, as shown in the next example.

Example 6 **(a)** Simplify $12p + 18q$.

Solution. The numbers 12 and 18 are both divisible by 6. Thus,

$$12p + 18q = 6 \cdot 2p + 6 \cdot 3q$$
$$= 6 \cdot (2p + 3q).$$

This last result is usually written without the multiplication dot, as follows:

$$12p + 18q = 6(2p + 3q).$$

To check this result, multiply 6 by $2p$ and 6 by $3q$.

(b) Simplify $15rs - 30rt + 5r$.

Solution. Each of $15rs$, $-30rt$, and $5r$ is divisible by $5r$. We have

$$15rs - 30rt + 5r = (5r) \cdot (3s) - (5r) \cdot (6t) + (5r) \cdot 1$$
$$= 5r\,(3s - 6t + 1).$$

We state the **general form of the distributive property:**

Let ☆ and ∘ be two operations defined for elements in the same set. Then ☆ is distributive over ∘ in case
$a ☆ (b ∘ c) = (a ☆ b) ∘ (a ☆ c)$
for *every* choice of elements a, b, and c from the set.

6.5 EXERCISES

Try examples to help you decide whether or not the operations, when applied to the real numbers, satisfy the distributive property.

1. multiplication over subtraction
2. addition over division
3. subtraction over multiplication
4. division over multiplication
5. addition over subtraction
6. subtraction over addition

Use the distributive property to work each problem.

7. $9 \cdot 52$	**11.** $5 \cdot 76$	**15.** $4 \cdot 374$
8. $6 \cdot 73$	**12.** $9 \cdot 88$	**16.** $5 \cdot 574$
9. $3 \cdot 95$	**13.** $7 \cdot 125$	**17.** $8 \cdot 9999$
10. $7 \cdot 59$	**14.** $9 \cdot 258$	**18.** $7 \cdot 99,999$

Use the distributive property, where possible, to simplify each algebraic expression. (*Hint:* m is the same as $1m$.)

19. $7p + 9p$	**29.** $11k - k$
20. $11r + 5r$	**30.** $32b - b$
21. $12z + 2z$	**31.** $12a - 18a$
22. $3k + 15k$	**32.** $25x - 35x$
23. $p + 5p$	**33.** $5k + 2r$
24. $12r + r$	**34.** $9z + 4y$
25. $a + a$	**35.** $7p + 8p - 9p - 2p + p$
26. $y + y$	**36.** $12z + 5z - 3z - 9z + 15z$
27. $14r - 3r$	**37.** $-2y + 9y - 8y - 7y + 6y - 14y$
28. $27m - 19m$	**38.** $-3k - 5k - 8k + 17k + 2k$

Factor each of the following.

39. $15m + 10k$	**43.** $60abc - 12bc + 10ab$
40. $20p + 70y$	**44.** $35mpq - 56mp + 28\mathit{l}$
41. $12ab + 30bp$	**45.** $17m + 21n - 5p$
42. $4mn - 8mp$	**46.** $9z + 16y + 23x$

Niels Henrik Abel *(1802–1829) of Norway was spotted in childhood as a mathematical genius, but never received in his lifetime the professional recognition his work deserved.*

At 16, influenced by a perceptive teacher, he read the works of Newton, Euler, and Lagrange. In only a few years he began producing original work of his own.

When his father died, Abel assumed responsibility for his family, and never escaped poverty. Even though a government grant enabled him to visit Germany and France, the leading mathematicians there failed to acknowledge his genius. He died of tuberculosis at age 27.

*6.6 GROUPS

We have studied mathematical systems and their properties. Not all systems satisfy all the basic properties—associative, commutative, identity, inverse, and closure. In order to study mathematical systems, it has been found helpful to classify systems according to the properties that they do satisfy. For example, a mathematical system satisfying the closure, associative, identity, and inverse properties is called a **group**. The set of integers together with the operation of addition satisfies all four of these properties, and thus is a group.

Groups have many applications, especially in mathematics itself, but also in chemistry and physics. What makes groups so appealing to mathematicians is group *structure*. Any system that is a group has a known structure— it satisfies the closure, associative, identity, and inverse properties. Thus, theorems and formulas can be proved for groups in general. Then, any group that we might encounter in our work will automatically satisfy all the formulas and theorems that have been developed. Groups are so important that the very term "algebra" to a mathematician means groups and other related structures.

Example 1   Is the system made up of the set of rational numbers and the operation of addition a group?

Solution.   This system satisfies the closure, associative, identity, and inverse properties, and thus is a group.

Example 2   Does the set of odd integers {. . . , −5, −3, −1, 1, 3, 5, . . .} and the operation of multiplication form a group?

Solution.   Here the closure, associative, and identity properties are satisfied, but not the inverse property. Thus, the system is *not* a group.

Example 3   Does the set {−1, 1} and the operation of multiplication form a group?

Solution.   Check the necessary four properties.
Closure   The given system leads to this multiplication table:

$\bullet$	−1	1
−1	1	−1
1	−1	1

All the entries in the body of the table are either −1 or 1, so that the system is closed.
Associative   Both −1 and 1 are integers, and we know that multiplication of integers is associative.
Identity   The identity for multiplication is 1, an element of the set of the system, {−1, 1}.
Inverse   Both −1 and 1 are their own inverses for multiplication.
    Since all four of the necessary properties are satisfied, the system is a group.

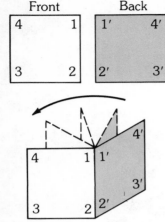

Front | Back

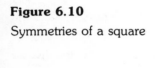

Figure 6.9 Original position of square

Group structure applies not only to sets of numbers. One common group is the group of **symmetries of a square,** which we will develop. First, cut out a small square, and label it as shown in Figure 6.9.

Make sure that 1 is in front of 1', 2 is in front of 2', 3 is in front of 3', and 4 is in front of 4'. Let the letter M represent a clockwise rotation of 90°. Let N represent a rotation of 180°, and so on. A complete list of the symmetries of a square is given in Figure 6.10.

We can combine symmetries as follows. We let *NP* represent *N* followed by *P*. Performing *N* and then *P* is the same as performing just *M,* so that *NP = M.* See Figure 6.11.

Example 4 Find *RT*.

Solution. First, perform *R* by flipping the square about a horizontal line through the middle. Then, perform *T* by flipping the result of *R* about a diagonal from upper left to lower right. The result of *RT* is the same as performing only *M,* so that *RT = M.*

Figure 6.10

Symmetries of a square

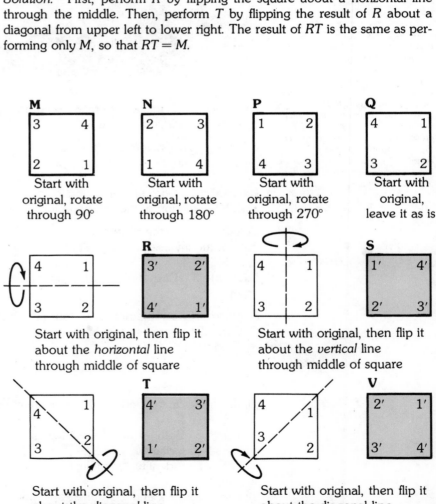

Original (Q)

Apply N

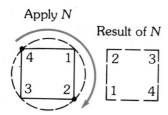

Result of N

Apply P

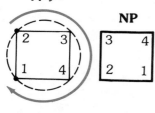

NP

Apply M

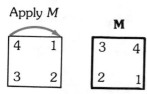

M

Figure 6.11 Think of N as advancing each corner two quarter turns clockwise. Thus 4 goes from upper left to lower right. To this result apply P, which advances each corner three quarter turns. Thus 2 goes from upper left to lower left. The result, NP, is the same as advancing each (original) corner one quarter turn, which M does. Thus, NP = M.

Using the method of Example 4, we can complete the following table for combining the symmetries of a square.

□	M	N	P	Q	R	S	T	V
M	N	P	Q	M	V	T	R	S
N	P	Q	M	N	S	R	V	T
P	Q	M	N	P	T	V	S	R
Q	M	N	P	Q	R	S	T	V
R	T	S	V	R	Q	N	M	P
S	V	R	T	S	N	Q	P	M
T	S	V	R	T	P	M	Q	N
V	R	T	S	V	M	P	N	Q

Example 5 Show that the system made up of the symmetries of a square is a group.

Solution. We must show that the system satisfies the closure, associative, identity, and inverse properties.

Closure All the entries in the body of the table come from the set $\{M, N, P, Q, R, S, T, V\}$. Thus, the system is closed.

Associative We can try examples:

$$P(MT) = P(R) = T.$$

Also,
$$(PM)\,T = (Q)\,T = T,$$
so that
$$P(MT) = (PM)\,T.$$

Other similar examples also work. (See Exercises 23–28 below.) Thus, the system has the associative property.

Identity The column at the left in the table is repeated under Q. Check that Q is indeed the identity element.

Inverse In the first row, Q appears under P. Check that M and P are inverses of each other. In fact, every element in the system has an inverse. (See Exercises 29–34 below.)

Since all four of the properties are satisfied, the system is a group.

Example 6 Form a mathematical system by using only the set $\{M, N, P, Q\}$ from the group of symmetries of a square. Is this new system a group?

Solution. The table for the elements $\{M, N, P, Q\}$ is just one corner of the table for the entire system.

	M	N	P	Q
M	N	P	Q	M
N	P	Q	M	N
P	Q	M	N	P
Q	M	N	P	Q

Verify that the system represented by this table satisfies all four properties and thus is a group. This new group is a *subgroup* of the original group of the symmetries of a square.

6.6 EXERCISES

Decide whether or not each system is a group. If *not a group,* tell which
property or properties is (are) not satisfied.

  1. odd integers; addition

  2. counting numbers; addition   *no*

  3. integers; multiplication

  4. integers; subtraction   *no*

  5. rational numbers; addition

  6. non-zero rational numbers; multiplication   *yes*

  7. non-zero rational numbers; division

  8. prime numbers; addition   *no*

 *9. $\{0, 1\}$; addition   *no   closure 1+1=2*

*10. $\{0, 1\}$; multiplication   *no - inverse*

*11. $\{0\}$; addition   *yes*

*12. $\{1\}$; multiplication   *yes*

*13. $\{1, -1\}$; addition   *no   closure*

*14. $\{1, -1\}$; division   *yes*

*15. List five different subgroups of the group of integers with addition.

The following exercises apply to the system of symmetries of a square
presented in the text.

Find each of the following.

  16. RV **18.** ST **20.** PT **22.** VV

  17. NR **19.** VM **21.** QV

Show that each statement is true.

  23. $N(PR) = (NP)R$ **26.** $P(RS) = (PR)S$

  24. $S(TV) = (ST)V$ **27.** $T(RV) = (TR)V$

  25. $V(MN) = (VM)N$ **28.** $V(MP) = (VM)P$

Find the inverse of each element.

  29. N **31.** R **33.** T

  30. Q **32.** S **34.** V

The following exercises apply to the *modulo systems* we discussed earlier
in this chapter. Do the following modulo systems form groups?

  35. addition modulo 3   **38.** addition modulo 6

  36. addition modulo 4   **39.** addition modulo 12

  37. addition modulo 5

Emmy Noether (1882–1935) was an outstanding mathematician in the field of abstract algebra. She studied and worked in Germany at a time when it was very difficult for a woman to do so. At the University of Erlangen in 1900, Noether was one of only two women. Although she could sit in on classes, professors could and did deny her the right to take the exams for their courses. It wasn't until 1904 that Noether was allowed to register officially. She completed her doctorate four years later.

In 1916 Emmy Noether went to Gottingen to work with David Hilbert on the general theory of relativity. But even with Hilbert's backing and prestige, it was three years before the faculty voted to make Noether a Privatdozent, the lowest rank in the faculty. In 1922 Noether was made an unofficial professor (or assistant). She received no pay for this post, although she was given a small stipend to lecture in algebra.

Noether's area of interest was abstract algebra, particularly structures called rings and ideals. (Groups are structures, too, with different properties.) One special type of ring bears her name, since she was the first to study its properties.

Forced to leave Germany in 1933, Noether taught at Bryn Mawr for the last 18 months of her life. There she became good friends with Anna Pell Wheeler, head of the mathematics department. Wheeler had studied at Gottingen and understood well the problems that Emmy Noether had faced as a woman scholar in Germany.

*40. Complete this statement: _____ modulo systems form a group for addition.

41. $\{1, 2, 3\}$ and multiplication modulo 4

42. $\{1, 2, 3, 4\}$ and multiplication modulo 5

43. $\{1, 2, 3, 4, 5\}$ and multiplication modulo 6

44. $\{1, 2, 3, 4, 5, 6\}$ and multiplication modulo 7

45. $\{1, 2, 3, 4, 5, 6, 7, 8\}$ and multiplication modulo 9

*46. Complete this statement: modulo systems form a group for multiplication only when the modulo number is _____.

Consider the set $\{1, 2, 3, 4\}$ and the operation of multiplication modulo 5. Find each of the following in this system.

47. 2^1 **48.** 2^2 **49.** 2^3 **50.** 2^4 **51.** 2^5

The answers you got should have been the elements in the set $\{1, 2, 3, 4\}$. (Not in that order.) Thus, the various powers of 2 lead to all the elements in the set of the system. For this reason, 2 is called a *generator* of the group, and the group itself is called a **cyclic group.** Check whether the following elements are generators of this same group.

52. 3   **53.** 4   **54.** 1

*55. Is the mathematical system made up of the set $\{1, 2, 3, 4, 5, 6\}$ and multiplication modulo 7 a cyclic group?

*56. Is the group of symmetries of the square a cyclic group?

CHAPTER 6 SUMMARY

Keywords

Mathematical system	Closure	Factoring
12-hour clock	Commutative	Group
Finite system	Associative	Subgroup
Congruent modulo k	Identity	Symmetries of a square
Casting out nines	Identity element	Generator
Operation table	Inverse	Cyclic group

Properties   Here a, b, and c represent elements from the set of a mathematical system, and $\circ$ represents the operation of the system.

Closure   The system is closed if for any two elements a and b,
$$a \circ b$$
is in the set of the system.

Commutative   The system has the commutative property if
$$a \circ b = b \circ a$$
for every choice of two elements a and b from the system.

Associative   The system has the associative property if
$$(a \circ b) \circ c = a \circ (b \circ c)$$
for every choice of three elements a, b, and c of the system.

Identity   The system has an identity element e (where e is in the set of the system) if
$$a \circ e = a \quad \text{and} \quad e \circ a = a,$$
for every element a in the system.

Inverse   The system satisfies the inverse property if, for every element a of the system there is an element b in the system such that
$$a \circ b = e \quad \text{and} \quad b \circ a = e,$$
where e is the identity element of the system.

Distributive   In addition to operation $\circ$, suppose another operation $\star$ is defined on the system. Then $\star$ is distributive over $\circ$ in case
$$a \star (b \circ c) = (a \star b) \circ (a \star c)$$
for every choice of elements a, b, and c from the set.

CHAPTER 6 TEST

Find each of the following, using a 12-hour clock.

1. $6 + 11$ **3.** $5 - 8$ **5.** $8 \cdot 6$

2. $9 + 10$ **4.** $7 - 11$ **6.** $9 \cdot 7$

Find each of the following, using a 7-hour clock.

7. $5 + 6$ **9.** $1 - 6$

8. $4 \cdot 3$ **10.** $3 - 5$

Answer *true* or *false* for each of the following.

11. $27 \equiv 3 \pmod 5$ **13.** $12,597 \equiv 12 \pmod 3$

12. $598 \equiv 7 \pmod{16}$

Find each sum or product.

14. $(38 + 49) \pmod 7$ **16.** $(25 \cdot 38) \pmod 2$

15. $(76 \cdot 3) \pmod 5$

17. Write an addition table for a modulo 8 system.

18. Write a multiplication table for a modulo 7 system.

Find all possible solutions for the following.

19. $x \equiv 7 \pmod 9$ **22.** $x \equiv 2 \pmod 5$

20. $x + 6 \equiv 8 \pmod 3$ $x \equiv 3 \pmod 7$

21. $7x \equiv 1 \pmod 5$ x is less than 50

Check each of the following by casting out nines.

23.
$$
\begin{array}{r}
7496 \\
8917 \\
+\quad 3258 \\
\hline
19,671
\end{array}
$$

25.
$$
\begin{array}{r}
76,928 \\
-\ 64,761 \\
\hline
12,167
\end{array}
$$

27.
$$
\begin{array}{r}
58 \\
76\overline{)4408}
\end{array}
$$

24.
$$
\begin{array}{r}
568,342 \\
+\ 971,897 \\
\hline
1,550,239
\end{array}
$$

26.
$$
\begin{array}{r}
897 \\
\times\quad 64 \\
\hline
57,408
\end{array}
$$

P	a	b	c	d	e
a	c	d	a	b	d
b	d	c	b	e	b
c	a	b	c	d	e
d	b	e	d	c	a
e	d	b	e	a	c

Let operation **P** be defined by the table, and evaluate items 28–30.

28. $a\,\boldsymbol{P}\,d$ **29.** $(b\,\boldsymbol{P}\,c)\,\boldsymbol{P}\,d$ **30.** $(c\,\boldsymbol{P}\,d)\,\boldsymbol{P}\,(a\,\boldsymbol{P}\,b)$

Which of the following properties are satisfied by the set $\{a, b, c, d, e\}$ and operation **P?**

31. Closure   **32.** Commutative   **33.** Associative

34. Identity   (What is the identity element?)

35. Inverse   (Give the inverse of each element that has one.)

Use the distributive property to simplify:

36. $8 \cdot 77$ **37.** $4x + 9x - 8x$

38. Factor $16k - 12z$.

Decide whether or not each system represents a group.

39. $\{\ldots, -6, -4, -2, 0, 2, 4, 6, \ldots\}$; addition

40. $\{\ldots, -7, -5, -3, -1, 1, 3, 5, 7, \ldots\}$; multiplication

Chapter 7 *Algebra*

Moslem scribes *copied ancient manuscripts, as did Christian and Jewish scribes. During the Middle Ages there was no less laborious way to duplicate texts. Printing was not invented until the 15th century.*

The Republic of Mali (northwest Africa) has issued a series of glowing stamps to honor Moslem culture. The scribe pictured in this stamp is from a Persian miniature painting. The words running vertically (in French and Arabic) command us to learn from the cradle to the grave.

Algebra is a generalized arithmetic. Algebra deals with numbers, as does arithmetic, but uses *variables*—letters which stand for numbers. This permits more complicated problems to be solved.

Algebra goes back to the Babylonians of 2000 BC. They developed methods of solving quadratic equations (those containing x^2). Multiplication of positive and negative numbers can be traced back to 300 BC.

The Egyptians also worked problems in algebra, but the problems were not as complex as those of the Babylonians. The Egyptians drew a pair of legs walking from right to left to represent "plus," with legs walking left to right for "minus."

Little further was done in algebra until the time of the Hindus, about the sixth century. The Hindus had methods for solving many types of practical problems, such as interest, discounts, and partnerships.

Many Hindu and Greek works on mathematics were saved for us today only because Moslem scholars made translations of them. These translations, made generally from 750 to 1250, were mostly done in Bagdad. The Arabs took the work of the Greeks and Hindus and greatly expanded it. For example, Mohammed ibn Musa al-Khowarizmi wrote books on algebra and on the Hindu numeration system (the one we use). His books had a tremendous influence in Western Europe; his name is remembered today in the word *algorithm* (see Chapter 3). Perhaps al-Khowarizmi's most famous book was *Hisab al-jabr, wal-mugabalah,* from whose title we get the very word "algebra."

Throughout the fifteenth and sixteenth centuries, a main interest of mathematicians was in solving more and more complicated equations. Gradually, algebra found much use in the development of other branches of mathematics, such as the calculus.

7.1 EQUATIONS AND FORMULAS

An algebraic expression is made up of three things: numbers, variables, and operation signs, such as $\cdot$ or $+$. A *variable* is a letter, such as x, y, or z, that stands for a number. A statement that says two such expressions are equal is called an **equation,** for example,

$$y + 8 = 15, \qquad 5m = 30, \qquad 8p - 2 = 3p + 8, \qquad t = 3.$$

To *solve* an equation, you must find all numerical values of the variable that make the equation true. The values of the variable that make the equation true are called the **solutions** of the equation.

For example, the number 3 is a solution to the equation $8y + 7 = 31$, since replacing the variable y with the number 3 makes the equation true:

$$
\begin{aligned}
8\,y + 7 &= 31 \\
8 \cdot 3 + 7 &= 31 \qquad \text{Let } y = 3 \\
24 + 7 &= 31 \\
31 &= 31 \qquad \text{True}
\end{aligned}
$$

In fact, 3 is *the* (one and only) solution for the equation $8y + 7 = 31$.

Our plan in solving an equation is to go through a series of steps, ending up with an equation where only the variable is on one side of the equals sign, and a number is on the other side. That is, we want to reduce the given equation to the form

$$x = \text{number.}$$

Think about the equation

$$x + 5 = 12.$$

We know that $x + 5$ and 12 represent the same number, since this is the meaning of the $=$ sign. We want to change the left-hand side from $x + 5$ to just x. We could do this by adding -5 to $x + 5$:

$$x + 5 + (-5) = x.$$

To keep the equality between the left and right sides of the equation, we must also add -5 to 12: $\quad 12 + (-5) = 7.$

By adding -5 to both sides of the equation $x + 5 = 12$, we get the streamlined equation

$$x = 7,$$

which gives the solution to the original equation, $x + 5 = 12$. The solution is 7. Our series of steps goes from the given equation to successively simpler ones, ending with the variable equal to a number:

$$
\begin{aligned}
x + 5 &= 12 \\
x + 5 + (-5) &= 12 + (-5) \\
x + 0 &= 7 \\
x &= 7.
\end{aligned}
$$

al-jabr. . .algebra. . .algebrista
In the title of Khowarizmi's book, written in the 9th century, jabr *("restoration") refers to transposing negative quantities across the equals sign in solving equations. From Latin versions of Khowarizmi's text, "al-jabr" became the broad term covering the arts of equation solving. (The prefix al means "the.")*

In Spain under Moslem rule, the word algebrista *referred to the person who restores (resets) broken bones. You would have seen signs outside barber shops saying Algebrista y Sangrador (bonesetter and bloodletter). Such services were part of the barber's trade. The red-and-white striped barber pole you can see today symbolizes blood and bandages (see page 248).*

In solving this equation, we used the following rule:

> The same number can be added to or subtracted from both sides of an equation.

A similar rule holds for multiplication and division:

> The same non-zero number can be used to multiply or divide both sides of an equation.

We have to say "non-zero number" since multiplying both sides of *any* equation by 0 gives $0 = 0$, which is true, but of no help in finding the solution.

Example 1   Solve each equation: (a) $x + 9 = 17$ (b) $7r = 84$ (c) $3k - 2 = 10$.

(a) We want to get x alone on one side of the equals sign. We would get x alone if the 9 could be made to disappear. To get rid of the 9 on the left-hand side, add -9 to *both* sides of the equation:

$$x + 9 = 17$$
$$x + 9 + (-9) = 17 + (-9)$$
$$x = 8 \qquad\qquad 9 + (-9) = 0$$

(b) To solve $7r = 84$, we can divide both sides by 7:

$$7r = 84$$
$$\frac{7r}{7} = \frac{84}{7} \qquad \frac{7}{7}r = 1r = r$$
$$r = 12$$

(c) To solve $3k - 2 = 10$, we need to go through two steps.
First, add 2 on both sides: *Second*, divide both sides by 3:

$$3k - 2 = 10$$
$$3k - 2 + 2 = 10 + 2$$
$$3k = 12.$$

$$\frac{3k}{3} = \frac{12}{3}$$
$$k = 4.$$

The solution is 4.

Solving equations usually involves more steps than in Example 1 above, because most practical problems involve equations that are not as simple as those in the example. For one thing, we often must **combine terms.** We do this with the distributive property of Chapter 5, as follows.

$$8x + 6x = (8 + 6)x \qquad \text{Distributive property}$$
$$8x + 6x = 14x. \qquad\qquad 8 + 6 = 14$$

Also, $9y - 4y + 3y - 11y = (9 - 4 + 3 - 11)y = -3y.$

Notice that $5x$ and $6y$ cannot be combined in this way.

Example 2 Solve the equation $4r + 5r - 3 + 8 - 3r - 5r = 12 + 8$.

Solution. First, simplify the equation by combining terms wherever possible. On the left side of the equals sign, we get

$$4r + 5r - 3 + 8 - 3r - 5r = (4r + 5r - 3r - 5r) + (-3 + 8)$$
$$= 1r + 5$$
$$= r + 5.$$

On the right of the equals sign, $12 + 8 = 20$. The equation is now simplified to

$$r + 5 = 20.$$

Then we can add -5 to both sides:

$$r + 5 + (-5) = 20 + (-5)$$
$$r = 15.$$

The solution of the given equation is 15. To check this, substitute 15 for r in the original equation.

Example 3 Solve $3r + 4 - 2r - 7 = 4r + 3$.

Solution. Combine terms as possible.

$$r - 3 = 4r + 3.$$

There are several possible ways to proceed now. One way is first to add 3 on both sides.

$$r - 3 + 3 = 4r + 3 + 3$$
$$r = 4r + 6.$$

Next add $-4r$ to both sides.

$$r + (-4r) = 4r + 6 + (-4r)$$
$$-3r = 6.$$

Finally divide both sides by -3.

$$\frac{-3r}{-3} = \frac{6}{-3}$$
$$r = -2.$$

The solution is -2. (*Recall:* The quotient of a positive number and a negative number is negative.)

Example 4 Solve the equation $4(k - 3) - k = k - 6$.

Solution. Before combining terms, use the distributive property to simplify $4(k - 3)$.

$$4(k - 3) = 4k - 4 \cdot 3 = 4k - 12.$$

Now combine terms.

$$4k - 12 - k = k - 6$$
$$3k - 12 = k - 6$$

Add 12 to both sides.

$$3k - 12 + 12 = k - 6 + 12$$
$$3k = k + 6.$$

Add $-k$ to both sides.

$$3k + (-k) = k + 6 + (-k)$$
$$2k = 6.$$

Divide both sides by 2.

$$k = 3.$$

François Viète (1540–1603) was a lawyer at the court of Henri IV of France and studied equations. Viète simplified the notation of algebra and was among the first to use letters to represent numbers. For centuries, algebra and arithmetic were expressed in a cumbersome way with words and occasional symbols. Since the time of Viète, algebra has gone beyond equation solving; the abstract nature of higher algebra depends on its symbolic language.

Solving equations often depends on knowledge of **formulas,** which give connections between certain dimensions, amounts, or quantities. For example, formulas exist for distance and for money earned on savings (interest, as discussed in Chapter 5). Scientists, engineers, and economists, among others, make use of formulas that are often very complicated. Geometric formulas for area and volume have been known since the ancient Babylonians solved agricultural and building problems. (We will discuss such formulas in the next chapter.)

The problems in Examples 5 and 6 below depend on the formula for the distance traveled by an object moving at a certain rate of speed. Distance in this case is the product of rate and time:

Words   distance = rate · time
Symbols $d = rt$

Example 5   Jim drove at the rate of 75 kilometers per hour for 7 hours. Find the total distance that he drove.

Solution.   Use the formula $d = rt$. Here $r = 75$ and $t = 7$, so that

$$d = 75 \cdot 7 = 525.$$

Jim drove 525 kilometers.

Example 6   Two cars start from the same point at the same time and travel in the same direction at constant speeds of 34 and 45 kilometers per hour, respectively. In how many hours will they be 33 kilometers apart?

Solution.   To work this problem, we use the formula $d = rt$. Let t represent the unknown number of hours. The distance traveled by the slower car is its rate multiplied by its time, or $34t$. The distance traveled by the faster car is $45t$. The numbers $34t$ and $45t$ represent different distances. From the information of the problem, we know that these distances differ by 33 kilometers, or $45t - 34t = 33$. We can now solve this equation to find t.

$$45t - 34t = 33$$
$$11t = 33$$
$$t = 3.$$

Figure 7.1

In 3 hours, the two cars will be 33 kilometers apart.

Example 7 How much simple interest will be earned on a deposit of $600 for 4 years at an interest rate of 6% per year?

Solution. The formula for simple interest is $I = prt$, where I represents interest, p represents principal, r is rate, and t is time in years. Using the given information, we have

$$I = prt$$
$$I = (600)(.06)(4) \qquad 6\% = .06$$
$$I = 144.$$

The deposit will earn interest of $144.

When you buy an item on the installment plan, you agree to pay a certain finance charge. If you pay off the loan early, you do not need to pay the entire finance charge. By one method, called the *rule of 78,* the amount of finance charge that you do not have to pay (called the *unearned interest*) is given by

$$u = f \cdot \frac{k(k+1)}{n(n+1)} \qquad \text{where} \quad \begin{aligned} &u = \text{unearned interest} \\ &f = \text{original finance charge} \\ &k = \text{number of payments remaining} \\ &\qquad \text{when the loan is paid off} \\ &n = \text{original number of payments} \end{aligned}$$

Example 8 Find the unearned interest if the total finance charge is $360, the loan was scheduled for 36 payments, and the loan is paid off after 24 payments (that is, 12 payments ahead of schedule).

Solution. Here, $f = 360$, $n = 36$, and $k = 12$. From the formula,

$$u = 360 \cdot \frac{12(12+1)}{36(36+1)} = 360 \cdot \frac{12(13)}{36(37)} = 42.16.$$

(We used a calculator.) Thus, $42.16 of the $360 finance charge need not be paid. The total finance charge is $360 - $42.16 = $317.84.

Note that the loan was paid off with 1/3 of the original life remaining. However, only $42.16 out of $360, or about 12%, of the finance charge was saved. Paying off an installment loan ahead of schedule may not be as beneficial as depositing the same money in a bank account.

Problem Solving

It is often necessary to translate the words of a problem into algebra before the solution can be found. This is done following these steps:

(1) Choose a variable to represent the value that you need to find — the unknown number.

(2) Translate the problem into an equation.

(3) Solve the equation.

(4) Check your solution in the words of the original problem.

Step (2) is usually the hardest. To translate the problem, go from facts stated in words to mathematical *expressions*. The problem also gives you some fact about equality, which leads to an *equation*.

What about words? You are likely to see certain words again and again in problems. Some of these words are listed here, along with the mathematical symbols that represent them. We use x here as the unknown, but any letter could have been used.

Addition   5 **plus** a number $5 + x$
add 20 to a number $x + 20$
7 **more than** a number $x + 7$
a quantity is **increased by** 3 $x + 3$

Subtraction   3 **less than** a number $x - 3$
a number **decreased by** 14 $x - 14$
subtract 6 **from** a number $x - 6$

Multiplication   the **product** *of* the number and 3 $3x$
three times a number $3x$
a number is **tripled** $3x$
two-thirds **of** a number $\frac{2}{3}x$ (fractions only)

Division   (Use a fraction bar instead of $\div$)

the **quotient** of a number and 2 $\dfrac{x}{2}$

half the number $\dfrac{x}{2}$ or $\dfrac{1}{2}x$

Since equal mathematical expressions are names for the same number, any words that mean *equality* or *sameness* translate as =.

Equality   Four times a number, decreased by 7 **is** 97. $4x - 7 = 97$

The quotient of a number plus 7, and 2, **is** 4. $\dfrac{x + 7}{2} = 4$

If you add 10 to a number, the **result is** 20. $x + 10 = 20$

Example 9   If five times a number is added to three times the number, the result is the sum of seven times the number, and 9. Find the number.

Solution.   We have to translate the words into mathematical symbols, using the essential facts to find an equation.

five times a number	added to	three times a number	result is	sum
$5x$	$+$	$3x$	$=$	$7x + 9$

We now have the equation $5x + 3x = 7x + 9$, which we can solve.

$$5x + 3x = 7x + 9$$
$$8x = 7x + 9$$
$$8x + (-7x) = 7x + 9 + (-7x)$$
$$x = 9.$$

The number we want is 9. Check in the words of the original problem: *Five times 9* [45] *added to three times 9* [27] *is seven times 9* [63] *added to 9;* or 45 + 27 = 63 + 9.

When trying to solve problems, you might find the following steps to be helpful. They are from *How to Solve It,* by George Polya, published by Princeton University Press. This book is a modern classic among mathematics books.

Understanding the problem

**First.
You have to *understand* the problem.**

What is the unknown? What are the data? What is the condition? Is it possible to satisfy the condition? Is the condition sufficient to determine the unknown? Or is it insufficient? Or redundant? Or contradictory?
Draw a figure. Introduce suitable notation.
Separate the various parts of the condition. Can you write them down?

Devising a plan

**Second.
Find the connection between the data and the unknown. You may be obliged to consider auxiliary problems if an immediate connection cannot be found. You should obtain eventually a a *plan* of the solution.**

Have you seen it before? Or have you seen the same problem in a slightly different form?
Do you know a related problem? Do you know a theorem that could be useful?
Look at the unknown! And try to think of a familiar problem having the same or a similar unknown.
Here is a problem related to yours and solved before. Could you use it? Could you use its result? Could you use its method? Should you introduce some auxiliary element in order to make its use possible?
Could you restate the problem? Could you restate it still differently? Go back to definitions.
If you cannot solve the proposed problem try to solve first some related problem. Could you imagine a more accessible related problem? A more general problem? A more special problem? An analogous problem? Could you solve a part of the problem? Keep only a part of the condition, drop the other part; how far is the unknown then determined, how can it vary? Could you derive something useful from the data? Could you think of other data appropriate to determine the unknown? Could you change the unknown or the data, or both if necessary, so that the new unknown and the new data are nearer to each other?
Did you use all the data? Did you use the whole condition? Have you taken into account all essential notions involved in the problem?

Carrying out the plan

**Third.
Carry out your plan.**

Carrying out your plan of the solution, *check each step.* Can you see clearly that the step is correct? Can you prove that it is correct?

Looking back

**Fourth.
Examine the solution obtained.**

Can you *check the result?* Can you check the argument?
Can you derive the result differently? Can you see it at a glance?
Can you use the result, or the method, for some other problem?

7.1 EXERCISES

Solve the equations in Exercises 1–36.

1.	$m + 2 = 6$	**19.**	$2p + 6 = 10 + p$
2.	$y + 5 = 12$	**20.**	$5r + 2 = -1 + 4r$
3.	$a - 8 = 7$	**21.**	$2k + 2 = -3 + k$
4.	$p - 14 = 2$	**22.**	$6 + 7x = 6x + 3$
5.	$2x = 30$	**23.**	$4x + 3 + 2x - 5x = 2 + 8$
6.	$5s = 75$	**24.**	$3x + 2x - 6 + x = 9 + 4 + 5x$
7.	$6m = 12$	**25.**	$3x + 9 = -3(2x + 3)$
8.	$10b = 70$	**26.**	$4z + 2 = -2(z + 2)$
9.	$-4k = 12$	**27.**	$3k - 5 = 2(k + 6) + 1$
10.	$-8p = 32$	**28.**	$4a - 7 = 3(2a + 5) - 2$
11.	$-7z = -28$	**29.**	$-4 - 3(2x + 1) = 11$
12.	$-2k = -40$	**30.**	$8 - 2(3x - 4) = 2x$
13.	$6z + 1 = 43$	**31.**	$5(2m - 1) = 4(2m + 1) + 7$
14.	$7y + 2 = 37$	**32.**	$3(3k - 5) = 4(2k - 5) + 7$
15.	$2a + 9 = 19$	**33.**	$-2(3s + 9) - 6 = -3(3s + 11) - 6$
16.	$9y - 21 = 87$	**34.**	$-3(5z + 24) + 2 = 2(3 - 2z) - 10$
17.	$2 + 3x = 2x$	**35.**	$6(2p - 8) + 24 = 3(5p - 6) - 6$
18.	$10 + r = 2r$	**36.**	$2(5x + 3) - 3 = 6(2x - 3) + 15$

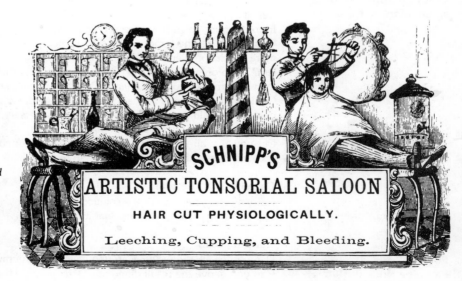

Blood and bandages *The barber pole was not yet a quaint relic in 19th century America, as this advertisement shows. Besides cutting hair, the barber also provided* bloodletting *services. (Doctors believed the practice had curative effects.) Blood was drawn ("let") by opening a vein and either attaching bloodsucking worms called* leeches *or placing a glass cup over the incision to create a partial vacuum and thus draw blood out. Let's hope that Schnipp did that with style.*

Solve each word problem. Use the following formulas.

$$d = rt \qquad I = prt \qquad u = f \cdot \frac{k(k+1)}{n(n+1)}$$

Approximate *true annual interest rate* for a loan paid off in monthly installments:

$$A = \frac{24f}{b(p+1)}$$

A = approximate true annual interest rate
f = total finance charge
b = amount borrowed
p = total number of payments

37. If Lupe goes 80 kilometers at 40 kilometers per hour, find the time it takes her.

38. Distance is 400 kilometers. Rate is 100 kilometers per hour. Find the time.

39. What amount of principal must be invested at 6% per year for 5 years to earn $60 interest?

40. Find the time if a deposit of $4000 earns $200 interest at a rate of 10%.

41. Find the approximate true annual interest rate for an installment loan to be paid off in 24 monthly payments. The finance charge is $200, and the original loan balance (amount borrowed) is $1920.

42. Find the approximate true annual interest rate for an automobile loan to be repaid in 36 monthly payments. The finance charge on the loan is $740 and the amount borrowed is $3600.

43. Joe bought a Honda and agreed to pay for it in 24 monthly payments. The total finance charge on the loan was $450. With 9 payments remaining, he decided to pay the loan in full. Find the amount of unearned interest.

44. Adrian bought a new Ford and agreed to pay it off in 36 monthly payments. The total finance charge is $700. Find the unearned interest if she pays it off 4 payments ahead of schedule.

45. From a point on a straight road, John and Fred ride ten-speed bicycles in opposite directions. John rides 10 miles per hour and Fred rides 12 miles per hour. In how many hours will they be 55 miles apart?

46. Two trains leave Charlotte at the same time. One travels north at 60 miles per hour and the other south at 80 miles per hour. In how many hours will they be 280 miles apart?

*47. Ms. Sullivan has $10,000 to invest. She wants to invest part at 5% interest and part at 6%. The total annual interest is $560. How much is invested at each rate?

*48. Mr. Jones received $16,000 from an insurance settlement. He invested part at 8% a year. He put $2000 less than twice that amount in a safe 5% bond. His total annual income from interest is $980. How much did he invest at each rate?

*49. A boat travels upstream for 3 hours. The return trip requires 2 hours. If the speed of the current is 5 miles per hour, find the speed of the boat in still water.

*50. In an automobile motocross, a driver was 120 miles from the finish line after 5 hours. Another driver, who was in a later heat, traveled at the same speed as the first driver. After 3 hours, she was 250 miles from the finish. What was the speed of each driver?

The local electric utility provided the explanation of the EER (Energy Efficiency Ratio) of an air conditioner.

Here's something to think about when you buy an air conditioner. Choose a unit that gives you more cooling with less electricity— a unit with a high EER (Energy Efficiency Ratio). To find the EER of a unit, just divide the BTU number by the watts. Your answer is the EER number of the unit. EER numbers range from 4.5 to 12. The higher the number, the more efficient the unit. Recommended numbers are 7 or higher. For example, a unit with EER number 8 uses 30% less electricity than a unit with EER 5.6.

51. Write a formula for EER. Let B represent the BTU rating of an air conditioner, and let W represent watts.

52. An air conditioner in a recent Montgomery Ward catalog had a BTU rating of 12,600, with a wattage rating of 1315. Find the EER for this machine.

Work the word problems in Exercises 53–60.

53. The sum of five times a number, three times the number, and 1, is 49. Find the number.

54. The difference of four times a number and three times a number, added to 8, is 11. Find the number.

55. On a recent trip to McDonald's, one of the authors noticed a total of five employees. The number of managers was one less than the number of other employees. Find the number of managers.

56. Joann is three times as old as Marilyn. The sum of their ages is 24. Find Joann's age.

57. Ann has saved $163 for a trip to Disney World. Transportation will cost $28, tickets for the park entrance and rides will cost $15 per day, and lodging and meals will cost $30 per day. How many days can she spend there?

58. Hamburgers cost 90 cents each, and a bag of french fries costs 40 cents. How many hamburgers and how many bags of fries can Ted buy with $8.80 if he wants twice as many hamburgers as bags of french fries?

*59. A grocer buys lettuce for $2.60 a crate. Of the lettuce she buys, 10% spoils and cannot be sold. If she charges 20¢ for each head she sells, and makes a profit of 5¢ on each head she buys, how many heads of lettuce are in a crate? (*Hint:* profit = income − cost.)

*60. A store has 39 quarts of milk, some in pint cartons and some in quart cartons. There are six times as many quart cartons as pint cartons. How many quart cartons are there?

7.2 EQUATIONS IN TWO VARIABLES; SYSTEMS OF EQUATIONS

The equations in the preceding section, such as $x + 5 = 12$ and $y = 7$, were all equations in one variable, whether $x, y, z, r, s, t,$ and so on. We can extend our methods of problem solving by studying equations in *two* variables, such as

$$y = 6x + 9 \qquad \text{or} \qquad 2x + 3y = 12.$$

An equation such as $y = 6x + 9$ shows a *relationship* between the variables x and y. Indeed, one of the main tasks of people using mathematics is finding relationships between variables. For example, we might be interested in the relationship between smoking and lung cancer, between temperature and cricket matings, or between hours of study and scores on a mathematics test.

Suppose we are doing research on this last relationship—between hours of study and scores on a test. We might begin by gathering data from a number of students. We might find, for example, that one student studied 5 hours and earned a score of 84, while another student studied 9 hours and earned a score of 91. As a shortcut way of expressing these relationships, we write (5, 84) to represent the fact that 5 hours of study led to a score of 84; in the same way (9, 91) shows that 9 hours led to a score of 91.

We call (5, 84) an **ordered pair** of numbers. The word "pair" is used because there are two numbers, while "ordered" means that 5 is to be considered the *first* number in (5, 84), and 84 the *second*. Usually, the first number in an ordered pair is called the *x-value*; the second number is the *y-value*.

A **relation** is defined as a set of ordered pairs. In the relationship between hours of study and test scores, we could form a relation by listing all the ordered pairs we get by considering each student taking the test. This relation would be a set of ordered pairs:

$$\{(5, 84), (9, 91), (4, 76), (5, 80), (8, 76), (9, 53), \ldots\}.$$

Henri Poincaré *(1854–1912) is often heralded as the greatest mathematician of his time. He was professor at the Sorbonne, Paris, for twenty-one years and lectured on topics ranging over mechanics, mathematical physics and astronomy, celestial mechanics, probability, and topology. He wrote several popular books about science, including* The Value of Science *and* Science and Hypothesis. *In* Foundations of Science *he talks about mathematical creativity and problem solving. He observed his own mind at work, as the notes on the facing page recount.*

Normally, we state a formula or rule that shows how to find a *y*-value from a given *x*-value. This rule is often an *equation*. For example, the set of ordered pairs

$$\{(x, y) \mid x + y = 2, x \text{ and } y \text{ are integers}\}$$

is a typical relation. The rule relating *x* and *y* is the equation $x + y = 2$. The elements of this relation are the ordered pairs (x, y) such that $x + y = 2$, where both *x* and *y* are integers. The ordered pair $(1, 1)$ belongs to this relation since $1 + 1 = 2$ and 1 is an integer. Other examples of ordered pairs belonging to this relation include $(2, 0)$, $(5, -3)$, and $(7, -5)$. This relation contains an infinite number of ordered pairs.

In the relation

$$\{(x, y) \mid y = 2x + 1, x = 0, 1, 3, \text{ or } 5\},$$

the variable *x* is restricted to the values 0, 1, 3, or 5. Using the rule for this relation, the equation $y = 2x + 1$, and the set of *x*-values $\{0, 1, 3, 5\}$, we can find the ordered pairs belonging to the relation. For example, if $x = 3$, we then have

$$\begin{aligned} y &= 2x + 1 \\ y &= 2 \cdot 3 + 1 \qquad \text{Let } x = 3. \\ y &= 6 + 1 \\ y &= 7, \end{aligned}$$

so that the ordered pair $(3, 7)$ belongs to the relation. Replace *x*, in turn, with the values 0, 1, and 5. This gives the additional ordered pairs $(0, 1)$, $(1, 3)$, and $(5, 11)$.

For simplicity, we shall abbreviate a relation such as

$$\{(x, y) \mid y = 2x + 1, x = 0, 1, 3, \text{ or } 5\}$$

by omitting all set symbols; instead, we write

$$y = 2x + 1, \qquad x = 0, 1, 3, \text{ or } 5.$$

Example 1   Find all the ordered pairs belonging to the relation

$$y = -3x + 9, \qquad x = -2, 0, 3, \text{ or } 5.$$

To find the ordered pairs belonging to the relation, replace *x*, in turn, with -2, 0, 3, and 5. Let $x = -2$ and $x = 0$:

If $x = -2$,
then $y = -3(-2) + 9$
$\quad y = 6 + 9$
$\quad y = 15$
Ordered pair: $(-2, 15)$

If $x = 0$,
then $y = -3(0) + 9$
$\quad y = 0 + 9$
$\quad y = 9$
Ordered pair: $(0, 9)$

In the same way, $x = 3$ leads to the ordered pair $(3, 0)$; while $x = 5$ leads to $(5, -6)$.

The equations of this section, such as $2x + 3y = 12$, are examples of **linear equations.** In general, a linear equation (in two variables) has the form

$$ax + by = c,$$

where a, b, and c are real numbers, with not both a and b equal to 0. When either $a = 0$ or $b = 0$, then a linear equation becomes $by = c$ (when $a = 0$), or $ax = c$ (when $b = 0$). For example, $3y = 17$ (where $0 \cdot x$ disappears) and $2x = 13$ are linear equations.

The name "linear" comes from the fact that the graph of a linear equation is a straight line, as we shall see in the next section.

Systems of equations   We can further extend our techniques of problem solving by studying *systems of two equations,* each with two variables, such as

$$3x + y = 5$$
$$2x - y = 10.$$

Each of these two equations is satisfied by an infinite number of ordered pairs. We want to see if there are any ordered pairs that make both equations true *at the same time.* Any ordered pair that does make both equations true is called a *solution* for the system.

Example 2   Decide whether or not $(-3, 1)$ is a solution for each system:

(a) $x + 5y = 2$ (b) $3x - 5y = -14$
     $2x + y = -5$ $5x + 4y = -10.$

(a) Replace x with -3 and y with 1 in each equation of the system.

$$\begin{array}{ll} x + 5y = 2 & 2x + y = -5 \\ -3 + 5(1) = 2 & 2(-3) + 1 = -5 \\ -3 + 5 = 2 & -6 + 1 = -5 \\ 2 = 2 \quad \text{True} & -5 = -5 \quad \text{True} \end{array}$$

Since $(-3, 1)$ makes both equations true, it is a solution of system (a).

(b) We again let $x = -3$ and $y = 1$:

$$\begin{array}{ll} 3x - 5y = -14 & 5x + 4y = -10 \\ 3(-3) - 5(1) = -14 & 5(-3) + 4(1) = -10 \\ -9 - 5 = -14 & -15 + 4 = -10 \\ -14 = -14 \quad \text{True} & -11 = -10 \quad \textbf{False} \end{array}$$

The ordered pair $(-3, 1)$ does *not* make both equations true, and thus is not a solution of the system (b).

How do we find a solution for a system such as this?

$$x + y = 5$$
$$x - y = 3$$

We can use an algebraic method called the **elimination method,** or **addition method.** The next examples show how this method is used to *eliminate* one of the two variables of the system by *adding together* the equations of the system.

Example 3 Solve the system

$$x + y = 5$$
$$x - y = 3.$$

Solution. Draw a line under the equations, and add. [*Recall:* $x + x = 2x$ and $y + (-y) = 0$.]

$$x + y = 5$$
$$\underline{x - y = 3}$$
$$2x \quad = 8$$

The result, $2x = 8$, is an equation with only one variable, which we can solve. Since $2x = 8$, then

$$x = 4.$$

Thus, $x = 4$ gives the x-value of the solution of the given system. To find the y-value, substitute 4 for x in either of the two equations of the system. If we use the first equation, $x + y = 5$, we get

$$x + y = 5$$
$$4 + y = 5 \qquad \text{Let } x = 4.$$
$$y = 1 \qquad \text{Add } -4 \text{ on both sides.}$$

Then $y = 1$ gives the y-value of the solution. The complete solution is the ordered pair $(4, 1)$.

Example 4 Solve the system

$$x + 3y = 7 \qquad \qquad (1)$$
$$2x + 5y = 12. \qquad \qquad (2)$$

Solution. If we draw a line under the two equations and add, we get $3x + 8y = 19$, which does not help us find the solution. What we need to do is to multiply both sides of either (or both) equations so that we can then add two equations, so that one variable drops out. Let us multiply both sides of equation (1) by -2:

$$-2(x + 3y) = -2(7)$$
$$-2x - 6y = -14 \qquad \qquad (3)$$

Now write equation (3) above equation (2), draw a line, and add:

$$-2x - 6y = -14$$
$$\underline{2x + 5y = 12}$$
$$-y = -2$$
$$y = 2 \qquad \text{Divide both sides by } -1.$$

To find x, substitute 2 for y in equation (1).

$$x + 3y = 7$$
$$x + 3(2) = 7 \qquad \text{Let } y = 2.$$
$$x + 6 = 7$$
$$x = 1.$$

The solution of the system is $(1, 2)$.

You might think that multiplying only one equation of the given system might result in a solution different from a solution obtained from the equations as given. This doesn't happen. As long as we use the common rules of algebra, the solution of our second system will also be a solution of the original system. What we have done is to produce an **equivalent system,** a system which has the same solution as the original system.

Example 5 Solve the system

$$2x + 3y = -15 \qquad\qquad (4)$$
$$5x + 2y = 1. \qquad\qquad (5)$$

Solution. Here we must multiply *both* equations. We need to multiply by two numbers that will cause the coefficients of x (or of y) in the two equations to be negatives of each other. We can do this here if we multiply both sides of equation (4) by 5, and multiply both sides of equation (5) by -2. This gives

$$10x + 15y = -75$$
$$\underline{-10x - 4y = -2}$$
$$11y = -77$$
$$y = -7$$

Check that $x = 3$, so that the solution is $(3, -7)$.

In the preceding section we solved word problems by letting the unknown number be represented by one variable, such as x, and then writing an equation from the information of the problem. In some cases, however, *two* unknown numbers are involved, and we must use *two variables* to represent them. This is shown in the next example.

Example 6 The sum of two numbers is 9. Their difference is 1. Find the numbers.

Solution. Here we can use two different variables—let x and y represent the numbers. "The sum of two numbers is 9" becomes $x + y = 9$. "Their difference is 1" is $x - y = 1$. We must now solve the system

$$x + y = 9$$
$$x - y = 1.$$

Use the elimination method to verify that the solution is $(5, 4)$. Thus, the numbers we want are 5 and 4.

Double Descartes *After the French postal service issued the above stamp in honor of René Descartes, sharp eyes noticed that the title of Descartes' most famous book was wrong. Thus a second stamp (see facing page) was issued with the correct title. The book in question, Discourse on Method, appeared in 1637. In it Descartes rejected traditional Aristotelian philosophy, outlining a universal system of knowledge that was to have the certainty of mathematics. He first adopted a skeptical view of everything, seeking "clear and distinct" ideas that any rational person could not doubt. One of these is his famous statement, "I think, therefore I am." For Descartes, method was analysis, going from self-evident truths step-by-step to more distant and more general truths. (Thomas Jefferson, also a rationalist, began the Declaration with the words, "We hold these truths to be self-evident.")*

Descartes first conceived his method about 1620, during his military service, when extreme winter conditions forced him to spend time shut up and without any diversion. From 1628 to 1649 Descartes deliberately lived in seclusion in the Netherlands. In 1649 he went to Sweden to tutor Queen Christina. She preferred working in the unheated castle in the early morning; Descartes was used to staying in bed until noon. Ironically, the rigors of Swedish winter proved too much for him, and he died less than a year later.

7.2 EXERCISES

In each equation, find y by replacing x with 2. Then find a second value of y by replacing x with -3. Use each answer to write an ordered pair.

1. $y = 4x + 5$

2. $y = 7x + 12$

3. $y = 8x + 1$

4. $y = 2x + 9$

5. $y = 3x - 2$

6. $y = 6x - 9$

7. $y = -8 + 5x$

8. $y = -7 + 9x$

9. $6x + y = 0$

10. $3x - y = 0$

11. $x + y = 6$

12. $x + y = 12$

13. $x - y = -2$

14. $y - x = -8$

15. $4x + 3y = 6$

16. $5x + 3y = 15$

Decide whether or not the given ordered pair is a solution of the system in Exercises 17–26.

17. $(2, -5)$ $\quad 3x + y = 1$
$2x + 3y = -11$

18. $(-1, 6)$ $\quad 2x + y = 4$
$3x + 2y = 9$

19. $(4, -2)$ $\quad x + y = 2$
$2x + 5y = 2$

20. $(-6, 3)$ $\quad x + 2y = 0$
$3x + 5y = 3$

21. $(2, 0)$ $\quad 3x + 5y = 6$
$4x + 2y = 5$

22. $(0, -4)$ $\quad 2x - 5y = 20$
$3x + 6y = -20$

23. $(5, 2)$ $\quad 4x + 3y = 26$
$3x + 7y = 29$

24. $(9, 1)$ $\quad 2x + 5y = 23$
$3x + 2y = 29$

25. $(6, -8)$ $\quad x + 2y + 10 = 0$
$2x - 3y + 30 = 0$

26. $(-5, 2)$ $\quad 3x - 5y + 20 = 0$
$2x + 3y + 4 = 0$

Use the elimination method to find the solution for each system.

27. $x - y = 3$
$x + y = -1$

28. $x + y = 7$
$x - y = -3$

29. $x + y = 2$
$2x - y = 4$

30. $3x - y = 8$
$x + y = 4$

31. $2x + y = 14$
$x - y = 4$

32. $2x + y = 2$
$-x - y = 1$

33. $x + 3y = 16$
$2x - y = 4$

34. $4x - 3y = 8$
$2x + y = 14$

35. $5x - 4y = -1$
$-7x + 5y = 8$

36. $6x + 2y = 0$
$-5x + 3y = 56$

37. $3x + 5y = 33$
$4x - 3y = 15$

38. $2x + 8y = 30$
$5x - 3y = 6$

*39. $8x + 12y = 13$
$16x - 18y = -9$

*40. $9x + 6y = -9$
$6x + 8y = -16$

Solve each word problem.

41. The sum of two numbers is 15. Their difference is 3. Find the numbers.

42. The sum of two numbers is 23. Their difference is 7. Find the numbers.

43. Two numbers have a sum of 18. If twice the second is subtracted from three times the first, the result is 14. Find the numbers.

Descartes wrote his Geometry *as an application of his method; it was published as an appendix to the* Discourse. *His attempts to unify algebra and geometry influenced the creation of what became coordinate geometry, and influenced the development of calculus by Newton and Leibniz in the next generation. Compare the following diagram by Descartes with the fully evolved "Cartesian" coordinate system of today, as in Figure 7.2 below.*

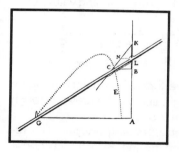

44. The sum of two numbers is 33. The second number is twice the first. Find the numbers.

45. The cashier at Mason's Pharmacy has some $10 bills and some $20 bills. The total value of the money is $1480. If there is a total of 85 bills, how many of each denomination are there?

46. A bank teller has 154 bills of $1 and $5 denominations. How many of each type of bill does he have if the total value of the money is $466?

47. A club secretary bought 8¢ and 10¢ pieces of candy to give to the members. She spent a total of $15.52. If she bought 170 pieces of candy, how many of each kind did she buy?

48. There were 311 tickets sold for a basketball game, some for students, and some for nonstudents. Student tickets were 25¢ each and nonstudent tickets were 75¢ each. The total receipts were $108.75. How many of each type of ticket were sold?

7.3 COORDINATE GEOMETRY

As we said in the introduction to this chapter, the development of algebra from Babylonian times involved studying a variety of types of equations. Progress was made more rapid by the introduction of symbols, gradually freeing algebra from words. Another boost to the development of algebra came in the seventeenth century, when mathematicians began to discover that algebra and geometry were not as separate as was then commonly believed. They began to discover that algebra could be used as an aid in studying geometry, and geometry could be used as an aid in studying algebra.

Much of the credit for the unification of algebra and geometry into the branch of mathematics called **coordinate geometry,** or **analytic geometry,** goes to the French mathematicians René Descartes (1596–1650) and Blaise Pascal (1623–1662). We will see in Chapter 9 that Pascal was also a co-founder of probability.

Coordinate geometry allows us to draw graphs of relations, such as the equations we studied in the previous section. We actually picture the *ordered pairs* of a relation as *points* on a coordinate system. To do this, begin with a horizontal and a vertical number line, which are placed perpendicular to each other, as in Figure 7.2. The horizontal line is called the *x-axis*, while the vertical line is called the *y-axis*. Together, the axes form a **Cartesian coordinate system,** named for Descartes. We can locate points on a co-ordinate system that correspond to the ordered pairs of a relation.

The point where the horizontal and vertical lines cross is called the **origin** of the coordinate system. To the right of the origin is a scale of positive values of *x*, with negative values to the left. The positive values of *y* are scaled above the origin, with negative values below.

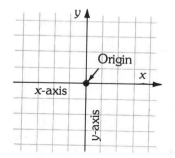

Figure 7.2 Cartesian coordinate system

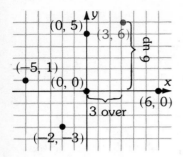

Figure 7.3 Some points are located (or "plotted")

To locate the point corresponding to the ordered pair $(3, 6)$, count 3 units to the right on the x-axis, and then count 6 units straight up, parallel to the y-axis. Figure 7.3 shows the point for $(3, 6)$ and other sample points.

Example 1 Graph the relation

$$y = 2x - 3, \qquad x = -2, -1, 0, 1, 2, \text{ or } 3.$$

Solution. Replace x, in turn, by $-2, -1, 0, 1, 2,$ and 3. Doing this leads to the ordered pairs $(-2, -7)$, $(-1, -5)$, $(0, -3)$, $(1, -1)$, $(2, 1)$, and $(3, 3)$. These ordered pairs have been graphed as points in Figure 7.4.

In the previous section we mentioned that linear equations (those with the form $ax + by = c$) are the equations of straight lines. Also, all graphs of linear equations are straight lines. To see how these equations lead to straight line graphs, let us graph the linear equation

$$y = 4 - 2x.$$

We begin by choosing several values of x, such as $x = -1, 0, 1, 2, 3,$ and 4. We then use the equation $y = 4 - 2x$ to find the corresponding values of y.

x	-1	0	1	2	3	4
y	6	4	2	0	-2	-4
(x, y)	$(-1, 6)$	$(0, 4)$	$(1, 2)$	$(2, 0)$	$(3, -2)$	$(4, -4)$

The six ordered pairs that we have found are located as points in Figure 7.5(a). Note the pattern that emerges. A straight line can be drawn through all the points on the graph, as we did in Figure 7.5(b).

Select any other x-value, and find the corresponding y-value. The resulting ordered pair will locate a point which is on the line. Also, if you choose any point on the line, the x-value and y-value for that point can be substituted into the equation $y = 4 - 2x$, producing a true statement.

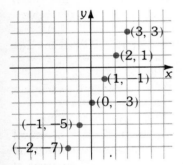

Figure 7.4

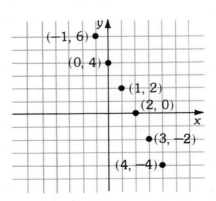

Figure. 7.5(a) Six ordered pairs are plotted as points

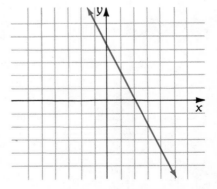

Figure 7.5(b) A straight line through the points plotted in figure (a) gives the graph of the linear equation $y = 4 - 2x$

The geometry of ethics
Descartes was a strong influence on Benedict Spinoza (1632– 1677), a Dutch-Jewish philosopher who made his living as a lens-grinder. Spinoza went much further than Descartes in viewing the world as a system of logical relations. He even took Euclid's Elements as the model for his own monumental treatise on moral philosophy, Ethic Demonstrated in Geometric Order (1677).

Example 2   Draw the graph of $4x - 5y = 20$.

Solution.   This equation has a straight line as its graph. (How do we know this?) We need only two distinct points to locate a straight line, but it's usually a good idea to plot three points, with the third point serving as a check.

We can choose any three x-values that we like, and then find the corresponding y-values. Some x-values are simpler to work with than others; here for example it is perhaps easiest to let $x = 0$, $x = 5$, and $x = 2\frac{1}{2}$.

x	0	5	$2\frac{1}{2}$
y	-4	0	-2
(x, y)	$(0, -4)$	$(5, 0)$	$(2\frac{1}{2}, -2)$

If we locate these points on a coordinate system, and draw a straight line through them, we get the graph in Figure 7.6.

Example 3   Graph $y = 3$.

As in the previous example, we need to select three different values of x and then find the corresponding values of y. But here, no matter which x-values we pick, the y-value is always the same, namely, $y = 3$. Therefore, if we choose -4, 0, and 3 for x, we get the ordered pairs $(-4, 3)$, $(0, 3)$, and $(3, 3)$. These ordered pairs have been graphed in Figure 7.7. The line that goes through these points is a horizontal line.

In general, $y = k$ (where k is a real number) has a graph which is a horizontal line. The line goes through the point $(0, k)$.

In the same way, the graph of $x = k$ (k is a real number) is a vertical line going through $(k, 0)$. Here x takes on only one value, $x = k$, while y can have any value. (To see this, rewrite $x = k$ as $x = k + 0 \cdot y$.) The graph of $x = -2$ in Figure 7.8 is an example.

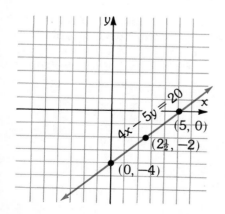

Figure 7.6   The graph of $4x - 5y = 20$

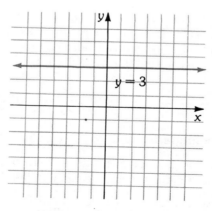

Figure 7.7   The graph of $y = 3$ is a horizontal line

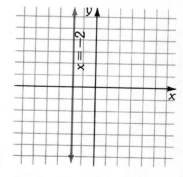

Figure 7.8   The graph of $x = -2$ is a vertical line

Slope   We often need a numerical measure of the "steepness" of a straight line. This measure is called the *slope* of the line, and is defined as follows. A line going through two distinct points (x_1, y_1) and (x_2, y_2) has **slope** given by m, where

$$m = \frac{\text{change in } y}{\text{change in } x} = \frac{y_2 - y_1}{x_2 - x_1}.$$

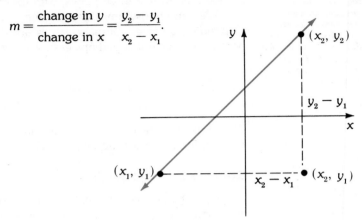

Figure 7.9

Think of the slope as a *vertical* change in the graph, $y_2 - y_1$, divided by the corresponding *horizontal* change, $x_2 - x_1$.

Example 4   Find the slope of the line through $(-2, 4)$ and $(4, 6)$.

Solution.   We can let (x_1, y_1) be either $(-2, 4)$ or $(4, 6)$. (It doesn't matter.) If we let $(x_1, y_1) = (-2, 4)$ so that $(x_2, y_2) = (4, 6)$, the slope is

$$m = \frac{y_2 - y_1}{x_2 - x_1} = \frac{6 - 4}{4 - (-2)} = \frac{2}{6} = \frac{1}{3}.$$

Example 5   Find the slope of the line $2x - y = 5$.

Solution.   To find the slope of $2x - y = 5$, we need to find two points that lie on the line. We can find these two points by choosing two values of x and finding corresponding values of y. If we choose $x = 0$ and $x = 5$, we get the ordered pairs $(0, -5)$ and $(5, 5)$. Now we can use the formula for slope:

$$m = \frac{y_2 - y_1}{x_2 - x_1} = \frac{5 - (-5)}{5 - 0} = \frac{10}{5} = 2.$$

Example 6   Find the slope of the horizontal line $y = 2$.

Solution.   Find two ordered pairs for $y = 2$, such as $(3, 2)$ and $(5, 2)$. Using these ordered pairs, we can find the slope:

$$m = \frac{2 - 2}{5 - 3} = \frac{0}{2} = 0.$$

Every horizontal line has a slope of 0.

Example 7   Find the slope of the vertical line $x = -3$.

Solution.   We find two ordered pairs that satisfy this equation, such as $(-3, 5)$ and $(-3, 8)$. The slope is then

$$m = \frac{8 - 5}{-3 - (-3)} = \frac{3}{0}.$$

But this is not a real number, since division by 0 is impossible. Thus, the slope of a vertical line is undefined; in other words,

A vertical line has no slope.

We have seen how to use the equation of a straight line to find the slope of the line. If we know the slope of a line and a point that the line goes through, we can work backwards to find the equation of the line. This is done with the following formula giving the **point-slope** form of the equation of a line:

The line with slope m
going through the point (x_1, y_1) has equation
$y - y_1 = m(x - x_1)$.

Slope warning   *Take special care when you see this traffic symbol. You're headed for a downgrade that may be long, steep, or sharply curved.*

Example 8   Find the equation of the line with slope $m = -2$, going through the point $(-3, 4)$.

Solution.   We have $m = -2$, $x_1 = -3$, and $y_1 = 4$. Substituting these numbers into the point-slope form of the equation of a line, we have

$$
\begin{aligned}
y - y_1 &= m(x - x_1) \\
y - 4 &= -2[x - (-3)] \\
y - 4 &= -2(x + 3) \\
y - 4 &= -2x - 6 \quad &\text{Distributive property} \\
y + 2x &= -2 \quad &\text{Add } 2x \text{ and } 4 \text{ to both sides.}
\end{aligned}
$$

Example 9   Find the equation of the line with slope $m = -3/5$ and going through $(4, -1)$.

Solution.   We have $m = -3/5$, $x_1 = 4$, and $y_1 = -1$.

$$
\begin{aligned}
y - y_1 &= m(x - x_1) \\
y - (-1) &= -\frac{3}{5}(x - 4) \\
y + 1 &= -\frac{3}{5}(x - 4) \\
5(y + 1) &= -3(x - 4) \quad &\text{Multiply both sides by 5.} \\
5y + 5 &= -3x + 12 \quad &\text{Distributive property} \\
3x + 5y &= 7 \quad &\text{Add } 3x \text{ and } -5 \text{ to both sides.}
\end{aligned}
$$

Maria Gaetana Agnesi *(1719–1799) did much of her mathematical work in coordinate geometry. She grew up in a scholarly atmosphere; her father was a mathematician on the faculty at the University of Bologna. In a larger sense she was an heir to the long tradition of Italian mathematicians.*

Algebraists such as Tartaglia, Cardano, and Bombelli in the 16th century produced the general solution of the cubic equation. (Cubic equations have variables to the third power, such as x³.) Galileo was as much a physicist and astronomer as mathematician. (See page 276 and in Chapter 8.) Cavalieri, a pupil of Galileo, had the gist of the calculus, as he showed in his Geometry of Indivisibles (1635), which was an important influence on Isaac Barrow, Newton's teacher.

Maria Agnesi was fluent in several languages by the age of 13, but she chose mathematics over literature. After many years of study and working alone, she was given the post her father had at the university.

7.3 EXERCISES

Exercises 1–10 give linear equations and some values for x (or y). Find the corresponding y-values and write ordered pairs. (Make a table for each equation, as in Exercise 1.) Then locate the ordered pairs for each equation on a separate coordinate system, and draw a straight line through the points.

1. $y = 2x$

x	-2	-1	0	1	2
y					
(x, y)					

2. $y = x + 3$; x-values: $-2, -1, 0, 1, 2$

3. $y = 4x - 3$; x-values: $-1, 0, 1, 2, 3$

4. $y = 5 - 2x$; x-values: $-2, -1, 0, 1, 2$

5. $2x + 3y = 6$; x-values: $-3, 0, 3, 6$

6. $4x + 5y = 20$; x-values: $-5, 0, 5, 10$

7. $6x - 5y = 20$; x-values: $-5, 0, 5$

8. $7x - 2y = 14$; x-values: $-2, 0, 2, 4$

9. $y = -5$; x-values: $-3, 0, 2, 5, 6$

10. $x = 2$; y-values: $-5, -3, 0, 1, 2$ (Be careful here!)

Draw the graph of each straight line.

11. $x - y = 2$

12. $x + y = 6$

13. $y = x + 2$

14. $y = x - 1$

15. $3x - 2y = 6$

16. $2x + 3y = 12$

17. $3x + 7y = 21$

18. $6x - 5y = 30$

19. $y = 2x$

20. $y = -3x$

21. $x = -2$

22. $y = 3$

23. $y = 6$

24. $x = 2$

25. $x = 0$

26. $y = 0$

Find the slope of the line through each pair of points.

27. $(6, 1)$ and $(5, 2)$

28. $(-9, 2)$ and $(3, -5)$

29. $(-5, 2)$ and $(8, -1)$

30. $(-3, 7)$ and $(-9, 2)$

31. $(8, 0)$ and $(6, 5)$

32. $(-2, 3)$ and $(0, 2)$

33. $(5, 1)$ and $(3, 1)$

34. $(7, 5)$ and $(12, 5)$

35. $(8, 3)$ and $(8, 7)$

36. $(17, 2)$ and $(17, -5)$

Find the slope of each line.

37. $x + y = 5$ **40.** $y = -5x$ **43.** $x = -5$

38. $x - y = 6$ **41.** $y = 6$ **44.** $x + 1 = 0$

39. $y = 3x$ **42.** $y - 4 = 0$

Find the equation of the line going through the given point and having the given slope.

45. $(-2, 3)$; $m = 2$ **50.** $(-7, 1)$; $m = -9/10$

46. $(8, -6)$; $m = -3$ **51.** $(-8, 9)$; $m = 0$

47. $(2, -8)$; $m = 3/4$ **52.** $(2, -5)$; $m = 0$

48. $(5, -2)$; $m = 2/3$ ***53.** $(3, -2)$; no slope

49. $(-1, 6)$; $m = -5/8$ ***54.** $(-1, 7)$; no slope

Find the equation of each line.

55. through $(-2, 7)$ and $(1, 5)$ **56.** through $(-2, 1)$ and $(6, 8)$

The number of times that a cricket chirps in one minute gives an estimate of the temperature. The following table shows the relationship between chirps and temperature.

Number of chirps in one minute	x	0	20	40	60	80	160
Temperature (in degrees Fahrenheit)	y	40	45	50	55	60	80

57. Graph the ordered pairs represented by this table.

58. Draw a straight line through them.

59. Choose two points on the line and find the slope of the line.

60. Find the equation of the line, using the point-slope formula.

From your equation in Exercise 60, find the temperatures related to:

61. 100 chirps   **62.** 120 chirps

Roughly estimated, the weight of a male human taller than about 60 inches is approximated by $y = 5.5x - 220$, where x is the height of the human in inches, and y is the weight in pounds. Estimate the weights for the given heights.

63. 62 inches   **64.** 64 inches   **65.** 68 inches

66. 72 inches

67. Graph $y = 5.5x - 220$. Use only the numbers 62 through 76 on the x-axis.

In economics, **supply and demand** are related in general: as the price of an item goes up, the demand for the item goes down. (Exceptions include cosmetics and pet food.) Also, as the price goes up, the supply goes up. For example, suppose that the demand for an item, D, is given by

$$D = -\frac{4}{3}x + 80,$$

where x is the price of the item in dollars. Find the demand for each of the following prices.

68. $0   **69.** $12   **70.** $30   **71.** $60

72. Graph D. (Graph only the portion where x is positive.)

Suppose now that the supply for this same item, S, is given by

$$S = \frac{4}{3}x,$$

where x is the price in dollars. Find the supply for each of the following prices.

73. $0   **74.** $12   **75.** $30   **76.** $45   **77.** $60

78. Graph S on the same coordinate system you used in Exercise 72.

79. Where do the lines for D and S cross?

80. The price where supply and demand are equal is called the *equilibrium price*. Find the equilibrium price here.

The demand and supply for a certain type of candy are given by

$$D = \frac{300 - 4x}{3} \quad \text{and} \quad S = \frac{2}{3}x,$$

where x is the price in dollars.

81. Graph D and S on the same coordinate system.

82. Find the equilibrium price.

7.4 LINEAR INEQUALITIES

A *linear inequality* contains one of the following symbols of inequality:

$<$ is less than	$\leq$ is less than or equal to
$>$ is greater than	$\geq$ is greater than or equal to

$\neq$ is not equal to

Examples of linear inequalities include

$$x + y \leq 6, \quad 2x + 3y \geq 12, \quad x < 4, \quad x \neq 6.$$

To graph the linear inequality $2x - y \leq 4$, we first graph the equation $2x - y = 4$; its graph is the straight line in Figure 7.10.

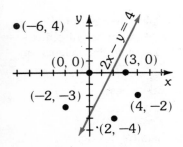

Figure 7.10 The graph of $2x - y = 4$,

The line drawn in Figure 7.10 divides the plane of the coordinate system into three parts: the line itself, and two *half-planes,* one on either side of the line. The line belongs to neither half-plane. We apply the general idea of a half-plane in graphing the linear inequality $2x - y \leq 4$ (and other inequalities).

The linear inequality $2x - y \leq 4$ involves *less than* as well as *equal to,* and we must find the set of all ordered pairs which make the inequality true. Figure 7.10 also shows six selected points, three on each side of the line. We can test that $(-6, 4)$, $(-2, -3)$, and $(0, 0)$ make $2x - y \leq 4$ true, while the other three points do not. For example, test the point $(-6, 4)$; let $x = -6$ and $y = 4$. Then the inequality becomes

$$2(-6) - 4 = -16 \leq 4.$$

It is true that $-16 \leq 4$, so $(-6, 4)$ makes the inequality true.

Since the three ordered pairs which satisfy the relation $2x - y \leq 4$ all lie in the same half-plane, we test other points in this half-plane. For example, if we test $(-8, -2)$ and $(-3, 2)$ we find that both make $2x - y \leq 4$ true. In fact, all the points in this half-plane make $2x - y \leq 4$ true. We show this by shading the entire half-plane, as in Figure 7.11.

Example 1   Graph the linear inequality $3x + 2y \leq 6$.

Solution.   To begin, we draw the straight line $3x + 2y = 6$, as shown in Figure 7.12. We know that the graph of $3x + 2y \leq 6$ is one of the half-planes determined by this line. To decide which half-plane it is, select any point *not* on the line and see if it makes the inequality true. Often a good choice is the point $(0, 0)$. Does $(0, 0)$ make $3x + 2y \leq 6$ true? To test $(0, 0)$, replace x with 0 and y with 0:

$$3x + 2y \leq 6$$
$$3(0) + 2(0) \leq 6$$
$$0 + 0 \leq 6$$
$$0 \leq 6 \quad \text{True}$$

Since $(0, 0)$ makes $3x + 2y \leq 6$ true, we shade the half-plane containing $(0, 0)$, as shown in Figure 7.12.

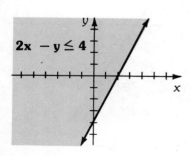

Figure 7.11

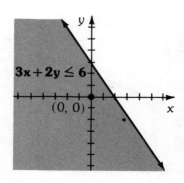

Figure 7.12

Example 2 Graph $2x - 5y > 10$.

Solution. The points of the line $2x - 5y = 10$ are not in the graph of $2x - 5y > 10$. (Why?) A common way to show that the line itself is omitted is to draw dashes, as in Figure 7.13. To decide which half-plane to shade, use $(0, 0)$ as a test point.

$$2x - 5y > 10$$
$$2(0) - 5(0) > 10$$
$$0 > 10 \quad \text{False}$$

Since $0 > 10$ is false, $(0, 0)$ is *not* in the graph of $2x - 5y > 10$, so we shade the half-plane that does *not* contain $(0, 0)$, as in Figure 7.13.

A **system of inequalities** is made up of two or more inequalities. The solution of such a system is the set of all ordered pairs that make all inequalities of the system true at the same time. To find the solution, we can graph all the inequalities of the system on the same coordinate system, and then note where the common points (the intersection) of all the graphs fall. This method is explained in the next example.

Example 3 Graph the system

$$3x + 2y \le 6$$
$$2x - 5y \ge 10.$$

Solution. First graph the linear inequality $3x + 2y \le 6$, as explained above. The graph of $3x + 2y \le 6$ is shown in grey in Figure 7.14. Then graph $2x - 5y \ge 10$ on the same coordinate system. This inequality is graphed in color. The solution of the given system is the heavily shaded portion, where the colors "intersect." The solution includes portions of both boundary lines. (Why?)

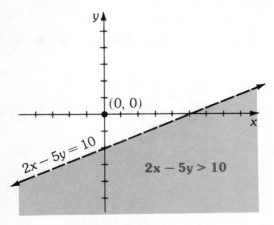

Figure 7.13 The graph of $2x - 5y > 10$.

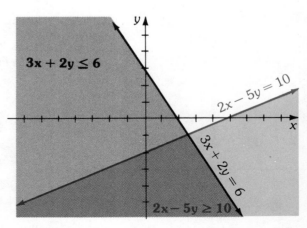

Figure 7.14 The graph of the system in Example 3

7.4 EXERCISES

In Exercises 1–9, each figure shows the first step in graphing the given inequality: the required straight line has been drawn. Complete each graph by shading the correct half-plane.

1. $x + y \leq 4$

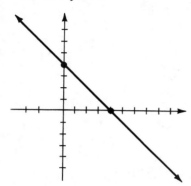

4. $2x + y \leq 5$

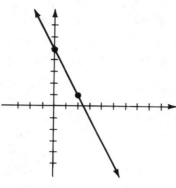

7. $5x + 3y > 15$

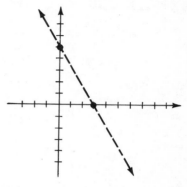

2. $x + y \geq 2$

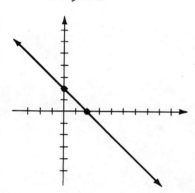

5. $-3x + 4y < 12$

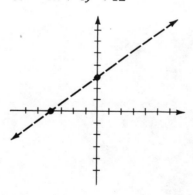

8. $x < 4$

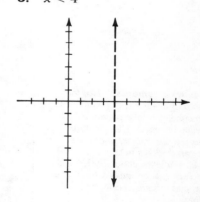

3. $x + 2y \leq 7$

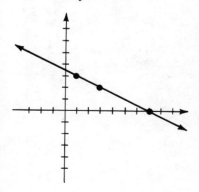

6. $4x - 5y > 20$

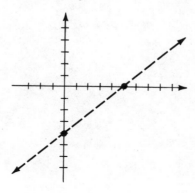

9. $y > -1$

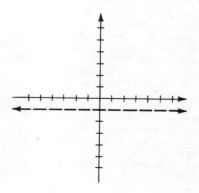

Graph each of the following inequalities.

10.	$x + y \leq 8$	**17.**	$2x + 3y > 6$	*$24.$	$x \leq 5y$
11.	$x + y \geq 5$	**18.**	$3x + 4y > 12$	*$25.$	$x \geq -2y$
12.	$x - y \leq -2$	**19.**	$3x - 4y < 12$	*$26.$	$x > -4y$
13.	$x - y \leq 3$	**20.**	$2x - 3y < -6$	**27.**	$x < 4$
14.	$x + 2y \geq 4$	**21.**	$3x + 7y \geq 21$	**28.**	$x \leq -2$
15.	$x + 3y \leq 6$	**22.**	$2x + 5y \geq 10$	**29.**	$y \geq -3$
16.	$4x + y \leq 8$	*$23.$	$x \leq 3y$	**30.**	$y \geq 1$

Graph each of the following systems of inequalities.

31.	$x + y \leq 6$ $x - y \leq 1$	**35.**	$x + 4y \leq 8$ $2x - y \leq 4$	**39.**	$x \geq 2$ $y \leq 3$
32.	$x + y \geq 2$ $x - y \leq 3$	**36.**	$3x + y \leq 6$ $2x - y \leq 8$	**40.**	$x \geq -1$ $y \geq 4$
33.	$x - 2y > 6$ $2x + y > 4$	**37.**	$x + 2y \leq 4$ $x + 1 \geq y$	*$41.$	$2x + 3y \leq 6$ $4x + y \leq 6$ $x \geq 0$ $y \geq 0$
34.	$3x + y < 4$ $x + 2y > 2$	**38.**	$x < 2y + 3$ $x + y > 0$		

Algebraic symbols? *No; the photo of chevrons, X's, and Y's is a magnetic bubble pattern circuit, greatly enlarged. The chevrons you see are actually about one micron thick—a micron is 1/1000 of a millimeter, about .00004 inch! Such circuits are etched by a new X-ray technique developed by IBM Corporation.*

Magnetic bubble circuits have been produced containing over 90,000 bits in a one-inch square device. (A bit is short for binary digit—the presence of a bubble represents 1, the absence 0. See page 113.) Hand-held calculators will become minicomputers, with memories in the millions of bits.

The new X-ray technique can be applied in biology, too. For example, biological specimens can be examined microscopically in much greater detail than before, and almost unchanged from their natural state, and it will be easier to map the arrangements of genes in chromosomes.

*7.5 LINEAR PROGRAMMING

One of the main applications of mathematics to business and social science is called *linear programming.* Linear programming is used to find minimum cost, maximum profit, the maximum amount of learning that can take place under given conditions, and so on. We shall explain the basic ideas of linear programming with an example.

The Smith Company makes two products, tape decks and amplifiers. Each tape deck gives a profit of $3, while each amplifier earns $7. The company must manufacture at least one tape deck per day to satisfy one of its customers, but no more than five because of production problems. Also, the number of amplifiers produced cannot exceed six per day. As a further requirement, the number of tape decks cannot exceed the number of amplifiers. **How many of each should the company manufacture in order to obtain the maximum profit?**

This problem is an example of a *linear programming problem,* in which we need to find the maximum value of an expression. The procedures for solving linear programming problems were developed in 1947 by George Dantzig. He was working on a problem of allocating supplies for the Air Force in a way that minimized total cost. Business quickly picked up on the utility of Dantzig's methods. Today, few problems on warehouse location, or design of factories, or utilization of resources (among other problems) are solved without linear programming.

Let's now return to the problem of the Smith Company. To begin, we translate the statements of the problem into symbols by assuming

$$x = \text{number of tape decks to be produced daily}$$

$$y = \text{number of amplifiers to be produced daily.}$$

According to the statement of the problem given above, the company must produce at least one tape deck (one or more), so that

$$x \geq 1.$$

No more than 5 tape decks may be produced:

$$x \leq 5.$$

Not more than 6 amplifiers may be made in one day:

$$y \leq 6.$$

The number of tape decks may not exceed the number of amplifiers:

$$x \leq y.$$

The number of tape decks and of amplifiers cannot be negative:

$$x \geq 0 \quad \text{and} \quad y \geq 0.$$

George B. Dantzig *is Chairman of the Operations Research Center at Stanford University.*

Operations research *(OR) is making significant contributions to business and industry. As a management science, OR is not a single discipline, but draws from mathematics, probability theory, statistics, and economics. The name given to this "multiplex" shows its historical origins in World War II, when operations of a military nature called forth the efforts of many scientists to research their fields for applications to the war effort, and to solve tactical problems. Applications of mathematical and statistical techniques were found to be feasible by Dantzig and other scientists.*

Operations research is an approach to problem solving and decision making. First of all the problem has to be clarified. Quantities involved have to be designated as variables, and the objectives as functions. Use of **models** *is an important aspect of OR. In fact, the Smith Company problem in the text illustrates three types of models: verbal, mathematical, and iconic. The statement of the Smith problem in words is a verbal model. The inequalities are mathematical models of the problem's conditions. The graphs of the inequalities are iconic models, and are basic to the linear programming method outlined here.*

Let us now summarize all the restrictions, or **constraints,** that are placed on production:

$$x \geq 1, \qquad x \leq 5, \qquad y \leq 6, \qquad x \leq y, \qquad x \geq 0, \qquad y \geq 0.$$

We need to find the maximum possible profit that the company can make, subject to these constraints. To begin, we sketch the graph of each constraint. The graphs of the constraints are shown in Figure 7.15. The only feasible values of x and y are those that satisfy all constraints; that is, the values which lie in the intersection of the graphs of the constraints. The intersection is shown in Figure 7.16. Any point lying inside the shaded region (boundaries also) of Figure 7.16 satisfies the restrictions as to the number of tape decks and amplifiers that may be produced.

Since each tape deck gives a profit of $3, the daily profit from the production of x decks is $3x$ dollars. Also, the profit from the production of y amplifiers will be $7y$ dollars per day. The total daily profit is thus

$$\text{Profit} = 3x + 7y.$$

The problem of the Smith Company may now be stated as follows: **Find values of x and y in the shaded region of Figure 7.16 that will produce the maximum possible value of $3x + 7y$.**

A basic idea of linear programming says that in a case such as ours the maximum (or minimum) profit will be found at a *corner* of the graph. By looking at Figure 7.16, we see that the corner points are $(1, 1)$, $(1, 6)$, $(5, 5)$, and $(5, 6)$. We now check each corner point to find the profit.

Figure 7.15

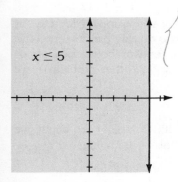

At least one tape deck

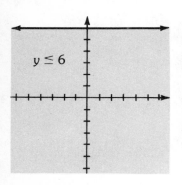

Five tape decks or less

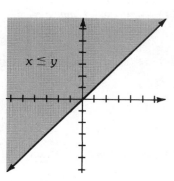

Six amplifers or less

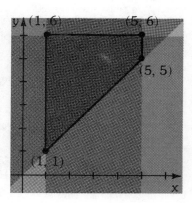

Number of tape decks, x, is no more than the number of amplifiers, y

Figure 7.16   Intersection of the four graphs in Figure 7.15

Point	Profit $= 3x + 7y$	
(1, 1)	$3(1) + 7(1) = 10$	← minimum
(1, 6)	$3(1) + 7(6) = 45$	
(5, 5)	$3(5) + 7(5) = 50$	
(5, 6)	$3(5) + 7(6) = 57$	← maximum

Maximum profit of $57 per day will be obtained if five tape decks and six amplifiers are made each day.

Example 1   Robin, who is ill, takes vitamin pills. Each day, she must have at least 16 units of Vitamin A, at least 5 units of Vitamin B_1, and at least 20 units of Vitamin C. She can choose between red pills, costing 10¢ each, which contain 8 units of A, 1 of B_1, and 2 of C; and blue pills, which cost 20¢ each, and contain 2 units of A, 1 of B_1, and 7 of C. **How many of each pill should she buy in order to minimize her cost and yet fulfill her daily requirements?**

Solution.   Let $x =$ number of red pills to buy,
$y =$ number of blue pills to buy.

The cost in pennies per day is given by

$$\text{Cost} = 10x + 20y,$$

since Robin buys x of the 10¢ pills and y of the 20¢ pills. Vitamin A comes as follows: 8 units from each red pill and 2 units from each blue pill. Altogether, she gets $8x + 2y$ units of A per day. Since she must get at least 16 units,

$$8x + 2y \geq 16.$$

Each red pill and each blue pill supplies 1 unit of Vitamin B_1. Robin needs at least 5 units per day:

$$x + y \geq 5.$$

For Vitamin C we get the inequality

$$2x + 7y \geq 20.$$

Also, $x \geq 0$ and $y \geq 0$, since Robin cannot buy negative numbers of the pills.
Again, we minimize total cost of the pills by finding the intersection of the graphs of the constraints. (See Figure 7.17.) We check the corners to find the lowest cost.

Point	Cost $= 10x + 20y$	
(10, 0)	$10(10) + 20(0) = 100$	
(3, 2)	$10(3) + 20(2) = 70$	← minimum
(1, 4)	$10(1) + 20(4) = 90$	
(0, 8)	$10(0) + 20(8) = 160$	

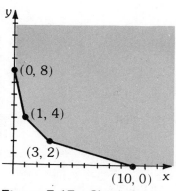

Figure 7.17 Check the corners

Robin's best bet is to buy 3 red pills and 2 blue ones, for a total cost of 70¢ per day. She receives just the minimum amounts of Vitamins B_1 and C, but an excess of Vitamin A. Even though she has an excess of A, this is still the best buy.

7.5 EXERCISES

Exercises 1–4 show regions of feasible solutions. Find the maximum and minimum values of the given expressions.

1. $3x + 5y$

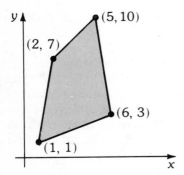

2. $6x + y$

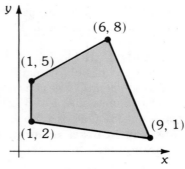

3. $40x + 75y$

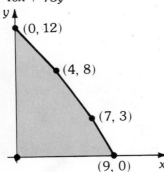

4. $35x + 125y$

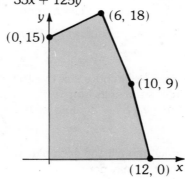

In Exercises 5–10, use graphical methods to find values of x and y satisfying the given conditions.

5. Find $x \geq 0$ and $y \geq 0$ such that
$$2x + 3y \leq 6$$
$$4x + y \leq 6$$
and $5x + 2y$ is maximized.
[*Hint:* One corner point is $(\frac{6}{5}, \frac{6}{5})$.]

6. Find $x \geq 0$ and $y \geq 0$ such that
$$x + y \leq 10$$
$$5x + 2y \geq 20$$
$$2y \geq x$$
and $x + 3y$ is minimized.

7. Find $x \geq 2$ and $y \geq 5$ such that
$$3x - y \geq 12$$
$$x + y \leq 15$$
and $2x + y$ is minimized.

8. Find $x \geq 10$ and $y \geq 20$ such that
$$2x + 3y \leq 100$$
$$5x + 4y \leq 200$$
and $x + 3y$ is maximized.

9. Find $x \geq 0$ and $y \geq 0$ such that
$$x - y \leq 10$$
$$5x + 3y \leq 75$$
and $4x + 2y$ is maximized.

10. Find $x \geq 0$ and $y \geq 0$ such that
$$10x - 5y \leq 100$$
$$20x + 10y \geq 150$$
and $4x + 5y$ is minimized.

11. Maximize $10x + 12y$ subject separately to each set of constraints.

(a) $x + y \leq 20$
$x + 3y \leq 24$
$x \geq 0$
$y \geq 0$

(b) $3x + y \leq 15$
$x + 2y \leq 18$
$x \geq 0$
$y \geq 0$

(c) $2x + 5y \geq 22$
$4x + 3y \leq 28$
$2x + 2y \leq 17$
$x \geq 0$
$y \geq 0$

12. Minimize $3x + 2y$ subject separately to each set of constraints.

(a) $10x + 7y \leq 42$
$4x + 10y \geq 35$
$x \geq 0$
$y \geq 0$

(b) $6x + 5y \geq 25$
$2x + 6y \geq 15$
$x \geq 0$
$y \geq 0$

(c) $x + 2y \geq 10$
$2x + y \geq 12$
$x - y \leq 8$
$x \geq 0$
$y \geq 0$

#13

Let $x = $ pigs

$y = $ geese

$x + y \leq 16$

$x \geq 0$

$y \geq 0$

$y \leq 12$

$5x + 2y \leq 50$

$P = 4x + 8y$

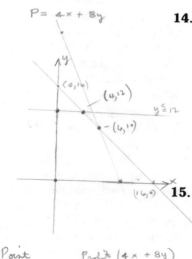

Point	Profit $(4x + 8y)$
$(0,0)$	0
$(0,12)$	96
$(10,0)$	40
$(4,12)$	$\$112 \leftarrow$ max.
$(6,10)$	104

Solve each of the following linear programming problems.

13. Farmer Jones raises only pigs and geese. He wants to raise no more than 16 animals with no more than 12 geese. He spends $5 to raise a pig and $2 to raise a goose. He has $50 available for this purpose. Find the maximum profit he can make if he makes a profit of $8 per goose and $4 per pig.

14. A wholesaler of party goods wishes to display her products at a convention of social secretaries in such a way that she gets the maximum number of inquiries about her whistles and hats. Her booth at the convention has 12 square meters of floor space to be used for display purposes. A display unit for hats requires 2 square meters; and for whistles, 4 square meters. Experience tells the wholesaler that she should never have more than a total of 5 units of whistles and hats on display at one time. If she receives three inquiries for each unit of hats and two inquiries for each unit of whistles on display, how many of each should she display in order to get the maximum number of inquiries?

15. An office manager wants to buy some filing cabinets. She knows that cabinet #1 costs $10 each, requires 6 square feet of floor space, and holds 8 cubic feet of files. On the other hand, cabinet #2 costs $20 each, requires 8 square feet of floor space, and holds 12 cubic feet. She can spend no more than $140 due to budgetary limitations, while her office has room for no more than 72 square feet of cabinets. She desires the maximum storage capacity within the limitations imposed by funds and space. How many of each type of cabinet should she buy?

16. The manufacture of "Smokey the Bear" ashtrays and "Superman" cufflinks requires two machines, a drill and a saw. Each ashtray needs one minute on the drill and two minutes on the saw; each cufflink set needs two minutes on the drill and one minute on the

saw. The drill is available at most for 12 minutes per day, and the saw is available no more than 15 minutes per day. Each product provides a profit of $1. How many of each should be manufactured in order to maximize profit?

17. Seeall Manufacturing Company makes color TV sets. They produce a bargain set that sells for $100 profit and a deluxe set that sells for $150 profit. The bargain set requires 3 hours on the assembly line, while the deluxe set takes 5 hours. The cabinet shop spends one hour on the cabinet for the bargain set and 3 hours on the cabinet for the deluxe set. Both sets take 2 hours for testing and packing. On a particular production run the Seeall Company has available 3900 worker hours on the assembly line, 2100 worker hours in the cabinet shop, and 2200 worker hours in the testing and packing department. How many sets of each type should they produce, and what is the maximum profit?

18. In a small town in South Carolina, zoning rules require that the window space (in square feet) in a house be at least one-sixth of the space taken up by solid walls. The cost to heat the house is 2¢ for each square foot of solid wall and 8¢ for each square foot of windows. Find the maximum total area (windows plus walls) if $16 is available to pay for heat.

*19. A candy company has 100 kilograms of chocolate-covered nuts and 125 kilograms of chocolate-covered raisins to be sold as two different mixtures. One mix will contain 1/2 nuts and 1/2 raisins, and will sell for $5 per kilogram. The other mix will contain 1/3 nuts and 2/3 raisins and will sell for $4 per kilogram. How many kilograms of each mix should the company prepare for maximum revenue? What is the maximum revenue?

*20. Ms. Coffman was given the following advice. She should supplement her daily diet with at least 6000 USP units of Vitamin A, at least 195 mg of Vitamin C, and at least 600 USP units of Vitamin D. Ms. Coffman finds that her local pharmacy carries blue vitamin pills at 5 cents each and red ones at 4 cents each. Upon reading the labels she sees that each blue pill contains 3000 USP units of A, 45 mg of C, and 75 USP units of D; the red pills contain 1000 USP units of A, 50 mg of C, and 200 USP units of D. What combination of vitamin pills should Ms. Coffman buy to obtain the least cost? What is the least cost per day?

Once on a hilltop in Vermont *stood the Smith-Putnam Wind Turbine, called* Grandpa's Knob Windmill *in honor of the Hubbardton farmers who owned the hill it stood on. October 19, 1941, for the first time in human history, power from the wind— via this windmill—fed into an electric utility's lines. Electricity was generated for Rutland and surrounding towns. The windmill operated when feasible for several years until March, 1945, when one of the blades broke off. Grandpa's Knob Windmill was shut down and later dismantled.*

Functions in the wind *The energy generated by a windmill is related functionally to the force of the wind—energy, in fact, varies as the cube (third power) of wind velocity. The mathematics of wind power is much more complex than that, since technological, economic, and ecological issues are involved.*

Experimental windmills are being designed to maximize the energy from wind power. Their blades rotate as fast as possible, catching the wind from all corners, even small amounts of moving air. New designs are being created by individuals at home, by university departments, business, industry, and governmental agencies such as the U.S. Energy Research and Development Administration (ERDA) and the National Aeronautics and Space Administration (NASA).

What's the wind like where you live? Does it blow more than 8 miles an hour? You may see more and more roofs with contraptions moving too fast to be TV antennas.

Vertical axes *in experimental wind turbines let the wind hit the blades from any direction. The turbine on the left is being tested by ERDA in Albuquerque. Each blade has the shape of a catenary (free-hanging rope). The "eggbeater" (right) is on exhibit in Toronto.*

7.6 FUNCTIONS

In Section 7.2 of this chapter, we defined a relation as a set of ordered pairs. In this section, we shall look at possibly the most important type of relation in mathematics, a *function*. Functional relationships occur often in everyday life too: an electric bill is a function of the amount of electricity used, a person's weight is a function of how much food the person eats, and so on. In the language of mathematics, a relation is a **function** if there is a rule or formula that yields exactly *one value of y for each value of x.*

For example, $y = 5x - 7$ is a function. For any value of x that we might choose, the rule $y = 5x - 7$ produces exactly one value of y. If we begin, say, with $x = 6$, we can find the value for y from the rule:

$$y = 5x - 7$$
$$y = 5(6) - 7$$
$$y = 23.$$

If $x = -3$, then $y = 5(-3) - 7 = -22$. For any value of x that we choose, the rule $y = 5x - 7$ will produce exactly one value of y.

On the other hand, the relation $x = y^2$ is not a function. If we start with the x-value $x = 16$, we have $16 = y^2$. There are *two* values of y that make this statement true: $y = 4$ and $y = -4$. Since one value of x leads to more than one value of y, the relation $x = y^2$ is not a function.

Letters such as f, g, or h are used to name functions. If we have a function f, we know that for each value of x there is exactly one value of y. Since

Galileo Galilei (1564–1642) *died in the year Newton was born; his work was important in Newton's development of the calculus. The idea of* function *is implicit in Galileo's analysis of the parabolic path of a projectile, where height and range are functions (in our terms) of the angle of elevation and the initial velocity.*

there is only one such y, we often call it $f(x)$, (read "f of x"), to indicate the result of applying the function f to the number x.

For example, if f is the function given by $f = \{(x, y) | y = 4x - 5\}$, we know that if $x = 3$, then $y = 4(3) - 5 = 7$. We express this in abbreviated form as

$$f(3) = 7 \qquad \text{(the value of the function when } x = 3\text{)}.$$

We also abbreviate the function $f = \{(x, y) | y = 4x - 5\}$ simply as

$$f(x) = 4x - 5.$$

Example 1   Find $f(2)$ and $f(-4)$ for each function.

(a) $f(x) = -3x + 9$

Solution.   Replace x, in turn, by 2 and by -4.

$$\text{If } x = 2, \qquad\qquad \text{If } x = -4,$$
$$f(2) = -3(2) + 9 \qquad f(-4) = -3(-4) + 9$$
$$f(2) = 3 \qquad\qquad f(-4) = 21$$

(b) $f(x) = x^2 + 2x - 5$

Solution.
$$f(2) = 2^2 + 2(2) - 5 \qquad f(-4) = (-4)^2 + 2(-4) - 5$$
$$\qquad\quad = 4 + 4 - 5 \qquad\qquad = 16 - 8 - 5$$
$$f(2) = 3 \qquad\qquad\qquad f(-4) = 3$$

For a particular function, the set of all possible values of x is called the **domain** of the function, while the set of all possible values of y is called the **range** of the function. For example, in the function $y = 6x + 8$, the variable x can take on any value at all, so that the domain is the set of all real numbers. Also, y can take on any value, so that the range is also the set of all real numbers. Functions with more restricted domains and ranges are shown in the examples below.

We can graph functions just as we did relations. To do this for a function f, we can let the horizontal axis represent x, with the vertical axis representing y or $f(x)$.

Example 2   Let the domain of the function $f(x) = 8 - 5x$ be the set $\{-1, 0, 1, 2, 3\}$. Draw the graph of the function. Find the range.

Solution.   Since the domain gives the possible values of x, we can draw the graph by finding the values of y corresponding to those x-values. By this process, we can find ordered pairs for graphing the function. For example, if $x = -1$, then $y = f(-1) = 8 - 5(-1) = 13$. The ordered pair is $(-1, 13)$. All the ordered pairs for this function are shown in the following table.

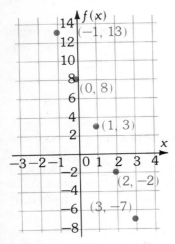

Figure 7.18

x	-1	0	1	2	3
$y = f(x)$	13	8	3	-2	-7
(x, y)	$(-1, 13)$	$(0, 8)$	$(1, 3)$	$(2, -2)$	$(3, -7)$

These ordered pairs are graphed in Figure 7.18. (How would this graph change if we changed the domain to include *all* real numbers?) The range, the set of all y-values, is $\{13, 8, 3, -2, -7\}$.

Example 3   Let the domain of the function $y = x^2$ be the set $\{-3, -2, -1, 0, 1, 2, 3,\}$. Graph the function. Find the range.

Solution.   To graph the function, we first find a y-value for each x-value in the domain.

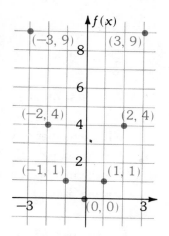

x	-3	-2	-1	0	1	2	3
$y = f(x)$	9	4	1	0	1	4	9
(x, y)	$(-3, 9)$	$(-2, 4)$	$(-1, 1)$	$(0, 0)$	$(1, 1)$	$(2, 4)$	$(3, 9)$

These ordered pairs yield the graph in Figure 7.19. The points of this graph do not lie on a straight line, but on a curve called a *parabola*, which is discussed in more detail in the next section. The range of this function is the set of y-values $\{0, 1, 4, 9\}$.

Figure 7.19

7.6 EXERCISES

Find all the ordered pairs belonging to each relation, with the given domain. Graph the relation, and give the range. Identify any relation which is also a function.

1. $y = x + 6$; $\{3, 4, 5, 6, 7\}$
2. $y = 2x + 1$; $\{0, 1, 2, 3, 4\}$
3. $2x + y = 4$; $\{-1, 0, 1, 2, 3, 4\}$
4. $4x + y = 9$; $\{-2, -1, 0, 1, 2, 3\}$
5. $y - 2x = 6$; $\{-1, 0, 1, 2, 3, 4, 5\}$
6. $y - x = -2$; $\{3, 4, 5, 6, 7\}$
7. $x + 4y = 8$; $\{-4, -2, 0, 4, 2\}$
8. $x + 2 = y$; $\{-2, -1, 0, 1, 2, 3\}$
9. $y \leq 2x + 1$; $\{-2, -1, 0, 1\}$
    (Assume $y \geq -4$.)
10. $y \leq x - 2$; $\{3, 4, 5, 6\}$
    (Assume $y \geq 0$.)

11. $y \leq x + 3$; $\{-1, 0, 1, 2\}$
    (Assume $y \geq -1$.)
12. $x \geq 2y$; $\{0, 2, 4, 6\}$
    (Assume $y \geq 0$.)
13. $x = 2y$; $\{-4, -2, 0, 2, 4\}$
14. $y = -3x$; $\{-2, -1, 0, 1, 2, 3\}$
15. $x + 3y = 0$; $\{-9, -6, -3, 0, 3, 6\}$
16. $y - 2x = 0$; $\{-2, -1, 0, 1, 2, 3\}$
*17. $x = 2$ (y is $-1, 0, 1$, or 2)
*18. $y = 3$; $\{0, 1, 2, 4, 5\}$
*19. $y + 1 = 0$; $\{-2, -1, 0, 1, 2, 3\}$
*20. $x + 4 = 0$ (y is $-1, 0, 1$, or 2)

In Exercises 21–34, find $f(-2)$, $f(0)$, and $f(3)$ for each function.

21. $f(x) = 2x + 1$
22. $f(x) = 5x + 3$
23. $f(x) = 9 - 2x$
24. $f(x) = 3 - 8x$
25. $f(x) = x^2 + 5x$
26. $f(x) = x^2 - 4x$

27. $f(x) = 2x^2 + x - 6$

28. $f(x) = 3x^2 - 2x + 1$

29. $f(x) = (x + 1)(x + 2)$

30. $f(x) = (x - 3)(x + 1)$

31. $f(x) = \dfrac{8 + x}{3 + x}$

32. $f(x) = \dfrac{6 - 2x}{x + 1}$

*33. $f(x) = \dfrac{3 + 4x}{x + 2}$

*34. $f(x) = \dfrac{5 - 5x}{x - 3}$

Write *function* or *not a function* for each relation. In Exercises 45–48, the domain is the set of all people.

35. $y = 12x - 3$

36. $y = x^2 + 1$

37. $x + 2y = 5$

38. $3x + 5y = 8$

39. $y^2 + 1 = x$

40. $x^2 + 1 = y$

41. $x + 3y < 6$

42. $y - x > -4$

*43. $x = 4 + 0y$

*44. $y = 0x - 3$

*45. y is the mother of x

*46. y is the sister of x

*47. x is the mother of y

*48. x is the sister of y

A chain-saw rental firm charges a fixed $4 to rent out a saw, plus $3 per day or any fraction of a day. Let $f(x)$ represent the cost of renting a saw for x days. Find each of the following values.

Example: $f(5)$ represents the cost for renting a saw 5 days. This cost is $4 + 3(5) = 19$ dollars.

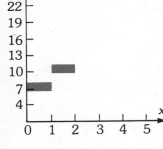

49. $f(1)$

50. $f(2)$

51. $f(3)$

52. $f(4)$

53. $f(\frac{1}{2})$

54. $f(\frac{3}{4})$

55. $f(3\frac{2}{3})$

56. $f(5\frac{1}{8})$

57. $f(2\frac{7}{8})$

58. Complete the graph of the chain-saw function f.

By the definition of function, for each value of x we must be able to find exactly one value of y. For the graph shown in the figure, the marked value of x leads to *two* values of y. The vertical line cuts the graph in two points. Thus, this graph is not the graph of a function. In general, a graph represents a function if any vertical line cuts the graph in no more than one point, Use this vertical line test to decide which of the following graphs in Exercises 59–61 represent functions.

59.

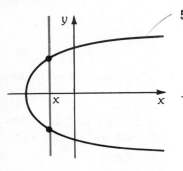

60.

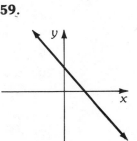

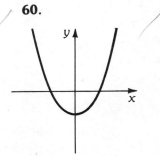

61.

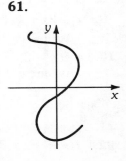

From the Tower of Pisa
Galileo (see page 276) did more than construct theories to explain physical phenomena—he set up experiments to test them. For one thing, Galileo delighted in confounding his critics, particularly believers in Aristotelian physics. Legend says that Galileo dropped objects of different weights from the Tower of Pisa to disprove the Aristotelian view that heavier objects fall to the ground faster than lighter objects.

Galileo said, "The book of nature is written in mathematical language." He applied mathematics to the motions of objects. Thus he developed a formula for freely falling objects (as in the tower legend):

$$d = 16t^2$$

where d is the distance that an object falls (discounting air resistance) in a given time, t, regardless of weight. Note that the formula is a second-degree relation, as discussed in the section beginning on this page.

*62. Does every straight line represent a function?

*63. Are there any linear inequalities which are functions?

A relation R is called an **equivalence relation** if it satisfies the following three properties:

Reflexive property   If r is an element of the domain of R, then (r, r) must belong to R.

Symmetric property   If (r, s) belongs to R, then (s, r) must belong to R.

Transitive property   If (r, s) and (s, t) belong to R, then (r, t) must also belong to R.

Determine whether the following relations are equivalence relations by checking for each of the three properties above.

*64. $x = y$ ***66.** $y < x$ ***68.** $y \geq x$

*65. $x < y$ ***67.** $y \neq x$ ***69.** $y \leq x$

*70. Consider the set $\{1, 2, 3, 4, \ldots, 9999\}$ and the relation R defined as follows: (x, y) belongs to R if x and y contain the same number of digits.

*71. Consider the set of all people and the relation R defined by: (x, y) belongs to R if x and y weigh the same.

*72. Consider the set of all people and the relation R defined by: (x, y) belongs to R if x and y weigh within two pounds of each other.

*7.7 SECOND DEGREE RELATIONS

Most of the equations we graphed thus far in this chapter involved linear equations, those of the form $ax + by = c$. The exponent on both variables is understood to be 1. The graph of a linear equation (first degree relation) is always a straight line.

In this section we shall look at some non-linear relations. In particular, we shall look at equations containing variables whose highest exponent is 2, such as $y = x^2 + 2x + 6$. These equations represent *second degree relations*.

The second degree equations that we shall study have the form

$$y = ax^2 + bx + c, \qquad (a \neq 0)$$

where a, b, and c are real numbers. The graphs of all such equations are *parabolas*. The simplest parabola is probably $y = x^2$, where $a = 1$, $b = 0$, and $c = 0$.

To graph $y = x^2$, we complete a table of ordered pairs, plot the points from the table, and draw a smooth curve through them, as in Figure 7.20. Is $y = x^2$ a function?

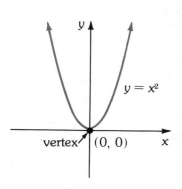

Figure 7.20 The graph of $y = x^2$ is a parabola with vertex at $(0, 0)$

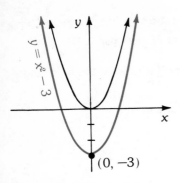

Figure 7.21   The graph of $y = x^2 - 3$ is shifted down three units

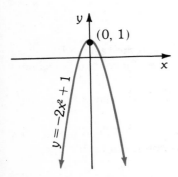

Figure 7.22   The graph of $y = -2x^2 + 1$ opens downward

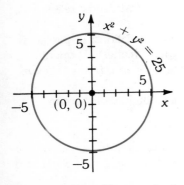

Figure 7.23   The graph of $x^2 + y^2 = 25$ is a circle centered at $(0, 0)$

Example 1   Graph $y = x^2 - 3$.

Solution.   This equation represents a parabola. (How do we know?) To find its graph, complete a table of ordered pairs, and then draw the graph. See Figure 7.21. As the graph shows, $y = x^2 - 3$ has the same *shape* as $y = x^2$, but is shifted 3 units down. The lowest point on this graph, $(0, -3)$, is called the **vertex** of the parabola.

In general, the graph of

$$y = a(x - h)^2 + k$$

is a **parabola** with vertex at (h, k). The parabola opens upward if a is positive, and downward if a is negative.

Example 2   Graph $y = -2x^2 + 1$.

Solution.   The minus sign in the term $-2x^2$ indicates that the graph opens downward. The 2 makes the parabola a little thinner than the "normal" $y = x^2$. Here, the vertex $(0, 1)$ is the *highest* point on the graph. See Figure 7.22.

The equation $x^2 + y^2 = r^2$

represents a **circle** with center at $(0, 0)$ and radius r.

Example 3   Graph $x^2 + y^2 = 25$.

Solution.   Since $5^2 = 25$, this circle has center at $(0, 0)$ and radius 5. See Figure 7.23. Does this graph represent a function?

The equation $\dfrac{x^2}{a^2} + \dfrac{y^2}{b^2} = 1$

represents an **ellipse** centered at $(0, 0)$ and going through the points $(a, 0)$ and $(-a, 0)$ on the x-axis and $(0, b)$ and $(0, -b)$ on the y-axis.

Example 4   Graph $\dfrac{x^2}{9} + \dfrac{y^2}{25} = 1$.

Solution.   This equation represents an ellipse going through $(3, 0)$ and $(-3, 0)$ on the x-axis, and $(0, 5)$ and $(0, -5)$ on the y-axis. The graph is shown in Figure 7.24. Does this graph represent a function?

Hyperbolas have equations of either form:

$$\frac{x^2}{a^2} - \frac{y^2}{b^2} = 1 \qquad \text{or} \qquad \frac{y^2}{a^2} - \frac{x^2}{b^2} = 1$$

hyperbola goes through   hyperbola goes through
$(a, 0)$ and $(-a, 0)$ $(0, a)$ and $(0, -a)$

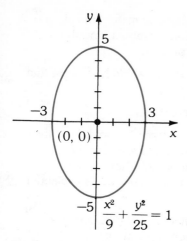

Figure 7.24 The graph of an ellipse

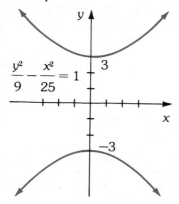

Figure 7.25 The graph of a hyperbola through $(0, 3)$ and $(0, -3)$

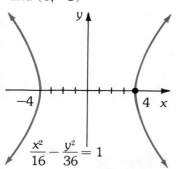

Figure 7.26 The graph of a hyperbola through $(4, 0)$ and $(-4, 0)$

Example 5 Graph $\dfrac{y^2}{9} - \dfrac{x^2}{25} = 1$.

Solution. Here we have a hyperbola going through $(0, 3)$ and $(0, -3)$. See Figure 7.25. Is this hyperbola a function?

Example 6 Graph $\dfrac{x^2}{16} - \dfrac{y^2}{36} = 1$.

Solution. This hyperbola goes through $(4, 0)$ and $(-4, 0)$, as shown in Figure 7.26. Is this hyperbola a function?

7.7 EXERCISES

Graph each parabola. Identify the vertex of each.

1. $y = 3x^2$
2. $y = -2x^2$
3. $y = -\frac{1}{4}x^2$
4. $y = \frac{1}{3}x^2$
5. $y = x^2 - 1$
6. $y = x^2 + 3$
7. $y = -x^2 + 2$
8. $y = -x^2 - 4$
9. $y = 2x^2 - 2$
10. $y = -3x^2 + 1$
11. $y = (x - 1)^2$
12. $y = (x + 2)^2$
13. $y = -2(x + 1)^2$
14. $y = 3(x - 3)^2$
15. $y = (x + 1)^2 - 2$
16. $y = (x - 2)^2 + 3$
17. $y = 2(x - 1)^2 - 3$
18. $y = -3(x + 4)^2 + 5$
19. $y = -3(x + 2)^2 + 2$
20. $y = -2(x + 1)^2 - 3$
*21. $3y = 2(x - 1)^2 - 6$
*22. $4y = 5(x - 2)^2 - 8$

Graph each curve (circle or ellipse).

23. $x^2 + y^2 = 16$
24. $x^2 + y^2 = 9$
25. $x^2 + y^2 = 81$
26. $x^2 + y^2 = 100$
27. $\dfrac{x^2}{4} + \dfrac{y^2}{9} = 1$
28. $\dfrac{x^2}{16} + \dfrac{y^2}{25} = 1$
29. $\dfrac{x^2}{9} + \dfrac{y^2}{16} = 1$
30. $\dfrac{x^2}{36} + \dfrac{y^2}{9} = 1$
31. $\dfrac{x^2}{16} + \dfrac{y^2}{4} = 1$
32. $\dfrac{x^2}{49} + \dfrac{y^2}{4} = 1$
*33. $\dfrac{x^2}{9} + y^2 = 1$

Graph each curve (hyperbola).

34. $\dfrac{x^2}{25} - \dfrac{y^2}{9} = 1$
35. $\dfrac{y^2}{16} - \dfrac{x^2}{16} = 1$
36. $\dfrac{y^2}{49} - \dfrac{x^2}{36} = 1$
37. $\dfrac{x^2}{144} - \dfrac{y^2}{49} = 1$
38. $\dfrac{x^2}{64} - \dfrac{y^2}{100} = 1$
39. $\dfrac{y^2}{9} - \dfrac{x^2}{25} = 1$

Parabolas are often useful in finding the maximum or minimum values of a function. In Exercises 59 and 60, graph the given parabola, find the vertex, and use your information to solve the problem.

*40. Harry owns a taco stand in downtown Phoenix. He has found that the profits of his stand are approximated by

$$P = -(x - 8)^2 + 20,$$

where P represents profit when x units of tacos are sold. Find the number of units of tacos that Harry should sell to make the maximum profit. What is the maximum profit? (Note that if Harry sells too many tacos, his profit will fall. This fact is explained in detail in economics classes.)

*41. Carrie runs a sandwich shop. By studying past results, she has found that the cost of operating her shop is given by

$$C = 2(x - 12)^2 + 50,$$

where C is the cost to produce x units of sandwiches. Find the number of units of sandwiches that Carrie should produce in order to keep cost at a minimum. What is the minimum cost?

Energy from wind and sun *Smokey Yunick operates a garage in Daytona Beach, Florida, the site of his "total energy system," as reported in* Popular Science *(August 1975). In the photo here you can see his bicycle-wheel wind turbine and solar panels over the roof. The diagram on the facing page shows the process of breaking water down into oxygen and hydrogen, then combining with carbon to yield methane and methanol. Smokey Yunick envisions the possibilities of his system down on the farm. It would furnish electricity, tractor fuel, fertilizers, refrigeration for preserving farm products, and steam for processing foods.*

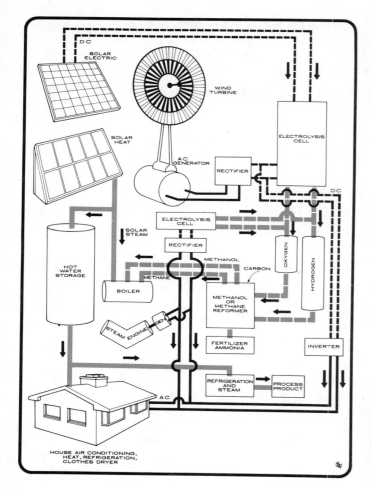

CHAPTER 7 SUMMARY

Keywords

It's a bird *The towering object at the upper right is not a bird, although it seems about to take off. It's NASA's experimental wind turbine near Sandusky, Ohio. It's the largest wind turbine in operation. Each aluminum blade is 62.5 feet long and weighs 2000 pounds. The turbine begins to generate power in an 8-mile an hour wind; blades rotate as fast as 40 rpm.*

Equation
Variable
Solution
Formulas

Coordinate (analytic)
  geometry
Cartesian coordinate
  system
Origin
x-axis
y-axis

Linear (first degree)
  equation
Slope of a line
System of equations
Elimination (addition)
  method

Linear inequality
Half-plane
System of inequalities
Linear programming
Constraint

Relation
Ordered pair
Equivalence relation
Function
Domain
Range
Second degree equation
Parabola
Vertex
Circle
Ellipse
Hyperbola

Formulas

$d = rt$

$I = prt$

d = distance I = simple interest u = unearned interest
r = rate p = principal f = original finance charge
t = time r = rate k = number of payments remaining
   t = time n = original number of payments

$u = f \cdot \dfrac{k(k+1)}{n(n+1)}$

$A = \dfrac{24f}{b(p+1)}$

A = approximate true annual interest rate for a loan paid off in monthly
   payments
f = total finance charge
b = amount borrowed
p = total number of payments

$m = \dfrac{y_2 - y_1}{x_2 - x_1}$

Slope of a line through the distinct points (x_1, y_1) and (x_2, y_2). Point-slope form of the equation of a line through (x_1, y_1) with slope m:

$$y - y_1 = m(x - x_1).$$

Second degree equations

$y = a(x - h)^2 + k$
is a **parabola**
with vertex at (h, k);
opens upward if $a > 0$

$x^2 + y^2 = r^2$
is a **circle**
centered at $(0, 0)$
with radius r

$\dfrac{x^2}{a^2} + \dfrac{y^2}{b^2} = 1$
is an **ellipse** through
$(a, 0)$, $(-a, 0)$ and
$(0, b)$, $(0, -b)$

$\dfrac{x^2}{a^2} - \dfrac{y^2}{b^2} = 1$
is a **hyperbola**
through $(a, 0)$
and $(-a, 0)$

$\dfrac{y^2}{a^2} - \dfrac{x^2}{b^2} = 1$
is a **hyperbola**
through $(0, a)$
and $(0, -a)$

Out of the well at last!
An old algebra problem asks how long it takes a serpent to crawl out of a well at a given rate. The above woodcut was among the first illustrations of algebra word problems—in a text of 1491 by Fillipo Calandri. It seemed fitting to end this chapter on algebra with a hiss of relief.

CHAPTER 7 TEST

Solve each of the following equations.

1. $9k + 3k + 2 = 6k + 8$ **2.** $4(m + 5) + 6m = 7m + 5$

3. Find the rate if $1000 earns $240 simple interest in 4 years.

Solve each of the following systems of equations.

4. $2x + y = -1$
   $4x - y = -11$

5. $5x - 3y = 3$
   $2x - y = 0$

6. $3x + 7y = 31$
   $4x - 2y = -4$

Graph each of the following straight lines.

7. $3x + 2y = 12$ **8.** $x - 4y = 8$ **9.** $y + 3 = 0$

Find the equation of each of the following straight lines.

10. through $(2, -3)$; $m = 2$ **11.** through $(-4, 6)$; $m = \dfrac{2}{3}$

Graph each of the following inequalities.

12. $2x + 5y \le 10$ **13.** $3x - 2y > 12$ **14.** $x \le -3$

Catastrophe Theory *The tempo of change for many of nature's processes is gradual. The distance a car travels, the amount of interest money earns, and biological aging usually change by a small amount over any short time interval. Such events are termed* continuous. *Mathematical functions which represent continuous events can be graphed by unbroken lines or curves. Calculus has refined our knowledge of continuous processes during the past 300 years.*

Until recently there has been no comparable mathematical tool for describing discontinuous events where abrupt change occurs. For example, the heartbeat is a discontinuous event: the resting heart suddenly beats, then rests, then suddenly beats again. The buckling of a beam, a stock-market crash, a riot, a sudden loss of self-control are other examples of discontinuous events.

In the 1960s Rene Thom, a French mathematician, began to investigate such processes using the methods of topology. Thom called discontinuous events **catastrophes** *to emphasize the feature of sudden change. Thom proved that all catastrophic events (in our four-dimensional world) are combinations of seven elementary catastrophes. (In higher dimensions the number quickly reaches infinity.)*

Each of the seven catastrophes has a characteristic topological shape—think of an elastic sheet bent, folded, twisted, deformed, but not torn. (More on topology in Chapter 8.)

Two examples of elementary catastrophes are at the right. The top figure is called a cusp. The bottom figure is an elliptic umbilicus (a belly button with a sort of oval cross section).

Graph each of the following systems of inequalities.

15. $3x + 4y \leq 24$
$\quad\ \ 2x - y \geq 12$

16. $4x + 5y > 20$
$\quad\ \ 3x + 6y < 18$

17. Find the maximum values of $4x + 5y$ subject to these constraints.

$$3x + 2y \leq 8, \qquad x + 2y \leq 4, \qquad x \geq 0, \qquad y \geq 0.$$

For each of the following functions, find $f(-2)$, $f(0)$, and $f(4)$.

18. $f(x) = 9x + 3$

19. $f(x) = -x^2 + 2x$

Write *function* or *not a function* for each of the following.

20. $y = x^2 - 8x$

21. $x = y^2$

Graph each second degree equation. Identify each curve.

22. $y = x^2 + 1$

24. $x^2 + y^2 = 121$

23. $\dfrac{x^2}{9} + \dfrac{y^2}{25} = 1$

25. $\dfrac{x^2}{16} - \dfrac{y^2}{25} = 1$

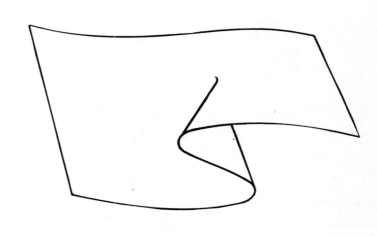

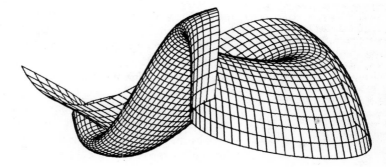

Chapter 8 *Geometry*

Geometria *The two close-ups here are from a bronze relief by the Renaissance sculptor Antonio del Pollaiolo (1433–1498). The complete sculpture is shown on page 289. Antonio's portrayal of geometry is one of ten reliefs depicting the Liberal Arts on the sloping base of the bronze tomb of Sixtus IV, pope from 1471 to 1484. The rectangular top of the tomb contains the prone effigy of Sixtus IV surrounded by reliefs of the seven virtues. Antonio worked on the tomb from 1484 to 1493, assisted by his younger brother Piero. Both this tomb and their bronze tomb of Pope Innocent VIII are in St. Peter's cathedral, Rome.*

Look at the front cover of this text before turning to page 289.

To the Greeks of 1000 BC mathematics meant geometry above all, a rigid kind of geometry from a twentieth-century point of view. They assumed that geometric figures could only be rotated or moved about without change—no deformations in figures were even considered. Perhaps it may have seemed perfectly natural to sculptors and architects working in wood or stone that shape and size were constant properties of objects. The Greeks studied the properties of figures identical in shape and size (congruent figures) as well as figures identical in shape but not necessarily in size (similar figures). They absorbed ideas about area and volume from the Egyptians and Babylonians, and established general formulas.

The Greek view of geometry (and other mathematical ideas) was summarized in the *Elements,* written by Euclid about 300 BC. The power this book exerted is extraordinary; it has been studied virtually unchanged to this day as a geometry textbook and as *the* model of deductive logic.

Euclid's *Elements* begins with definitions of basic ideas such as point, line, and plane. He then gives five postulates providing the foundation of all that follows.

Euclid then listed five axioms, which he viewed as general truths and not just facts about geometry. (To some of the Greek writers, postulates were truths about a particular field, while axioms were general truths. Today, "axiom" is used in either case.)

Euclid's postulates
1. Two points determine one and only one straight line.
2. A straight line extends indefinitely far in either direction.
3. A circle may be drawn with any given center and any given radius.
4. All right angles are equal.
5. Given a line *k* and a point *P* not on the line, there exists one and only one line *m* through *P* that is parallel to *k*.

Euclid's axioms
6. Things equal to the same thing are equal to each other.
7. If equals are added to equals, the sums are equal.
8. If equals are subtracted from equals, the remainders are equal.
9. Figures which can be made to coincide are equal.
10, The whole is greater than any of its parts.

Using only these ten statements and the basic rules of logic, Euclid was able to prove a large number of "propositions" about geometric figures.

By the time of the Renaissance, 17 centuries after Euclid, artists such as Leonardo da Vinci studied the images of three-dimensional objects on two-dimensional surfaces, and *projective geometry* emerged, with ideas distinct from those of Euclid. In the nineteenth century even the hallowed axioms of Euclid were challenged, yielding *non-Euclidean geometries* (discussed in Section 5 of this chapter). Modern geometry includes the important field of *topology,* in which figures can be stretched, twisted, and deformed (but not cut). (Topology and its applications are discussed in Section 6.)

8.1 BASIC IDEAS OF EUCLIDEAN GEOMETRY

The most basic ideas in geometry are *point, line,* and *plane.* In fact, it is not really possible to define them in other words. Euclid defined a point as "that which has no part," but this definition is so vague as to be meaningless. Do you think you could decide what a point is from this definition? But from your experience in saying "this point in time" or in sharpening a pencil, you have an idea of what he was getting at. Even though we don't try to define *point,* we do know some properties of points. For example, a point has no magnitude, or size. It has only position.

Euclid defined a line as "that which has breadthless length." Again, this definition is meaningless. Based on our experience, however, we know what Euclid meant, although his words tell us little about lines. We do know properties of lines: a line has no thickness or width; it extends indefinitely in two directions.

What do you visualize when you read Euclid's definition of a plane: "a surface which lies evenly with the straight lines on itself"? Do you think of a flat surface, such as a tabletop or a page in a book? That kind of thing is what Euclid intended.

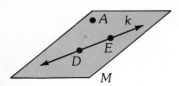

Figure 8.1

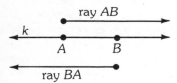

Figure 8.2

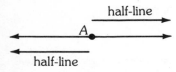

Figure 8.3

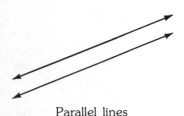

Parallel lines

Today we would not try to define the basic idea of plane, even though we study properties of planes. For example, if we are given any three points that are not in a straight line, then we can pass a plane through the points. That is why camera tripods have three legs — no matter how irregular the surface, the tips of the three legs determine a plane. On the other hand, a camera support with four legs would wobble unless we carefully extend each leg just the right amount.

There are certain conventions for naming points, lines, and planes. A capital letter names a point or a plane. A line is named either with a lowercase letter or with capitals for two points on the line. Figure 8.1 pictures plane M with point A in the plane. Plane M also contains a line, named either line k or line DE.

If we select any point on a line, we have divided the line into three parts: the point itself and two **half-lines,** one on each side of the point. Thus, in Figure 8.2, point A divides the line into three parts, A itself and two half-lines. Point A belongs to neither half-line. Note that each half-line extends indefinitely in the direction opposite to the other half-line.

If we include an initial point with a half-line, we get a **ray.** A ray is named with two letters, one for the initial point of the ray, and one for the point that the half-line goes through. For example, in Figure 8.3 ray AB starts at point A and goes through point B. On the other hand, ray BA starts at B and goes through A. Note that ray AB is not the same as ray BA. (Why?)

A finite portion of a line is called a **line segment.** A line segment includes both endpoints, and is named for its endpoints. Figure 8.3 shows line segment AB of line k.

The geometric meanings of "parallel" and "intersecting" apply to two or more lines or planes. (See Figure 8.4.) **Parallel lines** lie in the same plane and never meet, no matter how far extended. However, **intersecting lines** *do* meet or cross, and two distinct intersecting lines have one and only one point in common, the point of intersection.

Parallel planes also never meet, no matter how far extended. Two intersecting planes form a straight line, the one and only line they have in common. *Skew lines* do not lie in the same plane; they never meet, no matter how far extended.

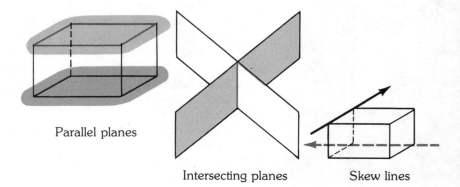

Intersecting lines   Parallel planes   Intersecting planes   Skew lines

Figure 8.4

Geometria *The complete depiction of Geometry by Pollaiolo has some elements in common with the front cover of this text. The important feature in both conceptions is the construction of a figure with compasses—one of two instruments allowed in classical geometry for constructions.*

The cover shows Euclid in action, as portrayed in School of Athens, a mural by Raphael in the Vatican palace. Raphael was commissioned by Pope Julius II, came to Rome in 1508, and finished the mural in 1511. Surely he had seen and studied Pollaiolo's work in St. Peter's—the resemblance is too close. Raphael used fellow artists as models for the faces of Euclid, Aristotle, Pythagoras, and others in the mural. Euclid was appropriately modeled after Bramante, painter and architect, who created the initial plans for the dome of St. Peter's. Bramante was the only Italian architect in the early 16th century who understood thoroughly the applications of mathematics and physics to the construction of vaults and domes.

There is a special and very important case of intersecting lines (or planes)—*perpendicular* lines (or planes). To explain this case we must first explore the meaning of "angle."

An **angle** is formed by two rays that have a common endpoint, as shown in Figure 8.5. Notice that "angle" is the first basic term in this section we have actually defined, using the previously defined terms *ray* and *endpoint*.

The rays forming an angle are called its *sides*. The common endpoint of the rays is the *vertex* of the angle. There are two common ways of naming angles. If no confusion would result, we can name an angle with the letter marking its vertex. Thus, the angles in Figure 8.5 can be named, respectively, angle *B*, angle *E*, and angle *K*. We also name angles with three letters. The first letter names a point on one side of the angle; the middle letter names the vertex; the third names a point on the other side of the angle. In this system, the angles in the figure can be named angle *ABC,* angle *DEF,* and angle *JKL.*

In Figure 8.6(a) angle *ABC* is again the union of rays *BC* and *BA*, but the angle can be thought of as an amount of rotation. Let *BA* first coincide with *BC*—as though they were the same ray. Then rotate *BA* (the endpoint doesn't move) in a counterclockwise direction to form the angle *ABC*.

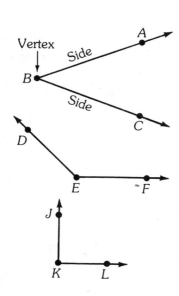

Figure 8.5

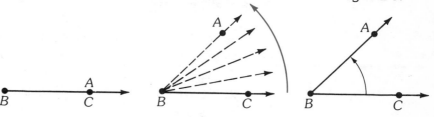

Figure 8.6 (a)

Angles are measured by the amount of rotation, in a system that goes back to the Babylonians some 2000 years before Christ. The Babylonian astronomers said that 360 represents the amount of rotation of a ray back onto itself. The number 360 may have been chosen because it is close to the number of days in a year and is conveniently divisible by 2, 3, 4, 5, 6, 8, 9, 10, 12, and other numbers. Note also that the Babylonians used a base 60 number system (see Chapter 3).

Figure 8.6(b) shows various angles measured in **degrees.** We use a degree sign ° to write, for example, angle $N = 45°$. (The same sign ° indicates degrees of temperature.)

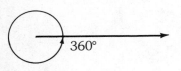

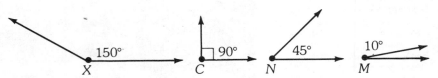

Figure 8.6(b)

Angles are classified and named with reference to their degree measure. An angle measuring more than 0° but less than 90° is called an **acute angle** (angles M and N in Figure 8.6). Angles more than 90° but less than 180° are called **obtuse angles** (angle X). An angle measuring 180° is a **straight angle;** its sides form a straight line.

Angles measuring exactly 90° are **right angles** (angle C). Now we can define "perpendicular": **perpendicular lines** intersect to form right angles. Also, intersecting lines that form right angles are perpendicular. Our sense of *vertical* depends on perpendicularity. Carpenters and stonemasons through the ages have built amazing structures with as few tools as a plumb line and a right angle (or T square).

In Figure 8.7 the sides of angle NMP have been extended to form sides of another angle, RMQ. The pair NMP and RMQ are called **vertical angles.** Another pair of vertical angles is formed at the same time, NMQ and PMR.

In the figure, the pair of angles NMQ and NMP have a common vertex, M, and a common side, NM. They are called **adjacent angles.** Actually, the figure contains four pairs of adjacent angles.

Angle pairs called *complementary* or *supplementary* are explained in Exercises 27–34 below.

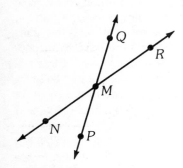

Figure 8.7

8.1 EXERCISES

Exercises 1–10 name portions of line k (see figure). For each exercise, draw a figure showing just the portion named.

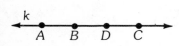

1. line segment AB
2. ray BC
3. ray CB
4. line segment AD

5. half-line BC
6. half-line AD
7. ray BA
8. ray DA

9. line segment CA
10. line segment DA

Skysurfers *look for a good glide angle in choosing wings so they can fly on gentler slopes and stay up longer.*

In Exercises 11–16, use a protractor (as below) to measure each angle. Identify each angle as either *acute, right, obtuse,* or *straight.*

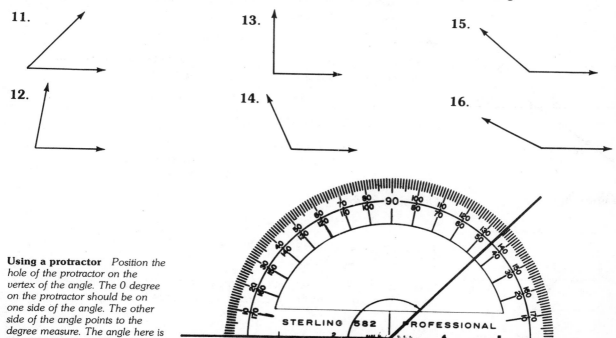

11.

12.

13.

14.

15.

16.

Using a protractor *Position the hole of the protractor on the vertex of the angle. The 0 degree on the protractor should be on one side of the angle. The other side of the angle points to the degree measure. The angle here is found to measure 135°.*

STERLING 582 PROFESSIONAL

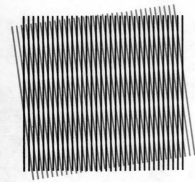

Moiré patterns *A set of parallel lines with equidistant spacing intersects an identical set, but at a small angle. The result is a moiré pattern, named after the fabric moiré ("watered") silk. You often see similar effects looking through window screens with bulges. Moiré patterns are related to periodic functions, which describe regular recurring phenomena (wave patterns such as biorhythms or business cycles). Moirés thus apply to the study of electromagnetic, sound, and water waves, to crystal structure, etc.*

Use a protractor to draw angles having the following measures.

17. 30°   **19.** 60°   **21.** 95°

18. 45°   **20.** 82°   **22.** 121°

Name all pairs of adjacent angles and of vertical angles in each figure.

23.

24.

25.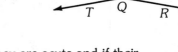

26.

Two angles are called *complementary* if they are acute and if their measures add to 90°. Find the angle complementary to each angle with the given measure.

27. 47°   **28.** 59°   **29.** 82°   **30.** 19°

Supplementary angles are two angles whose sum is 180°. Find the angle supplementary to each angle with the given measure.

31. 41°   **32.** 96°   **33.** 75°   **34.** 112°

We can use equation solving as in Section 7.1 to find the measure of angles in a complementary or supplementary pair.

Example: Find the measures of the angles in the figure.

Solution.   The figure shows two supplementary angles. The sum of their measures is 180°, a fact expressed by the equation

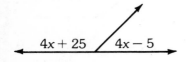

$$(4x - 5) + (4x + 25) = 180.$$

Simplify on the left-hand side, and then solve the equation.

$$\begin{aligned}
8x + 20 &= 180 &&(4x + 4x = 8x) \\
8x &= 160 &&(\text{Add } -20 \text{ to both sides.}) \\
x &= 20 &&(\text{Divide both sides by 8.})
\end{aligned}$$

One angle of the figure is $4x - 5$, so by substitution,

$$4x - 5 = 4(20) - 5 = 80 - 5 = 75.$$

The other angle is $4x + 25 = 4(20) + 25 = 105$. The two angles measure 75° and 105°, respectively.

Exercises 35–40 picture pairs of complementary or supplementary angles. Use algebra to find the measures of angles in each pair.

35.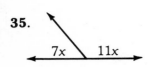

36.

37.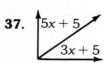

38.

39.

40.

8.2 FIGURES IN THE PLANE AND IN SPACE

The basic undefined term describing figures in the plane is **curve.** We use this term without any attempt to define it. We can, however, define common types of curves. (See the examples in Figure 8.8.)

A **simple** curve can be drawn without lifting the pencil from the paper, with no point of the paper touched twice.

A **closed** curve has its starting and ending points the same, and is also drawn without lifting the pencil from the paper.

Among the most common types of curves in mathematics are those which are both simple and closed. Perhaps the most important of these are **polygons:** simple closed curves made up of only straight line segments. Polygons are classified according to the number of line segments making their sides. Some common types of polygons are shown in Figure 8.9. Polygons with all sides equal and all angles equal are **regular polygons.**

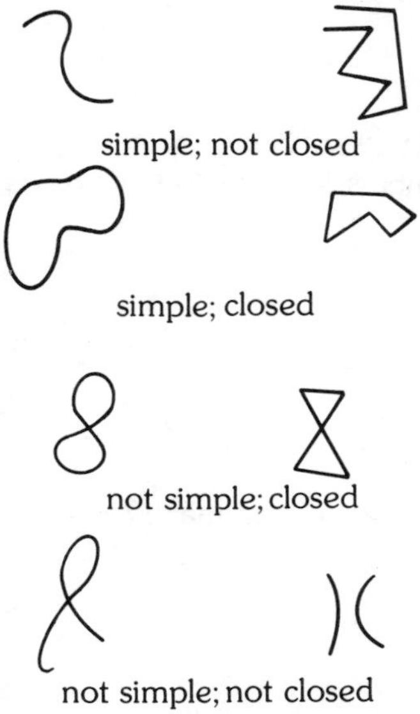

Figure 8.9

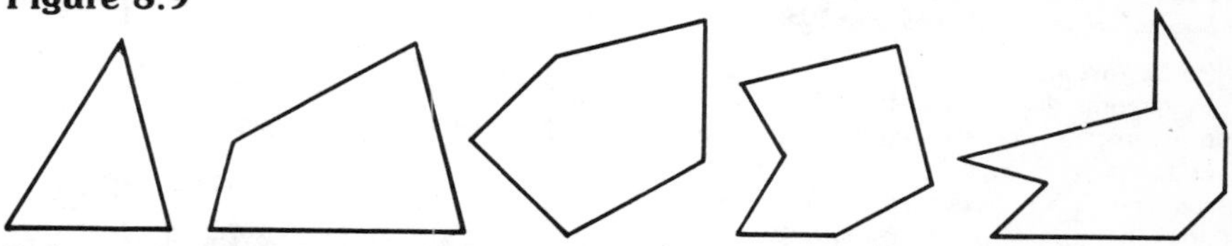

Polygons are simple closed curves made up of straight line segments

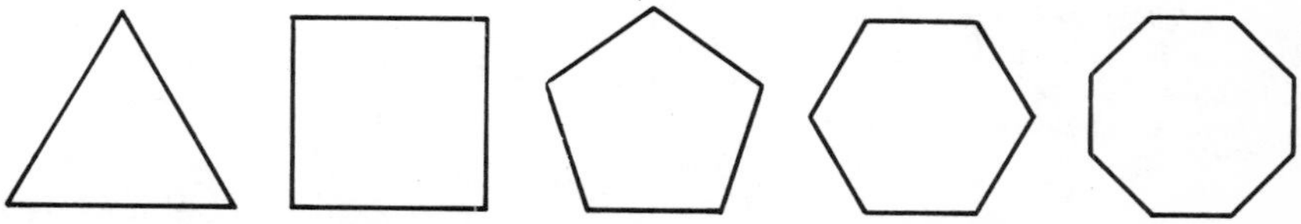

Regular polygons have equal sides and equal angles

simple; not closed

simple; closed

not simple; closed

not simple; not closed

Figure 8.8

The minimum number of straight line segments forming a polygon is three. A polygon of three sides is called a **triangle** (rather than "trigon"). Triangles are classified by angles as well as by sides, as shown in Figure 8.10.

	all acute	one right angle	one obtuse angle

Angles

 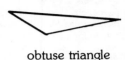

	acute triangle	right triangle	obtuse triangle

	all sides equal	two sides equal	no sides equal

Sides

Figure 8.10   equilateral triangle   isosceles triangle   scalene triangle

While triangles have been studied *extensively* in mathematics, one of the most famous results in all of geometry concerns *right* triangles:

Pythagorean Theorem

If c is the longest side (hypotenuse) of a right triangle having a and b as the shorter sides, then
$$a^2 + b^2 = c^2.$$

Example 1   Find the missing side of each right triangle:

(a)
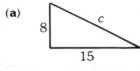

(b)

(a) In a right triangle, $a^2 + b^2 = c^2$. In the right triangle of figure (a), we have $a = 8$ and $b = 15$. If we substitute these numbers into the Pythagorean Theorem, then

$$a^2 + b^2 = c^2$$
$$8^2 + 15^2 = c^2.$$

We know $8^2 = 8 \cdot 8 = 64$, and $15^2 = 15 \cdot 15 = 225$. Thus,

$$64 + 225 = c^2$$
$$289 = c^2.$$

From the square root table in the back of the book, we find that $17 \cdot 17 = 289$, and so $c = 17$.

(b)
$$a^2 + b^2 = c^2$$
$$a^2 + 36^2 = 39^2$$
$$a^2 + 1296 = 1521$$
$$a^2 = 225 \qquad \text{Subtract 1296 from both sides.}$$
$$a = 15. \qquad \text{Since } 15 \cdot 15 = 225.$$

Pythagorean theorem
Pythagoras did not discover the theorem that was named after him, although legend tells that he sacrificed a hundred oxen to the gods in gratitude for the discovery. The Babylonians knew the theorem very well. The first proof may have come from Pythagoras, however. The Greek stamp above illustrates the theorem with a sort of tile pattern, suggesting a simple way to demonstrate the theorem.

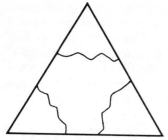

Another demonstration
*Construct any triangle, such as the
one above. Cut it out, and tear it
into three "corner" pieces. You
should be able to rearrange the
pieces so that the three angles
make a straight angle (as seen at
the right). A straight angle
measures 180° by definition.*

Another basic result that the Greeks proved about triangles is the
following:

The sum of the angles of any triangle is 180°.

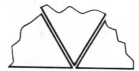

For example, suppose that we know that two of the angles of a triangle
are 48° and 61°. We can find the third angle, call it x, by using the fact that
all three angles add to 180°:

$$48 + 61 + x = 180$$
$$109 + x = 180$$
$$x = 71.$$

The third angle of the triangle is 71°.

Polygons with four sides are called **quadrilaterals.** The variety of
quadrilaterals can be classified by sides and angles, as Figure 8.11 shows. An
important distinction is whether one or more pairs of opposite sides are
parallel. Thus a **parallelogram** is a quadrilateral with both pairs of opposite
sides parallel. Parallelograms with all right angles are the most familiar—
rectangles and squares. Both the square and the rhombus have all sides
equal, but only the square is a regular polygon. (Why?)

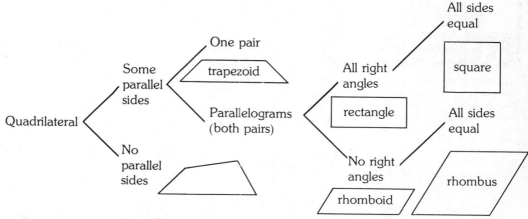

Figure 8.11   Quadrilateral tree   All squares are rectangles are parallelograms
are quadrilaterals are polygons—but the reverse order is not true.
The square is the only regular polygon among quadrilaterals.

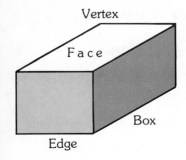

Vertex

Face

Box

Edge

Figure 8.12

FRANCE   POSTES 1977

VASARELY   3,00

What would Euclid say about this? *This French stamp reproduces a painting by Victor Vasarely, who works in the style called* op art *(using visual tricks). Vasarely here links hexagonal shapes to make a superhexagon — or is it about to burst into a sphere? The bulge is particularly tricky in the context of this page: the text says that six triangles fit perfectly in the plane around a point; no space figure results. Well, see here!*

Thus far in this chapter we have studied plane figures — figures which can be drawn completely in the plane of a piece of paper. However, it takes three dimensions of space to truly represent the solid world around us. The space figures we shall study are made up of plane figures. For example, Figure 8.12 shows a "box" (a *rectangular parallelepiped* in mathematical terminology). The *faces* of a box are rectangles and/or squares. The faces meet at *edges;* the "corners" are called *vertices* (plural of vertex — the same word as for the "corner" of an angle).

Boxes are one kind of space figure belonging to an important group called **polyhedrons,** the faces of which are made of only polygons.

Perhaps the most interesting polyhedrons are the **regular polyhedrons.** Recall that a *regular polygon* is a polygon with all sides equal and all angles equal. A *regular polyhedron* is a space figure, the faces of which are only of one kind of regular polygon.

It turns out that there are only five different regular polyhedrons. To see why, begin with a regular triangle. Three regular triangles meeting at one point lead to the space figure called a **tetrahedron.** (See Figure 8.13.) The faces of a tetrahedron are four equilateral triangles.

Four regular triangles meeting at a point lead to an **octahedron.** An octahedron has eight faces and twelve edges.

We get an **icosahedron** when five regular triangles meet at a point. It's a little hard to check from a drawing, but an icosahedron has 20 faces, 12 vertices, and 30 edges.

We don't get anywhere with six regular triangles meeting at a point. Each angle of a regular triangle is 60°. Six of these angles make 360°, the number of degrees around a point. Thus, we can put six triangles together only in a plane. No space figure can result. Only tetrahedrons, octahedrons, and icosahedrons can be made from regular triangles.

A regular quadrilateral is called a square. If three squares meet at a point, we get a *hexahedron,* or **cube.** Four squares completely fill the space around a point, so that no space figure is possible from four squares meeting at a point.

Three regular pentagons meeting at a point produce a **dodecahedron,** which has 12 faces, 20 vertices, and 30 edges. Four or more pentagons will not fit around a point, so that no further regular polyhedrons can be made from pentagons.

Three regular hexagons (six-sided polygons) completely fill the space around a point. Thus, there can be no regular space figure made only of hexagons. Also, there are no regular polyhedrons made up of regular polygons of more than six sides. Therefore, the five regular polyhedrons shown in Figure 8.13 are the only ones that can exist.

Figure 8.13 The five regular polyhedrons

Tetrahedron

Hexahedron (Cube)

Octahedron

Dodecahedron

Icosahedron

Power *The pyramid has been considered mystical since the Egyptians built them 4000 years ago. No one is sure why this shape was chosen; one idea is that when the sun's rays broke through the clouds from the occasional desert rainstorm a pyramid shape was suggested. Since the Egyptians worshipped the sun god, Ra, they may have felt that pyramid shapes were required by divine power.*

Today, pyramids are again popular, this time for their alleged properties: it is claimed that a rusty razor blade placed in a pyramid will become sharp again, a nearly dead plant placed in a pyramid will come back to life, tired people who sit in pyramids are refreshed, a pyramid placed over the heads of dental patients reduces pain, vegetables in a pyramid don't decay, cars run better with pyramids under the hood, and so on.

If you want to test these claims, you can make your own pyramid. Use almost any material, even string. The pyramid need not have solid sides; the shape is what is important. In deciding on the shape, first choose a convenient height (the distance from the center of the base to the point, or apex, of the pyramid).

According to The Secret Power of the Pyramids, *by Bill Schul and Ed Pettit, the length of each side of the base should be 1.5708 times as long as the height, while the distance from a corner of the base to the apex should be 1.4945 times as long as the height.*

The following table, from Schul and Pettit's book, shows dimensions for some pyramids.

Height	Base	Side
6	9.42	8.97
8	12.57	11.96
10	15.70	14.95
12	18.85	17.93

Two other types of polyhedrons are familiar space figures: pyramids and prisms. **Pyramids** are made of triangular sides and a polygonal base. **Prisms** have two faces in parallel planes; both are polygons equal in size and shape. The remaining faces of a prism are all parallelograms. (See Figure 8.14.) Thus, a box is also a prism.

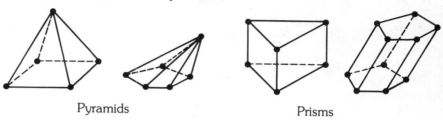

Pyramids   Prisms

Figure 8.14

Concentric circles *This rock carving from Hawaii may symbolize the firstborn (dot) in a family (circles).*

The circle, although a plane figure, is not a polygon. (Why?) Figure 8.15 shows space figures made up in part of circles, including circular cones and circular cylinders. The figure also shows how circles can generate toruses (the torus is a doughnut-shaped solid that has interesting topological properties—see Section 8.6).

Circular cone

Circular cylinder

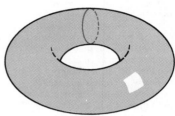

A rotating circle generates a torus

Figure 8.15

Monster of the deep *Someday you may see this out in the ocean, at least the top of it. This is a drawing of the Ocean Thermal Energy Conservation system (OTEC) by Lockheed in California. Plans call for a structure 250 feet in diameter, 1600 feet long, weighing 300,000 tons.*

Can it really generate power as it floats under the surface of the ocean? Yes, and it would do so by using solar energy. The sun's radiation on the ocean is stored as heat energy in surface waters. A difference then exists between temperature of the surface waters and that of deeper waters. In the tropics, for example, the temperature at the surface is from 30° to 35°F higher than in the deep. The difference (called thermal gradient) is enough so that a volatile liquid (ammonia, for example) could be vaporized at the higher temperature and condensed at the lower—recycling. The vapor would turn a turbine (as steam can do), which drives a generator to produce electricity. OTEC is designed to produce power for a city of 100,000 people. The turbines and pumps are attached on the outside; a maintenace crew would be housed on the top.

Figure 8.16 Conic sections

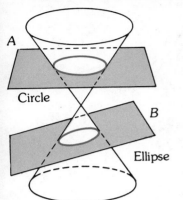

Circle

Ellipse

Some important plane figures are obtained when a cone is cut through by a plane, a fact known to the ancient Greeks. (Note that a mathematical cone is a double cone.) Such **conic sections** are illustrated in Figure 8.16. Plane *A* cuts the cone straight across, producing a section that is a **circle**. Plane *B*, tilted away from the horizontal, produces an **ellipse** (commonly called an *oval*). Plane *C* is tilted so that it is parallel to the edge of the cone, and a **parabola** is the section produced. Plane *D* cuts both parts of the cone, yielding a **hyperbola**. The algebraic equations of conic sections were discussed in Section 7.7, where these figures turned out to be graphs of second-degree equations (the exponent on one or more variables is 2).

8.2 EXERCISES

In Exercises 1–12, identify each curve as either *simple, closed, both,* or *neither.* Identify any that are polygons.

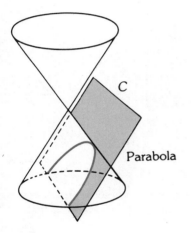

Parabola

1.

2.

3.

4.

5.

6.

7.

8.

9.

10.

11.

12.

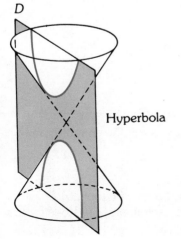

Hyperbola

In Exercises 13–24, *a* and *b* represent the two shorter sides of a right triangle, while *c* represents the longest side. Find the length of any missing sides.

13. 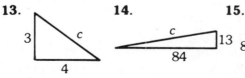   3   *c*   4

14.   *c*   84

15. 13   8   17   *b*

16. *a*   26   24

17. $a = 24$ cm; $c = 25$ cm

18. $a = 14$ m; $b = 48$ m

19. $a = 28$ km; $c = 100$ km

20. $b = 48$ feet; $c = 52$ feet

21. $b = 28$ inches; $c = 35$ inches

22. $a = 30$ m; $b = 16$ m

23. $b = 63$ cm; $c = 65$ cm

24. $a = 80$ km; $c = 82$ km

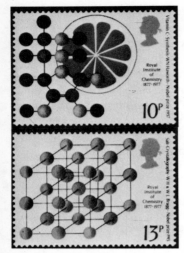

Geometry in nature   *These stamps are from a set of four honoring British achievement in chemistry. The molecular "plan" of Vitamin C (top) is a unique geometric structure. In the course of "proving" that its structure is indeed unique, Sir Norman Haworth synthesized Vitamin C, which occurs naturally in citrus fruit and some other foods. Haworth's work was also important in making the vitamin available to consumers whatever their diet—the vitamin is not produced in the human body.*

The salt crystal (bottom) is revealed under X-ray photography to have a cubic structure, with alternating ions of sodium and chlorine. Sir Lawrence Bragg pioneered in the study of crystal structure by discovering a law of X-ray diffraction in 1912. He and his father, Sir William Henry, a pioneer in solid state physics, jointly received a Nobel Prize in 1915.

Polyhedral structures other than cubic are found in certain crystals and in viruses, too. Many viruses that infect animals were thought to have spherical shapes until X-ray analysis revealed that their shells are icosahedral in shape.

Find the missing angles in each triangle.

25.   **26.**   **27.**   **28.**

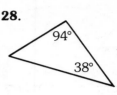

29. $A = 74°$; $B = 29°$ **32.** $M = 39°$; $N = 58°$

30. $A = 58°$; $C = 102°$ **33.** $R = 114°$; $S = 21°$

31. $B = 64°$; $C = 42°$ **34.** $Z = 38°$; $X = 38°$

Classify each triangle in Exercises 35–46 as either *acute, right,* or *obtuse;* and also as either *equilateral, isosceles,* or *scalene.*

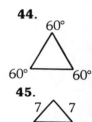

35.

38.

41.

44.

36.

39.

42.

45.

37.

40.

43.

46.

Complete the following table for the numbers of faces, vertices, and edges of each given figure. Compute $E + 2$ for each figure. (*Hint:* Many of the answers are given in the text.)

Figure	Faces (F)	Vertices (V)	Edges (E)	$E + 2$
47. Tetrahedron				
48. Hexahedron				
49. Octahedron				
50. Dodecahedron				
51. Icosahedron				

52.   By studying the table, you should find a pattern: the sum of F and V is always _____ more than E, or

$$F + V = \underline{\quad}.$$

This formula is called *Euler's formula,* in honor of Leonhard Euler (1707–1783), who developed it. Euler's solution of the Koenigsburg Bridge Problem is discussed in Section 8.6.

Parabola power   *The red dot in the above parabola marks a point called the focus of the parabola. Suppose a reflective material like polished metal were made into a dish or trough with parabolic cross section. Then incoming light rays, parallel to the axis of the parabola, would be reflected and brought together at the focus, as the drawing shows. This is the optical or reflective property of parabolas. It can be applied to collecting solar energy. The lower photo shows a large solar collector, trough-shaped, with parabolic cross section. The sun's rays are focused on tubes in the center. Light energy is converted to heat energy and to electrical energy. The process also heats up fluid used to heat and cool a nearby office building. The scene is Sandia Laboratories in New Mexico, part of the U.S. Energy Research and Development Administration.*

The top photo illustrates an idea many steps further. This is an artist's conception of a satellite-to-Earth energy system. The "kite" object in the upper righthand corner is a Solar Power Satellite, orbiting along with the Earth, 36,000 miles above it. The satellite beams solar energy to receivers, massed in circular formation, five miles across. The city in the distance would be supplied with electrical energy.

Geometry and Astronomy

Observations and orbits

Through the ages human beings have observed the planets and stars, recorded their motions, and tried to work out systems to account for regular motions and apparent irregular motions of the heavenly bodies. One practical result is a calendar, such as the Piedra del Sol, *pictured on the above stamp. This "stone of the sun," nearly 12 feet across, was carved for Aztec astronomer-priests in the early 1500s.*

About the same time, far away in Central Europe, a revolution in thinking about the sun was in the making. This Copernican Revolution, named after Nicolas Copernicus, abandoned what was obvious to everybody—that the planets, sun, and stars all revolved about the Earth, which was fixed in space.

The traditional system went back over 1500 years to the Egyptian port of Alexandria, a center of Greek science in late Antiquity. (See page 325.) There, in the second century A.D., Claudius Ptolemeus, called Ptolemy, compiled his Almagest *("the great"). That is what Arabian astronomers called the work they translated and introduced into medieval Europe.*

Ptolemy *The* Almagest *contains what the previous centuries knew about astronomy as well as Ptolemy's own geometric model of the solar system. The model was fairly complex, with the Earth at center, the planets and sun moving in circular orbits and on epicycles. Ptolemy meant it to be a means of predicting accurately the motions of the planets, not a true picture of reality. However, Ptolemy's system was accepted on authority during the Middle Ages and into the time of Copernicus.*

In fact, medieval astronomers also held that the planets were carried on rotating crystal spheres—an idea invented by Pythagoras and adopted by no less an authority than Aristotle himself.

Copernicus worked out his own system with the sun at center (heliocentric) instead of Earth at center (geocentric). A small change, you think—but it marked a revolution in thinking and the beginning of the modern world.

Mikolaj Kopernik *(1473–1543), called Copernicus, was born in Torun, on the Poland-Prussia border. (This stamp honors the 500th year since his birth.) He studied law, medicine, Greek, and mathematics in Italy, and returned to Poland to take an administrative post at the cathedral of Heilsberg. He practised medicine, read the works of pre-Ptolemaic writers, and even wrote a treatise on the coining of money. (In fact, Copernicus was the first to say that bad money drives out good (so-called Gresham's law).*

Above all, Copernicus made astronomical observations from his tower in the cathedral. By 1512 he was already working on his heliocentric model. In the early 1540s he decided to publish the theory. His work was titled De revolutionibus orbium coelestium *("on the revolutions of the heavenly spheres"). The book came out just in time for him to see it before he died—it was brought to him on his deathbed.*

Johannes Kepler *(1571–1630) was a mathematician and astronomer who avidly accepted the heliocentric theory of Copernicus, and went so far as to defend it in public while still a student. Such rashness displeased his theology professors, who told him that he was unfit for the Protestant ministry. Thus it was that Kepler came to be teaching high school mathematics and astronomy.*

He was as much a number mystic as a mathematician. In 1596 Kepler described the solar system with reference to the five regular polyhedrons (see page 296). Kepler's model is shown below. He fitted the five regular polyhedrons between the spheres of the six known planets. From the innermost planet out, the model gives the following sequence: Mercury, octahedron, Venus, icosahedron, Earth-Moon, dodecahedron, Mars, tetrahedron, Jupiter, cube, Saturn. In Harmony of the Worlds (1619), Kepler composed musical motifs for the harmony of the celestial spheres, an idea that goes back to Pythagoras. Mercury's motif, for example, runs up the scale and descends with an arpeggio. In contrast, Earth's motif is a two-note drone, mi, fa, mi ("misery, famine, misery").

In 1598 Kepler had to flee to Prague because of anti-Protestant feeling (this was during the Counter-Reformation). He was taken on as assistant by Tycho Brahe (see page 6). Tycho had devised his own planetary system, with Earth at center, but the other planets revolving about the sun. His observations of comets led him to drop the idea of crystal spheres. After Tycho's death in 1601, Kepler had the use of his teacher's records, and became court mathematician to the Emperor Rudolph.

Not circles, but ellipses!

Using Tycho's records, Kepler began an intensive study of the motion of planets, especially that of Mars. In 1609 Kepler published the New Astronomy, *which contains his first two laws of planetary motion. (The third law appeared in* Harmony of the Worlds.*)*

The first law is particularly interesting from a geometric viewpoint. The law states that the planets revolve around the sun in elliptical orbits—not circular orbits. This was a complete break with tradition. Kepler himself struggled and suffered trying to preserve the circular orbits since circles, he felt, are more regular and "harmonious." But the data forced him as scientist to throw away wrong ideas. One of Kepler's famous diagrams of the elliptical orbit of Mars is shown above on a commemorative stamp.

Note that both the ellipse and the circle are conic sections (see page 299). The paths of projectiles on Earth and bodies in the solar system (including comets and space satellites) are conic sections.

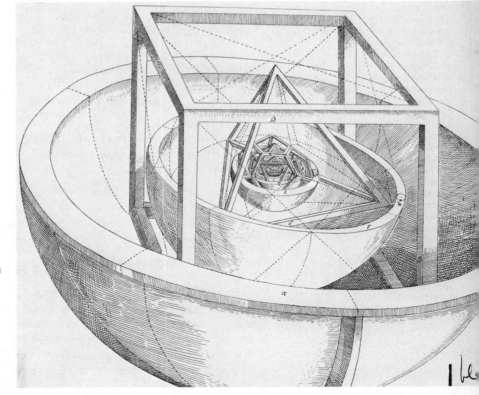

8.3 GEOMETRIC FORMULAS
FOR PROBLEM SOLVING

Geometry is not only seeing, but it is also measuring. We are surrounded with objects whose shapes can be approximated by the figures studied in the previous section. Suppose you had to fence in a field, tile a floor, cover a sofa, fill a tank, climb a pyramid, or make a dunce hat. We saw the shapes of many geometric figures in the previous section. In this section we discuss formulas for the perimeter, area, and volume of such figures. These formulas are very useful in problem solving.

The **perimeter** of a plane figure is the distance around the figure. (Exception: the perimeter of a circle is called its *circumference*.) To find the perimeter of a rectangle, we use the formula

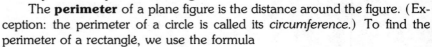

$$P = 2L + 2W,$$

where L represents the length of the rectangle and W represents the width.

Example 1 The length of a rectangle is 9 meters, and the width is 6 meters. Find the perimeter.

Solution.   We substitute 9 for L and 6 for W in the formula:

$$P = 2L + 2W$$
$$P = 2 \cdot 9 + 2 \cdot 6$$
$$= 18 + 12$$
$$P = 30.$$

The perimeter of the rectangle is 30 meters.

Example 2 The perimeter of a rectangle is 96 cm. The length is 32 cm. Find the width.

Solution.   Again we use the formula $P = 2L + 2W$. Here, $P = 96$ and $L = 32$. We solve the resulting equation for W.

$$P = 2L + 2W$$
$$96 = 2 \cdot 32 + 2W$$
$$96 = 64 + 2W$$
$$32 = 2W \qquad \text{Subtract 64 from both sides.}$$
$$16 = W \qquad \text{Divide both sides by 2.}$$

The width of the rectangle is 16 cm.

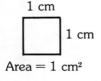

Area = 1 cm²

To define **area,** we need a basic *unit of area.* One that is commonly used is the *square centimeter,* abbreviated cm². One square centimeter, or 1 cm², is the area of a square one centimeter on a side. In place of 1 cm², our basic unit of area could have been 1 in², 1 ft², 1 m², or any appropriate unit.

What is the area of the rectangle in Figure 8.17? We find the area using our basic 1 cm² unit. By inspecting Figure 8.17, we see that 4 squares, each 1 cm on a side, can be laid off vertically, while 6 such squares can be laid

Figure 8.17

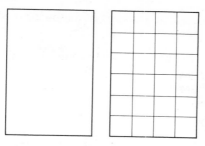

off horizontally. A total of $24 = 4 \cdot 6$ of the small squares are needed to cover the large rectangle. Thus, we say that the area of the large rectangle is 24 cm². In general, the area of a rectangle is given by the formula

$$\text{Area} = \text{length} \cdot \text{width} \qquad \text{(rectangle)}$$

or

$$A = LW. \qquad \text{(rectangle)}$$

Using the formula for the area of a rectangle, we can find formulas for the areas of other figures. For example, a square is a rectangle having all sides equal. The area of a square can be found by the formula for the area of a rectangle, $A = LW$. Let us use the letter s to represent the equal lengths of the sides of the square. Then,

$$A = LW \qquad \text{(rectangle)}$$
$$A = s \cdot s$$
$$A = s^2. \qquad \text{(square)}$$

Thus, $A = s^2$ is the usual formula for the area of a square. For example, a room 4 meters on a side would have a floor area of

$$A = s^2$$
$$A = 4^2 \qquad (\text{Let } s = 4)$$
$$A = 16 \text{ m}^2 \qquad (4^2 = 4 \cdot 4 = 16).$$

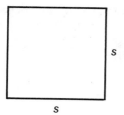

s

Square

The 10 mathematical formulas that changed the face of Earth

Nicaragua issued this series of ten stamps, including the one you see here, with The Law of Pythagoras, $A^2 + B^2 = C^2$.

Compasses dominate this stamp as they dominate geometry. In fact, the ordinary constructions of elementary geometry can be done with compasses alone. This was demonstrated in the late 18th century by Lorenzo Mascheroni.

The application of geometry to architecture is epitomized in the pictures of the Parthenon and its floor plan. The basic idea is the right angle—the angle of pillar to ground, of pillar to roof; the angle where the walls meet; the angle of the triangle where $A^2 + B^2 = C^2$ is the law, binding on all right triangles within the Euclidean system.

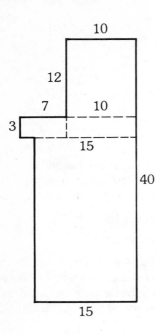

Example 3 Find the area of the building whose floor plan is shown in the figure.

Solution. In the figure, we have used dotted lines to break up the floor area into rectangles and squares. The areas of the various rectangles and squares that result are

$$10 \cdot 12 = 120 \text{ m}^2$$
$$3 \cdot 10 = 30 \text{ m}^2$$
$$3 \cdot 7 = 21 \text{ m}^2$$
$$15 \cdot 25 = 375 \text{ m}^2.$$

The total area is thus $120 + 30 + 21 + 375 = 546$ m².

As we said earlier in this chapter, a *parallelogram* is a four-sided figure having both pairs of opposite sides parallel. A parallelogram need not be a rectangle, so we cannot directly use the formula for the area of a rectangle. However, we can use this formula indirectly, as shown in Figure 8.18. Cut off the triangle in color, and attach it at the right. The resulting figure is a rectangle having the same area as the original parallelogram. The width of the rectangle equals the *height* of the parallelogram, and the length of the rectangle is the base, so that

$$A = \text{length} \cdot \text{width} \quad \text{(rectangle)}$$
$$A = \text{base} \cdot \text{height}$$
$$A = \mathbf{bh}. \quad \text{(parallelogram)}$$

Thus, $A = bh$ is the formula for the area of a parallelogram.

Figure 8.18

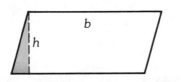
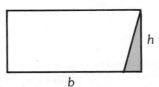

Example 4 Find the area of each of the following parallelograms:

(a)   (b)

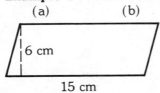
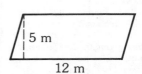

(**a**) Here the base has a length of 15 cm, while the height is 6 cm. Thus, $b = 15$ and $h = 6$:

$$A = bh$$
$$A = 15 \cdot 6$$
$$A = 90.$$

The area is 90 cm².

(**b**) In the second figure, $b = 12$ m and $h = 5$ m, so $A = 12 \cdot 5 = 60$. Thus, the area is 60 m².

A *trapezoid* is a four-sided figure having only one pair of opposite sides parallel. The formula for the area of a trapezoid is a little more complicated than the others of this section. Figure 8.19 shows how to convert a trapezoid into a rectangle by cutting off triangles and moving them around.

Figure 8.19

By studying trapezoids such as the one in Figure 8.19, we get the following formula:

Area = (average of length of parallel bases) · height;

or, by letting B and b represent the lengths of the parallel bases,

$$\mathbf{A} = \tfrac{1}{2}\,(\mathbf{B} + \mathbf{b})\mathbf{h}. \qquad (\text{trapezoid})$$

Example 5 Find the area of each of the following trapezoids:

(a)   **(b)**

(a) Here $h = 6$, $b = 3$, and $B = 9$.
Thus, $A = \tfrac{1}{2}(B + b)\,h$
$A = \tfrac{1}{2}(9 + 3)\,6$
$A = \tfrac{1}{2}(12)\,6$
$A = 36$ cm².

(b) We use the same formula:
$A = \tfrac{1}{2}(17 + 12)\,9$
$A = \tfrac{1}{2}(29)\,9$
$A = 130.5$ ft².

Even the formula for the area of a triangle can be found from the formula for the area of a rectangle. In Figure 8.20, the triangle whose vertices are at A, B, and C has been broken into two parts, one shown in color and one shown in gray. By repeating the part shown in color and the part in gray, we get a rectangle. The area of this rectangle is $A = \text{base} \cdot \text{height}$, or $A = bh$. However, the rectangle has *twice* the area of the triangle; in other words, the area of the triangle is *half* the area of the rectangle. Thus, the formula for the area of a triangle is

$$\mathbf{A} = \tfrac{1}{2}\,\mathbf{bh}. \qquad (\text{triangle})$$

Figure 8.20

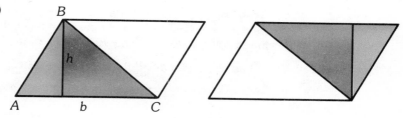

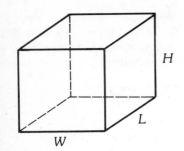

The important measure associated with space figures is their **volume.** Think of the volume of a figure as the amount of air or water that the figure could hold. Volumes are found by using the correct formula: the formula for the volume of a box is

$$V = \text{length} \cdot \text{width} \cdot \text{height}$$
$$\mathbf{V = LWH.} \qquad \text{(box)}$$

Example 6   Find the volume of the box shown in the figure.

Solution.   Use the formula $V = LWH$:

$$V = LWH$$
$$V = 14(5)(7)$$
$$V = 490,$$

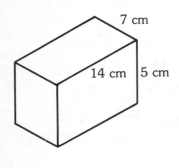

so that the volume is 490 cm³, where "cm³" represents "cubic centimeters."

Not only does a box have volume; it also has a **surface area.** Think of the surface area as analogous to the amount of paint that would be needed to paint the outside of the space figure.

Example 7   Find the surface area of the figure in Example 6.

Solution.   The box has six faces, each a rectangle, and opposite faces are equal.

Size of face	Area of rectangle	Number of faces	Subtotals
7 cm by 14 cm	$7 \cdot 14 = 98$	2	$2 \cdot 98 = 196$
5 cm by 14 cm	$5 \cdot 14 = 70$	2	$2 \cdot 70 = 140$
5 cm by 7 cm	$5 \cdot 7 = 35$	2	$2 \cdot 35 = 70$

Total surface area $= 406$ cm²

There are few space figures made with rectangles as useful as the box. However, there is one useful space figure, the *pyramid,* made from triangles and one rectangle. The formula for the volume of a pyramid is

$$\mathbf{V = \tfrac{1}{3} Bh} \qquad \text{(pyramid)}$$

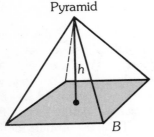

Pyramid

B is the area of the base

where B represents the area of the base, and h represents the height, or the distance from the top of the pyramid (called the *apex*) to the center of the base.

Example 8   The Transamerica Tower in San Francisco is a pyramid with a square base. Each side of the base has a length of 52 m, while the height of the building is 260 m. Find the volume of the building.

Solution.   We first find the area of the base:

$$B = 52 \cdot 52 = 2704.$$

Now use the formula $V = \frac{1}{3}Bh$ to find the volume:

$$V = \frac{1}{3} \cdot 2704 \cdot 260 = 234{,}000. \qquad \text{(rounded off)}$$

The volume of the building is about 234,000 m³.

One common figure, the circle, has a circumference (perimeter) and an area which cannot be found by the methods we used above. In fact, the correct formulas for a circle of radius r are

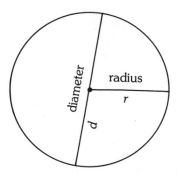

$$C = 2\pi r \qquad \text{(circumference of a circle)}$$
$$A = \pi r^2 \qquad \text{(area of a circle)}$$

where π is the irrational number approximated by 3.14. (We discussed π in more detail in Chapter 5.)

Example 9 A circle has a radius of 5.2 m. Find its circumference and area. Use 3.14 to approximate π.

Solution. Let $r = 5.2$, and substitute this value in the formulas.

Circumference | Area
$C = 2\pi r$ | $A = \pi r^2$
$C = 2\pi(5.2)$ | $A = \pi(5.2)^2$
$\quad = 2(3.14)(5.2)$ | $\quad = 3.14(5.2)(5.2)$
$C = 32.66$ | $A = 84.91 \qquad$ (rounded off)

The circumference is 32.66 m and the area is 84.91 m².

Rectangular solids *make up this vertical concrete building, the Johnson Museum at Cornell University, Ithaca, New York. The architect is I. M. Pei.*

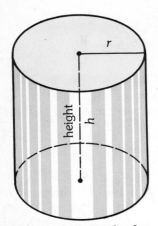

Right circular cylinder
Volume $= \pi r^2 h$
Surface
    area $= 2\pi rh + 2\pi r^2$

Example 10   A can of Pringle's Potato Chips measures 7.6 cm in diameter and 21.9 cm in height. Find the volume and surface area of the can.

Solution.   The diameter is 7.6 cm. However, the formulas call for the radius, and not the diameter. The radius is half the diameter, or

$$r = \frac{\text{diameter}}{2} = \frac{7.6}{2} = 3.8.$$

Now we can find the volume:

$$V = \pi r^2 h$$
$$V = \pi (3.8)^2 (21.9).$$

Using 3.14 as an approximation for π, we have

$$V = 3.14(3.8)(3.8)(21.9)$$
$$V = 992.98. \qquad \text{(rounded)}$$

Thus, the volume of the can is 992.98 cm³. The surface area is

$$S = 2\pi r^2 + 2\pi rh$$
$$S = 2\pi (3.8)^2 + 2\pi (3.8)(21.9)$$
$$= [2(3.14)(3.8)(3.8)] + [2(3.14)(3.8)(21.9)]$$
$$= 90.68 + 522.62$$
$$S = 613.30.$$

The surface area is 613.30 cm².

Example 11   Find the volume of a sphere having a radius of 4.2 m. Find the surface area, too.

Solution.   Use the formula given above.

$$V = \tfrac{4}{3}\pi r^3$$
$$V = \tfrac{4}{3}\pi (4.2)^3 \qquad \text{(Let } r = 4.2)$$
$$= \tfrac{4}{3}(3.14)(4.2)(4.2)(4.2)$$
$$V = 310.18$$

The volume is 310.18 m³. The surface area of this same sphere is

$$S = 4\pi r^2$$
$$S = 4(3.14)(4.2)(4.2)$$
$$S = 221.56.$$

The surface area is 221.56 m².

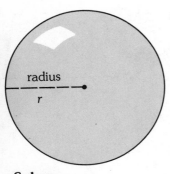

Sphere
Volume $= \tfrac{4}{3}\pi r^3$
Surface area $= 4\pi r^2$

    We can also use the formulas of this section to help solve word problems, as the next example shows.

Example 12   The length of a rectangle is 2 more than the width. The perimeter is 40 mm. Find the width and length.

Solution.   Let x represent the unknown width of the rectangle. Then $x + 2$ represents the length. The formula for the perimeter of a rectangle is $P = 2L + 2W$. Substitute x for W, $x + 2$ for L, and 40 for P:

$$P = 2L + 2W$$
$$40 = 2(x + 2) + 2x$$
$$40 = 2x + 4 + 2x \qquad \text{(distributive property)}$$
$$40 = 4x + 4$$
$$36 = 4x$$
$$9 = x.$$

The width of the rectangle is 9, and the length is $x + 2 = 9 + 2 = 11$.

8.3 EXERCISES

All the formulas of this section are concisely listed on page 333.

Find the area of each figure.

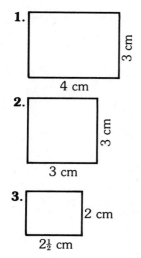

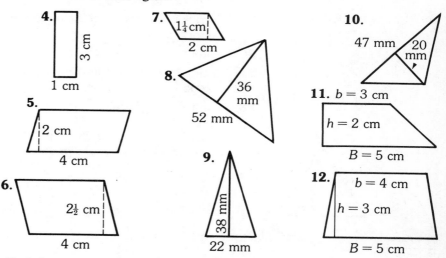

Find the areas of rectangles measuring:

13. 4 cm by 5 cm   **15.** 12 cm by 15 cm

14. 8 cm by 10 cm   **16.** 16 cm by 20 cm

Fill in the blanks in the following problems.

17. The rectangle of Exercise 14 had sides twice as long as the sides of the rectangle of Exercise 13. Divide the larger area by the smaller. By doubling the sides, the area increased by _____ times.

18. To get the rectangle of Exercise 15, we multiplied each side of the rectangle of Exercise 13 by _____. This made the larger area _____ times the size of the smaller area.

19. To get the rectangle of Exercise 16, we multiplied each side of the rectangle of Exercise 13 by _____. This made the area increase by _____ times what it was originally.

*20. In general, if the length of each side of a rectangle is multiplied by n, the area is multiplied by _____.

Use the results of Exercise 20 to answer Exercises 21–24.

21. A ceiling 9 feet by 15 feet can be painted for $80. How much would it cost to paint a ceiling 18 feet by 30 feet?

22. How much to paint a ceiling 27 feet by 45 feet?

23. A cake 41 cm by 58 cm costs $39. What should be the cost of a cake 82 cm by 116 cm?

24. A carpet cleaner charges $150 to do an area 31 feet by 31 feet. What would be the charge for an area 93 feet by 93 feet?

Circles follow the same type of results as rectangles: if the radius of a circle doubles, the area is multiplied by four. In general, if we multiply the radius of a circle by n, the area must be multiplied by n^2. Use these ideas to answer Exercises 25–28.

25. A pizza with a radius of 9 cm sells for $3. What should the charge be for a pizza of radius 18 cm?

26. For a pizza of radius 27 cm?

27. A circular area 49 m in diameter can be painted for $280. Find the charge for an area 98 m in diameter.

28. For an area 147 m in diameter.

In Exercises 29–34, break each figure into polygons and half-circles as needed. Then find the total area of each figure.

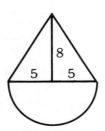

Example: The figure shows a half-circle and a triangle. The half-circle has an area half that of a circle with the same radius, that is,

$$A = \tfrac{1}{2}\pi(5)^2 = \tfrac{1}{2}(3.14)(5)(5) = \tfrac{1}{2}(78.5) = 39.25.$$

The triangle has an area $\tfrac{1}{2}(10)(8) = 40$. The total area of the original figure is thus $39.25 + 40 = 79.25$ cm².

29.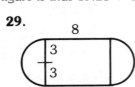

31.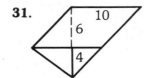

33.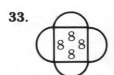

30.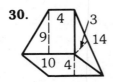

32.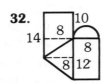

34.

Find the volume of each figure in Exercises 35–42.

35.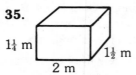

36.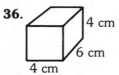

37.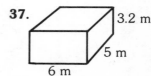

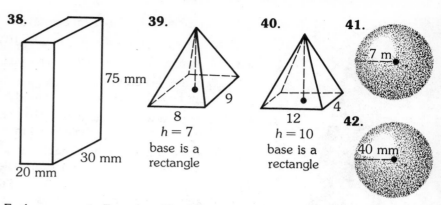

38. 75 mm, 30 mm, 20 mm

39. 8, 9, $h = 7$, base is a rectangle

40. 12, 4, $h = 10$, base is a rectangle

41. 7 m

42. 40 mm

Sphere of influence *Questions about "liveliness" of baseballs received a scientific answer in the 1977 season. That year saw a 47 percent rise in home runs in the majors, and new baseballs in use were suspected. Baseballs manufactured by Rawlings Sporting Goods Company were said to have more spring than the older balls manufactured by Spaulding.*

Various theories were given: the yarn layer was wound tighter; the inner cork was bouncier; or moisture and heat affected the older baseballs en route to Haiti, where their covers were sewn on. (The newer balls were made from scratch in Haiti.)

To settle the issue, tests on baseball resiliency were made by mathematicians at the University of Missouri at Rolla. They shot baseballs from air cannons against wooden backstops, and used radar to compute rebound speeds. Results showed that the new baseballs did have more liveliness than the older ones—about 5 percent more.

Each measure in Exercises 43–46 is the radius of a sphere. Find the volume of each sphere.

43. 1 m   **44.** 2 m   **45.** 3 m   **46.** 4 m

47. When a sphere of radius 1 m grows to become a sphere of radius 2 m, the volume goes from 4.19 m³ to 33.49 m³. The radius has doubled, while the volume has become how many times the original volume? (Divide 33.49 by 4.19 and round to the nearest whole number.)

48. If the radius triples, from 1 m to 3 m, the volume goes from 4.19 m³ to 113.04 m³. The radius tripled; the volume has become how many times the original volume?

49. How many times greater than the original volume does the volume become when the radius goes from 1 m ($V = 4.19$ m³) to 4 m ($V = 267.95$ m³)?

50. When the radius goes from 1 m to 5 m ($V = 523.33$ m³)?

Use the results of Exercises 47–50 to answer Exercises 51–54.

51. A spherical storage tank of radius 2 m can be filled with a liquid for $90. How much would it cost to fill a spherical tank of radius 4 m with the same liquid?

52. How much for a spherical tank of radius 6 m?

*53. A perfectly elastic balloon has a radius of 1.5 m when a certain volume of gas is pumped in. Find the radius of the balloon if a total of eight times as much gas is pumped in.

*54. Find the radius if 125 times as much gas is pumped in.

Find the volume and surface area in Exercises 55–66. If we do not give the measurements of an object, measure the necessary dimensions yourself to the nearest millimeter.

55. Nerf ball, radius 58 mm

56. beachball, 30 cm in diameter

57. tennis ball

58. 12-ounce soda pop can

59. 46-ounce fruit juice can

60. one-pound box of saltines

Atlanta's Peachtree Plaza Hotel

61. An ice-cream cone, 8 cm across, and 15 cm high.

62. A conical filter for a coffee maker, 10 cm across and 13 cm high.

*63. A pyramid 7 m high with a rectangular base, 8 m by 4 m. (The two triangles on the 8 m sides have a height of 7.3 m. The other two triangles have a height of 8.1 m.)

*64. A pyramid with a square base, 10 m on a side, and a height of 4 m. The height of each triangle is 6.4 m.

*65. Find the volume of the Great Pyramid of Cheops, near Cairo. The base is square, 230 m on a side, while the height is 137 m.

*66. Find the volume and surface area of the Peachtree Plaza Hotel in Atlanta, a cylinder with a radius of 46 m, and a height of 220 m.

Use the formulas of this section to work out Exercises 67–70.

67. The shorter base of a trapezoid is 16 cm and the longer base is 20 cm. The height is 6 cm. Find the area.

68. The circumference of a circle is five times the radius, increased by 2.56 m. Find the radius.

69. The perimeter of a rectangle is 16 times the width. The length is 12 cm more than the width. Find the length and width of the rectangle.

70. The width of a rectangle is one less than the length. The perimeter is five times the length, decreased by 5. Find the length.

8.4 SIMILAR TRIANGLES

If two triangles have exactly the same shape, they are called **similar** *triangles.* Figure 8.21 shows three pairs of similar triangles.

The two triangles on the very bottom have exactly the same shape and thus are similar. Also, these two triangles are exactly the same *size.* Triangles which are both the same size and the same shape are called **congruent** *triangles.* If two triangles are congruent, then it will be possible to pick one of them up and place it exactly upon the other so that they coincide. If two triangles are congruent, then they must be similar. However, two similar triangles need not be congruent.

An example of congruent triangles would be the triangular supports for a child's swing, machine produced with exactly the same dimensions each time. An example of similar triangles would be the supports of a long bridge, all the same shape, but decreasing in size toward the center of the bridge.

There are two basic properties of similar triangles that we shall use in this section:

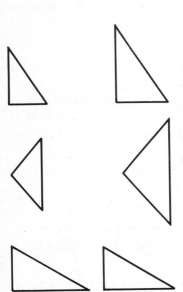

Figure 8.21 Three pairs of similar triangles

Corresponding angles of similar triangles are equal.

Corresponding sides of similar triangles are in proportion.

Triangular arrangement *of struts makes a strong support for the glass "eye" of the student center at the University of Rochester in New York. The architect is I. M. Pei.*

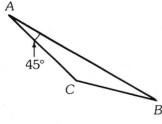

Example 1   Find the missing angles in the larger triangle. Assume that the two triangles are similar.

Solution.   If the triangles are similar, corresponding angles are equal. By looking at the figure, we see that angle C corresponds to angle P, so that angle C is 104°. Also, angle B corresponds to angle M. Thus, angle B is 31°.

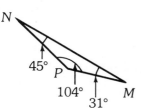

Example 2   Find the lengths of the missing sides of the smaller triangle in the figure. Assume the two triangles are similar.

Solution.   As we said above, similar triangles have corresponding sides in proportion. We use this fact to find the missing sides in the triangle on the right. Side DF of the small triangle corresponds to side AB of the larger one, and sides DE and AC correspond. This leads to the proportion

$$\frac{8}{16} = \frac{DF}{24}.$$

If we solve this equation for DF, we can find the length of the missing side. To solve the equation, multiply both sides by the least common multiple of 16 and 24, namely, 48. (We discussed least common multiple in Chapter 4.)

$$48 \cdot \frac{8}{16} = \frac{DF}{24} \cdot 48$$
$$24 = 2(DF)$$
$$12 = DF. \qquad \text{Divide both sides of the equation by 2.}$$

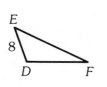

Side DF has a length of 12.
     Side EF corresponds to side BC.
Thus, we have another proportion:

$$\frac{8}{16} = \frac{EF}{32}.$$

Side EF has a length of 16.

Multiply both sides by 32.

$$32 \cdot \frac{8}{16} = \frac{EF}{32} \cdot 32$$
$$16 = EF.$$

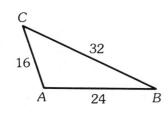

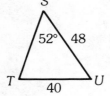

Example 3 Find the missing parts of the similar triangles in the figure.

Solution. Here angles X and T correspond, as do angles Y and U, and angles Z and S. Since angles Z and S correspond, and angle S is 52°, angle Z must also be 52°. The sum of the angles of any triangle is 180°. In the triangle on the left $X = 71°$ and $Z = 52°$. To find Y, set up an equation and solve for Y.

$$X + Y + Z = 180$$
$$71 + Y + 52 = 180$$
$$123 + Y = 180$$
$$Y = 57.$$

Angle Y is 57°. Since angles Y and U correspond, then $U = 57°$ also.

Now we can find the missing sides. Since SU and ZY correspond, as do XZ and TS, we have the proportion

$$\frac{48}{144} = \frac{ST}{126}.$$

Multiply both sides of this equation by 1008. (Why 1008?)

$$1008 \cdot \frac{48}{144} = \frac{ST}{126} \cdot 1008$$
$$336 = 8(ST)$$
$$42 = ST.$$

Also,
$$\frac{40}{XY} = \frac{48}{144}.$$

Multiply both sides by $144(XY)$:

$$144(XY) \cdot \frac{40}{XY} = \frac{48}{144} \cdot 144(XY)$$
$$5760 = 48(XY)$$
$$120 = XY.$$

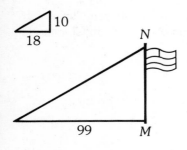

Example 4 The fire people at the Arcade Fire Station need to measure the height of the station flagpole. They notice that at the instant when the shadow of the station is 18 meters long, the shadow of the flagpole is 99 meters long. The station is 10 meters high. Find the height of the flagpole.

Solution. The figure represents the information of the problem. The two triangles shown there are similar (why?), so that corresponding sides are in proportion. Thus,   Multiply both sides by 90:

$$\frac{MN}{10} = \frac{99}{18}.$$ $$90 \cdot \frac{MN}{10} = \frac{99}{18} \cdot 90$$
$$9(MN) = 495$$
$$MN = 55.$$

The flagpole is 55 meters high.

8.4 EXERCISES

Write *similar* or *not similar* for each pair of triangles.

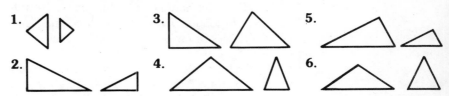

1.   **3.**   **5.**

2.   **4.**   **6.**

Find all missing angles in each pair of similar triangles.

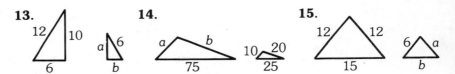

7. C ←78° A, B 46°   P M N Q   **9.** 74°→ X, Y Z B 30°   28° T V U, N 30° M K 106°   **11.** P ←20° 64° Q R, V T U   T M 90° N 38° P

8. A ←42° 90° C B   Q R P   **10.** A C   **12.** 90° X Y 38° Z, M 90° N 38° P

Find the lengths of the missing sides in each pair of similar triangles.

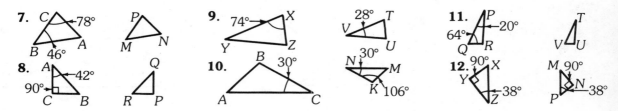

13. 12 10 6   a 6 b

14. a b 75   10 20 25

15. 12 12 15   6 a b

Find the measure of each side labeled with a letter in Exercises 16–18.

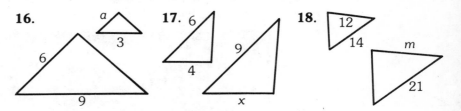

16. a 3 6 9   **17.** 6 4 9 x   **18.** 12 14 m 21

Solve the problems posed in Exercises 19–22.

19. On a photograph of a triangular piece of land, the lengths of the three sides measure 4 cm, 5 cm, and 7 cm, respectively. The shortest side of the actual piece of land is 400 m in length. Find the lengths of the other two sides.

20. By drawing lines on a map, a triangle can be formed by the cities of Phoenix, Tucson, and Yuma. On the map, the distance between Phoenix and Tucson is 8 cm, the distance between Phoenix and Yuma is 12 cm, and the distance between Tucson and Yuma is 17 cm. The actual straight-line distance from Phoenix to Yuma is 230 km. Find the distances between the other pairs of cities.

21. A tree casts a shadow 45 m long. At the same time, the shadow cast by a vertical 2-meter stick is 3 m long. Find the height of the tree.

22. The Santa Cruz lighthouse is 14 m tall and casts a shadow 28 m long at 7 PM. At the same time, the shadow of the lighthouse keeper is 3.5 m long. How tall is she?

Find the missing measurement in each of the following. (*Hint:* In the sketch for Exercise 23, the side 100 m long in the small triangle corresponds to a side of length $100 + 120 = 220$ m in the larger triangle.)

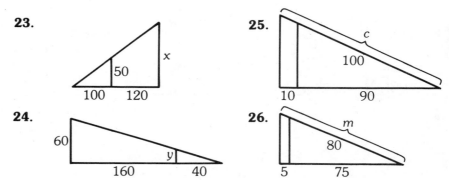

23.

24.

25.

26.

27. The stamp shows Stone Mountain, a memorial to Civil War leaders located near Atlanta. General Lee's head is 40 feet tall on the monument. A man 6 feet tall has a head about 3/4 feet high. If General Lee had been carved standing, to the same scale, how tall would he have been?

*8.5 NON-EUCLIDEAN GEOMETRY

At the beginning of this chapter we listed the axioms and postulates from Euclid's famous *Elements*. Actually, we substituted Playfair's Axiom on parallel lines for Euclid's own fifth postulate. However, to understand why the fifth bedeviled so many mathematicians so long, we must examine the original formulation.

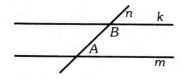

Figure 8.22

Euclid's Fifth Postulate ("Parallel Postulate")

If two lines (k and m in Figure 8.22) are such that a third line, n, cuts them so that the sum of the two interior angles (A and B) on one side of line n is less than two right angles, then the two sides, if produced far enough, will meet on the same side of n that has the sum of the interior angles less than two right angles.

John Playfair *(1748–1819) wrote his* Elements of Geometry *in 1795. Playfair's Axiom is: Given a line k and a point P not on the line, there exists one and only one line m through P that is parallel to k. Playfair was a geologist, who fostered "uniformitarianism," the doctrine that geological processes long ago gave Earth its features, and processes today are the same kind as those in the past. He was the first to state that a river cuts its own valley.*

Euclid's parallel postulate is quite different from the other nine axioms we listed. The others are simple statements that seem in complete agreement with our experience of the world around us. But the parallel postulate is long and wordy, and hard to understand without a sketch.

The difference between the parallel postulate and the other axioms was noted by the Greeks, as well as later mathematicians. It was commonly believed that this postulate was not an axiom at all, but a theorem to be proved. For over 2000 years mathematicians tried repeatedly to prove it.

The most dedicated attempt came from an Italian Jesuit, Girolamo Saccheri (1667–1733). He attempted to prove the parallel postulate in an indirect way, so-called "reduction to absurdity." He would assume the postulate to be false and then show that the assumption leads to a contradiction of something true (an absurdity). Such a contradiction would thus prove the postulate true.

Saccheri began with a quadrilateral, as in Figure 8.23. He assumed angles A and B to be right angles, and sides AC and BD to be equal. His plan is as follows:

1) To assume that angles C and D are obtuse angles, and to show that this leads to a contradiction.

2) To assume that angles C and D are acute angles, and to show that this also leads to a contradiction.

3) Then if C and D can be neither acute nor obtuse angles, they must be right angles.

4) If C and D are both right angles, then it can be proved that the fifth postulate is true.

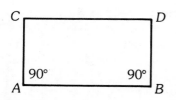

Figure 8.23 $AC = BD$

Saccheri had no trouble with part 1. However, he did not actually reach a contradiction in the second part 2, but produced some theorems so "repugnant" that he convinced himself he had vindicated Euclid. In fact, he published a book called in English "Euclid Freed of Every Flaw."

Farkas Bolyai *(note that on the stamp the family name is first— the Hungarian custom) was a mathematician who spent years trying to prove Euclid's fifth postulate, without success. His son Janos also became interested in the problem, but Janos began to see that the fifth postulate could not be proven. He then developed a form of non-Euclidean geometry based on a postulate different from Euclid's. His father was shocked, and wrote to the leading mathematician of the day, C. F. Gauss, for advice. Gauss told the father that his son's ideas were sound—in fact, Gauss himself had thought of them years earlier.* **Lobachevski,** *son of a Russian peasant, showed his mathematical genius early. He became a professor of mathematics at the University of Kazan at age twenty-one. He made known his non-Euclidean ideas of geometry in lectures there as early as 1826, but did not publish his ideas until 1830.*

Today we know that Saccheri was wrong—there is no contradiction to be found by assuming angles C and D acute angles. Thus, the fifth postulate is indeed an axiom, and not a theorem. Mathematicians had tried for 2000 years to do the impossible.

The ten axioms of Euclid describe the world around us with remarkable accuracy. We now realize that the fifth postulate is necessary in Euclidean geometry to establish *flatness*. That is, the axioms of Euclid describe the geometry of *plane surfaces*. By changing the fifth postulate, we can describe the geometry of other surfaces. So, other geometric systems exist as much as Euclidean geometry exists, and they even can be demonstrated in our world. They are just not as familiar. A system of geometry in which the fifth postulate is changed is called a **non-Euclidean geometry.**

One non-Euclidean system was developed amazingly by three separate thinkers at about the same time. Early in the nineteenth century Karl Friedrich Gauss, one of the great mathematicians, worked out a consistent geometry replacing Euclid's fifth postulate. He never published his work, however, because he feared the ridicule of people who could not free themselves from habitual ways of thinking. Gauss first used the term "non-Euclidean." Nikolai Ivanovich Lobachevski (1793–1856) published a similar system in 1830 in the Russian language. At the same time, Janos Bolyai (1802–1860), a Hungarian army officer, worked out a similar system, which he published in 1832, not knowing about Lobachevski's work. Bolyai never recovered from the disappointment of not being the first, and did no further work in mathematics.

Lobachevski replaced Euclid's fifth postulate with:

Angles C and D in the quadrilateral of Saccheri are acute angles.

This postulate of Lobachevski can be rephrased as follows:

Through a point P off a line k, at least two different lines can be drawn parallel to k.

Figure 8.24

Compare this form of Lobachevski's postulate to the geometry of Euclid. How many lines can we draw through P and parallel to k in Euclidean geometry? At first glance, the postulate of Lobachevski does not agree with what we know about the world around us. But this is only because we think of our immediate surroundings as being flat.*

Many of the theorems of Euclidean geometry are valid for the geometry of Lobachevski, but many are not. For example, in Euclidean geometry, the sum of the angles in any triangle is 180°. In Lobachevskian geometry, the sum of the angles in any triangle is *less* than 180°. Also, triangles of different sizes can never have equal angles.

---

*The International Flat Earth Research Society, with headquarters in Lancaster, California, publishes *Flat Earth News, Last Iconoclast,* and *Plane Truth.*

Georg Friedrich Bernhard Riemann *(1826–1866) was a German mathematician. Though he lived a short time and published few papers, his work forms a basis for much modern mathematics. He made significant contributions to the theory of functions (see Chapter 7) and the study of complex numbers (see Chapter 5) as well as to geometry. Most calculus books today use the idea of a "Riemann sum" in defining the integral.*

A second non-Euclidean system was developed by Georg Riemann (1826–1866). He pointed out the difference between a line that continues indefinitely and a line having infinite length. For example, a circle on the surface of a sphere continues indefinitely, but does not have infinite length. Riemann developed the idea of geometry on a sphere, and replaced Euclid's fifth postulate with:

Angles *C* and *D* of the Saccheri quadrilateral are obtuse angles.

In terms of parallel lines, Riemann's postulate becomes

Through a point *P* off a line *k*, no line can be drawn which is parallel to *k*.

Riemannian geometry is important in navigation. "Lines" in this geometry are really *great circles,* or circles whose centers are at the center of the sphere. The shortest distance between two points on a sphere lies along an arc of a great circle. Great circle routes on a globe don't look at all like the shortest distance when the globe is flattened out to form a map, but this is part of the distortion that occurs when this is done.

 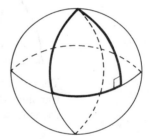

Figure 8.25

We can now summarize what we know about Euclidean and non-Euclidean geometries.

	non-Euclidean	
Euclidean (about 300 BC)	Lobachevskian (about 1830)	Riemannian (about 1850)
Angles *C* and *D* of a Saccheri quadrilateral are *right* angles.	Angles *C* and *D* are *acute* angles.	Angles *C* and *D* are *obtuse* angles.
Given point *P* off line *k*, exactly *one* line can be drawn through *P* and parallel to *k*.	*More than one* line can be drawn through *P* and parallel to *k*.	*No* line can be drawn through *P* and parallel to *k*.
The sum of angles of a triangle is *180°*.	*less than 180°*	*more than 180°*
Two triangles can have same size angles but different size sides. (similarity as well as congruency)	Two triangles with the same size angles must have same size sides. (congruency only)	
Lines have *infinite* length.		Lines have *finite* length.
Geometry on a plane	Geometry on a surface like a pseudosphere	Geometry on a sphere

8.5 EXERCISES

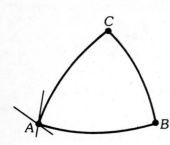

The figure at the side shows how to measure angles of a spherical triangle. The triangle is distorted—it is drawn the way it looks to us *away* from the surface of the sphere. The sides would seem straight to a person living *on* the sphere (where great circles are straight lines). To measure angle *A* in the figure, draw two tangent lines to the sides of angle *A* so that they cross. Then the angle between the tangents is defined as the measure of angle *A*.

Use a protractor to answer the following questions.

1. Find the measure of angle *A*. **2.** Of angle *B*. **3.** Of angle *C*.

4. Find the sum of angles *A*, *B*, and *C*.

Find the sum of the angle measures in each of the following triangles from the surface of a sphere.

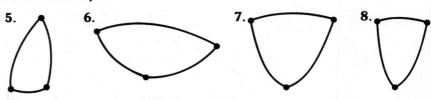

Triangles from Lobachevskian geometry look as shown below when transferred to a flat piece of paper. The angles of this triangle are measured with tangents, just as above. Find the measure of each angle.

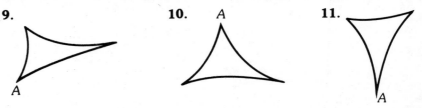

12. What is the sum of angles *A*, *B*, and *C*?

Find the sum of the angles in each of the following triangles from Lobachevskian geometry.

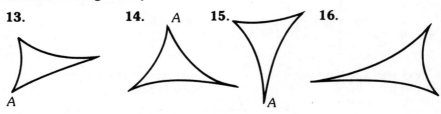

On a globe, let A represent New York City, B represent Cairo, and C the North Pole. Imagine "lines" drawn on the globe from A to B, from B to C, and from C to A, forming a triangle. Estimate each angle.

17. A   **18.** B   **19.** C   **20.** Find the sum of A, B, and C.

Now let M represent Nairobi, N represent Quito, and Q represent Philadelphia. Draw lines as above. Find the size of each angle.

21. M   **22.** N   **23.** Q

24. What is the sum of M, N, and Q?

In the studio *Albrecht Durer's woodcut (about 1527) pictures an artist using various devices in making a perspective drawing.*

Projective geometry   Artists must have a sound knowledge of geometry in order to paint realistically a world of solids in space. Our three-dimensional world must be represented in a convincing way on canvas or some other plane surface. For example, how should an artist paint a realistic view of railroad tracks going off to the horizon? In reality, the tracks are always at a constant distance apart, but they cannot be drawn that way except from overhead. The artist must make the tracks converge at a point. Only in this way will the scene look "real."

Beginning in the fifteen century, artists led by Leone Battista Alberti, Leonardo da Vinci, and Albrecht Dürer began to study the problems of representing three dimensions in two. They found geometric methods of doing this. What artists initiated, mathematicians developed into a geometry different from that of Euclid—*projective geometry*. Some of its key theorems are worked out in the remaining exercises in this set.

East vs West   *These side-by-side interiors are similar, yet strangely different. Which one seems more "natural"?*

Well, the drawing at left uses parallel perspective as in the classic art of China. The drawing at right uses the convergent perspective ("disappearing railroad tracks") that dominated art in the West from the Renaissance to the twentieth century.

Gerard Desargues *(1591–1661), architect and engineer, was a technical advisor to the French government. He met Descartes at the siege of La Rochelle in 1628, and in the 1630s was a member of the Parisian group that included Descartes, Pascal, Fermat, and Mersenne. Desargues was a music lover and even wrote about musical composition. In 1636 and 1639 he published a treatise and proposals about the perspective section—Desargues had invented projective geometry. His works unfortunately were too difficult for the times; his terms were from botany and he did not use Cartesian symbolism. Desargues' geometric innovations were hidden for nearly two hundred years. It was after Jean-Victor Poncelet rediscovered projective geometry in the 1810s that a handwritten manuscript of Desargues was found—that was in 1845.*

Desargues' theorem states that in a plane, if two triangles are placed so that lines joining corresponding vertices meet in a point, then corresponding sides, when extended, will meet in three collinear points. (Collinear points are defined as lying on the same line.)

Theorem of Desargues in a Plane   The figure shows triangles ABC and $A'B'C'$. Lines joining corresponding vertices have been drawn, the lines AA', BB', CC', which meet in point O.

25.   Extend sides AC and $A'C'$ of the two triangles to meet in a point M.

26.   Extend sides AB and $A'B'$ to meet in a point N.

27.   Extend BC and $B'C'$ to meet in a point P.

28.   Draw a straight line through M, N, and P. (If your drawing is correct, you should be able to do so.)

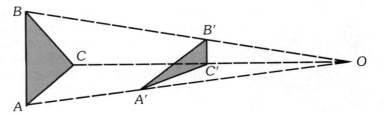

Theorem of Desargues in Space   The figure shows triangles ABC and $A'B'C'$ on separate planes K and R, respectively. Lines joining corresponding vertices have been drawn, lines AA', BB', CC', which meet in a point O.

29.   Extend AB and $A'B'$ to meet in a point P.

30.   Let AC and $A'C'$ meet in M.

31.   Let BC and $B'C'$ meet in N.

32.   You should be able to draw a straight line through points M, N, and P. Try it.

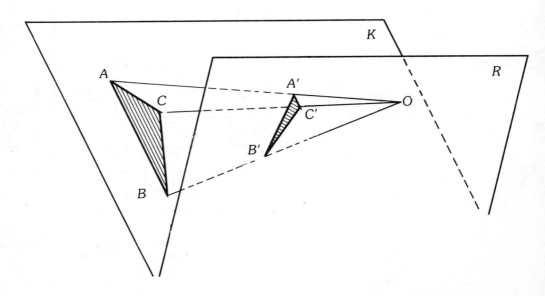

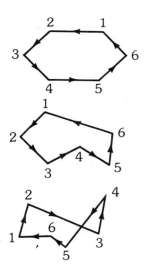

Pappus, a Greek mathematician in Alexandria about 320 AD, wrote a commentary on the geometry of the times. We will work out a theorem of his about a hexagon inscribed in two intersecting lines.

First we need to define an old word in a new way: a *hexagon* is any six lines in a plane, no three of which meet in the same point. As the figure in the margin shows, the vertices of several hexagons are labeled with numbers. Thus $1-2$ represents the line segment joining vertices 1 and 2. Segments $1-2$ and $4-5$ are *opposite sides* of a hexagon, as are $2-3$ and $5-6$, and $3-4$ and $1-6$.

33. Draw an angle less than 180°.

34. Choose three points on one side of the angle. Label them 1, 5, 3 in that order, beginning with the point nearest the vertex.

35. Choose three points on the other side of the angle. Label them 6, 2, 4 in that order, beginning with the point nearest the vertex.

36. Draw line segments $1-6$ and $3-4$. Draw lines through the segments so they extend to meet in a point, call it N.

37. Let lines through $1-2$ and $4-5$ meet in point M.

38. Let lines through $2-3$ and $5-6$ meet in P.

39. Draw a straight line through points M, N, and P.

*40. Formulate in your own words a theorem generalizing your result in Exercises $33-39$.

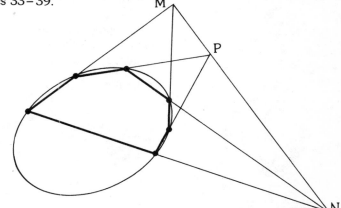

*8.6 TOPOLOGY AND NETWORKS

We began this chapter by suggesting that Euclidean geometry might seem rigid from a twentieth-century point of view. The plane and space figures we studied in the Euclidean system are carefully distinguished by differences in size, shape, angularity, and so on. For a given figure such properties are permanent, and thus we can ask sensible questions about congruence and similarity. Suppose we studied "figures" made of rubber bands, as it were, "figures" that could be stretched, bent, or otherwise distorted without tearing or scattering?

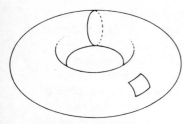

Torus *One of the most useful figures in topology is the torus, a doughnut-like or tire-shaped surface. Its properties are different from those of a sphere, for example. Imagine a sphere covered with hair. You cannot comb the hairs in a completely smooth way—one fixed point remains, as you can find on your own head. In the same way, on the surface of Earth the winds are not a smooth system—there is a calm point somewhere. However, a torus covered with hair can be combed smooth. The winds would be blowing smoothly on a planet shaped like a torus.*

Topology, an important twentieth-century geometry, does just that. Topological questions concern the basic structure of objects rather than size or arrangement. For example, a typical topological question has to do with the number of holes in an object, a basic structural property that does not change during deformation. You cannot deform a rubber ball to get a rubber band without tearing it—making a hole in it. Thus the two objects are not topologically equivalent. On the other hand, a rubber sheet *is* topologically equivalent to a rubber ball. Have you changed the basic structural property of a rubber ball by stretching it into a sheet?

Example 1 Decide whether or not the figures in each pair are topologically equivalent.

(a) A football and an oatmeal box

Solution. If we assume that a football is made of a perfectly elastic substance like rubber or dough, it could be twisted or kneaded into the same shape as an oatmeal box. Thus, the two figures are topologically equivalent.

(b) A doughnut and an unbuttoned blouse

Solution. A doughnut has one hole, while a blouse has two (the sleeve openings). Thus, a doughnut could not be stretched and twisted into the shape of a blouse without tearing another hole in it. Because of this, a doughnut and a blouse are not topologically equivalent.

In topology, figures are classified according to their **genus,** that is, the number of cuts that can be made without cutting the figure into two pieces. The genus of an object is the number of holes in it. See Figure 8.26.

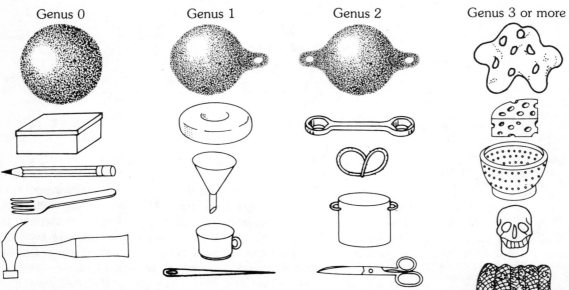

Figure 8.26 Genus of some common objects

Euler in Petrograd *It was only fitting that the USSR linked the 250th anniversary of Euler's birth with the Russian Academy. The Academy was founded by Peter the Great in 1725, and Euler went there two years later. It was there he heard about the bridges in Koenigsburg. By this time Catherine the Great was Empress. Euler died in Petrograd in 1783. He talked a little mathematics, had dinner, a cup of tea, a smoke—and suddenly the great man was no more. (Petrograd was renamed Leningrad after the revolution.)*

Some of the most important applications of topology come from the study of networks, for example, systems of transportation and communication. The interrelationships of people can also be described as networks, from families and clans to corporations and larger social organization.

A **network** is a diagram showing the various paths (or *arcs*) between points (called *vertices,* or *nodes*). A network can be thought of as a set of arcs and vertices. The study of networks began formally with the so-called Koenigsburg Bridge Problem as solved by Leonhard Euler (1707–1783). (Euler's formula for polyhedrons was discussed in Section 8.3.) In the town of Koenigsburg, Germany (now Kaliningrad, USSR), the River Pregel flows through the middle of town. There are two islands in the river. During Euler's lifetime, there were seven bridges connecting the islands and the two banks of the river. See Figure 8.27. (An eighth bridge was added since Euler's time.)

The people of the town loved Sunday strolls among the various bridges. Gradually, a competition developed to see if anyone could find a route that crossed each of the seven bridges exactly once. The problem concerns what topologists today call the *traversibility* of a network. No one could find a solution. The problem became so famous by 1735 it reached Euler, who was then at the court of the Russian empress Catherine the Great. In trying to solve the problem, Euler began by drawing a network representing the system of bridges, as in Figure 8.28.

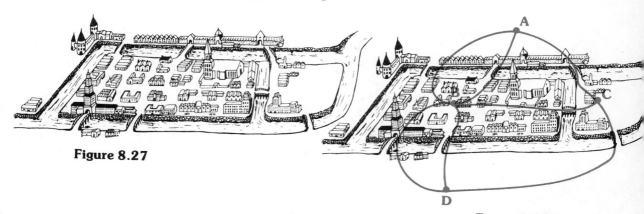

Figure 8.27

Figure 8.28

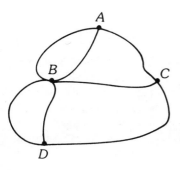

Euler first noticed that three routes meet at vertex A. Since 3 is an odd number, he called A an *odd vertex.* Since three routes meet at A, it must be a starting or ending point for any traverse of the network. This must be true; otherwise when you get to A on your second trip there would be no way to get out. Three paths also meet at C and D, with five paths meeting at B. Thus, B, C, and D are also odd vertices. An odd vertex must be a starting or ending point of a traverse. Thus, all four vertices A, B, C, and D must be starting or ending points. Since a network can have only two starting or ending points (one of each), this network cannot be traversed. The residents of Koenigsburg were trying to do the impossible.

Euler's results can be summarized as follows:

(1) The number of odd vertices of any network is *even*. (That is, a network must have 2*n* odd vertices, where *n* = 0, 1, 2, 3, ...)

(2) A network with exactly two odd vertices can be traversed. Start at one odd vertex and end at the other.

(3) A network with more than two odd vertices cannot be traversed.

Example 2   Are the following networks traversible?

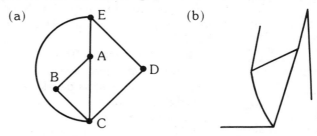

(a) One way to traverse the network of the figure is to start at *A*, go through *B* to *C*, then back to *A*. (It is acceptable to go through a vertex as many times as needed.) Then go to *E*, to *D*, to *C*, and finally back to *E*.

(b) Because of all the "dead ends" here, the network of figure (b) cannot be traversed.

A typical example of an application of topology comes from the research of the psychologists Lewin and Bavelas. They wanted to test the effects of various types of communication patterns on the behavior of small groups of people. To do this, they seated five people at a circular table. They then assigned various rules of communication among the people. For example, with the rules of communication diagrammed in Figure 8.29, person 1 can talk only with person 2, while person 2 can talk either with person 1 or with person 3, and so on. Other representations of communications rules are shown in Figure 8.30.

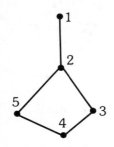

Figure 8.29

Figure 8.30   Some communication channels

After setting up the rules for communication, the researchers gave each member of the group one card from a set of six cards, numbered from 1 to 6. Each person was then asked to use only the established rules to decide which card was left. The performance of each group at this task was related to the communication rules in use. For example, any person seated in the center of the group tended to be a leader. A cyclic arrangement, such as in the left graph of Figure 8.30, was less likely to produce a leader. Satisfaction declined as distance from the center increased.

A.

B.

C.

D.

E.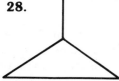

8.6 EXERCISES

In Exercises 1–10, each figure may be topologically equivalent to none or some of the objects grouped at the left, labeled A–E. List all topological equivalences (by letter) for each figure.

1.

2.

3.

4.

5.

6.

7.

8.

9.

10.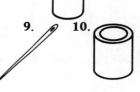

11. Group the capital letters of the alphabet into separate sets so that all the letters in each set are topologically equivalent.

12. Do the same thing with the set of digits 0, 1, 2, 3, 4, 5, 6, 7, 8, 9.

13. Do the same with the roman numerals I, II, III, IV, V, VI, VII, VIII, IX, X.

Give the the genus of each object in Exercises 14–29.

14. pencil	**19.** wedding band	**24.** the number 0
15. straw	**20.** the letter A	**25.** the number 3
16. apple	**21.** the letter B	**26.** the number 6
17. pretzel	**22.** the letter C	**27.** the number 8
18. phonograph record	**23.** the letter P	

For each of the following networks, decide if each vertex is even or odd. Count the number of odd vertices.

28.　　**29.**　　**30.**　　**31.**　　**32.**

Decide whether or not each network in Exercises 33–38 is traversible. If a network is traversible, show how it can be traversed.

33.　　**34.**　　**35.**　　**36.** 　**37.**　　**38.**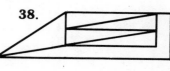

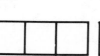
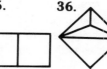

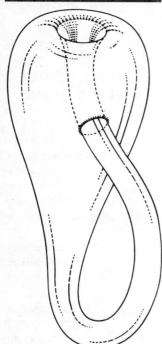

The **Moebius strip** is a figure with several interesting properties. To make one, cut a rectangle of paper, perhaps 3 cm by 25 cm. Paste together the two 3-cm ends, after giving the paper a half-twist.

39. This Moebius strip has only one surface. To see this, mark an x on the strip, and then mark another x on what appears to be the other "side." Begin at one of the x's you have drawn, and trace a path along the strip. You will eventually come to the other x, without crossing the edge of the strip.

40. Cut the strip lengthwise and see what happens.

41. Cut the strip from Exercise 40 lengthwise again.

42. Make a new strip, and start cutting about 1/3 of the way from one edge.

Moebius strip *decorates a stamp honoring the sixth mathematicial colloquium held in Brazil. This topological entity was named after August Ferdinand Möbius (1790–1868), a pupil of Gauss. Möbius anticipated the theory of vectors with his barycentric calculus of 1827, by which he could "add" forces and points.*

Klein bottle *A famous limerick sheds some light on it:*

> *A mathematician named Klein*
> *Thought the Moebius band was divine.*
> *Said he, If you glue*
> *The edges of two,*
> *You'll get a weird bottle like mine.*

The Klein bottle, like the Moebius strip, has one side only: its inside and outside are one and the same. Felix Klein (1849–1925) is famous for his classification of the fields of mathematics using the concept of group (see Chapter 6). He outlined the idea in his inaugural address on becoming professor at Erlangen University in 1872—the so-called "Erlangen Program." Each geometry, he said, studies its elements under certain limitations to change. To put it positively, each geometry has certain invariant aspects. In Euclidean geometry, lengths of line segments and degrees of angles, for example, are preserved in figures. Topology shrinks and distorts, but it stops short of tearing, or scattering points.

Appendix: The Four Color Theorem

Although computers appear often in our daily lives, they have made little impact on pure mathematics. Recently, however, computers were used to solve the four color problem, previously considered one of the classic unsolved problems of mathematics.

Look at any map of the world. Countries sharing a common border have to be colored with different colors to distinguish them. For example, the map shown in Figure 1 can be colored with three colors. The map in Figure 2 requires an additional color. The four color problem considers how to color a map of any conceivable arrangement of countries on a plane surface, not only actual countries with existing borders.

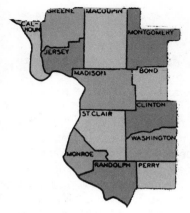

Figure 1 *The map of counties in downstate Illinois takes only three colors.*

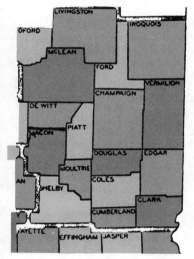

Figure 2 *A group of counties to the northeast of Figure 1 (the University of Illinois is in Champaign County) requires four colors.*

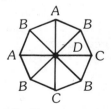

Figure 3 *The map of Shelby County (in the fourth color) and its neighbors, converted to a graph.*

Mathematicians proved without difficulty that five colors are sufficient to color any map of countries. Cartographers, however, never needed more than four colors in practice. This led Francis Guthrie to conjecture in 1853 that any map drawn on the plane can be colored with four colors. Interestingly, results were obtained more quickly for maps drawn on more complicated types of surfaces such as Moebius strip.

It turns out to be easier to color the points of a graph than to color countries. Thus the first step is to convert any map of countries to a graph, by replacing each country by a point. If two countries share a common border, the points which represent them are connected in the graph by a straight line (see Figure 3). The four color conjecture then could be phrased as

The points of any graph in the plane
can be colored with four colors, in such a way that
no points connected by a line are the same color.

The first proposed proof was given in 1879 by A. B. Kempe. Unfortunately, a flaw was discovered in his proof. In 1976 a method was finally found by which Kempe's proof could be completed, and a correct proof using a computer was given by Kenneth Appel and Wolfgang Haken of the University of Illinois. According to Appel and Haken, their method of proof for the four color theorem could not have worked previously because no computer was available to carry out the calculations needed to analyze a large number of maps. Appel and Haken estimate that these calculations involve about 10 billion logical decisions and take about 300 hours on a fast computer.

Appel and Haken found a collection of 1936 special graphs, with the property that every plane graph must contain one of these special graphs, called *reducible configurations*. (An analogy can be made with prime numbers: all numbers are composed of primes, and in a somewhat similar way all plane graphs are built from reducible configurations.) It already had been established that every graph containing one of these configurations is four-colorable. Together these two results yield the four color theorem.

The Haken-Appel technique consisted of "brute-force" enumeration and proof that each of their 1936 special graphs was indeed a reducible configuration. Such enumerative techniques are not possible except with computers, in view of the extremely large number of calculations needed.

The case-by-case method employed by Appel and Haken has been received with mixed feelings by the mathematical community. Many would have preferred a shorter, more elegant proof. Nevertheless, the validity of the Haken-Appel proof has not been seriously questioned. One can legitimately ask whether additional computer attacks on unsolved mathematical problems will be forthcoming.

Any computer-related solution would involve a necessary first step, that of "finitizing" the problem: instead of proving an assertion for an infinite number of possibilities, an argument is given to show that it is sufficient to prove the assertion for only a finite number of these. A computer then would be used to enumerate a proof of each of these possibilities. The four color problem was amenable to a computer approach because plane maps of one size look very much like plane maps of a slightly larger size. For example, adding one country to a map of the earth produces only a slight change in the map. However, many important mathematical problems are not so constituted. Fermat's famous conjecture, unsolved for 350 years, appears to be one example: There are no integers x, y, and z with $x^n + y^n = z^n$, if n is an integer > 2.

The Haken-Appel proof does raise a fascinating question: are there important mathematical problems for which *only* computer-assisted solutions exist? The four color theorem stands as a tentative example. It would be difficult, though, to show that no simpler proof ever will be formulated. Appel and Haken suggest that mathematician-and-computer exploration of such problems will be a major challenge. Haken has even remarked, "This work has changed my view of what mathematics is. I hope it will do the same for others."

Their own words   *Appel and Haken discuss the event in "The solution of the four-color-map. problem" (Scientific American, October 1977, pages 108–121). The article pictures the postage meter stamp ("four colors suffice"). used by the mathematics department at the University of Illinois, Urbana, in honor of the solution.*

CHAPTER 8 SUMMARY

Keywords

Euclidean geometry	Closed curve	Polyhedron
Axioms	Polygon	Face
Postulates	Regular polygon	Edge
Parallel postulate	Triangle	Vertex
Plane figure	Acute triangle	Regular polyhedron
Straight line	Obtuse triangle	Tetrahedron
Half-line	Right triangle	Octahedron
Ray	Equilateral triangle	Hexahedron (Cube)
Line segment	Isosceles triangle	Dodecahedron
Parallel lines	Scalene triangle	Pyramid
Perpendicular lines	Quadrilateral	Prism
Intersecting lines	Trapezoid	Sphere
Angle	Parallelogram	Cone
Side	Pentagon	Cylinder
Vertex	Hexagon	Volume
Acute angle	Conic section	Surface area
Obtuse angle	Circle	
Right angle	Ellipse	Projective geometry
Straight angle	Parabola	non-Euclidean geometry
Degrees of an angle	Hyperbola	Great circles
Protractor	Perimeter	Moebius strip
Adjacent angles	Area	Torus
Complementary angles	Similar polygons	Network
Supplementary angles	Congruent polygons	Arc
Vertical angles	Space figure	Node
Curve	Plane	Odd vertex
Simple curve	Skew lines	Traversibility

Formulas   **Perimeter**   **Area**

Rectangle $P = 2L + 2W$ $A = LW$
Square $P = 4s$ $A = s^2$
Parallelogram $P = 2a + 2b$ $A = bh$
Trapezoid $P = a + b + B + c$ $A = \frac{1}{2}(b + B)h$
Triangle $P = a + b + c$ $A = \frac{1}{2}bh$
Circle $C = 2\pi r$ $A = \pi r^2$

                        Volume   **Surface area**

Cylinder $V = \pi r^2 h$ $S = 2\pi r^2 + 2\pi rh$
Sphere $V = \frac{4}{3}\pi r^3$ $S = 4\pi r^2$
Cone $V = \frac{1}{3}\pi r^2 h$ $S = \pi r\sqrt{r^2 + h^2}$
Pyramid $V = \frac{1}{3}Bh$
Box $V = LWH$

CHAPTER 8 TEST

Draw angles having the following measures.

1. 29°   **2.** 172°

Find each of the following supplementary angles.

3.
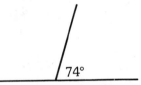
74°

4.
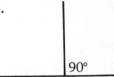
90°

Identify each of the following curves as *simple, closed, both,* or *neither.*

5.

6.

7. A right triangle has shorter sides of 16 cm and 30 cm. Find the length of the longest side.

Find the area of each of the following figures.

8.   **9.**   **10.**   **11.**

12
6

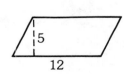
5
12

8
17

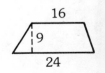
16
9
24

12. A wall 12 feet by 8 feet can be painted for $38. Find the cost to paint a wall 24 feet by 16 feet.

13. A small tank has a radius of 3 m. A large tank has a radius of 9 m. If the small tank can be filled with an industrial chemical for $2000, find the cost to fill the large tank.

Find the volume of each of the following figures.

14

6

15.

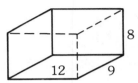

8

12   9

16.

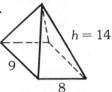

$h = 14$

9

8

17.

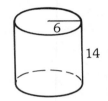

6

14

Find the missing sides or angles in each of the following pairs of similar triangles.

18.

12

10

11

y

x

22

19.

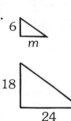

6

m

18

24

20.

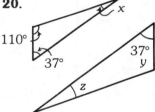

110°

37°

x

37°

y

z

Give the genus of the following objects.

21. desk calendar (See photograph at left.)

22. a coffee cup

Are the following networks traversible?

23.

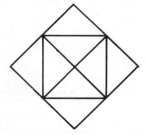

24.

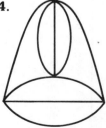

The Batcolumn *stands in a concrete base at right angles to the flat Chicago earth, all 21 tons of welded steel, 101 feet high. It was created by sculptor Claes Oldenberg. The photograph here and those on pages 309 and 315 are by Howard Kaplan.*

9 *Probability*

Try to determine each of the following:

(1) The area of a square one centimeter on a side.

(2) The number of days in next year if last year was leap year.

(3) The result of tossing an ordinary coin.

(4) The result when five children draw straws to determine the captain of the team.

(5) The number of air molecules that will bounce off the bottom inside an empty sixteen-ounce Coke bottle in 1 second at sea level if the temperature is 16°C.

(6) The number of buildings that will be destroyed in the next major California earthquake.

Questions (1) and (2) involve phenomena where the answers (namely, 1 square centimeter and 365) can be determined for certain. We call them *deterministic phenomena*.

Questions (3) and (4) involve phenomena that have long been recognized as unpredictable. In fact, their very uncertainty has made them useful in games of chance and in making certain kinds of decisions. Evidence of similar phenomena, such as casting lots and rolling dice, is found in the literature and remains of the ancient world.

Question (5) is a type that science had hopes of answering at one time. However, the study of molecules and other tiny particles has shown us that this question also involves an unpredictable phenomenon, and really belongs in the same category as questions (3) and (4).

Question (6), which seems similar at first glance, actually presents a different kind of problem altogether. Whereas coin-tossing, straw-drawing, and moving air molecules are commonplace, and can be observed whenever and as often as we please, major earthquakes in California do not occur that often. Questions (3)–(6) all involve uncertainty, but only (3), (4), and (5) have the following two characteristics:

Brownian motion *(or movement) of a particle suspended in a liquid or a gas describes a random path due to bombardment by molecules of the medium, themselves moving randomly as they collide with each other, bounce off, veer away, collide again. Everything is in flux, said Heraclitus 2500 years ago; so it seems.*

(a) Observation of the phenomenon under a given set of circumstances does not always lead to the same outcome.

(b) In a very large number of independent observations of the phenomenon, the relative number of occurrences of each possible outcome tends toward a particular fraction.

Any phenomenon with these two characteristics is called a **random phenomenon.** An ordinary coin can come up either heads or tails (two possible outcomes), and we cannot predict with any certainty the result of a particular toss. However, since heads and tails are equally likely (that is, one is just as apt to occur as the other), we can predict that in a large number of tosses, we will observe heads about half the time. California earthquakes are not random phenomena since the required "large number of observations" is not possible.

Probability theory is concerned with random phenomena. The relative frequency fraction in characteristic (b) above represents the *likelihood,* or *probability,* of a given outcome. In this chapter we will look at various ways of finding and using this fraction.

Some mathematics involved in games and gambling was studied in Italy as early as the fifteenth and sixteenth centuries, but a systematic mathematical theory of chance was not begun until 1654. In that year two French mathematicians, Pierre de Fermat (about 1601–1665) and Blaise Pascal (1623–1662), corresponded with each other regarding a problem posed by the Chevalier de Méré, a gambler and member of the aristocracy. *If the two players of a game are forced to quit before the game is finished, how should the pot be divided?* Pascal and Fermat solved the problem by developing basic methods of determining each player's chance, or probability, of winning.

About all the published work in probability theory for the next 150 years or so was based on dice and other games of chance. The man usually credited with being the "father" of probability theory was the French mathematician Pierre Simon de Laplace (1749–1827); he was one of the first to apply probability to matters other than gambling.

Despite the work of Laplace and the urgings of a few important mathematicians in the nineteenth century, the general mathematical community failed to get interested in probability.

One event that eventually helped to bring probability theory into prominence was the botanist Robert Brown's observations in 1828 of the irregular motion of pollen grains suspended in water. Such "Brownian motion," as it came to be called, was not understood until 1905 when Albert Einstein gave a complete explanation based on the treatment of molecular motion as a random phenomenon. (See Section 9.7, Exercises.)

The development of probability as a mathematical theory was mainly due to a line of remarkable scholars in Russia, including P. L. Chebyshev (1821–1894), his student A. A. Markov (1856–1922), and finally Andrei Nikolaevich Kolmogorov (born in 1903). Kolmogorov's *Foundations of the Theory of Probability* (1933) is an axiomatic treatment.

Pierre de Fermat *chose to write about his mathematical findings in letters to Pascal, Descartes, and others, rather than publish them. His correspondence with Pascal about probability was to have been expected. There are many unexpected things about him. Fermat was not a mathematician by profession, but a government official. He did not interest himself in mathematics until he was past 30. Fermat is best known for his work in number theory, notably on primes. He was fascinated by the* Arithmetica *of Diophantus, which became accessible to the reading public (in Latin) rather late—1621. Its title refers to number theory, not reckoning, and its emphasis is on solutions of various types of equations, now called* Diophantine equations. *(Diophantus probably lived in the third century AD in Alexandria—not much else is known about him.)*

Fermat wrote many marginal notes in his copy of Diophantus. One of them has confounded many mathematicians for over 300 years. Fermat claimed to have discovered a proof of the theorem: $x^n + y^n = z^n$ *is impossible for positive integers x, y, z, and n, where n is greater than 2. "But," Fermat wrote, "the margin is too narrow to contain it." Since then, no one has been able to disprove or prove Fermat's Last Theorem, as it is called.*

9.1 PROBABILITY EXPERIMENTS

In probability theory, we refer to the observation of a random phenomenon as an **experiment.** Each repetition of the experiment is called a *trial,* and on each trial the possible results (of which there are two or more) are called **outcomes.**

An example of a probability experiment is the tossing of a coin. On each trial of this experiment (each toss) there are two possible outcomes, heads and tails (which we abbreviate *h* and *t*). If the coin is not loaded to favor one side over the other, we say it is a "fair" coin. This means we assume that the two possible outcomes, *h* and *t,* are **equally likely,** or *equiprobable.* For a fair coin, this "equally likely" assumption is made on each trial.

Since there are *two* equally likely outcomes possible, heads and tails, and just *one* of them is heads, we say that the probability of heads is 1 divided by 2:

$$\text{Probability (heads)} = \frac{1}{2}, \qquad \text{or simply} \qquad P(h) = \frac{1}{2}.$$

Other ways to say the same thing are:

(1)   The chances of heads are one in two.

(2) $P(h) = 0.5$.

(3)   There is a "fifty percent chance" of heads.

Notice that the probability of tails is also $\frac{1}{2}$.

Example 1   If the spinner shown here is spun, find the probability that it will point to: (a) the number 1;   (b) the number 2.

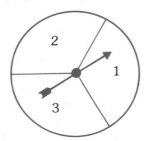

(a) Since there are three equally likely outcomes possible on a given spin, and one of them is the number 1, then $P(1) = \frac{1}{3}$.

(b) Also, the number 2 is one of three possible outcomes; so $P(2) = \frac{1}{3}$.

An ordinary die is a cube whose six faces show the numbers 1, 2, 3, 4, 5, and 6, respectively. If the die is not "loaded" to favor certain faces over others, then any one of the six faces is equally likely, and we call the die "fair." (The plural of "die" is "dice.")

REPUBLIQUE FRANÇAISE

30F +9F

POSTES

LAPLACE
1749-1827

I had no need of that hypothesis, *said Laplace to Napoleon, when the latter remarked that God was not mentioned in* Celestial Mechanics, *Laplace's great work on the solar system. Laplace was not denying God; he had succeeded in working out mathematically the interacting gravitational forces in the solar system and proving its stability. Over a century before, Newton had declared that God must intervene from time to time to keep the system's clockwork moving; it then seemed impossible to describe all intricate workings mathematically.*

In 1773 Laplace began to solve the problem of why Jupiter's orbit seems to shrink and Saturn's to expand. Eventually Laplace worked out the whole system. Celestial Mechanics *resulted from almost a lifetime of work. In five volumes, it was published between 1799 and 1825, and gained for Laplace the reputation "Newton of France."*

Laplace's work on probability was actually an adjunct to his celestial mechanics. He needed to demonstrate that probability is useful in interpreting scientific data. He also wrote a popular exposition of the system, which contains (in a footnote!) the "nebular hypothesis" that the sun and planets originated together in a cloud of matter, which then cooled, and condensed into separate bodies. (See also page 346.)

Example 2   If a single fair die is rolled, find the probability that the face turned up will show: (a) 4;   (b) 6.

(a) Since one out of six faces shows 4, $P(4) = \frac{1}{6}$.   **(b)** Also, $P(6) = \frac{1}{6}$.

Sometimes when an experiment is performed, we are interested in a result that is satisfied by more than one of the possible outcomes. To find the probability that the spinner in Example 1 will point to an odd number, we observe that two of the three possible outcomes are odd numbers, 1 and 3. Therefore

$$P(\text{odd}) = \frac{2}{3}.$$

The results of experiments we have discussed above, such as heads, tails, a die showing 6 or an odd number, are called *events*. In general, any outcome of a trial is called an **event**. In the experiment of coin tossing, for example, heads is such an event. We found the probability of the event *heads* to be $P(h) = \frac{1}{2}$.

Laplace, in his famous *Analytic Theory of Probability* (1812), provided us with a general formula for computing the **probability of an event:**

$$P(\text{event}) = \frac{\text{number of favorable outcomes}}{\text{total number of outcomes}}$$

We shall refer to this formula as the "classical definition of probability." It must be remembered that its use requires that all the outcomes be equally likely.

Example 3   When a single fair die is rolled, what is the probability of getting: (a) an even number;   (b) a number greater than 2;   (c) some number other than 6;   (d) the number 7?

Solution.   The probability of each of these events is found by counting the number of favorable outcomes.

(a) Three outcomes are even (2, 4, and 6), so

$$P(\text{even}) = \frac{3}{6} = \frac{1}{2}.$$

(b) The numbers 3, 4, 5, and 6 are all greater than 2, so

$$P(\text{greater than 2}) = \frac{4}{6} = \frac{2}{3}.$$

(c) Five of the possible outcomes are different than 6, so

$$P(\text{other than 6}) = \frac{5}{6}.$$

(d) None of the possible outcomes is the number 7, so

$$P(7) = \frac{0}{6} = 0.$$

Figure 9.1 *The set of 52 cards in the standard bridge deck has four 13-card subsets (or suits):* ♠ *spades,* ♥ *hearts,* ♦ *diamonds, and* ♣ *clubs.* Ace *is the unit card. The three face cards are* King, Queen, *and* Jack.

Example 4   If a single playing card is drawn at random from an ordinary 52-card bridge deck (Figure 9.1), find the probability of each of the following events: (a) ace;   (b) face card;   (c) spade;   (d) spade or heart.

(a) There are four aces in the deck, so

$$P(\text{ace}) = \frac{4}{52} = \frac{1}{13}.$$

(b)   Since there are 12 face cards,

$$P(\text{face card}) = \frac{12}{52} = \frac{3}{13}.$$

(c) The deck contains 13 spades, so

$$P(\text{spade}) = \frac{13}{52} = \frac{1}{4}.$$

(d) Besides the 13 spades, the deck contains 13 hearts, so

$$P(\text{spade or heart}) = \frac{26}{52} = \frac{1}{2}.$$

In the experiment of tossing a single fair coin, there are just two possible outcomes, heads and tails. If we call heads "success," and tails "failure," then

$$P(\text{success}) = \frac{1}{2} \quad \text{and} \quad P(\text{failure}) = \frac{1}{2}.$$

Even when an experiment offers more than two possible outcomes on each trial, we can still apply the terms "success" and "failure."

Example 5   If, in rolling a single fair die, we consider an even number to be success and an odd number to be failure, then find the probability of success and the probability of failure.

Solution.

$$P(\text{success}) = \frac{3}{6} = \frac{1}{2} \quad \text{and} \quad P(\text{failure}) = \frac{3}{6} = \frac{1}{2}.$$

Example 6   If a single card drawn from an ordinary 52-card deck is considered success if red and failure if black, then find the probabilities of success and failure.

Solution.

$$P(\text{success}) = \frac{26}{52} = \frac{1}{2} \quad \text{and} \quad P(\text{failure}) = \frac{26}{52} = \frac{1}{2}.$$

Success and failure do not always have probabilities of $\frac{1}{2}$, as illustrated by Examples 7 and 8.

Example 7   Suppose, in rolling a single fair die, we count it a success in the event that the number rolled is greater than 4, and a failure otherwise. Find the probabilities of success and failure.

Solution.   Since just two outcomes are greater than 4, namely, 5 and 6, we have

$$P(\text{success}) = \frac{2}{6} = \frac{1}{3} \quad \text{and} \quad P(\text{failure}) = \frac{4}{6} = \frac{2}{3}.$$

Example 8   In drawing a card from an ordinary deck, if hearts is considered success and anything else (any other suit) is failure, find the probabilities of success and failure.

Solution.

$$P(\text{success}) = \frac{13}{52} = \frac{1}{4} \quad \text{and} \quad P(\text{failure}) = \frac{39}{52} = \frac{3}{4}.$$

Actor Bruce Dern plays cards with a poker-faced droid in the sci-fi film, Silent Running.

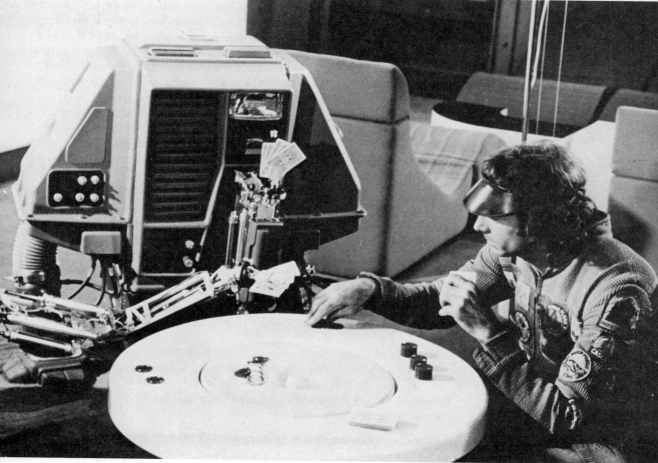

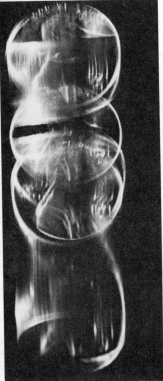

A trial of any experiment where the outcomes are classified as either success or failure is referred to as a *Bernoulli trial* (after the seventeenth-century mathematician, James Bernoulli). There are many applications, especially of *repeated* Bernoulli trials. We will discuss the application to Brownian motion in Section 9.7 (see Exercises 42 and 43 in that section).

A brief look at repeated Bernoulli trials here will help to clarify what probability is and what it is not. When we say that the probability of heads on a single coin toss is $\frac{1}{2}$, we do *not* mean that when the coin is tossed many times half the tosses will (necessarily) be heads. We mean that the fraction of times heads occurs should not be far from $\frac{1}{2}$ if we toss the coin enough times.

A series of 50 actual tosses of a coin produced this sequence of outcomes:

tthhh *ttthh* *hthtt* *hhthh* *ttthh* *thttt* *hhthh* *ththh* *tthhh* *hhtht*.

After two tosses the fraction of heads was 0 (far from $\frac{1}{2}$), and after eight tosses it was $\frac{3}{8}$ (approaching $\frac{1}{2}$). Checking the cumulative fraction of heads after each two tosses shows that 16 of these 25 values are $\frac{1}{2}$ exactly. Heads occurred on two of the first four tosses, on three of the first six, on six of the first twelve, and so on. After 42 tosses the fraction was $\frac{1}{2}$; and even though the next four tosses were all heads, the fraction after 46 was $\frac{25}{46}$, only $\frac{1}{23}$ higher than $\frac{1}{2}$. The effect here is that the fraction seems to stabilize around $\frac{1}{2}$ as the number of tosses increases, obeying the important principle known as Bernoulli's Theorem, or the **Law of Large Numbers.** (It is also commonly known as the "law of averages.")

There is a popular misconception about what this principle says. For example, if a fair coin is to be tossed five times and the first four tosses are all tails, many people think that, because of the law of averages, the fifth toss is more likely to be heads than tails. But this is not so. On any given toss the probability of heads is still $\frac{1}{2}$.

Closely related to the idea of probability is that of **odds.** If the probability of an event is described as the ratio of favorable outcomes to *total* outcomes, then the "odds in favor of an event" would be the ratio of favorable outcomes to *unfavorable* outcomes.

Example 9 A single marble is drawn at random from a jar containing 6 blue and 4 white marbles. Find the odds in favor of blue.

Solution. Since six outcomes are favorable and four are unfavorable, the required odds are $\frac{6}{4} = \frac{3}{2}$, which would usually be expressed as "3 to 2."

When there are more outcomes unfavorable to a given event than favorable, we refer to "odds against" rather than "odds in favor."

Example 10 A single card is selected from an ordinary deck. Find the odds against its being a face card.

Solution. There are twelve face cards and forty non-face cards in the deck, so the odds against a face card are 40 to 12, or 10 to 3.

Example 11 Find the odds in favor of a red card being drawn from an ordinary deck.

Solution. Since 26 cards are red and 26 are black, the odds are 26 to 26, or "1 to 1." Odds of 1 to 1 are commonly referred to as "even odds" or as a "50-50 chance."

9.1 EXERCISES

Identify each phenomenon as either *random* or *deterministic:*

1. The winner of a football game between the Steelers and the Raiders, if the Steelers lead 34 to 6 one second before the end of the game.

2. The result of drawing one card from a standard, shuffled deck.

3. The number of heads resulting when three fair coins are tossed.

4. The number which, when added to 6, gives the answer 9.

5. The number resulting when a roulette wheel is spun.

6. The number of auto accidents to occur in a given week on Interstate 5 between Portland and Seattle.

7. The result of tossing a coin that has heads on both sides.

8. The number of cosmic rays in an hour to be incident upon a 10 square centimeter detector at an elevation of 4000 feet above sea level at the equator.

When a single fair die is rolled, find the probability of:

9. the number 2 **10.** odd number **11.** number less than 5

A die is rolled. If a number greater than 4 is considered success, find:

12. $P(\text{success})$ **13.** $P(\text{failure})$

When a single card is drawn from an ordinary deck find the probability of:

14. spade **15.** queen **16.** black jack **17.** ace of spades

Find the odds against the card drawn being:

18. face card **19.** king **20.** red card **21.** three of hearts

Sometimes the assumption of equal likelihood, which is necessary to the classical determination of probability, is not justified.

22. Toss an ordinary coin fifty times and record the sequence of heads and tails. Calculate the cumulative fraction of heads after each ten tosses. Does this experiment verify the assumption that $P(h) = \frac{1}{2}$?

23. Toss a paper (or styrofoam) cup into the air thirty times, recording how it lands each time. Does this experiment verify the equal likelihood of *top*, *side*, and *bottom*?

There are many random phenomena whose possible outcomes are not equally likely. Sometimes the best (or even the only) way to find the probability for a given outcome is by experiment. It is then called an *empirical probability* (or sometimes *experimental*, or *statistical*) as opposed to the classical probability which assumes equally likely outcomes. Generally, the larger the number of trials, the truer the empirical probability.

24.   From your experimental results in Exercise 23, calculate empirical probabilities for *top, side,* and *bottom.*

25.   Toss a thumbtack fifty times, observing the number of times it lands point up (U) and point down (D). Calculate, for the particular tack you used, the empirical probabilities $P(U)$ and $P(D)$.

We often hear probabilities (or odds) stated which are neither classical nor empirical, but are someone's judgment, based upon his particular information and experience. An example is: "The probability is $\frac{7}{10}$ that the Dow Jones average will rise next week." This is a *subjective probability.* Two different people might well disagree on the appropriate value. Identify the probabilities in Exercises 26–32 as *classical, empirical,* or *subjective.*

26.   A weather forecaster allows a 70% chance of rain tomorrow.

27.   A gambler claims that on a roll of a fair die, $P(\text{even}) = \frac{1}{2}$.

28.   A surgeon gives a patient a 90% chance of full recovery.

29.   A boy notices that a certain coin landed tails 35 times out of 50, so for that coin he claims that $P(\text{tails}) = 0.7$.

30.   A bridge player has a probability of $\frac{1}{4}$ of being dealt a diamond.

31.   A scientist says the odds against a catastrophic core "meltdown" in a nuclear energy facility are five billion to one.

32.   A forest ranger states that the probability of a short fire season this year is only three in ten.

*33.   If, in a given experiment, we call the occurrence of a certain event a success, and its non-occurrence a failure, state a general relationship that always holds between $P(\text{success})$ and $P(\text{failure})$.

9.2 SAMPLE SPACES AND COMPOSITE EVENTS

As we discuss probability experiments that are more complicated than those in Section 9.1, it will be worthwhile to use set notation (see Chapter 1). The set of all possible outcomes is called the **sample space,** which is denoted by S. For example, the sample space for the experiment of tossing a single fair coin would be

$$S = \{h, t\}.$$

Example 1   Show the sample space for the experiment of rolling a single fair die.

Solution. $S = \{1, 2, 3, 4, 5, 6\}$.

An **event** is some particular subset of the sample space, and is often labeled with the letter E.

Example 2   For the experiment in Example 1, use set notation to show the event of an even number.

Solution. $E = \{2, 4, 6\}$.

For any experiment with all outcomes equally likely, if $n(S)$ is the number of outcomes possible (that is, the number of elements in the sample space S), and $n(E)$ is the number of outcomes favorable to some event E (that is, $n(E)$ is the number of elements in the subset E), then Laplace's classical probability formula can be written

$$P(E) = \frac{\text{favorable outcomes}}{\text{total outcomes}} = \frac{n(E)}{n(S)}.$$

We now repeat Example 3 of the previous section, using the new notation.

Example 3   When a single fair die is rolled, what is the probability of getting: (a) an even number;   (b) a number greater than 2;   (c) some number other than 6;   (d) the number 7?

Solution.   We give labels to these four events, as follows:

E_1 is the event of an even number
E_2 is the event of a number greater than 2
E_3 is the event of a number other than 6
E_4 is the event of a 7

Then we can express the events as subsets of the sample space $S = \{1, 2, 3, 4, 5, 6\}$.

$E_1 = \{2, 4, 6\}$
$E_2 = \{3, 4, 5, 6\}$
$E_3 = \{1, 2, 3, 4, 5\}$
$E_4 = \varnothing$

The required probabilities are

$$P(E_1) = \frac{n(E_1)}{n(S)} = \frac{3}{6} = \frac{1}{2} \qquad P(E_3) = \frac{n(E_3)}{n(S)} = \frac{5}{6}$$

$$P(E_2) = \frac{n(E_2)}{n(S)} = \frac{4}{6} = \frac{2}{3} \qquad P(E_4) = \frac{n(E_4)}{n(S)} = \frac{0}{6} = 0.$$

Mary Somerville *(1780–1872)*
is associated with Laplace because
of her brilliant exposition of his
Celestial Mechanics. She combined
a deep understanding of science
with the ability to communicate its
concepts to the general public.

In her childhood Somerville was
free to observe nature. She
studied Euclid thoroughly and
perfected her Latin so she could
read Newton's Principia. About
1816 she went to London, and
soon became part of its literary
and scientific circles. She also
corresponded with Laplace and
other Continental scientists.

Somerville's book on Laplace's
theories came out in 1831 with
great acclaim—her reputation was
made. Then followed a panoramic
book, Connection of the Physical
Sciences *(1834). A statement in*
one of its editions suggested that
irregularities in the orbit of
Uranus might indicate that a
farther planet, not yet seen,
existed. This caught the eye of the
scientist who worked out the
calculations for Neptune's orbit.

Notice in the above example that getting a number other than 6 is the same as not getting a 6. If we take S as the "universal" set, then the complement of E_3, denoted E'_3, is $\{6\}$.

Also, $P(E'_3) = \dfrac{n(E'_3)}{n(S)} = \dfrac{1}{6}$. Therefore, $P(E_3) + P(E'_3) = \dfrac{5}{6} + \dfrac{1}{6} = 1$,

or equivalently $P(E_3) = 1 - P(E'_3)$. In general, we have the "complements formula"

$$P(E) = 1 - P(E'),$$

which allows us to find probabilities indirectly when it is easier. (See Exercise 48.) If the event E is considered success, then this formula is saying

$$P(\text{success}) = 1 - P(\text{failure}).$$

Example 4   When a single card is drawn from an ordinary 52-card deck, what is the probability of the event of drawing a card other than a king?

Solution.   It is easier to count the kings than the non-kings, so we write

$$P(E) = 1 - P(E') = 1 - \frac{n(E')}{n(S)} = 1 - \frac{4}{52} = \frac{48}{52} = \frac{12}{13}.$$

So far our experiments have involved just one action, or task: toss a single coin, roll a single die, draw a single card, and so on. If an experiment consists of two or more distinct tasks, we call it a *composite experiment*, and any event that may occur is called a **composite event.**

For example, suppose the experiment is to toss *two* fair coins (or to toss one fair coin twice, an equivalent experiment). The first coin can yield either heads or tails, and so can the second. Figure 9.2 shows a two-by-two "product" table (similar to the operation tables in Chapter 6) that illustrates this experiment. In the body of the table we find the four equally likely outcomes, so the sample space is

$$S = \{hh,\ ht,\ th,\ tt\}.$$

Figure 9.3 shows a tree diagram for this same experiment. The outcomes are found by reading, left to right, along the tree's branches. The branch shown in color gives the outcome *ht*, one of the four equally likely possibilities.

		2d coin	
		h	*t*
1st	*h*	*hh*	*ht*
coin	*t*	*th*	*tt*

Figure 9.2

1st coin	2d coin	Outcome
h	*h*	hh
	t	ht
t	*h*	th
	t	tt

Figure 9.3   Tree diagram for tossing two fair coins

ESP   A deck of twenty-five cards, five per each symbol shown here, is used to test extrasensory perception (ESP). The cards were developed by J. B. Rhine at the Parapsychology Laboratory, Duke University. The Duke experiments are famous pioneering efforts to apply scientific methods to the investigation of clairvoyance or telepathy.

Rhine and other investigators believed that ESP does exist in certain individuals because their scores on ESP tests are higher than can be accounted for by mere chance. The probability that a test subject will correctly guess the symbols on cards hidden from the subject is 1/5. For example, in 800 trials a test subject may have a score of 160 correct. But suppose a subject's score is 207? Is the "extra" 47 only a matter of chance? Rhine did not think so. He calculated the probability of scoring the "extra" 47 in 800 trials to be 1/250,000 — too small to be just chance. What do you think?

Odds are given for success in ESP card tests. Odds against fair success are 20 to 1; odds against excellent results are 250 to 1. The following table lists number of trials (four runs through the 25-card deck are minimum in a serious experiment); number of correct responses by chance alone (probability is 1/5); number of responses considered fair success; number considered good.

Trials	Chance	Fair	Good
100	20	28	32
250	50	63	69
1250	250	279	293
2500	500	540	560

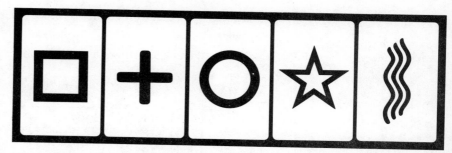

Example 5   For the experiment of tossing two fair coins, use set notation to show the event of two heads, and find the probability of this event.

Solution.   If we denote this event by E, then $E = \{hh\}$. Since $S = \{hh, ht, th, tt\}$, we have

$$P(E) = \frac{n(E)}{n(S)} = \frac{1}{4}.$$

Example 6   For the experiment of tossing one fair coin and rolling one fair die: (a) construct both a product table and a tree diagram;   (b) find the sample space S and the event E of getting tails on the coin and a number less than 3 on the die;   (c) find the probability of event E.

(a) A product table and a tree diagram for the experiment of tossing one fair coin and rolling one fair die are shown in Figures 9.4 and 9.5 respectively.

(b) $S = \{h1, h2, h3, h4, h5, h6, t1, t2, t3, t4, t5, t6\}$

$E = \{t1, t2\}$

(c) $P(E) = \dfrac{2}{12} = \dfrac{1}{6}.$

	Coin	
	Die	

Coin		Die					
	1	2	3	4	5	6
h	h1	h2	h3	h4	h5	h6
t	t1	t2	t3	t4	t5	t6

Figure 9.4

Coin   Die   Outcome

- h
  - 1 → h1
  - 2 → h2
  - 3 → h3
  - 4 → h4
  - 5 → h5
  - 6 → h6
- t
  - 1 → t1
  - 2 → t2
  - 3 → t3
  - 4 → t4
  - 5 → t5
  - 6 → t6

Figure 9.5   Tree diagram for tossing one fair coin and rolling one fair die

Example 7   What is the sample space for the experiment of rolling two fair dice?

Solution.   Figure 9.6 shows a product table.

		Second die					
		1	2	3	4	5	6
	1	(1, 1)	(1, 2)	(1, 3)	(1, 4)	(1, 5)	(1, 6)
	2	(2, 1)	(2, 2)	(2, 3)	(2, 4)	(2, 5)	(2, 6)
First	3	(3, 1)	(3, 2)	(3, 3)	(3, 4)	(3, 5)	(3, 6)
die	4	(4, 1)	(4, 2)	(4, 3)	(4, 4)	(4, 5)	(4, 6)
	5	(5, 1)	(5, 2)	(5, 3)	(5, 4)	(5, 5)	(5, 6)
	6	(6, 1)	(6, 2)	(6, 3)	(6, 4)	(6, 5)	(6, 6)

Figure 9.6

The sample space consists of all 36 number pairs in the body of the table. A tree diagram here would contain six branches for the first die, *each of* which split into six branches for the second die. Compare such a tree with Figure 9.5.

If three dice are rolled, tree diagrams become very impractical because of the size and complexity of the tree. Furthermore, a product table cannot be used unless it is done in three dimensions. Fortunately, though, most probability problems don't require an actual listing of the sample space so long as we can determine the number of outcomes it would contain. To do that, we can apply the following rule:

Product rule for counting
In a composite experiment, the total number of outcomes
can be found by determining the number of ways
of performing each of the separate tasks,
and then multiplying these numbers together.

To review Example 6, the total number of outcomes was $2 \cdot 6 = 12$, since each of the *two* branches for the coin split into *six* branches for the die.
The product rule for counting gives rise to the following rule which applies more directly to probabilities (see Exercises 17–22).

Product rule for probability
In a composite experiment, the probability of an event
can be found by determining the probability of the necessary result
on each of the separate tasks,
and then multiplying these probabilities together.

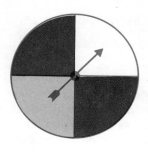

To review Example 6 again, the probability of event E (tails on the coin and a number less than 3 on the die) could have been found by this rule. The probability of tails on the coin is $\frac{1}{2}$; the probability of a number less than 3 on the die is $\frac{2}{6}$, or $\frac{1}{3}$. So the probability of the composite event is

$$P(E) = \frac{1}{2} \cdot \frac{1}{3} = \frac{1}{6}.$$

Example 8 If the spinner shown here is spun three times, find the probability of the composite event gray, then black, then white.

Solution. The probabilities of the separate events are

$$P(\text{gray}) = \frac{1}{4}, \qquad P(\text{black}) = \frac{1}{2}, \qquad P(\text{white}) = \frac{1}{4}.$$

So, $\quad P(\text{gray, then black, then white}) = \frac{1}{4} \cdot \frac{1}{2} \cdot \frac{1}{4} = \frac{1}{32}.$

9.2 EXERCISES

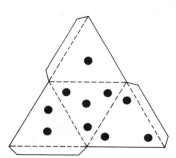

Die cut *Use this model to make your own tetrahedral dice. Cut along the solid outline. Fold on the dashed lines, with the three flaps toward the inside. Glue the flaps to respective triangles.*

Dice can be made from other regular solids, too, such as the dodecahedron (see Chapter 8). When all is said and done, the cube works best. Tetrahedrons don't really roll; on the other hand, dodecahedrons roll too well (off a tabletop, say).

A tetrahedral "die" has four faces, numbered 1, 2, 3, and 4, each equally likely to fall face down on a given toss. For the experiment of tossing a tetrahedral "die" once, use set notation for Exercises 1–4.

1. Write the sample space S.

2. Write the event E of a number greater than 1.

3. Find the probability of E.

4. Find the odds in favor of E.

For the experiment of tossing two tetrahedral "dice":

5. Construct a product table.

6. Construct a tree diagram.

7. Write the sample space.

For the experiment of tossing four tetrahedral "dice":

8. Find the number of outcomes possible.

9. Find the number of ways of getting four even numbers.

10. Find the probability of four even numbers.

For composite card-drawing experiments, it is important to know if the drawing is done "with replacement" (that is, each drawn card being shuffled back into the deck before the next card is drawn) or "without replacement." If two cards are drawn from a 52-card deck *with* replacement, find the probability of:

11. face card on the first draw.

12. face card on the second draw

13. face cards on both draws

Gregor Johann Mendel *(1822–1884) came from a peasant family who managed to send him to school. By 1847 he had been ordained and was teaching at the Abbey of St. Thomas in Brunn, present-day Brno, Czechoslovakia. He finished his education at the University of Vienna, and returned to the abbey to teach mathematics and natural science. Mendel began to carry out experiments on plants in the abbey garden, notably pea plants, whose distinct traits* (unit characters) *he had puzzled over. In 1865 he published his results* (the Czech stamp above commemorates the centennial). *His work was not appreciated at the time even though he had laid the foundation of classical genetics. Mendel had established the basic laws of heredity: law of unit characters, of dominance, and of segregation. The exercises on the facing page involve the* unit character *of flower color in pea plants, where red is* dominant *over white in the first generation of hybrids. In the second generation of hybrids, however, white is "segregated"; it shows up again, as the WW combination in Figure 9.7.*

Mendel had to give up his experiments after 1868, when he became Abbot of St. Thomas.

If two cards are drawn *without* replacement, find the probability of:

14.   heart on the first draw   ***16.**   hearts on both draws

*15.   heart on the second draw

For a two-task composite experiment, suppose the first task can be done in n_1 ways, S_1 of which are successful; and the second task can be done in n_2 ways, S_2 of which are successful. Event E requires success on both tasks. Exercises 17–22 will establish the equivalence of the two product rules in this section, the one for counting and the one for probability.

Use the product rule for counting:

17.   Find the total number of outcomes for the composite experiment.

18.   Find the number of outcomes favorable to event E.

19.   From your answers to 17 and 18, find $P(E)$.

20.   Find the probability of success on the first task.

21.   Find the probability of success on the second task.

22.   Using the product rule for probability, find $P(E)$.

*23.   For an experiment with sample space S and some event E, we know by the complements formula that if $n(E)$ is difficult to find directly, we can get $P(E)$ as follows:

$$P(E) = 1 - P(E') = 1 - \frac{n(E')}{n(S)}.$$

Derive a "complements formula" for finding the *odds* in favor of E.

Suppose we want to form three-digit numbers using the set of digits {0, 1, 2, 3, 4, 5}. For example, 201 and 334 are such numbers, but 052 is *not*.

*24.   How many such numbers are possible?

*25.   How many of these numbers are multiples of 5?

*26.   If one three-digit number is chosen at random from all those that can be made from the above set of digits, find the probability that the one chosen is not a multiple of 5.

Gregor Mendel, an Austrian monk, was the first to allow for randomness in the study of genetics. In an effort to understand the mechanism of character transmittal from one generation to the next in plants, he counted the number of occurrences of various characteristics. Mendel found that the flower color in certain pea plants obeyed this scheme:

Pure-red crossed with pure-white produces red.

Now, the red offspring received from its parents genes for both red (R) and white (W), but in this case red is "dominant" and white "recessive," so the offspring exhibits the color red. However, the offspring still carries both genes, and when two such offspring are crossed, several

More on genetics   *In the early 1900s, shortly after Mendel's work received the recognition it missed in 1865, a new branch of genetics was established — population genetics. The basis is the Hardy-Weinberg law, which concerns the genetic stability of a population. For example, if hybrid pea plants reproduce in nature randomly, the ratio of red to white characters may stay the same, that is, red does not necessarily drive out white, even though red is dominant. The Hardy-Weinberg law involves the formula $p^2 + 2pq + q^2 = 1$, where p and q are alternative unit characters, such as red and white. (For more on the mathematics involved, see Joe Dan Austin's article, "The Mathematics of Genetics," in The Mathematics Teacher, November, 1977, pages 685–690.)*

In the 1960s, experiments in transferring genetic material from cell to cell founded another branch of genetics, called molecular genetics. Genes are molecules of a substance called DNA for short, arranged usually in a double helix (a sort of spiral shape) and sometimes in a ring called a plasmid. In the 1970s controversy centered around experiments with plasmid rings, "recombinant DNA research," in which genes from higher organisms could be transplanted into the plasmids of bacteria and studied. However, scientists and others have raised questions about the safety of such research, and about the social effects of what is called "genetic engineering."

things can happen in the third generation as shown in the table of Figure 9.7, which is called a *Punnet square.*

		2nd parent	
		R	W
1st	R	RR	RW
parent	W	WR	WW

Figure 9.7

The body of this table shows the possible combinations of genes. Recalling that red is dominant over the recessive white character (one or more red genes leads to a red-flowered offspring), find:

*27. $P(\text{red})$ ***28.** $P(\text{white})$

Mendel found no dominance in snapdragons, with one red gene and one white producing pink-flowered offspring. These second generation pinks, however, still carry one red and one white gene, and when they are crossed, the next generation still yields the Punnet square, Figure 9.7. Find:

*29. $P(\text{red})$ ***30.** $P(\text{pink})$ ***31.** $P(\text{white})$

32. Mendel verified these probability ratios experimentally with large numbers of observations, and did the same for many character units other than flower color. The importance of his work, published in 1865, was not recognized until 1900. Explain the connection between the Punnet square (Figure 9.7) and the product table for tossing two fair coins (Figure 9.2, page 346).

33. Construct a product table for the experiment of rolling two fair dice, but in place of each of the 36 number pairs write the sum of the two numbers in the pair.

34. How many different sums are possible when rolling two dice?

Find the probabilities of the following sums:

35. 1	**38.** 4	**41.** 7	**44.** 10
36. 2	**39.** 5	**42.** 8	**45.** 11
37. 3	**40.** 6	**43.** 9	**46.** 12

47. For the experiment of rolling two dice, let success be associated with the sum 3. Then failure would mean a sum "not-3." If the event of success is denoted by E, how would you denote the event of failure? We see that *negation* of a statement in logic (see Section 2.2) corresponds to the *complement* of a set in set theory (see Section 1.2) and to the *failure* of an event to occur in probability theory.

48. If S is the sample space for an experiment, find $P(S)$.

Perhaps the real importance of coin-tossing, die-rolling, and similar games of chance is that they suggest real life situations. If we assume that a newborn baby is as likely to be a girl as a boy (which is actually not quite true), then we can perform a repeated coin-tossing experiment which will imitate, or **simulate,** an actual series of births. To find out, for example, how often in a series of 30 births two consecutive babies will be girls, we toss a coin 30 times, and a record *g* (girl) for each head and *b* (boy) for each tail:

bbggg   bbbgg   bgggg   gbbgg   ggbbb   bbggg

49. How many pairs of consecutive girls occurred?

*50. In a series of 30 tosses, there are 29 consecutive pairs of tosses. Based on the above series, what is the empirical probability of two consecutive babies both being girls? Give your answer in decimal form, to two places.

51. Find the classical probability (using the product rule) that two consecutive babies will be both girls. How does it compare with the experimental value of Exercise 50?

52. In the pattern of *b*'s and *g*'s above, how many consecutive "triples" (three in a row) of *tosses* are there?

53. How many of the triples are all boys?

54. Find the empirical probability, based on the pattern above, that three consecutive babies will all be boys.

*55. Toss a coin 30 times to extend the pattern above to 60 babies. Based on all 60, find a new empirical probability that two consecutive babies will both be girls. Is this value closer to the classical value than your answer to Exercise 50?

56. The tossing of two fair dice can be simulated by spinning the small pair of spinners shown here. Explain how the single larger spinner can simulate the same thing (provided the sum is all we are interested in on each roll).

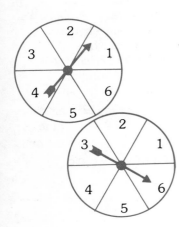

Of the 360° in the total circle of the large spinner below, find the angles associated with the regions for these numbers:

57. 5   **58.** 7   **59.** 10   **60.** 12

61. If there are actually 105 boys born for every 100 girls, explain how a spinner could be made to simulate births more accurately than coin-tossing.

*62. Explain how a fair die could be used to simulate the advice of a stock broker who has a record of advising wisely two out of three times on the average.

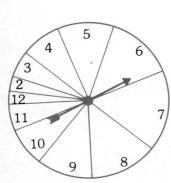

Roll a die 50 times and find an empirical probability that this stock broker will advise:

*63. wisely three times in a row   ***64.** unwisely three times in a row

9.3 MORE ON COMPOSITE EVENTS

Consider the experiment of randomly selecting one number from the set

$$S = \{1, 2, 3, 4, 5\},$$

and let a and b denote logical statements, as follows:

  a: The outcome is greater than 3.

  b: The outcome is odd.

If A is the event that a is satisfied and B is the event that b is satisfied, then the situation can be pictured in a Venn diagram (Figure 9.8).

Now, the logical condition "greater than 3 or odd" (symbolized $a \vee b$) is satisfied by the numbers 1, 3, 4, and 5. The diagram shows that these are the outcomes that make up the event (set) $A \cup B$. Likewise, the logical condition "greater than 3 and odd" ($a \wedge b$) is satisfied only by the number 5, and corresponds to the event $A \cap B$. The Venn diagram shows that $n(A) = 2$, $n(B) = 3$, $n(A \cap B) = 1$, and $n(A \cup B) = 4$, so the following relationship from set theory is satisfied:

$$n(A \cup B) = n(A) + n(B) - n(A \cap B).$$

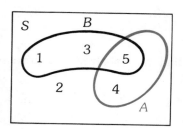

Figure 9.8

In words, we have the following:

> **Addition rule for counting**
> To find the number of ways that at least one
> (that is, one or the other or both) of two events could occur,
> add the number of ways the first could occur
> to the number of ways the second could occur,
> and then subtract the number of ways they could both occur.

Example 1 When a single fair die is rolled, how many ways could the outcome be "even or a multiple of 3"?

Solution. Let E be the event "even," and M the event "multiple of 3." Then $E = \{2, 4, 6\}$, $M = \{3, 6\}$, and $E \cap M = \{6\}$. So the number of outcomes that would be one or the other is

$$\begin{aligned} n(E \cup M) &= n(E) + n(M) - n(E \cap M) \\ &= 3 + 2 - 1 \\ &= 4. \end{aligned}$$

Specifically, the four outcomes are 2, 3, 4, 6.

Just as in the case of products (in Section 9.2), the addition rule of counting gives us a corresponding probability rule (see Exercises 22 and 23):

> **Addition rule for probability**
> If A and B are any two events for the same experiment,
> then $P(A \cup B) = P(A) + P(B) - P(A \cap B)$.

Example 2   If a single card is drawn from an ordinary deck, find the probability that it will be "red or a face card."

Solution.   Let R and F be the events "red" and "face card," respectively. Then

$$P(R) = \frac{26}{52}, \qquad P(F) = \frac{12}{52}, \qquad \text{and} \qquad P(R \cap F) = \frac{6}{52}$$

(there are 6 red face cards in the deck: the king, queen, and jack of hearts; the king, queen, and jack of diamonds), so

$$P(R \cup F) = P(R) + P(F) - P(R \cap F)$$
$$= \frac{26}{52} + \frac{12}{52} - \frac{6}{52} = \frac{32}{52} = \frac{8}{13}.$$

Example 3   When two fair dice are rolled, find the probability that either "the first die shows 2" or "the sum is 6 or 7."

Solution.   The sample space is shown in Figure 9.9. The two events, labeled A and B, are shaded.

Figure 9.9   Event A is "the first die shows 2" and event B is "the sum is 6 or 7."

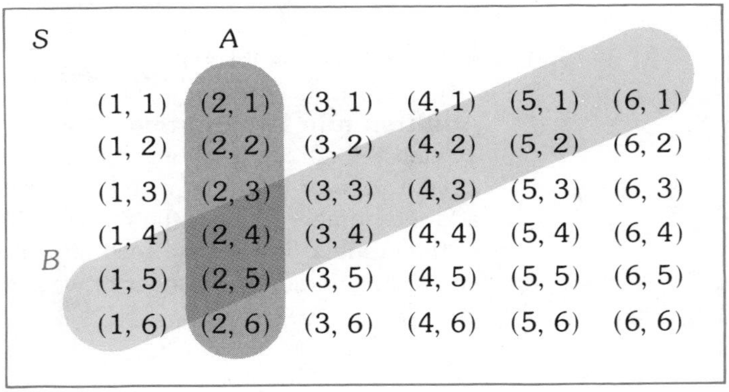

Since $P(A) = \frac{6}{36}$, $P(B) = \frac{11}{36}$, and $P(A \cap B) = \frac{2}{36}$, we get

$$P(A \cup B) = \frac{6}{36} + \frac{11}{36} - \frac{2}{36} = \frac{15}{36} = \frac{5}{12}.$$

Example 4   Of the 20 television programs to be aired this evening, Marc plans to watch one, which he will pick at random by throwing a dart at the TV schedule. If 8 of the programs are educational, 9 are interesting, and 5 are both educational and interesting; find the probability that the show he watches will have at least one of these attributes.

Solution.   If E denotes "educational," and I denotes "interesting," then $P(E) = \frac{8}{20}$, $P(I) = \frac{9}{20}$, and $P(E \cap I) = \frac{5}{20}$. So

$$P(E \cup I) = \frac{8}{20} + \frac{9}{20} - \frac{5}{20} = \frac{12}{20} = \frac{3}{5}.$$

If two events cannot happen at the same time, they are called **mutually exclusive.** For example, rolling a die cannot produce both "odd" and "even." The equivalent idea in set theory is that of *disjoint sets* (that is, sets whose intersection is the empty set). Two events are mutually exclusive whenever their representations as sets are disjoint. In general, if a set of events has the property that no *two* events can occur simultaneously, the events are called "pairwise mutually exclusive." The addition rule of probability is most useful in this case, and can be written as follows:

Addition rule for mutually exclusive events
If $E_1, E_2, \ldots, E_n$ are n pairwise mutually exclusive events, then
$$P(E_1 \cup E_2 \cup \cdots \cup E_n) = P(E_1) + P(E_2) + \cdots + P(E_n).$$

The probability that any one of the events will occur is found by adding the probabilities of the individual events.

Example 5 Alice figures there is only a slight chance she can finish her homework in just 1 hour tonight. In fact, she assigns probabilities to the various numbers of hours as shown in the table of Figure 9.10. Find the probability that it will take her: (a) fewer than 3 hours; (b) more than 2 hours; (c) more than 1 but no more than 5 hours.

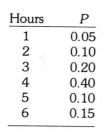

Hours	P
1	0.05
2	0.10
3	0.20
4	0.40
5	0.10
6	0.15

Figure 9.10

(a) "fewer than 3" means 1 or 2. So $P(\text{fewer than } 3) = 0.05 + 0.10 = 0.15$.

(b) $P(\text{more than } 2) = P(3) + P(4) + P(5) + P(6)$
$= 0.20 + 0.40 + 0.10 + 0.15 = 0.85$

(c) $P(\text{more than 1 but no more than } 5) = P(2) + P(3) + P(4) + P(5)$
$= 0.10 + 0.20 + 0.40 + 0.10 = 0.80.$

The following example shows how the addition rule of this section and the product rule of the last section are sometimes both necessary to find a single probability.

Example 6 Let two cards be drawn from a standard deck *with replacement* (the first card is shuffled back into the deck before the second is drawn). Find the probability of the event E that one card is a 5 and one is a face card.

Solution. There was nothing said here about order, so there are two (mutually exclusive) ways to succeed.

A: 5 and then face card; *B:* face card and then 5

By the product rule,
$$P(A) = \frac{4}{52} \cdot \frac{12}{52} = \frac{3}{169} \quad \text{and} \quad P(B) = \frac{12}{52} \cdot \frac{4}{52} = \frac{3}{169}.$$

So
$$P(E) = P(A) + P(B) = \frac{3}{169} + \frac{3}{169} = \frac{6}{169}.$$

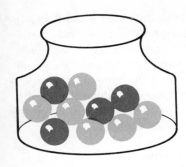

Example 7   Suppose 4 balls are drawn at random (with replacement) from the jar shown here. Find: (a) the probability of the event E of getting 3 colored and 1 white;   (b) the odds against event E.

(a) Since the 1 white ball could occur on any of the four draws, we have 4 ways to succeed:

E_1:   white — colored — colored — colored

E_2:   colored — white — colored — colored

E_3:   colored — colored — white — colored

E_4:   colored — colored — colored — white

Also, $P(E_1) = \dfrac{6}{10} \cdot \dfrac{4}{10} \cdot \dfrac{4}{10} \cdot \dfrac{4}{10} = \dfrac{384}{10000} = \dfrac{24}{625}$

$P(E_2) = \dfrac{4}{10} \cdot \dfrac{6}{10} \cdot \dfrac{4}{10} \cdot \dfrac{4}{10} = \dfrac{24}{625}$

$P(E_3) = \dfrac{4}{10} \cdot \dfrac{4}{10} \cdot \dfrac{6}{10} \cdot \dfrac{4}{10} = \dfrac{24}{625}$

$P(E_4) = \dfrac{4}{10} \cdot \dfrac{4}{10} \cdot \dfrac{4}{10} \cdot \dfrac{6}{10} = \dfrac{24}{625}$

and so $P(E) = P(E_1) + P(E_2) + P(E_3) + P(E_4) = 4 \cdot \dfrac{24}{625} = \dfrac{96}{625}.$

(b)   Odds against E are $(625 - 96)$ to 96, or 529 to 96.

9.3   EXERCISES

If a single card is drawn from an ordinary deck, find the number of ways of getting

1.   ace

3.   black ace

2.   black card

4.   card that is black or an ace

Find the probability that the card drawn is

5.   club

7.   6 and a club

6.   6

8.   6 or a club

Find the odds in favor of its being

9.   black or an ace

10.   6 or a club

When two fair dice are rolled, find the probability that the resulting sum is

11.   greater than 7

13.   even and greater than 7

12.   even

14.   even or greater than 7

below 60	0.05
60–64	0.08
65–69	0.15
70–74	0.28
75–79	0.22
80–84	0.08
85–89	0.06
90–94	0.04
95–99	0.02
100 or over	0.02

The table gives a certain golfer's probabilities of scoring in various ranges on a par-70 course. In a given round, find the probability that his score will be:

15. 90 or higher   **18.** not in the 60's

16. below par   **19.** not in the 60's, 70's or 80's

17. in the 70's

20. What are the odds against his shooting below par?

*21. Prove that if $E_1, E_2, \ldots, E_n$ are pairwise mutually exclusive and if their union is the whole sample space, then the sum of their individual probabilities is 1.

For a certain experiment, let A be the set of all outcomes that satisfy the logical statement a, and B the set of all outcomes that satisfy the logical statement b, and suppose $A \cap B = \emptyset$.

*22. What expression using sets and set operations is equivalent to the logical expression "a or b"?

*23. What expression using probabilities and arithmetic operations is equivalent to the logical expression "a or b"?

If you are dealt two cards successively (without replacement) from a standard deck, find the probability of getting

*24. two red cards   ***26.** one card of each color

*25. red on the first; black on the second

If you are dealt a 5-card poker hand (this implies *without replacement*) from a standard deck, find the probability of getting

*27. a heart flush (all hearts)   ***29.** a flush

*28. a red flush   ***30.** three clubs and two hearts

31. Toss a pair of dice 50 times keeping track of the number of times the sum is "even or greater than 7" (that is, 2, 4, 6, 8, 9, 10, 11, or 12). Calculate an empirical probability for this event, and compare with the classical answer in Exercise 14.

Mendel's experiments in genetics (see Section 9.2 Exercises) established that when two pea plants, each carrying a gene for white and a gene for red, are crossed, their offspring have a probability of $\frac{1}{4}$ of having white flowers. (Recall that red is dominant, white is recessive.) Any other random phenomenon with a recognizable event having probability $\frac{1}{4}$ can be used to simulate the pea plant genetic process. For example, if two fair coins are tossed the probability of two heads is $\frac{1}{4}$. Hence, we can repeatedly toss two coins, where two heads means white (W) and anything else means red (R). (Exercises 32–37)

32. Toss a pair of coins forty times, recording a sequence of R's and W's. Is the number of W's close to 10?

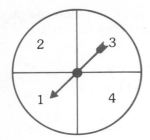

33. Estimate from your experiment the probability that any two plants observed will both have white flowers.

34. Estimate the probability that five successive plants will be observed to have red flowers.

35. Explain how flower color reproduction in pea plants could be simulated with a deck of cards.

36. If a fair die is tossed and 5's and 6's are ignored, then each of the other four numbers are still equally likely. Explain how to use a die to simulate the genetic process in pea plants.

37. Explain how to simulate flower color reproduction in peas with the spinner shown here (which is a kind of roulette wheel).

9.4 CONDITIONAL PROBABILITY

If two cards are drawn successively from a standard deck *with* replacement, then the probability of a spade in the second draw is $\frac{13}{52}$ or $\frac{1}{4}$, regardless of the result on the first draw. But if the drawing is done *without* replacement, then the probability of a spade on the second draw depends on what happened on the first. It will be $\frac{13}{51}$ if the first was not a spade, and $\frac{12}{51}$ if the first was a spade.

When two events are related such that knowing whether or not one of them occurred has no effect on the probability of the other one, then we call the events **independent.** When two or more events are all mutually independent, the probability that they all will occur is found by the product

Monte Carlo *is one of the three cities in Monaco, a principality ruled by Prince Rainier III and Princess Grace (the former Grace Kelly of the United States). The stamps show Monaco's coastline on the Mediterranean, a famous resort area. Its three other sides are bounded by France.*

The equally famous gambling casino has been in operation since 1866. François Blanc (pictured above) was the first to own the gambling concession. A private company later held it; in 1967 Rainier took control by buying up the shares owned by Aristotle Onassis.

Movie images of the casino give us the faces of roulette players— tense, hopeful, doomed, triumphant, mad—and the stony faced croupier who rakes in the money with irreversible gesture. Everything is random action. It was fitting to code secret work on nuclear reactions with "Monte Carlo." (See page 386.)

rule of Section 9.2. If two cards are drawn from a standard deck with replacement, the probability they will both be spades can be found by defining these two events:

$$E_1: \text{spade on first draw} \qquad E_2: \text{spade on second draw}$$

Then

$$P(E_1 \cap E_2) = P(E_1) \cdot P(E_2) = \frac{13}{52} \cdot \frac{13}{52} = \frac{169}{2704} = .0625.$$

However, if the drawing is done without replacement, then the second probability is computed on the assumption that the first event has occurred. The notation $P(E_2|E_1)$ is read "the probability of E_2, given E_1." We then have

$$P(E_1 \cap E_2) = P(E_1) \cdot P(E_2|E_1) = \frac{13}{52} \cdot \frac{12}{51} = \frac{156}{2652} = .0588.$$

A probability of the form $P(E_2|E_1)$ is called a **conditional probability** since it is the probability of one event, assuming the "condition" that the other event occurs. This idea plays an important role whenever we have partial information about a situation, that is, when we have some but not all of the relevant facts. There are many everyday situations where this is the case. An insurance company, when setting its premium rates, cannot predict when a given individual will die, but having certain information will certainly improve its accuracy. Age is an important condition. For example, by studying past records, we have approximately

$$P(\text{a man will die this year}|\text{he is } 21) = 0.001,$$

whereas

$$P(\text{a man will die this year}|\text{he is } 66) = 0.025.$$

Age and likelihood of dying are clearly not independent conditions.

Conditional probabilities are often found by rearranging the formula given above:

$$P(E_2|E_1) = \frac{P(E_1 \cap E_2)}{P(E_1)}.$$

Example 1 Two fair coins were tossed, and it is known that at least one was heads. Find the probability they were both heads.

Solution. The sample space has four equally likely outcomes: $S = \{hh, ht, th, tt\}$. We define two events:

$$E_1: \text{at least one head} \qquad E_2: \text{both heads.}$$

By the formula,

$$P(E_2|E_1) = \frac{P(E_1 \cap E_2)}{P(E_1)} = \frac{1/4}{3/4} = \frac{1}{3}.$$

Conditional probabilities can often be reasoned without any special formulas.

Example 2   If a single fair die is rolled, find the probability that the number will be: (a) odd, given that it was between 2 and 5;   (b) even, given that it was greater than 5;   (c) a 7, given that it was odd.

(a) Since there are two equally likely possibilities between 2 and 5, namely 3 and 4, and one of them is odd, we have

$$P(\text{odd}|\text{between 2 and 5}) = \frac{1}{2}.$$

(b) The only possibility greater than 5 is 6, and it is even, so

$$P(\text{even}|\text{greater than 5}) = 1.$$

(c) A 7 is impossible under any condition, so $P(7|\text{odd}) = 0$.

The following example shows how a tree diagram can be used in answering several kinds of questions about composite events.

Example 3   The local garage employs two mechanics, A and B, but the service manager never lets you know which one worked on your car. All you know is the following set of facts, discovered by a wily newspaper reporter: A does twice as many jobs as B, A does a satisfactory job three out of four times, and B does satisfactorily on only two out of five jobs. Suppose you take your car in for work. Find the probabilities of the following events:
(a) Mechanic A does the work with satisfactory results.
(b) Mechanic B does the work with satisfactory results.
(c) The results are satisfactory.
(d) The results are unsatisfactory.
(e) Either A does the work or the results are satisfactory (or both).
(f) B did the work, given that the results are unsatisfactory.
(g) A did the work, given that the results are satisfactory.

Solution.   We first construct a tree diagram to illustrate this composite experiment. (See Figure 9.11.) Since A does twice as many jobs as B, the (unconditional) probabilities of A and B are, respectively, $\frac{2}{3}$ and $\frac{1}{3}$, as shown on the first stage of the tree. The second stage of the tree shows four different conditional probabilities. The job ratings are S (satisfactory) and U (unsatisfactory). For example, along the branch from B to S we have

$$P(S|B) = \tfrac{2}{5}$$

since mechanic B does a satisfactory job on two out of five attempts.

Each of the four composite branches in the tree is numbered, and its probability is given on the right.

(a) The event A $\cap$ S is associated with branch 1, so $P(A \cap S) = \tfrac{1}{2}$.

(b) The event B $\cap$ S is shown on branch 3. $P(B \cap S) = \tfrac{2}{15}$.

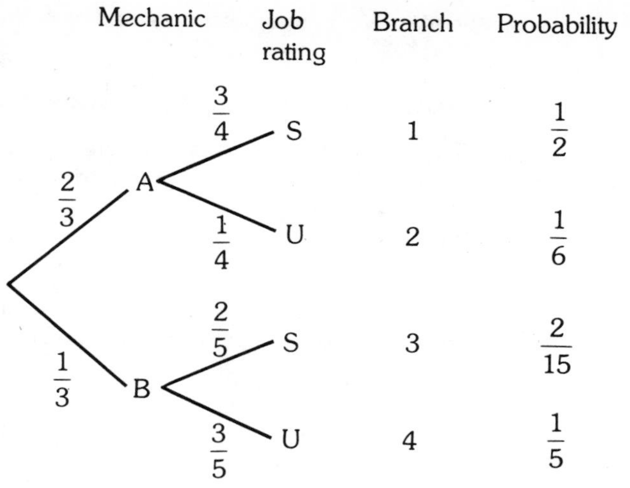

Figure 9.11 Garage mechanics experiment

Mechanic	Job rating	Branch	Probability

(c) The event S combines branches 1 and 3, so $P(S) = \frac{1}{2} + \frac{2}{15} = \frac{19}{30}$.

(d) The event U combines branches 2 and 4, so $P(U) = \frac{1}{6} + \frac{1}{5} = \frac{11}{30}$.

(e) Event A combines branches 1 and 2, while event S combines branches 1 and 3. Thus we use branches 1, 2, and 3. $P(A \cup S) = \frac{1}{2} + \frac{1}{6} + \frac{2}{15} = \frac{4}{5}$.

(f) Here we need the conditional probability

$$P(B|U) = \frac{P(B \cap U)}{P(U)}$$

B ∩ U is branch 4, while U combines branches 2 and 4. Hence

$$P(B|U) = \frac{\frac{1}{5}}{\frac{1}{6} + \frac{1}{5}} = \frac{6}{11}.$$

(g) $P(A|S) = \dfrac{P(A \cap S)}{P(S)} = \dfrac{\frac{1}{2}}{\frac{1}{2} + \frac{2}{15}} = \dfrac{15}{19}$.

Example 4 Twenty single people live in an apartment house, including eight men and twelve women. Three of the men and eight of the women have full-time jobs. All of the others are full-time college students. If one of the residents is chosen at random, find the probability it will be: (a) a woman; (b) a college student; (c) a woman student; (d) a man, given it is a worker; (e) a student, given it is a woman.

The table below classifies the 20 people, giving totals for both rows and both columns.

	College student (C)	Person with job (J)	
Man (M)	5	3	8
Woman (W)	4	8	12
	9	11	20

Solution.   We count outcomes and use the classical probability formula:

(a) $P(W) = \dfrac{12}{20} = \dfrac{3}{5}$ **(b)** $P(C) = \dfrac{9}{20}$

(c) The favorable outcomes here are shown in the box which is the intersection of the row for women and the column for college students.

$$P(W \cap C) = \dfrac{4}{20} = \dfrac{1}{5}$$

(d) The given condition here, serves to reduce the sample space only to those (11) people who have full-time jobs. Of these, the table shows that 3 are men.

$$P(M|J) = \dfrac{3}{11} \qquad\qquad \textbf{(e)}\ \ P(C|W) = \dfrac{4}{12} = \dfrac{1}{3}$$

9.4 EXERCISES

If a single fair die is rolled, find the probability the number is:

  1.   2, given that it was odd   **3.**   even, given that it was a 6

  2.   4, given that it was *even*

If two fair dice are rolled, find the probability the result is (recall the 36-outcome sample space):

  4.   a sum of 8, given the sum was greater than 7

  5.   a sum of 6, given the roll was a "double" (two identical numbers)

  6.   a double, given that the sum was 9

If two cards are drawn without replacement from a standard deck, find the probability that:

  7.   the second is a heart, given that the first is a heart

  8.   they are both hearts

  9.   the second is black, given that the first is a spade

10.   the second is a face card, given that the first is a jack

*11.   the first is a king, given that the second is a king

*12.   the second is a queen, given that the first is a face card

*13.   the first is a face card, given that the second is a queen

*14.   the second is a spade, given that the first is black

If five cards are drawn without replacement from a standard deck, find the probability that *all* the cards are:

15.   diamonds

16.   diamonds, given that the first and second were diamonds

17.   diamonds, given that the first four were diamonds

18. clubs, given that the third was a spade

*19. hearts, given that the first three were red   20. the same suit

21. Both of a certain pea plant's parents had a gene for red and a gene for white flowers. (See Section 9.2, Exercises 27–31.) If the offspring has red flowers, find the probability that it combined a gene for red and a gene for white (rather than two for red or two for white).

Assume that girl and boy babies are equally likely. Fill in the remaining probabilities on the three-stage tree diagram below, and use the information to find the probability that a family with three children has all girls, given that:

22. the first is a girl   25. at least two are girls

23. the third is a girl   26. at least one is a girl

24. the second is a girl

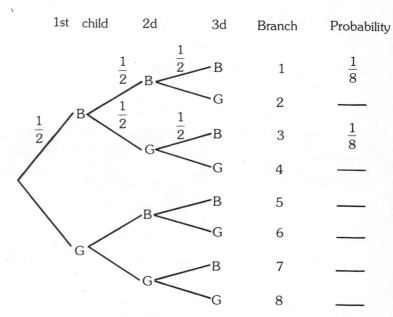

1st child	2d	3d	Branch	Probability

A bicycle factory runs two assembly lines, say A and B. If 95% of line A's products pass inspection, while only 90% of line B's products pass inspection, and 60% of the factory's bikes come off assembly line B (the rest off A), find the probability that one of the factory's bikes which did not pass inspection came off

27. assembly line A   28. assembly line B

29. A box contains 6 red and 4 white balls. If three are drawn out at random with replacement, find the probability that all are red, given that the third is red.

30. Repeat Exercise 29 if the drawing is done without replacement.

*31. Explain how the random phenomenon of Exercise 30 can be simulated with playing cards.

*32. Carry out the simulation experiment you described (using at least 50 trials), and find an empirical probability of "all red," given "third red."

*33. Describe and use a simulation scheme to find an empirical value for the probability of Example 4(a).

A pet shop has 10 puppies, 6 of them males. There are 3 beagles (1 of them a male), 1 cocker spaniel (a male), and the rest are poodles. Construct a table similar to the one in Example 4, and find the probability that one of these puppies, chosen at random, will be:

34. a beagle

35. a beagle, given that it is a male

36. a male, given that it is a beagle

37. a cocker spaniel, given that it is a female

38. a poodle, given that it is a male

39. a female, given that it is a beagle

40. Recall that, in general, $P(A) + P(A') = 1$. Comparing the answers to Exercises 36 and 39 above, do you think that conditional probabilities obey this same relationship? (That is, $P(A|B) + P(A'|B) = 1$.)

Big deal for whom? *What do you think about state lotteries? Are the ticket sellers selling dreams? If so, what do you think a dream is worth?*

9.5 EXPECTED VALUES AND SIMULATION WITH RANDOM NUMBERS

Hours	P
1	0.05
2	0.10
3	0.20
4	0.40
5	0.10
6	0.15

We repeat Figure 9.10 from Example 5, Section 9.3, which shows the probabilities assigned by Alice to the numbers of hours spent on homework on a given night.

Notice that each of the possible values for hours is paired with a probability value, and that the six probability values add up to 1. This is an example of a *probability distribution*. If Alice's friend Fred asks her how many hours her studies will take, what would be her best guess? Six different time values are possible, some more likely than others. Alice could calculate a "weighted average" by multiplying each possible value by its probability and adding the six products:

$$1(0.05) + 2(0.10) + 3(0.20) + 4(0.40) + 5(0.10) + 6(0.15)$$
$$= 0.05 + 0.20 + 0.60 + 1.60 + 0.50 + 0.90 = 3.85.$$

Thus 3.85 hours is the *mathematical expectation* (or **expected value**) of the quantity of time to be spent. In general, if a given quantity x can have any of the values $x_1, x_2, x_3, \ldots, x_n$, and the corresponding probabilities of these values occurring are $P(x_1), P(x_2), P(x_3), \ldots, P(x_n)$, then the expected value of the quantity x is defined as

$$x_1 \cdot P(x_1) + x_2 \cdot P(x_2) + x_3 \cdot P(x_3) + \cdots + x_n \cdot P(x_n).$$

Example 1 Find the expected number of boys for a three-child family (that is, the expected value of the number of boys).

Solution. The sample space for this experiment is

$$S = \{ggg, ggb, gbg, bgg, gbb, bgb, bbg, bbb\}.$$

The probability distribution is shown in Figure 9.12, along with the products and their sum, which gives the expected value. The expected number of boys is $\frac{12}{8}$, or 1.5.

Notice that the expected value for the number of boys in the family is itself an impossible value. So, in general, we don't necessarily *expect* the *expected value* ever to occur. Many times, as in the above example, the expected value actually cannot occur. It is only a kind of average of the various values that could occur. (See Section 10.2 for a more detailed discussion of averages.) If we recorded the number of boys in lots of different three-child families, then by the Law of Large Numbers the greater the number of families observed, the surer we could be that the observed average would be close to the expected value.

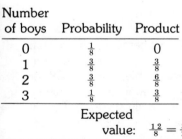

Number of boys	Probability	Product
0	$\frac{1}{8}$	0
1	$\frac{3}{8}$	$\frac{3}{8}$
2	$\frac{3}{8}$	$\frac{6}{8}$
3	$\frac{1}{8}$	$\frac{3}{8}$
	Expected value:	$\frac{12}{8} = \frac{3}{2}$

Figure 9.12

The idea of expected value is important in analyzing some games, as illustrated in Examples 2 through 6.

Example 2 A player pays $1 for the privilege of rolling a single fair die once. If he rolls a 6, he receives a "payoff" of $5 (*net winnings* would be $4, which is payoff minus cost to play). If he rolls anything other than 6, he receives nothing back (*net winnings*: payoff minus cost is $0 - \$1 = -\$1$). Find the expected net winnings of this game.

Solution.

Number rolled	Payoff	Net winnings	Probability	Product
6	$5	$4	$\frac{1}{6}$	$\$\frac{4}{6}$
1–5	$0	−$1	$\frac{5}{6}$	$-\$\frac{5}{6}$
			Expected value:	$-\$\frac{1}{6}$

The expected net winnings are

$$\frac{-\$1}{6} = -17\cancel{c} \quad \text{(approximately)}.$$

The player would not lose 17¢ on any single play. But, if he plays this game repeatedly, then in the long run, he should expect to lose about 17¢ per play *on the average.* If he played, say 100 times, he would win sometimes and lose sometimes, but should come out in the end about $100 \cdot (17\cancel{c}) = \$17$ in the hole.

A game in which the expected net winnings is zero is called a **fair game.** The game in Example 2 has negative expected net winnings, and so is unfair against the player. A game with positive expected net winnings is unfair in favor of the player.

Example 3   A game that costs $1 per play involves rolling a single fair die and pays off as follows: $3 for a 6, $2 for a 5, $1 for a 4, no payoff otherwise. Is this game fair?

Solution.

Number rolled	Payoff	Net winnings	Probability	Product
1–3	$0	−$1	$\frac{3}{6}$	−$\frac{3}{6}$
4	$1	$0	$\frac{1}{6}$	$0
5	$2	$1	$\frac{1}{6}$	$\frac{1}{6}$
6	$3	$2	$\frac{1}{6}$	$\frac{2}{6}$
			Expected value:	$0

Since the expected net winnings are $0, the game is fair.

Example 4   *Numbers* is an ever popular (but illegal) game. In one version of this game, the player selects a three-digit number and gives it to the numbers runner, with a dollar. Any three digits are acceptable, such as 008, or 041. A winning number is chosen by the game operators in some highly honest manner, such as the last three digits of the daily volume of stock transactions on Wall Street. Anyone who selected the correct number receives a payoff of $700. Find the expected net winnings for the player.

Solution.   The probability of selecting all three digits correctly is

$$\frac{1}{10} \cdot \frac{1}{10} \cdot \frac{1}{10} = \frac{1}{1000}, \quad \text{and incorrectly is} \quad 1 - \frac{1}{1000} = \frac{999}{1000}.$$

So the expected net winnings are

$$(\$699)\frac{1}{1000} + (-\$1)\frac{999}{1000} = -\$0.30.$$

The expected *loss* is 30¢ per play, which adds up to a fantastic profit for those collecting the money.

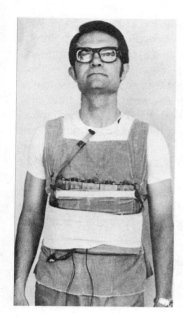

Games in a gambling casino are usually set up so that they are (at least slightly) unfair in favor of the house.

Example 5 One simple type of *roulette* is played with an ivory ball and a wheel set in motion. The wheel contains thirty-eight compartments. Eighteen of the compartments are black, eighteen red, one is labeled "zero," and one is labeled "double zero." (These last two are neither black nor red.) The player places $1 on either red or black. If the player picks the correct color of the compartment in which the ball finally lands, the payoff is $2; otherwise the payoff is zero. Find the expected net winnings.

Solution. By the expected value formula, expected net winnings are

$$(\$1)\frac{18}{38} + (-\$1)\frac{20}{38} = -\$\frac{1}{19}.$$

The expected loss here is $\$\frac{1}{19}$, or about 5.3¢, per play.

Example 6 In *keno*, the house has a pot containing 80 balls, numbered 1 through 80. In a simple version of keno, a player buys a ticket for $1 and marks one number on it (from 1 to 80). The management then selects 20 numbers at random. If the number selected by the player is among the 20 selected by the management, the player is paid $3.20. Find the expected net winnings.

Solution.
$$(\$2.20)\frac{20}{80} + (-\$1.00)\frac{60}{80} = -\$0.20,$$

so the expected loss is 20¢ per play.

A time to lose *This table comes from an article by Andrew Sterrett in* The Mathematics Teacher, *March, 1967. Sterrett discusses rules for some games, and calculates their expected values. He uses expected values to find expected times it would take to lose $1000 if you played continually at the rate of $1 per play and one play per minute.*

Game	Expected value	Days	Hours	Minutes
Roulette (with one 0)	−$0.027	25	16	40
Roulette (with 0 and 00)	−$0.053	13	4	40
Chuck-a-luck	−$0.079	8	19	46
Keno (one number)	−$0.200	3	11	20
Numbers	−$0.300	2	7	33
Football Pool (4 winners)	−$0.375	1	20	27
Football Pool (10 winners)	−$0.658	1	1	19

Most of the exercise sets of this chapter have included some exercises in **simulation,** where one random phenomenon is studied by actually observing another one with similar characteristics. But in every case, the simulation technique utilized something like tossing coins, rolling dice, drawing cards, or spinning pointers (generalized roulette wheels). In order to get accurate probabilities by simulation, we need to perform large numbers of repetitions of these experiments, so large, in fact, that the method becomes totally impractical for many kinds of problems. Here we introduce *random numbers* as a partial solution to this problem. (Even so, simulation would not enjoy its widespread application today, except for the development of computers.)

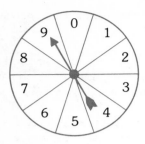

Figure 9.13

Figure 9.14
Random numbers
Group 1

51592	94928
77876	15709
36500	39922
40571	96365
04822	14655
53033	65587
92080	76905
01587	12369
36006	54219
63698	89329
17297	90060
22841	06975
91979	05050
96480	69774
74949	78351
76896	11464
47588	84086
45521	51497
02472	12307
55184	68009
40177	39687
84861	45062
86937	43752
20931	05477
22454	28636
73219	44070
55707	77653
48007	39931
65191	04744
06772	

A **random number** is one that has been selected in some random way. Using a spinner, like the one in Figure 9.13, we could spin the pointer a large number of times and record the results to get a list of random numbers, as in Figure 9.14. Actually, even the numbers in Figure 9.14 were obtained (more efficiently) on a computer, by a process that could be repeated to produce identical results (the same numbers in the same order) unlike the spinner. (These numbers should really be called "pseudo-random" numbers.) At any rate, they pass the generally accepted tests of randomness and give us the following advantage: we can read a sequence of numbers from the table to represent the actual experiment, rather than actually spinning a pointer, and so on.

Example 7   Assuming boy and girl babies are equally likely, use random number simulation to estimate the probability that a five-child family will have all boys.

Solution.   We use Group 1 in Figure 9.14, taking an odd digit to represent a boy and an even digit to represent a girl. Of the fifty-nine families represented by Group 1, only two have all boys (91979 and 39931), so our estimated probability is $\frac{2}{59} = 0.034$. Compare this with the classical probability: $\frac{1}{2} \cdot \frac{1}{2} \cdot \frac{1}{2} \cdot \frac{1}{2} \cdot \frac{1}{2} = \frac{1}{32} = 0.031$.

Example 8   Use simulation with random numbers to estimate the probability that two cards drawn from a standard deck with replacement will both be of the same suit.

Solution.   Here, we go down the columns in Group 2, reading only the first digit of each entry and using the following scheme: 0 and 1 mean clubs; 2 and 3 mean diamonds; 4 and 5 mean hearts; 6 and 7 mean spades; 8 and 9 are disregarded. The sequence 0–3–1–2–7–9– and so on gives us club–diamond–club–diamond–spade–and so on. Of the fifty-eight successive pairs we get using the first digits from the second group, twelve of them give both the same suit (check this for yourself). So, the estimated probability is $\frac{12}{58} = 0.21$. (For comparison, the classical probability is 0.25.)

9.5   EXERCISES

1.   If a single fair die is rolled once, what is the expected number to come up?

Suppose you pay $1 to roll a fair die with the understanding that you will get back $3 for a 1 or a 6, nothing otherwise.

2.   What are your expected net winnings?   **3.**   Is this game fair?

4.   A certain game involves tossing 3 fair coins, and pays 10¢ for 3 heads, 5¢ for 2 heads, and 3¢ for 1 head. Is 5¢ a fair price to pay to play this game? (That is, does the 5¢ cost to play make the game fair?)

5. In a form of roulette slightly different than in Example 5 above, a more generous management supplies a wheel having only thirty-seven compartments, with eighteen red, eighteen black, and one zero. Find the expected net winnings if you bet on red in this game.

6. Suppose you buy one of 1000 tickets at 10¢ each in a lottery where the prize is to be $50. What are your expected *net* winnings?

7. If 5 apples in a barrel of 25 apples are known to be rotten, what is the expected number of rotten apples in a sample of 2 (drawn without replacement)?

8. If two cards are drawn at one time from a standard deck of 52, what is the expected number of diamonds?

Figure 9.14
Random numbers
Group 2

05409	55741
36431	31249
16309	82319
28158	09359
70116	78731
98859	18804
88676	58230
93663	38927
24654	04606
25989	42553
29748	55470
01089	06250
25875	01132
41199	51403
05575	39635
73132	20344
63242	09843
63906	73378
14355	44437
39661	39674
89610	42846
54344	81641
56274	84131
42679	88906
17567	52109
65549	58436
65723	86333
20246	88249
87395	37991
31033	

9. Suppose someone offers to pay you $5 if you draw two diamonds in the experiment of Exercise 8. He says that you should pay 50¢ for the chance to try. Is this a fair game?

10. Jack must choose at age 20 either to inherit $25,000 at age 30 if he is alive then, or $30,000 at age 35 if he is alive then. If the probabilities for a person aged 20 to live to be 30 and 35 are .9 and .7 respectively, which choice should he make to secure the largest expected inheritance?

11. An insurance company has written one hundred policies of $10,000, five hundred policies of $5,000, and one thousand policies of $1,000, all on people of age 20. If experience shows that the probability of dying during age 20 is .001, how much can the company expect to pay out during the year?

12. A builder is considering a job which promises a profit of $30,000 with a probability of .7, or a loss (due to bad weather conditions, labor problems, and such) of $10,000 with a probability of .3. What is her expected profit (or loss)?

13. Experience has shown that a ski lodge which accommodates 160 guests will be full during the Christmas holidays if there is a good snow pack in December, while a light snowfall means it will have only 90 guests. What is the expected number of guests if the probability for heavy snow in December is .40?

In a game costing 10¢, two fair dice are rolled with a payoff of 50¢ for a 7 or an 11, nothing otherwise.

14. What are the expected net winnings?

15. Is the game fair?

16.

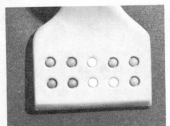

17.

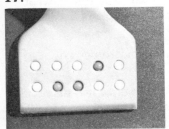

18.

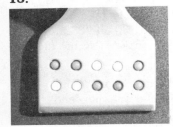

The photo shows the authors (Heeren is on the right) drawing samples from a bowl containing 10,000 beads of various colors. The actual probability of randomly drawing a bead of a given color is the number of beads of that color in the bowl divided by 10,000. Samples can be used to empirically estimate the probability. Exercises 16–18 show three samples of size 10. From these samples, find three estimates of the probability of drawing a white bead from the bowl.

19. Find the average of the three values in Exercises 16–18.

Exercises 20–22 make use of larger samples (size 100). Find three more estimates for the probability of drawing a white bead.

20. **21.** **22.**

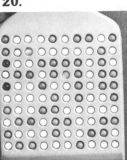 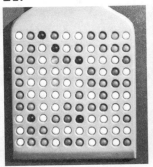 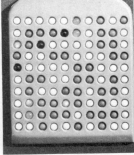

23. Find the average of the values in Exercises 20–22.

24. Would you be more likely to use the answer to Exercise 19 or the answer to Exercise 23 as your best estimate for the probability of a white bead? Why?

25. Use all six samples together to get a final estimate for the probability of a white bead.

26. Based on Exercise 25, what is the (approximate) total number of white beads in the bowl?

1	2	3	4	5	6	7
Account number	Existing volume	Potential additional volume	Probability of getting additional volume	Expected value of additional volume	Existing volume plus Expected value of additional volume	Classification
1	$15,000	$ 10,000	.25	$2,500	$17,500	
2	40,000	0	——	——	40,000	
3	20,000	10,000	.20	2,000		
4	50,000	10,000	.10	1,000		
5	5,000	50,000	.50			
6	0	100,000	.60			
7	30,000	20,000	.80			

The table above illustrates how a salesman for Levi Strauss & Co. rates his accounts by considering the existing volume of each account plus potential additional volume.*

*27. Compute the missing expected values in column **5**.

*28. Compute the missing amounts in column **6**.

*29. In column **7**, classify each account according to this scheme: Class A if the column **6** value is $55,000 or more; Class B if the column **6** value is at least $45,000 but less than $55,000; Class C if the column **6** value is less than $45,000.

*30. Considering all seven of this salesman's accounts, compute the total additional volume he can "expect" to get.

Appendix: The Typical Expected Value of a Slot Machine

Slot machines are a popular game for those who want to lose their money with very little mental effort. In this appendix we shall calculate the typical expected value of a slot machine. We say "typical expected value," since it is not possible to calculate an expected value applicable to all slot machines — payoffs vary from machine to machine.

A player operates a slot machine by pulling a handle after inserting a coin. Three reels inside the machine then rotate, and come to rest in some random order. Assume that the three reels show the pictures listed in Table 1. For example, of the 20 pictures on the first reel, 2 of them are cherries, 5 are oranges, 5 are plums, 2 are bells, 2 are melons, 3 are bars, and 1 is the number 7.

A picture of cherries on the first reel, but not on the second, leads to a payoff of 3 coins (*net* winnings: 2 coins); a picture of cherries on the first two reels, but not the third, leads to a payoff of 5 coins (*net* winnings: 4 coins). All other possible payoffs are as shown in Table 2.

Table 1

Pictures	Reels		
	1	2	3
Cherries	2	5	4
Oranges	5	4	5
Plums	5	3	3
Bells	2	4	4
Melons	2	1	2
Bars	3	2	1
7s	1	1	1
Totals	20	20	20

*This example was provided by James McDonald of Levi Strauss & Co., San Francisco.

Table 2

Winning combinations	Number of ways	Probability	Number of coins received	Net winnings (in coins)	Probability times Winnings
1 cherry (on first reel)	$2 \times 15 \times 20 = 600$	600/8000	3	2	1200/8000
2 cherries (on first two reels)	$2 \times 5 \times 16 = 160$	160/8000	5	4	640/8000
3 cherries	$2 \times 5 \times 4 = 40$	40/8000	10	9	360/8000
3 oranges	$5 \times 4 \times 5 = 100$	100/8000	10	9	900/8000
3 plums	$5 \times 3 \times 3 = 45$	45/8000	14	13	585/8000
3 bells	$2 \times 4 \times 4 = 32$	32/8000	18	17	544/8000
3 melons (jackpot)	$2 \times 1 \times 2 = 4$	4/8000	100	99	396/8000
3 bars (jackpot)	$3 \times 2 \times 1 = 6$	6/8000	200	199	1194/8000
3 7s (jackpot)	$1 \times 1 \times 1 = 1$	1/8000	500	499	499/8000
	Totals: 988				6318/8000

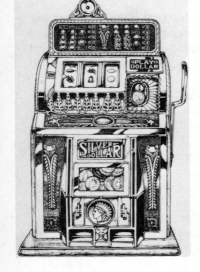

Since, according to Table 1, there are 2 ways of getting cherries on the first reel, 15 ways of *not* getting cherries on the second reel, and 20 ways of getting anything on the third reel, we have a total of $2 \times 15 \times 20 = 600$ ways of getting a net payoff of 2. Since there are 20 pictures per reel, there are a total of $20 \times 20 \times 20 = 8000$ possible combinations. Hence, the probability of receiving a net payoff of 2 coins is 600/8000. Table 2 takes into account all *winning* combinations, with the necessary products for computing expectation added in the last column. However, since a *non-winning* combination can occur in $8000 - 988 = 7012$ ways (with winnings of -1 coin), the product $(-1) \cdot \frac{7012}{8000}$ must also be included. Hence, the expected value of this particular slot machine is

$$\frac{6318}{8000} + (-1) \cdot \frac{7012}{8000} = -0.087 \text{ coin}.$$

On a machine costing one dollar per play, the expected *loss* (per play) is

$$(.087)(1 \text{ dollar}) = 8.7 \text{ cents}.$$

The expected value of this game shows that it rates between one-number keno and chuck-a-luck (see the chart on page 365).

Actual slot machines in Nevada vary in expected loss per dollar of play from about 6 cents to about 50 cents, with the better payoffs coming in the larger establishments. But you still lose.

To a slot machine *You've come a long way, bandit, since 1895, when the Fey Manufacturing Company invented the 3-reel, automatic payout machine. Fey Company's first Silver Dollar Slot Machine (pictured here) came along in 1929. These days, bandit, you're an electronic product, along with pinball and video machines. So you're trying to turn legit, eh? Well, see you someday at the supermarket, and we can talk about old times in Nevada.*

9.6 PERMUTATIONS AND COMBINATIONS

In order to use the classical probability formula

$$P(E) = \frac{\text{number of favorable outcomes}}{\text{total number of outcomes}},$$

we need to be able to determine the numbers that go in the numerator and denominator of the fraction. Our main methods so far have been to list the sample space (sometimes with the aid of a tree diagram) and to use the product rule for counting from Section 9.2. Here we develop some more sophisticated methods of counting.

As an illustration, we consider a club, N, with five members:

$$N = \{\text{Abbot, Babbit, Cynthia, Dillon, Ellen}\}, \quad \text{or} \quad N = \{A, B, C, D, E\}.$$

Example 1   The club N has a dinner at which 5 door prizes are to be given. If there are no restrictions on one person winning more than one prize, how many ways might the prizes be distributed among the members?

Solution.   The first prize could go to any one of the 5 members. So could the second, third, fourth, and fifth prizes. Hence, the product rule gives $5 \cdot 5 \cdot 5 \cdot 5 \cdot 5 = 3125$ possible ways. One of these ways might be written D, A, C, B, E (meaning D won the first prize, A won the second, C won the third, B won the fourth, and E won the fifth). Another possibility would be B, B, C, E, C (B won the first and second, C won the third and fifth, E won the fourth). Notice that no restriction against multiple prizes per person is equivalent to saying "repetitions are allowed."

Example 2   In how many different ways can the members of club N arrange themselves in a row for a photograph?

Solution.   Any one of the five could occupy the left end position. Then any one of the remaining four could be next. Any one of three could be next, and so on. The product rule gives $5 \cdot 4 \cdot 3 \cdot 2 \cdot 1 = 120$ different ways.

Notice that since no person can occupy more than one spot, we have "no repetitions allowed." The arrangement D, C, B, E, A is allowed, but B, B, D, B, A is not.

The product $5 \cdot 4 \cdot 3 \cdot 2 \cdot 1$ is an example of a **factorial.** In general, for any counting number n, the product of all counting numbers from n down through 1 is called n *factorial,* and is denoted $n!$.

Example 3   Evaluate: (a) 2!;   (b) 3!;   (c) 4!;   (d) 7!.

(a) $2! = 2 \cdot 1 = 2$ **(c)** $4! = 4 \cdot 3 \cdot 2 \cdot 1 = 24$

(b) $3! = 3 \cdot 2 \cdot 1 = 6$ **(d)** $7! = 7 \cdot 6 \cdot 5 \cdot 4 \cdot 3 \cdot 2 \cdot 1 = 5040$

So that factorials will be defined for all whole numbers, including zero,

$0! = 1$
$1! = 1$
$2! = 2$
$3! = 6$
$4! = 24$
$5! = 120$
$6! = 720$
$7! = 5040$
$8! = 40,320$
$9! = 362,880$
$10! = 3,628,800$
Short table of factorials
Their numerical value increases rapidly. The value of 100! is a number with 158 digits.

we make the special definition

$$0! = 1.$$

Any time we need to know the number of ways of *arranging n things in a row,* the answer is n! For reference, a table of factorial values is included in the margin.

Example 4 Asssuming no person can hold more than one office, in how many ways can club N select a slate of three officers (say president, vice-president, and secretary)?

Solution. First task: select a president (5 choices).
Second task: select a vice-president (only 4 choices remaining).
Third task: select a secretary (only 3 choices remaining).
By the product rule, we have $5 \cdot 4 \cdot 3 = 60$ ways.

Example 4, like Example 2, involved arrangements, but in this case, we were interested in arrangements of three numbers at a time rather than all five. The real question was: How many different arrangements are there of five things taken three at a time? Notice that since no member was allowed to hold more than one office, we again have "no repetitions allowed." The slate of officers B, A, D (B president, A vice-president, D secretary) is allowed, but B, A, B is not allowed.

In counting theory we usually call an arrangement a **permutation**. In general, the number of permutations of n things taken r at a time is denoted P(n, r).* Following the pattern of Example 4, we see that

$$P(n, r) = n(n-1)(n-2) \cdots (n-(r-1))$$
$$= n(n-1)(n-2) \cdots (n-r+1).$$

The factors in this product begin at n and descend downward until the total number of factors is r.

Example 5 Evaluate: (a) $P(4, 2)$; (b) $P(5, 2)$; (c) $P(7, 3)$; (d) $P(8, 5)$; (e) $P(5, 5)$.

(a) $P(4, 2) = 4 \cdot 3 = 12$

(b) $P(5, 2) = 5 \cdot 4 = 20$

(c) $P(7, 3) = 7 \cdot 6 \cdot 5 = 210$

(d) $P(8, 5) = 8 \cdot 7 \cdot 6 \cdot 5 \cdot 4 = 6720$

(e) $P(5, 5) = 5 \cdot 4 \cdot 3 \cdot 2 \cdot 1 = 120$

Notice that $P(5, 5)$ is equal to 5!. It is true in general that $P(n, n) = n!$.

Change ringing, *the English way of ringing church bells, combines mathematics and music. Bells are rung first in sequence, 1, 2, 3, ... Then the sequence is permuted ("changed"). On six bells, 720 different "changes"— different permutations of tone— can be rung:* $P(6, 6) = 6!$.
Composers work out changes so that musically interesting and harmonious sequences occur regularly.

*Alternate notations are $_nP_r$ and P_r^n.

The art of pulling *The bells are swung by means of ropes attached to the wheels beside them. One ringer swings each bell, listening all the while, watching the other ringers closely. If you get lost and stay lost, the rhythm of the ringing cannot be maintained — all the ringers have to stop.*

A ringer can spend weeks just learning to keep a bell going, months learning to make the bell ring exactly in the right place. Errors of 1/4 second mean that two bells are ringing at the same time. Even errors of 1/10 second can be heard.

Example 6 Again consider the club $N = \{A, B, C, D, E\}$. In how many ways can they elect a slate of three officers if: (a) the first office must be held by a woman and the second and third must be held by men; (b) all three must be held by men; (c) there are no restrictions; (d) the third office must be held by a woman?

(a) There are two choices for the first office (Cynthia or Ellen), three for the second (Abbot, Babbit, or Dillon), and two for the third (either of the two remaining men). Hence, by the product rule, there are $2 \cdot 3 \cdot 2 = 12$ ways.

(b) $P(3, 3) = 3! = 3 \cdot 2 \cdot 1 = 6$ ways.

(c) $P(5, 3) = 5 \cdot 4 \cdot 3 = 60$ ways.

Note that permutations apply in both part (b) and part (c). We need, respectively, the number of arrangements of 3 things taken 3 at a time, and of 5 things taken 3 at a time.

(d) In this case, the restriction is placed only on the third position. Consider that position first. There are two choices, Cynthia and Ellen. Once the third officer is determined, four choices remain for the first, and then three for the second. So we get $2 \cdot 4 \cdot 3 = 24$ ways. If we had considered the first office first here, then the number of possibilities remaining for the second and third would have been harder to determine. So, in general, we can avoid difficulties by dealing first with tasks having special restrictions.

Example 7 How many different 3-digit numbers are possible if: (a) no restrictions are made; (b) no repetitions are allowed?

(a) The available digits are 0, 1, 2, 3, 4, 5, 6, 7, 8, 9. Arrangements like 068 and 003 are not 3-digit numbers, so there are only 9 choices for the first position. There are 10 choices for the second and 10 for the third. The total number is $9 \cdot 10 \cdot 10 = 900$.

(b) There are 9 choices for the first position (0 not available), and 9 for the second (0 available but the digit used first not available), and 8 for the third: $9 \cdot 9 \cdot 8 = 648$.

Now suppose that club N decides to appoint a committee of three members to oversee future elections. In how many ways can the members of this committee be selected?

It may seem at first that the answer to this question is $P(5, 3)$. But notice the following important distinction between selecting three officers and selecting three committee members. A, D, E is a *different* slate of officers than D, A, E. But A, D, E is the *same* committee as D, A, E. That is, when officers are being designated, the order in which the three are listed is important (the order makes a difference). But when the committee members are being designated, the order of listing is not important (the order does not make a difference). The possible number of committees is not the number of arrangements of size 3. Rather, it is the number of *subsets* of size 3 (since the order of listing elements in a set makes no difference).

Permutational patterns *Below
is a pattern for ringing seven bells
from the Ringers' Notebook and
Diary. The changes are written in
columns, and the wavy line traces
the "path" of the seventh bell. To
ring the pattern, each ringer
would learn the shape of the line
and follow it, each one starting at
a different point. Note that no
bell may shift more than one place
in the sequence at a time.*

*To reach the 5040 (or 7!)
possible changes, the ringers
would use further variations in
this basic pattern.*

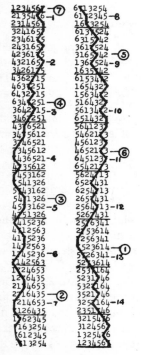

② STARTING POSITION

Subsets in this new context are called **combinations.** In general, the number of combinations of n things taken r at a time (that is, the number of size r subsets, given a set of size n) is denoted $C(n, r)$.*

Here is a list of all the size 3 committees (subsets) of the club (set) $N = \{A, B, C, D, E\}$:

$$\{A, B, C\}, \quad \{A, B, D\}, \quad \{A, B, E\}, \quad \{A, C, D\}, \quad \{A, C, E\},$$

$$\{A, D, E\}, \quad \{B, C, D\}, \quad \{B, C, E\}, \quad \{B, D, E\}, \quad \{C, D, E\}$$

There are ten size-3 subsets, so ten is the number of 3-member committees possible. Just as with permutations, repetitions are not allowed. For example, $\{E, E, B\}$ is not a valid 3-member subset any more than it is a valid 3-member arrangement.

To see how to find the number of such subsets without listing them all, notice that each size-3 subset (combination) gives rise to 6 size-3 arrangements (permutations). For example, the single combination A, D, E yields these 6 permutations:

$$A, D, E \quad D, A, E \quad E, A, D \quad A, E, D \quad D, E, A \quad E, D, A$$

Then there must be 6 times as many size-3 permutations as there are size-3 combinations, or in other words, one-sixth as many combinations as permutations. Therefore

$$C(5, 3) = \frac{P(5, 3)}{6} = \frac{5 \cdot 4 \cdot 3}{6} = 10.$$

The 6 appears in the denominator because there are 6 different ways to arrange a set of 3 things (since $3! = 3 \cdot 2 \cdot 1 = 6$). In general, r things can be arranged in $r!$ different ways, so we get the formula

$$C(n, r) = \frac{P(n, r)}{r!}.$$

Since $P(n, r) = n(n-1)(n-2) \cdots (n-r+1)$, we have

$$C(n, r) = \frac{P(n, r)}{r!} = \frac{n(n-1)(n-2) \cdots (n-r+1)}{r(r-1)(r-2) \cdots 2 \cdot 1}.$$

Example 8 Find the number of different subsets of size 2 that can be found in the set $\{a, b, c, d\}$. List them all to check the answer.

Solution. $\qquad C(4, 2) = \frac{P(4, 2)}{2!} = \frac{4 \cdot 3}{2 \cdot 1} = 6.$

They are $\{a, b\}, \{a, c\}, \{a, d\}, \{b, c\}, \{b, d\}, \{c, d\}$.

---

*Alternate notations include $_nC_r$, C_r^n, and $\binom{n}{r}$.

Example 9   How many different 5-card hands could possibly be dealt from a 52-card deck?

Solution.   The order is unimportant since a given hand depends only on the cards it contains, and not on the order in which they were dealt or the order in which they are listed. Thus we use combinations:

$$C(52, 5) = \frac{P(52, 5)}{5!} = \frac{52 \cdot 51 \cdot 50 \cdot 49 \cdot 48}{5 \cdot 4 \cdot 3 \cdot 2 \cdot 1} = 2,598,960.$$

In this case, we will refrain from listing them all.

Example 10   Melvin wants to buy 10 different books but can afford only 4 of them. In how many ways can he make his selection?

Solution.   The order of the 4 chosen seems to have no bearing here, so we use combinations:

$$C(10, 4) = \frac{P(10, 4)}{4!} = \frac{10 \cdot 9 \cdot 8 \cdot 7}{4 \cdot 3 \cdot 2 \cdot 1} = 210 \text{ ways.}$$

Example 11   Susie wants to give books to 4 of her friends for Christmas. If there are 10 different books to choose from, how many choices does she have?

Solution.   In this case, a different order of the same 4 books means a different choice since a given friend might then get a different book. Thus, we use permutations:

$$P(10, 4) = 10 \cdot 9 \cdot 8 \cdot 7 = 5040 \text{ choices.}$$

Example 12   What is the probability of being dealt a heart flush in 5-card poker? (A heart flush is a 5-card hand of all hearts.)

Solution.   The total number of possible outcomes is found just as in Example 9. The number of favorable outcomes is $C(13, 5)$ since there are 13 hearts in the deck and a favorable outcome is any combination of 5 of those 13. (As in nearly all card-drawing situations, the order is unimportant.) We obtain a probability of

Making history in Chicago—1975   *On March 22 of that year, six ringers at the University of Chicago rang 5040 changes on six of the ten bells in Mitchell Tower. It took two hours and fifty-two minutes to run through the 720 possible permutations seven times (varying the sequence each time). This was the first peal (5000 or more changes rung without break, without substitutions, without paralyzing errors) on the tower bells. That is just 1/8 of the record 8! changes rung in Loughborough, England in 1963, but for the Chicago Six the feeling of achievement outlasted the blisters.*

$$\frac{C(13, 5)}{C(52, 5)} = \frac{\dfrac{P(13, 5)}{5!}}{\dfrac{P(52, 5)}{5!}} = \frac{13 \cdot 12 \cdot 11 \cdot 10 \cdot 9}{52 \cdot 51 \cdot 50 \cdot 49 \cdot 48} = \frac{33}{66,640} = 0.0005$$

(approximately).

Example 13   In 5-card poker: (a) What is the probability of being dealt a full house of aces and 8s (three aces and two 8s)?   (b) What are the odds against such a hand?

Number of poker hands in 5-card poker; nothing wild

Event E	Number of outcomes favorable to E
Royal flush	4
Straight flush	36
Four of a kind	624
Full house	3,744
Flush	5,108
Straight	10,200
Three of a kind	54,912
Two pairs	123,552
One pair	1,098,240
No pair	1,302,540
Total	2,598,960

(a) Success here requires the completion of two tasks: getting three aces (out of four in the deck) and getting two 8s (out of four in the deck). By the product rule, the number of ways of succeeding is $C(4, 3) \cdot C(4, 2)$, so the probability of success is

$$\frac{C(4, 3) \cdot C(4, 2)}{C(52, 5)} = \frac{\frac{P(4, 3)}{3!} \cdot \frac{P(4, 2)}{2!}}{\frac{P(52, 5)}{5!}} = \frac{P(4, 3)}{3!} \cdot \frac{P(4, 2)}{2!} \cdot \frac{5!}{P(52, 5)}$$

$$= \frac{(4 \cdot 3 \cdot 2)(4 \cdot 3)(5 \cdot 4 \cdot 3 \cdot 2 \cdot 1)}{(3 \cdot 2 \cdot 1)(2 \cdot 1)(52 \cdot 51 \cdot 50 \cdot 49 \cdot 48)} = \frac{1}{108,290}.$$

(b) This probability indicates a ratio of 1 chance of success in 108, 290 tries, which is equivalent to 1 chance of success for every 108,289 chances of failure. Therefore, the odds against an aces and 8s full house are 108,289 to 1.

In summary, we outline the similarity of permutations and combinations as well as their differences.

Permutations	**Combinations**
Number of ways of selecting r items out of n items	
Repetitions are not allowed	
Order is important	Order is **not important**
Arrangements of n items taken r at a time	**Subsets** of n items taken r at a time
$P(n, r) =$ $n(n-1)(n-2) \cdots (n-r+1)$	$C(n, r) =$ $\dfrac{n(n-1)(n-2) \cdots (n-r+1)}{r(r-1)(r-2) \cdots 2 \cdot 1}$

It should be stressed that not all counting problems lend themselves to either of these techniques. Whenever a tree diagram or the product rule can be used directly, use them.

9.6 EXERCISES

Evaluate the following factorials, permutations, and combinations.

1. $P(4, 2)$ **4.** $7!$ **7.** $4!$

2. $3!$ **5.** $P(8, 1)$ **8.** $P(4, 4)$

3. $C(8, 3)$ **6.** $C(8, 1)$

A child is allowed to choose two items from a bowl which contains a candy bar, a lollipop, bubble gum, and a licorice stick.

9. List three of the ways he can make his choices.

10. Is there any reason to think that the *order* of choosing makes any difference? Explain why.

11. Are *repetitions* allowed here? Explain why.

12. Which one(s) of these counting methods would be appropriate?
(a) Product rule; (b) Factorials; (c) Permutations;
(d) Combinations; (e) None of these.

13. Find the number of different ways the child can choose.

You are to construct a three-digit number using digits from the set {1, 2, 3, 4, 5}.

14. List four of the possible numbers.

15. Does *order* make a difference here? Show why.

16. Are *repetitions* allowed here? Explain why.

17. Which counting method(s) is appropriate? (a) Product rule;
(b) Factorials; (c) Permutations; (d) Combinations; (e) None of these.

18. How many different numbers are possible?

Five runners, say A, B, C, D, and E, enter a race in which first, second and third place medals will be awarded.

19. List three different ways the medals could be won.

20. Does *order* make a difference? Explain why.

21. Are *repetitions* allowed? Explain why.

22. Which counting methods are appropriate? (a) Product rule; (b) Factorials; (c) Permutations; (d) Combinations; (e) None of these.

23. How many different ways can the medals be won?

24. In how many different ways can a 7-member rock group arrange themselves in 7 stage positions?

Find out, by listing them, how many subsets each of the following sets has. (Don't forget the empty subset.)

25. {a}   **26.** {a, b}   **27.** {a, b, c}   **28.** {a, b, c, d}

29. How many subsets are there of a set with 5 elements?

*30. Give a formula for the number of subsets there are of a set with *n* elements.

31. Using the digits in the set {0, 1, 2, 3, 4}, how many 2-digit numbers can be formed?

32. How many of the numbers of Exercise 31 are even?

33. How many of the numbers of Exercise 31 do not have a repeating digit?

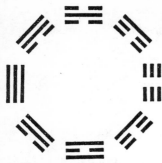

I Ching (Book of Changes) *is an ancient Chinese art of divination based on dual symbols: unbroken line (yang) and broken line (yin). Yang–yin lines form 3-line units (trigrams) in $2^3 = 8$ ways, as you see above. In the "circular" arrangement each trigram is opposite its complement (yang interchanged with yin).*

Yang–yin lines form 6-line units (hexagrams) in $2^6 = 64$ ways. Divination rituals begin with choosing one of the 64 hexagrams at random. One way is to toss six identical coins, where heads is assigned to be either yang or yin. The outcome is keyed to one of the crucial numbers 6, 7, 8, or 9 (as noted on page 132), which determines the bottom line of a hexagram. Repeated tosses give successive lines from bottom to top.

In the I Ching the hexagrams are in the so-called King Wen sequence, paired so that the second in each pair is either the complement of the first or its inverse (upside-down image). No one has yet found a pattern in the seemingly random order of pairs. Here is a pair of inverses:

The later Fu Hsi sequence can be seen as the integers 0–63 in binary notation. Let each yang line be 0, and each yin line be 1. The first four hexagrams in Fu Hsi are (right to left, as is traditional):

000011 000010 000001 000000

How many 7-digit telephone numbers are possible if the first digit cannot be zero and:

34.   only odd digits may be used

35.   the number must be a multiple of 10

36.   the number must be a multiple of 100

37.   the first 3 digits are 484

38.   no repetitions are allowed

39.   How many different license numbers are possible consisting of 3 letters followed by 3 digits?

40.   If 5 points are drawn in a plane with no 3 of them collinear (on a common line), how many lines do they determine? (Any 2 points determine a line.)

41.   If three 2-digit counting numbers are written down and concealed, and a person guesses one 2-digit number at random, what is the probability he will guess one of the numbers written down?

42.   What are the odds against the event of Exercise 41?

In a club with 8 men and 11 women members, find the probability that a 5-member committee chosen randomly will have

43.   all men   **45.**   3 men and 2 women

44.   all women   ***46.**   more men than women

If 5 cards are drawn at random without replacement from a standard 52-card deck, find the probability that

47.   all are queens   **49.**   none is a face card

48.   all are face cards   **50.**   2 are face cards, 3 are not

51.   1 is a heart, 2 are diamonds, and 2 are clubs

In bridge, each of 4 players are dealt 13 of the 52 cards in the deck. Suppose I get a hand containing 8 hearts, 2 clubs, and 3 spades

52.   How many hands are now available for distribution to the other players?

53.   How many of the hands available to the other players would contain the other 5 hearts?

54.   What is the probability that my partner's hand contains the other 5 hearts?

55.   If a baseball coach has 5 good hitters and 4 poor hitters on the bench and chooses 3 of these at random, find the probability that he will choose at least 2 good hitters.

56.   If your college offers 400 courses, 20 of which are in mathematics, and your counselor arranges your schedule of 4 courses by random selection, what is the probability that you will not get a math course?

The Great Art of Combination
The above diagram is one of many devised by Ramon Lull (1232–1315), a Spanish theologian and writer. Lull worked out a geometric method for obtaining truth in his Ars Magna (Great Art), about 1375. He believed that God had revealed the art to him in a vision. Lull asserted that combinations of basic principles would yield new truths. He symbolized principles by letters, such as A for God, B for Bonitas (Bounty), and so on. Lull marked concentric circles with such symbols, and rotated the inner circle(s) to obtain an exhaustive list of combinations. (The lines connecting the symbols also indicate the combinations.) For example, the combination BK tells us that God's Bounty (B) is glorious (K for Gloria).

Lull's art fascinated people, particularly in the Renaissance, but received strong criticism as well. The young Leibniz was influenced by Ars Magna to write a Dissertation on the Art of Combinations (1666). Lull's symbolism had an affinity with his own universal characteristic (see page 36). Incidentally, Leibniz, who invented a binary arithmetic, was amazed to learn about the Fu Hsi sequence (see opposite) from a missionary to China.

See the biography by E. Allison Peers, Fool of Love: The Life of Ramon Lull (West, 1973). Peers also translated Blanquerna, an allegory comparable to Pilgrim's Progress, which Lull wrote in the Catalan language.

*57. If the digits 1, 2, 3, 4, 5, 6, 7, 8 are randomly arranged in a row, what is the probability that 4 and 5 will occur together in the order 4–5? (For example, one such arrangement would be 6 8 4 5 7 3 1 2.)

*58. In 5-card poker, what is the probability of being dealt a full house, of any type? (Refer to Example 13.)

*59. How many 3-digit numbers are there which are multiples of 5 and have no repeating digits?

*60. Recall that in the game keno (see Example 4, Section 9.5) the house randomly selects 20 numbers from the counting numbers 1–80. In the variation called 6-spot keno, the player pays 60¢ for his ticket and marks 6 numbers of his choice. If the 20 numbers selected by the house contain at least 3 of those chosen by the player, he gets a payoff according to this scheme:

3 of the player's numbers among the 20	$ .35
4 of the player's numbers among the 20	2.00
5 of the player's numbers among the 20	60.00
6 of the player's numbers among the 20	1,250.00

Find the player's expected net winnings in this game.

*61. In bridge, the ace, king, queen, jack, and 10 of the trump suit are called "honors." If it is known that a certain hand (of 13 cards) contains 6 trump cards, what is the probability that at least 2 of those are honors?

*62. Begin with the formula $P(n, r) = n(n-1)(n-2) \cdots (n-r+1)$, and show that

$$P(n, r) = \frac{n!}{(n-r)!}.$$

*63. Derive a formula for $C(n, r)$ which involves factorials. (*Hint:* Use the formula of Exercise 62.)

A man tips his waiter by choosing 3 coins randomly from his pocket, which contains 1 half-dollar, 1 quarter, 1 dime, 1 nickel, and 1 penny.

64. How many different amounts are possible for this tip?

65. What is the greatest amount possible?

66. What is the least amount possible?

Now suppose the man has 3 of each of the 5 types of coins in his pocket, and again selects 3 coins at random for the tip.

*67. How many different amounts are now possible?

68. What is the greatest possible amount?

69. What is the least possible amount?

Pascal's search *for mathematical truth began early in his life. A child prodigy, he worked out by himself the fundamental ideas of Euclidean geometry—without the aid of books. By 16 he formulated a theorem about a hexagon inscribed in a conic section (called the "mystic hexagon"), a result in projective geometry (see page 323). A few years later he invented an adding machine, which "carried" digits from one column to the next by means of geared wheels.*

Pascal summed up his experiments on air and water pressure in a treatise on atmospheric weight (1647). He applied the principles and invented a hydraulic press. His study of the number triangle named after him summed up what was known about it and more—but Pascal did not discover the array.

One of Pascal's "thoughts" (in the Pensées) distinguishes between the spirit of geometry and the spirit of "finesse" (or intuition, or instinct). His life seems to illustrate that deductive reasoning, however satisfying it is, is not enough to satisfy the human hunger for truth.

9.7 PASCAL'S TRIANGLE

The triangular array in Figure 9.16 has many interesting properties, some of which are useful in probability. It is named for Pascal, who wrote a treatise about it in 1653; but there is definite evidence that it was known as early as about 1100, and may have been studied in China or India still earlier.

The array in Figure 9.16 can be extended indefinitely, each row being one member longer than the row above it, and each row beginning and ending with 1. Each entry other than the 1's down the edges can be found by adding the two numbers just above it. The entry 20 in row 6, for example, is the sum of 10 and 10. Each row is symmetric with respect to its center: every even-numbered row has a unique center number, while every odd-numbered row has two identical numbers at center.

The feature of Pascal's triangle that helps us in probability is that its entries are none other than the combination values $C(n, r)$. You can verify this fact for as many values as you like by comparing Figure 9.17 with Figure 9.16. For example,

$$C(4, 2) = 6; \qquad C(5, 3) = 10; \qquad C(6, 2) = 15; \qquad \text{and so on.}$$

If, in any counting problem, we need to know the number of combinations of n things taken r at a time (that is, the number of subsets of size r in a set of size n), we simply read entry number r of row number n. Note that the *first* row given is *row number* **0**. Also the first entry of each row can be called entry number 0. It represents the number of subsets of size 0 (which is always 1 since there is only one empty set).

Example 1 Three marbles are chosen randomly (without replacement) from a bag containing 6 red and 4 green marbles. Find the probability of getting: (a) 2 red and 1 green; (b) all red; (c) at least 2 red; (d) at least 1 red.

Row number	Pascal's triangle	Row sum
0	1	1
1	1 1	2
2	1 2 1	4
3	1 3 3 1	8
4	1 4 6 4 1	16
5	1 5 10 10 5 1	32
6	1 6 15 20 15 6 1	64
7	1 7 21 35 35 21 7 1	128
	and so on	

Figure 9.16

(a) Since 3 are being chosen from a total of 10, the total number of choices is $C(10, 3)$. Success here requires the completion of two tasks: getting 2 red, which can happen in $C(6, 2)$ ways; and getting 1 green, which can happen in $C(4, 1)$ ways. By the product rule, the number of favorable outcomes is $C(6, 2) \cdot C(4, 1)$, so the probability of success is

$$\frac{C(6, 2) \cdot C(4, 1)}{C(10, 3)} = \frac{15 \cdot 4}{120} = \frac{1}{2}.$$

The value of $C(10, 3)$ was taken from an extended Pascal's triangle.

(b) All red means 3 red (out of 6) and no green (out of 4), so the number of ways to succeed is $C(6, 3) \cdot C(4, 0)$. The probability of success is

$$\frac{C(6, 3) \cdot C(4, 0)}{C(10, 3)} = \frac{20 \cdot 1}{120} = \frac{1}{6}.$$

(c) $P(\text{at least 2 red}) = P(\text{exactly 2 red}) + P(\text{all 3 red})$

$$= \frac{1}{2} + \frac{1}{6} \qquad [\text{from (a) and (b)}]$$

$$= \frac{2}{3}.$$

(d) We can think of "at least 1 red" as the complement of "no reds." So

$$P(\text{at least 1 red}) = 1 - P(\text{no reds})$$
$$= 1 - P(\text{all 3 green})$$

$$= 1 - \frac{C(6, 0) \cdot C(4, 3)}{C(10, 3)}$$

$$= 1 - \frac{1 \cdot 4}{120} = 1 - \frac{1}{30} = \frac{29}{30}.$$

Row number														
0								$C(0, 0)$						
1							$C(1, 0)$		$C(1, 1)$					
2						$C(2, 0)$		$C(2, 1)$		$C(2, 2)$				
3					$C(3, 0)$		$C(3, 1)$		$C(3, 2)$		$C(3, 3)$			
4				$C(4, 0)$		$C(4, 1)$		$C(4, 2)$		$C(4, 3)$		$C(4, 4)$		
5			$C(5, 0)$		$C(5, 1)$		$C(5, 2)$		$C(5, 3)$		$C(5, 4)$		$C(5, 5)$	
6		$C(6, 0)$		$C(6, 1)$		$C(6, 2)$		$C(6, 3)$		$C(6, 4)$		$C(6, 5)$		$C(6, 6)$
7	$C(7, 0)$		$C(7, 1)$		$C(7, 2)$		$C(7, 3)$		$C(7, 4)$		$C(7, 5)$		$C(7, 6)$	$C(7, 7)$

and so on

Figure 9.17

Up front and before Pascal
This version of the arithmetic triangle appeared on the title page of an arithmetic book by Petrus Apianus, published 1527 in Germany. Pascal's treatise was not published until 1665, after his death.

One of the most common uses of Pascal's triangle is in studying repeated trials with two equally likely outcomes possible at each step, such as tossing a fair coin. This is an example of "repeated Bernouilli trials." To illustrate, suppose we toss 4 fair coins. What is the probability of getting exactly 3 heads? We can think of this experiment as involving 4 "cells" in a row, each of which will be filled with either *h* (heads) or *t* (tails). The favorable outcomes are shown in Figure 9.18.

$$\boxed{h\,|\,h\,|\,h\,|\,t} \qquad \boxed{h\,|\,h\,|\,t\,|\,h} \qquad \boxed{h\,|\,t\,|\,h\,|\,h} \qquad \boxed{t\,|\,h\,|\,h\,|\,h}$$

Figure 9.18

The number of successful outcomes possible is just the number of ways of choosing 3 cells out of 4 to be filled with *h*'s, but this is just the number of size 3 subsets of a set of size 4, or $C(4, 3)$, which we read (from the triangle) to be 4. This will be the numerator of the probability fraction. To get the denominator, notice that in 4 cells there could be either 0, 1, 2, 3, or 4 *h*'s, and the numbers of ways of achieving each of these possibilities are just the entries in row **4** of the triangle. Hence the *total* number of outcomes for this experiment is the sum of all entries in row **4**, which we read (from the triangle) as 16. (See also Exercise 44.) Hence

$$P(3 \text{ heads}) = \frac{4}{16} = \frac{1}{4}.$$

Example 2   If 6 fair coins are tossed, find the probability of getting: (a) exactly 2 heads;   (b) exactly 4 heads;   (c) at least 4 heads.

Solution.   Row **6** of Pascal's triangle is:

$$1 \quad 6 \quad 15 \quad 20 \quad 15 \quad 6 \quad 1 \qquad \text{Sum} = 64$$

(a) $P(\text{exactly } 2) = \dfrac{\text{row 6, entry 2}}{\text{sum of row 6}} = \dfrac{15}{64}.$

(b) $P(\text{exactly } 4) = \dfrac{\text{row 6, entry 4}}{\text{sum of row 6}} = \dfrac{15}{64}.$

(c) $P(\text{at least } 4) = P(4, \text{ or } 5, \text{ or } 6)$

$$= P(4) + P(5) + P(6) = \frac{15}{64} + \frac{6}{64} + \frac{1}{64} = \frac{11}{32}.$$

Notice that $P(\text{exactly 4 heads}) = P(\text{exactly 2 heads})$. This makes sense because in filling 6 cells, selecting 4 cells to receive *h*'s is really the same as selecting 2 cells to receive *t*'s. Since one of these selections can be done in $C(6, 4)$ ways and the other in $C(6, 2)$ ways, we have $C(6, 4) = C(6, 2)$. This helps to explain the symmetry we see in the rows of Pascal's triangle. In general, we always have

$$C(n, r) = C(n, n - r).$$

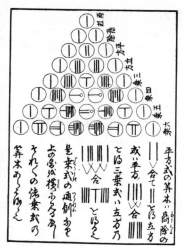

Japanese version *of the triangle dates from the 18th century. The "stick numerals" evolved from bamboo counting pieces used on a ruled board. In China the triangle was known early in the 14th century. Possibly Omar Khayyam, 12th century Persian mathematician and poet, may have divined its patterns in pursuit of algebraic solutions. (The triangle lists the coefficients of the binomial expansion.)*

This is a help in evaluating combinations when the triangle values are not handy. For example, instead of writing out

$$C(10, 7) = \frac{10 \cdot 9 \cdot 8 \cdot 7 \cdot 6 \cdot 5 \cdot 4}{7 \cdot 6 \cdot 5 \cdot 4 \cdot 3 \cdot 2 \cdot 1},$$

we can simplify the arithmetic by observing that

$$C(10, 7) = C(10, 10 - 7) = C(10, 3) = \frac{10 \cdot 9 \cdot 8}{3 \cdot 2 \cdot 1} = 120.$$

9.7 EXERCISES

Read the following combination values directly from Pascal's triangle.

1. $C(4, 2)$ 4. $C(7, 4)$ 7. $C(10, 3)$
2. $C(5, 3)$ 5. $C(8, 3)$ 8. $C(10, 8)$
3. $C(6, 5)$ 6. $C(9, 6)$

A bag contains 5 black, 1 red, and 3 yellow jelly beans; you reach in and take 3 at random. Find the probability of getting:

9. all black   12. 2 black, 1 red   15. 2 yellow, 1 red
10. all red   13. 2 black, 1 yellow   16. 1 of each color
11. all yellow   14. 2 yellow, 1 black

Find the number of 2-member committees that could be selected from the club {Abbot, Babbit, Cynthia, Dillon, Ellen} if:

17. all club members are eligible
18. both must be women
19. there must be 1 woman and 1 man
20. there must be at least 1 man

A couple plans to have 6 children. Assuming boys and girls are equally likely, find the probability that the couple will have:

21. exactly 2 boys   24. more girls than boys
22. exactly 2 girls   25. either all boys or all girls
23. exactly 1 boy   26. at least 1 of each

If 5 cards are drawn *with* replacement from a standard deck of playing cards, find the probability of getting

27. exactly 2 red   30. exactly 2 black
28. all 5 red   31. at least 3 red
29. exactly 4 red

32. Jill is hiding somewhere in a 7-room house, but Jack only has time to check 3 rooms. If he picks 3 at random, what is the probability he will find her?

33. A fisherman has 6 trout in his creel; 2 of them shorter than the legal size limit. If a game warden picks 3 fish at random from the creel and measures them, what is the probability that he will detect the violation? (*Hint:* First consider the event of no detection.)

Random walk   Harvey is standing on a corner flipping his last quarter and trying to decide what to buy to eat. In order to work up a good appetite first, he decides to take a walk, following these rules: he will flip the quarter 10 times, each time walking 1 block north if it comes up heads, and 1 block south if it falls tails. Find the probability that when he finishes his walk he will be exactly

34. 10 blocks north of his corner

35. 6 blocks north of his corner

36. 6 blocks south of his corner

37. 5 blocks south of his corner

38. 2 blocks north of his corner

39. at least 2 blocks north of his corner

40. at least 2 blocks from his corner

41. *on* his corner

Exercises 34–41 illustrate what we call a "random walk." It is a simplified model of Brownian motion (see the chapter introduction). Atomic particles released in nuclear fission also move in a random fashion. During World War II, John von Neumann and Stanislas Ulam used simulation with random numbers to study particle motion in nuclear reactions. Von Neumann coined the name "Monte Carlo" for the methods used, and since then the terms "Monte Carlo methods" and "simulation methods" have often been used with very little distinction.

   The figure below suggests a model for random motion in two dimensions. Assume that a particle moves in a series of 1-unit "jumps," each one in a random direction, any one of 12 equally likely possibilities. One way to choose directions is to roll a fair die and toss a fair coin. The die determines one of the directions 1–6, coupled with heads on the coin. Tails on the coin reverses the direction of the die, so that the die coupled with *tails* gives directions 7–12. Thus 3h (meaning 3 with the die and heads with the coin) gives direction 3; 3t gives direction 9 (opposite to 3); and so on.

42. Simulate the motion described above with 10 rolls of a die (and a coin). Draw the 10-jump path you get. Make your drawing accurate enough so you can estimate (by measuring) how far from its starting point the particle ends up.

43. Repeat the experiment of Exercise 42 four more times. Measure distance from start to finish for each of the 5 "random trips." Add these 5 distances and divide the sum by 5, to arrive at an "expected distance" for such a trip.

Over the years, many interesting patterns have been discovered in Pascal's Triangle.

44. Identify the pattern of the sums of the rows in the triangle, and use it to find the sum of row number **11** and of row number **12.**

45. Complete the sequence of sums on "diagonal" rows, as the figure shows.

Stanislaw Ulam *was born and educated in Poland. He was one of a group of distinguished scientists who came to the United States in the 1930s. Professor Ulam is interested in applications of mathematics in fields such as biology and space technology.*

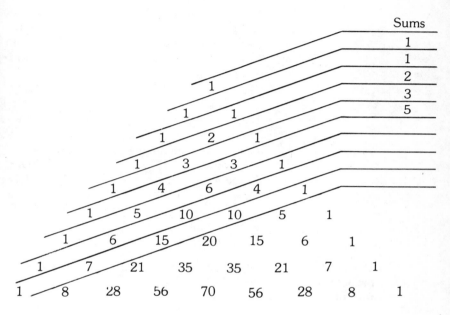

What pattern do these sums make? What is the name of this important sequence of numbers? (Compare Section 4.7.) The presence of this sequence in the triangle apparently was not recognized by Pascal.

46. Construct another "triangle" by replacing every number in the first 5 rows of Pascal's triangle by its remainder when divided by 2. What special property is shared by rows **2** and **4** of this new triangle?

47. What is the next row that would have the same property as rows **2** and **4** in Exercise 46?

*48. How many even numbers are there in row number **256** of Pascal's triangle? (Do Exercises 46 and 47 first.)

49. Identify other properties or patterns you can see in Pascal's triangle.

---

*For example, see the article "Serendipitous Discovery of Pascal's triangle" by Francis W. Stanley in *The Mathematics Teacher*, February 1975.

CHAPTER 9 SUMMARY

Keywords

Random phenomena
Probability (likelihood)
Odds
Classical probability
Equally likely (equiprobable)
Empirical probability
Subjective probability
Conditional probability
Probability distribution

Bernoulli trial
Fair coin; fair die
Law of Large Numbers
With replacement
Without replacement
Mutually exclusive
Pairwise mutually exclusive
Expected value
Net winnings
Fair game

Experiment
Trial
Outcome
Sample space S
Event E
Composite experiment
Composite event
Success
Failure

Simulation
Monte Carlo method
Random number

Factorial
Factorial n $n!$
Permutation $P(n, r)$
Combination $C(n, r)$
Pascal's triangle

Product table
Punnet square

Classical probability formula

When all outcomes in a sample space S are equiprobable, an event E has probability

$$P(E) = \frac{n(E)}{n(S)}.$$

Complements formula

$P(E) = 1 - P(E')$ or $P(\text{success}) = 1 - P(\text{failure}).$

Product rule for counting

In a composite experiment, the total number of outcomes can be found by determining the number of ways of performing each of the separate tasks, and then multiplying these numbers together.

Product rule for probability

In a composite experiment, the probability of an event can be found by determining the probability of the necessary result on each of the separate tasks, and then multiplying these probabilities together.

Addition rule for counting

To find the number of ways that at least one (that is, one or the other or both) of two events could occur, add the number of ways the first could happen to the number of ways the second could happen, and then subtract the number of ways they could both happen.

Addition rule for probability

If A and B are any two events for the same experiment, then
$$P(A \cup B) = P(A) + P(B) - P(A \cap B).$$

Addition rule for mutually exclusive events	If $E_1, E_2, \ldots, E_n$ are n pairwise mutually exclusive events, then $$P(E_1 \cup E_2 \cup \cdots \cup E_n) = P(E_1) + P(E_2) + \cdots + P(E_n).$$
Conditional probability formula	$$P(E_2 \vert E_1) = \frac{P(E_1 \cap E_2)}{P(E_1)}$$
Expected value	If a quantity x can take on the values $x_1, x_2, x_3, \ldots, x_n$, with probabilities, respectively of $P(x_1), P(x_2), P(x_3), \ldots, P(x_n)$, then the expected value of x is defined as $$x_1 \cdot P(x_1) + x_2 \cdot P(x_2) + x_3 \cdot P(x_3) + \cdots + x_n \cdot P(x_n).$$
Factorials	$n! = n(n-1)(n-2) \cdots 2 \cdot 1$
Permutations (arrangements)	$$P(n, r) = n(n-1)(n-2) \cdots (n-[r-1])$$ $$= n(n-1)(n-2) \cdots (n-r+1)$$
Combinations (subsets)	$$C(n, r) = \frac{P(n, r)}{r!} = \frac{n(n-1)(n-2) \cdots (n-r+1)}{r(r-1)(r-2) \cdots 2 \cdot 1}$$

Joker: *"It's been a wild chapter!"*

CHAPTER 9 TEST

1. If a single fair die is rolled, and 1 and 6 are considered "success," find $P(\text{failure})$.

When a single card is drawn from an ordinary deck, find the probability that it will be

2. heart
3. red queen
4. face card
5. black or face card

6. red, given it was a queen
7. jack, given it was a face card
8. face card, given it was a king

Find the odds against a single card drawn from an ordinary deck being:

9. a club   10. black jack   11. red face card or a queen

Identify each of the following proabilities as *classical, empirical,* or *subjective:*

12. The probability that my alarm clock will awaken me tomorrow morning is 0.8, since it has worked 80 percent of the time in the past.

13. The probability of a dry year again next year is 0.7.

14. The probability of drawing a king from an ordinary deck of cards is 1/13.

Two cards are drawn successively from an ordinary deck *with* replacement. Find the probability that:

15. both are red   **16.**   *not* both are red

17. both are red, given that the second is red

Now assume that the two cards are drawn *without* replacement. Find the probability that:

18. both are red   **19.**   *not* both are red

20. they are not the same color

21. both are red, given that the second is red.

22. The Punnet square below represents the four possible (equiprobable) cell-matching combinations when both parents are carriers of the Sickle Cell Anemia trait. Notice that each carrier parent has normal cells (N) and trait cells (T). Fill in the missing combinations in the table.

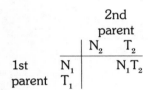

23. If the disease occurs only when two trait cells combine, find the probability that a child born to these parents will have Sickle Cell Anemia.

24. The child will carry the trait but not have the disease if a normal cell combines with a trait cell. Find the probability of this occurrence.

When two fair dice are rolled, find the probability that the resulting sum will be:

25. odd and greater than 8   **26.**   12, given that it was greater than 10.

27. A certain coin has its edges ground so that $P(\text{heads}) = \frac{3}{4}$. Draw a spinner that could be used to simulate the tossing of this coin.

x	P(x)
0	0
1	
2	
3	

28. Three checkers are drawn without replacement from a bag containing three red and two black checkers. Complete this probability distribution table, where x denotes the number of reds drawn.

29. Find the probability that the three checkers are not all red.

30. Find the "expected" number of reds.

If numbers are formed (with no repetition of digits allowed) from the set of digits {1, 2, 3, 4}, then:

31. How many different two-digit numbers are possible?

32. How many three-digit numbers can be formed?

33. How many different numbers are possible altogether?

34. Construct the first six rows of Pascal's triangle.

35. How many different subsets does a size 5 set have?

Chapter 10 Statistics

"*Tonight, we're going to let the statistics speak for themselves.*"

Drawing by Koren; © 1974 The New Yorker Magazine, Inc.

Aspects of the census *The top stamp refers to the use of data processing techniques in handling the information obtained by the census. Information about the growth of a population is nicely symbolized in the middle stamp. The curve running above the heads of the people is an exponential curve, indicating that there is a quickening rather than a steady growth rate. The bottom stamp shows a histogram and a frequency polygon—matters of descriptive statistics as discussed in this section of the text.*

Governments collect and analyze an amazing quantity of "statistics"; the census, for example, is a vast project for gathering data. The census is not a new idea—two thousand years ago Mary and Joseph traveled to Bethlehem to be counted in a census. Long before that the Egyptians recorded numerical information that is still being studied. For a long time, in fact, "statistics" referred to information about political territories. The word itself comes from the Latin *statisticus,* meaning "of the state." The term easily transferred during the nineteenth century to numerical information of other kinds and then to the methods for analyzing data.

The development of current statistical theories, methods, tests, experimental designs, and so on took centuries and is due to many people rather than to one or two inspired minds. John Gaunt, for example, analyzed the weekly Bills of Mortality that recorded deaths in London in the first half of the 17th century. He published his *Observations* in 1662, noting that male births were more numerous than female, but eventually the numbers of both sexes came to be about equal. From these beginnings developed insurance companies, founded on the predictability of deaths via mortality tables. From the field of biology, another example, Sir Francis Galton and Karl Pearson in the 19th century made important contributions to statistical theory. William S. Gossett, a student of Pearson, produced some of the first results concerning small samples. He felt obliged to publish under the name "Student"; his basic findings came to be known as Student's *t*-test—an important statistical tool.

In this chapter we can only begin to explore the ideas and methods of statistics. Much of the chapter discusses topics from *descriptive statistics,* a broad area dealing with the organization of data to reveal its tendencies.

The chapter ends with consumer applications of statistics, inasmuch as we are at the mercy of "statistics" in the news, in advertising, and almost everywhere, it seems. We need ways to defend ourselves against claims and propaganda that use (and often abuse) statistics.

10.1 FREQUENCY DISTRIBUTIONS AND GRAPHS

Suppose you are offered the following deal: Shady Sam will roll a die. If 6 comes up, you pay him $10. If any other number comes up, he pays you $10. The probability that, on a normal die, 1, 2, 3, 4, or 5 comes up is $\frac{5}{6}$. The probability of 6 coming up is $\frac{1}{6}$. The chances of Sam's losing are 5 out of 6. However, these probabilities assume that the die is fair. The only way that Sam can win is if 6 comes up a lot more often than it should.

How do you check and see if the die is "loaded"? One way is to roll it a large number of times and record the numbers that come up. Suppose you roll the die 60 times and record the numbers on the face of the die as they turn up.

```
6  6  4  6  5  6  6  6  3  3  6  5  2  4  6
3  6  5  6  6  5  6  4  2  6  4  6  6  4  6
6  5  4  6  1  6  6  1  5  6  1  6  6  6  1
2  6  6  2  3  5  6  4  6  6  1  6  1  6  2
```

Your list may be an array of numbers as above, conveniently arranged on the page – but the numbers are "raw" data. They are not yet arranged to show the patterns they may hold. For example, the pattern you are seeking is how many times 6 occurs. At first glance, it *appears* that 6 came up much more often than expected. To make sure of the pattern, we can construct a table showing *how often each number appears,* that is, the **frequency distribution.**

Now it is clear from the frequency distribution that 6 turned up more often than you would expect from the probabilities. You can conclude that Shady Sam's die is not fair.

Often the comparison of different frequencies can be shown best in a graph. One of the most popular types in statistics is the histogram, somewhat similar in appearance to a bar graph. In a histogram, the frequency is shown by a vertical bar – the higher the bar, the greater the frequency. The numbers along the left-hand side of the graph give the scale. A histogram for the frequency distribution of Shady Sam's die is shown in Figure 10.1.

Another graphic method of showing frequency distributions is a frequency polygon. While a frequency polygon can be plotted directly from a frequency distribution, the easiest way is to construct a histogram first. Then connect the midpoints of the tops of the bars with line segments. To complete the polygon, connect the midpoints of the first and last bars to the horizontal line of the graph (the horizontal axis). Choose points on the axis at a distance from the first and last bars equal to one-half the width of a bar. Figure 10.2 should make this clear.

Shady Sam's die

Number	Frequency
1	6
2	5
3	4
4	7
5	7
6	31

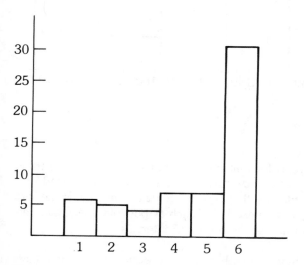

Figure 10.1

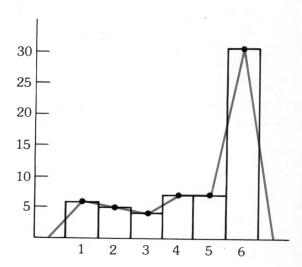

Figure 10.2

Example 1   The raw data below gives the number of sets of encyclopedias sold in a recent year by the twenty-one people on a local sales staff. (a) Make a table for the frequency distribution.   (b) From the table, construct a histogram.

$$
\begin{array}{ccccccc}
120 & 130 & 144 & 132 & 147 & 158 & 174 \\
135 & 142 & 155 & 174 & 162 & 151 & 178 \\
145 & 151 & 139 & 128 & 147 & 134 & 146
\end{array}
$$

(a) The frequency distribution alongside Figure 10.3 does not show each number separately. Instead, the numbers are grouped in **class intervals,** 120–129, 130–139, and so on. It is more practical to use intervals when many numbers make up the raw data and would take an extensive table to show their frequencies. For "grouped data" the frequency tells you how many times any numbers within the given interval occur.

(b) The histogram derived from the table is shown in Figure 10.3

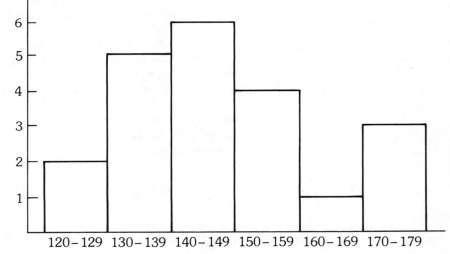

Intervals	Frequency
120–129	2
130–139	5
140–149	6
150–159	4
160–169	1
170–179	3

Figure 10.3

There are no fixed rules for making up a frequency distribution using intervals. There are, however, a few rules of thumb that can be used. For example, always choose intervals that accommodate all the data. Make sure each item goes into only one interval. As much as possible, make all intervals of equal width. As a rule, use from 6 to 15 intervals.

We now describe a type of graph that is appropriate when the data is expressed in percents of the total. A **circle graph** is useful here.

For example, a student itemized her expenses in certain categories over a period of time in order to work out a budget. She found what percent of total expenses each category was. She decided to show the percents in a circle graph by dividing the circle into sectors proportional to the percents. This is like cutting a pie into wedges; *pie chart* is an alternate name for this kind of graph. The cutting can be done fairly accurately, as follows.

Start at any point on a circle and trace out the circle until you come back to the starting point. You have gone a total of 360 degrees (written 360°). Let 360° represent total expenses. To find a percent of 360°, multiply 360 by the decimal fraction equivalent to the percent. For example, Food is 30% of total expenses, and 30% of 360 is

$$360 \cdot 30\% = 360 \cdot (.30) = 108.$$

Thus 30% of the circle is 108°, so a slice of the circle equaling 108° is labeled Food. In the same way, if Rent is 25% of expenses, then 25% of the circle is

$$360 \cdot 25\% = 360 \cdot (.25) = 90,$$

or 90°. On the graph, a 90° slice is labeled Rent.

After the student figured out how many degrees of the circle represented each expense, she drew a circle graph, as in Figure 10.4.

Expenses	Percent of total
Food	30%
Rent	25%
Entertainment	15%
Clothing	10%
Books	10%
Other	10%

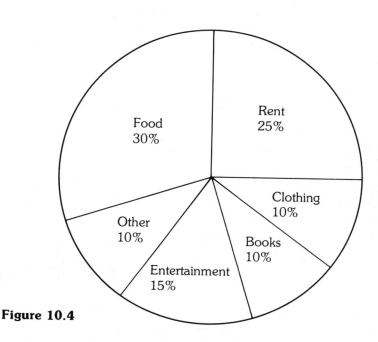

Figure 10.4

To compare different items, we sometimes use a **comparison graph.** A comparison graph shows values of several items on one graph. We can then make comparisons among the various items.

Example 2 The comparison graph in Figure 10.5 shows the average annual consumption per person of four different foods. (a) Find the average annual consumption of potatoes in 1970. (b) Find the consumption of coffee in 1976.

(**a**) Locate 1970 on the horizontal line of the graph. Then draw a line directly up, until it hits the line for potatoes. Read across over to the line marked "average annual consumption per person." You will have to estimate the answer; we get about 118 pounds.

(**b**) Read the graph as above. We get about 12.0 pounds of green coffee.

Figure 10.5

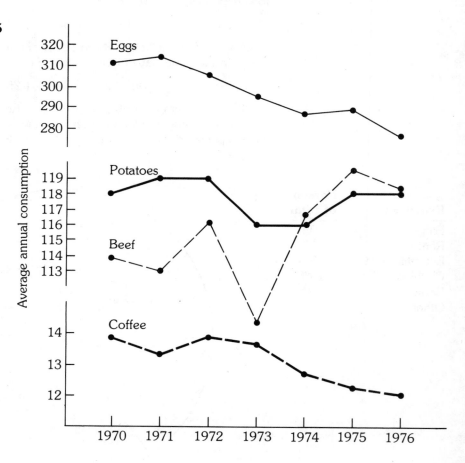

10.1 EXERCISES

1. Toss five coins all at once. Keep track of the number of heads. Repeat this experiment 64 times. Make a histogram of your results.

2. On the histogram from Exercise 1, draw a frequency polygon.

Francis Galton *(1822–1911) learned to read at age three, was interested in mathematics and machines, but was an indifferent mathematics student at Trinity College, Cambridge. After several years as a traveling gentleman of leisure, he became interested in researching methods of predicting weather. It was during this research on weather that Galton developed early intuitive notions of correlation and regression, and posed the problem of multiple regression.*

Galton was interested in eugenics (his term), the study of human inheritance, as a basis for improving human stock. He founded what became the Galton Laboratories of National Eugenics, over the years a source of much statistical theory and method. An important influence on Galton and his interest in the statistics of human inheritance came from his cousin, Charles Darwin, through the epochal work Origin of the Species.

Galton's key statistical work is Natural Inheritance. *In it, he set forth his ideas on regression and correlation. Examining the relative heights of fathers and sons led him to seek a unit-free measure of association. He discovered it, the correlation coefficient, while pondering Alphonse Bertillon's scheme for classifying criminals by physical characteristics. It was at the time a major contribution to statistical method.*

Galton's work led to the science of biometry, the application of mathematics to the problems of biological inheritance. He generously underwrote the beginnings of the journal Biometrika, *begun by two of his followers. It was the first journal devoted to the theory and practice of statistics.*

Many honors came Galton's way in his later life, but he maintained a modest humility, remarking upon receiving a medal, "Well, I am very pleased except that I stand in the way of younger men."

3. The following frequency distribution gives the *theoretical* results of tossing 5 coins 64 times. Use the distribution to draw a frequency polygon. Compare it to the one you drew in Exercise 2.

Number of heads observed	0	1	2	3	4	5
Frequency of occurrence	2	10	20	20	10	2

4. Toss *six* coins a total of 64 times. Keep track of the number of heads. Make a histogram of your results.

5. Draw a frequency polygon using the results of Exercise 4.

6. The following frequency distribution gives the theoretical results of tossing 6 coins 64 times. Use the distribution to draw a frequency polygon. Compare it to the one you drew in Exercise 5.

Number of heads observed	0	1	2	3	4	5	6
Frequency of occurrence	1	6	15	20	15	6	1

The following data gives the number of college units completed by 30 of the employees of the EZ Life Insurance Company.

74	133	4	127	20	30
103	27	139	118	138	121
149	132	64	141	130	76
42	50	95	56	65	104
4	140	12	88	119	64

Use the data to complete the following distribution.

	Number of units	Frequency		Number of units	Frequency
7.	0–24	_____	**10.**	75–99	_____
8.	25–49	_____	**11.**	100–124	_____
9.	50–74	_____	**12.**	125–149	_____

13. Make a histogram using the frequencies that you found.

The daily high temperatures in Phoenix for the month of July in a certain year were as follows:

79	84	88	96	102	
104	99	97	92	94	
85	92	100	99	101	
104	110	108	106	106	
90	82	74	72	83	
107	111	102	97	94	92

Use the data to complete the following distribution.

Temperature	Frequency		Temperature	Frequency
14. 70–74	_____	**19.**	95–99	_____
15. 75–79	_____	**20.**	100–104	_____
16. 80–84	_____	**21.**	105–109	_____
17. 85–89	_____	**22.**	100–114	_____
18. 90–94	_____			

23. Make a histogram showing the results that you found.

24. Make a frequency polygon from the histogram from Exercise 23.

The following graph adapted from *Business Week* magazine, shows that more Fords are stolen than any other make, considering the relative sales of the various makes.

Which cars do auto thieves prefer? *The bars shaded gray represent the numbers of cars stolen during September/October, 1974. The bars shaded in color represent the market shares of new car sales, 1969–1974. The names of auto manufacturers are abbreviated as follows: AM, American Motors; C, Chrysler; GM, General Motors; and Impts. stands for "all imports."*

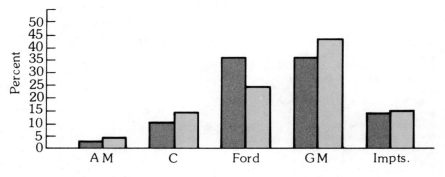

The reason for the higher rate of thefts of Ford cars seems to be the poor steering column lock used in those years. (This problem has now been corrected.) However, a Ford spokesman said that more Fords are stolen because they are more desirable: "The Mark IV is an extremely popular car among the criminal element. We joke about the number of bodies found in the trunks of Mark IV's."

25. About what percent of all cars stolen are Fords?

26. About what percent of all cars sold are Fords?

27. What makes of cars have sales and thefts about equal?

The following table (also from *Business Week*) shows the cost to repair various types of cars in recent years.

| | Average insured repair cost | | |
Type of car	2 years ago	Last year	This year
Subcompact	$500	$542	$684
Compact	488	513	596
Full-size	491	478	550
Luxury	581	585	684

28. Make a comparison graph for this data, using 4 different types of lines.

An article in *Ms* magazine discusses various ways of setting values on the dollar worth of people in today's societies. The scale below is based on expected salaries and expected lifespans.

Age	Value of a man	Value of a woman
20	$45,000	$40,000
30	48,000	38,000
40	38,000	30,000
50	22,000	16,000
60	9,000	5,000
70	−3,000	−3,000

*29. Draw a comparison graph for these numbers. The vertical scale should go from −$10,000 to $50,000. The horizontal scale should have marks for each of the given ages in the table.

*30. Locate the points from the table on the graph of Problem 29.

Over a period of 1 year, a student had $1400 in expenses, as shown in the following table. Find all the missing numbers.

	Expense	Dollar amount	Percent of total	Degrees of a circle
	Food	350	25	90
31.	Rent	280	_____	72
32. **33.**	Clothing	210	_____	_____
34.	Books	140	10	_____
35.	Entertainment	210	_____	54
36. **37.**	Savings	70	_____	_____
38. **39.**	Other		_____	36

40. Draw a circle graph using this information.

The following data from a major record distributor is based on a record with a list price of $5.98.

Manufacturer's costs

Artist advance	.30
Payment to publisher	.24
Musician's trust fund fee	.08
Manufacturing Cost	.35
Jacket, inner sleeve	.15
Artist royalty	.68
Freight to distributor	.03

Total: $1.83

41. Draw a circle graph representing this information. (*Hint:* The arithmetic is tedious. Use a hand calculator if possible.)

42. The manufacturer sells the record to the distributor for $2.51. How much does the manufacturer make on each record?

Signature verification. *John F. Morrisey, left, and Noel M. Herbst display how their signature verification method can distinguish between true and forged signatures despite what appears to be a close visual match. The motions of signing one's name seem to become imprinted in the nervous system. The pattern of accelerations of the pen during a signature reflect the muscle forces corresponding to these imprinted motions. A statistical analysis was made of such patterns to identify the distinctive characteristics. These are then stored in a computer, which can compare them with a new signature to determine if it is a forgery.*

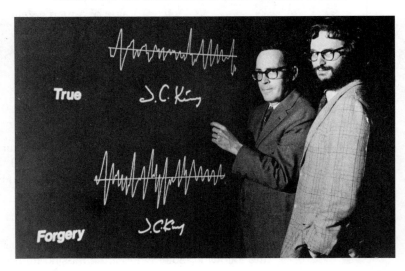

10.2 MEASURES OF CENTRAL TENDENCY

Suppose the daily sales at the Eastside Plant Boutique last week were

$86, $103, $118, $117, $126, $158, $149.

It would be desirable to have a single number to serve as a kind of representative value for this whole set of numbers, that is some value around which all the numbers in the set tend to cluster, a kind of "middle" number. In short, we are seeking a **measure of central tendency**. There are three such measures that we discuss in this section: the mean, the median, and the mode.

The **mean** is commonly called "average," and we speak of an "average person" as being fairly representative of current tendencies. In statistics, the mean is a number. It is important enough to have a special symbol: $\bar{x}$ (read "x bar"). The mean is found by the formula

$$\textbf{mean} = \bar{x} = \frac{\text{sum of all numerical values}}{\text{number of values}}$$

Now we can find the central tendency of the seven Eastside Plant Boutique sales figures by computing the mean according to the formula:

$$\text{mean} = \bar{x} = \frac{86 + 103 + 118 + 117 + 126 + 158 + 149}{7}$$

$$= \frac{857}{7}$$

$$= 122.43 \qquad \text{(rounded to nearest hundredth)}$$

Thus the mean value for the sales that week is $122.43.

Some calculators find the mean automatically when you enter the numbers given in the problem. To recognize these calculators, look for a key marked "$\bar{x}$."

Example 1   Find the mean for each list of values.

(a) $25.12, $42.58, $76.19, $32, $81.11, $26.41, $19.76, $59.32, $71.18, $21.03

Solution.   Add the numbers and divide by 10 since there are 10 numbers. Check that the sum of the 10 numbers is $454.70. From the formula,

$$\bar{x} = \frac{454.70}{10} = 45.47.$$

The mean is $45.47.

(b) The sales for one year at 8 different Denny's Restaurants were

$374,910	$321,872	$242,943	$351,147
$382,740	$412,111	$334,089	$262,900.

Solution.   The sum is $2,682,712.

$$\bar{x} = \frac{2,682,712}{8} = 375,339.$$

The mean is $375,339. (This mean is for restaurants actually owned by the Denny's Company, according to *Business Week.*)

The following table gives the annual salaries received by the players of a certain professional football team.

As the first column shows, 2 players were each paid $16,000, 5 were each paid $20,000, and so on. To find the average salary paid to the team members, we cannot merely add up the salaries since the different salaries are earned by different numbers of players. To find the mean annual salary, we must first multiply each annual salary by the number of players receiving it. The results are in the third column below:

Salary	Number of players	Salary times number of Players
$16,000	2	$32,000
$20,000	5	$100,000
$25,000	3	$75,000
$36,000	12	$432,000
$45,000	9	$405,000
$48,000	4	$192,000
$55,000	2	$110,000
$80,000	2	$160,000
$148,000	1	$148,000
Totals:	40	$1,654,000

The team has a total of 40 players. The mean is found by dividing total salary, $1,654,000, by 40:

$$\bar{x} = \frac{\$1,654,000}{40} = \$41,350.$$

(This mean is taken from a recent issue of *Harper's* magazine.)

Suppose you own a small company, and employ five persons. Assume that you are cheap, and pay annual salaries as follows:

$2,500, $3,000, $3,200, $4,000, $5,000.

The mean salary you pay your employees is

$$\bar{x} = \frac{\$2,500 + \$3,000 + \$3,200 + \$4,000 + \$5,000}{5} = \frac{\$17,700}{5} = \$3,540.$$

Suppose now that your employees go on strike, demanding a raise. To get public support, they go on television and talk about their miserable salaries, where the mean is only $3,540 per year.

The television station sends a reporter to interview you. Before the interviewer arrives, you decide to find the mean salary of *all* workers, including yourself. To do this, you add up the five salaries given above and your own salary of $117,000. The mean now is:

$$\bar{x} = \frac{\$2,500 + \$3,000 + \$3,200 + \$4,000 + \$5,000 + \$117,000}{6} = \frac{\$134,700}{6} = \$22,450.$$

When the television reporter arrives, you are all ready to report that there is no reason for your employees to be on strike since the company pays a mean salary of $22,450.

There are two points to this story. First, both averages are correct, depending on what you are measuring. This shows how easy it can be to manipulate statistics. Also, the mean is sometimes a poor indicator of the "middle" of a list of numbers. In fact, when the mean was computed by the owner, it turned out that 5 out of the 6 salaries were actually considerably less than the mean.

The mean is often affected by an extreme value. When this is the case, the "middle" number is more descriptive of the general tendency. This measure of central tendency is known as the **median.** In general, the median divides a group of numbers into two groups, with half the numbers below the median and half above it.

Example 2   Find the median for each list of numbers.

(a) 6, 7, 12, 15, 18, 23, 24

Solution.   Here the median is actually the middle number, 15; three numbers are less than 15, and three numbers are greater than 15. (Note that the numbers are arranged in numerical order.)

(b) 17, 15, 9, 13, 21, 32, 41, 7, 12

Solution.   First, place the numbers in numerical order from smallest to largest:

$$7, \quad 9, \quad 12, \quad 13, \quad 15, \quad 17, \quad 21, \quad 32, \quad 41$$

The middle number can now be picked out: the median is 15.

If there is an even number of data values, then there is no single middle number. In this case, the median is the average of the two middle numbers.

Example 3   Find the median for each list of numbers.

(a) 7, 13, 15, 25, 28, 32, 47, 59, 68, 74

Solution.   There are 10 numbers in this list. There is no single middle number. To find the median, we must take the average of the two middle numbers. The two middle numbers are 28 and 32. The average of these is

$$\frac{28 + 32}{2} = \frac{60}{2} = 30.$$

(b) 147, 159, 132, 181, 174, 253

Solution.   First write the numbers in numerical order:

$$132, \quad 147, \quad 159, \quad 174, \quad 181, \quad 253$$

The two middle numbers are 159 and 174. The median is the average of these two numbers:

$$\frac{159 + 174}{2} = \frac{333}{2} = 166\tfrac{1}{2}.$$

The third important statistical measure of central tendency is called the **mode.** The mode is the value that occurs most often in a list of data. For example, if ten students earned scores on a business law examination of

74,   81,   39,   74,   82,   80,   100,   92,   74,   85

then the mode is 74 since this score was earned by more students than any other.

Example 4   Find the mode for each list of numbers.

(a) 51, 32, 49, 49, 74, 81, 92

Solution.   The number 49 occurs more often than any other. Therefore, 49 is the mode. Note that it is not necessary to place the numbers in numerical order when looking for the mode.

(b) 482, 485, 483, 485, 487, 487, 489

Solution.   Both 485 and 487 occur twice. This list is said to have *two* modes, or is said to be *bimodal.*

(c) 10,708,   11,519,   10,972,   17,546,   13,905,   12,182

Solution.   No number here occurs more than once. This list has no mode.

10.2 EXERCISES

Find the mean in each list of data. Round it to the nearest tenth.

1. 6, 8, 14, 19, 23
2. 51, 48, 32, 43, 74, 58
3. 40, 51, 59, 62, 68, 73, 49, 80
4. 31, 37, 43, 51, 58, 64, 79, 83
5. 21,900,   22,850,   24,930,   29,710,   28,340,   40,000
6. 38,500,   39,720,   42,183,   21,982,   43,250
7. 9.4, 11.3, 10.5, 7.4, 9.1, 8.4, 9.7, 5.2, 1.1, 4.7
8. 30.1, 42.8, 91.6, 51.2, 88.3, 21.9, 43.7, 51.2
9. .06, .04, .05, .08, .03, .14, .18, .29, .07, .01
10. .31, .09, .08, .22, .46, .51, .48, .42, .53, .42

Many Kentucky Fried Chicken outlets are owned by the company itself, while others are franchised by individual operators. To see which outlets produced the largest annual sales, the company found the average sales for each. (The results were given in *Business Week* magazine.) Find the mean for each of the following. (Round to the nearest thousand dollars.)

11. Six company-owned outlets had sales of $240,000, $320,000, $300,000, $340,000, $250,000, $350,000.

12. Seven franchised outlets had sales of $240,000, $220,000, $300,000, $320,000, $260,000, $250,000, $270,000.

Denny's Restaurants did the same for its restaurants. Find each of the following means. (Round to the nearest thousand dollars.)

13. Eight company-owned outlets had sales of $382,520, $321,710, $308,512, $371,519, $382,710, $297,413, $314,725, $303,603.

14. Five franchised restaurants had sales of $217,941, $223,825, $234,818, $239,513, $238,403.

15. At Marriott's Great America, the "Turn of the Century" ride goes 2000 feet in 92 seconds. Find the mean number of feet traveled in one second. (Round to the nearest foot.)

16. One of the most popular rides in amusement parks today is the log flume ride. The company that makes these rides has sold 46 flumes, for a total of $34,500,000. Find the mean price of these flumes.

In Exercises 17–26, find the median of each list. (Don't forget to first place the numbers in numerical order, if necessary.)

17. 12, 18, 32, 51, 58, 92, 106
18. 596, 604, 612, 683, 719
19. 100, 114, 125, 135, 150, 172
20. 298, 346, 412, 501, 515, 521, 528, 621
21. 32, 58, 97, 21, 49, 38, 72, 46, 53
22. 1072, 1068, 1093, 1042, 1056, 1005, 1009
23. 576, 578, 542, 551, 559, 565, 525, 590
24. 7, 15, 28, 3, 14, 18, 46, 59, 1, 2, 9, 21
25. 28.4, 9.1, 3.4, 27.6, 59.8, 32.1, 47.6, 29.8
26. .6, .4, .9, 1.2, .3, 4.1, 2.2, .4, .7, .1

In Exercises 27–36, find the mode or modes for each list.

27. 4, 9, 8, 6, 9, 2, 1, 3
28. 21, 32, 46, 32, 49, 32, 49
29. 80, 72, 64, 64, 72, 53, 64
30. 97, 95, 94, 95, 94, 97, 97
31. 74, 68, 68, 68, 75, 75, 74, 74, 70
32. 158, 162, 165, 162, 165, 157, 163
33. 5, 9, 17, 3, 2, 8, 19, 1, 4, 20
34. 12, 15, 17, 18, 21, 29, 32, 74, 80
35. 6.1, 6.8, 6.3, 6.3, 6.9, 6.7, 6.4, 6.1, 6.0
36. 12.75, 18.32, 19.41, 12.75, 18.30, 19.45, 18.33

A chemistry student working on a lab experiment copied the following meter readings into a notebook: 3, 4, 5, 2, 3, 2, 2, 30, 3

37.   Find the mean.   **38.**   Find the median.

The student later decided that 30 was way out of line. The student was sure that the meter must have been misread. The number 30 was omitted from the list, producing   3, 4, 5, 2, 3, 2, 2, 3

39.   Find the mean.   **40.**   Find the median.

41.   Which seems better — the mean or the median — if you want to avoid the distortion caused by a single extreme value?

42.   Suppose you own a hat shop, and decide to order hats in one size only for the coming season. To decide on which size to order, you look at last year's sales figures, which are itemized for size, style, and so on. Should you find the mean, median, or mode for the data?

Exercises 43–48 give frequency distributions for sets of data values. For each set find the mean to the nearest tenth.

43.

Value	Frequency
3	4
5	2
9	1
12	3

44.

Value	Frequency
9	3
12	5
15	1
18	1

45.

Value	Frequency
12	4
13	2
15	5
19	3
22	1
23	5

46.

Value	Frequency
25	1
26	2
29	5
30	4
32	3
33	5

47.

Value	Frequency
104	6
112	14
115	21
119	13
123	22
127	6
132	9

48.

Value	Frequency
246	2
291	4
295	3
304	8
307	9
319	2

Find the average salary of selected players on each team. (The averages come from a recent *Harper's* magazine article.)

49.

Baseball salaries	Number of players
$16,000	3
$24,000	2
$32,000	5
$41,000	3
$45,000	4
$59,000	3
$63,000	2
$81,000	1

50.

Hockey salaries	Number of players
$15,000	1
$25,000	2
$37,000	4
$51,000	3
$82,000	3
$104,000	3
$108,000	2
$130,000	2

Find the average cost to buy one minute of advertising time for each sporting event in Exercises 51–56. (These averages come from a recent article in *Harper's* magazine.)

51. Seven minutes on a Saturday afternoon baseball game cost $67,500.

52. Four minutes on the Super Bowl cost $920,000.

53. Three minutes on a night World Series game cost $390,000.

54. Eleven minutes on Sunday football cost $748,000.

55. Five minutes on a basketball championship game cost $320,000.

56. Eight minutes on the Indianapolis 500 cost $544,000.

57. A 7-pound ham is sold by one high-priced store for $52. Find the average cost per pound.

58. The same store sells a 4-pound turkey breast for $25. Find the average cost per pound.

Sakowitz is an exclusive department store in Houston. Every year it publishes a Christmas catalog. To bring attention to its catalog, the firm includes some extremely expensive gifts in it. The problems in Exercises 59–62 are based on a recent catalog.

59. You can take 5000 friends (including yourself) for a day at Six Flags Over Texas amusement park for $47,500. Find the mean cost per friend.

60. Suppose you take 4500 friends (including yourself) to Astroworld, for $50,000. Find the mean cost per friend.

61. A total of 720 people can have an expense-paid weekend at the Hyatt at Palmetto Dunes, South Carolina—the cost, $286,125. Find the mean cost per person.

62. For $220,000 a famous interior decorator will redo your bedroom. Suppose you used this bedroom for 10 years (of 360 days each) before getting tired of it. Find the mean decorating cost of one night in the room.

An athlete's official times, in seconds, for the 100-yard dash during a season are

$$10.3, 9.9, 10.4, 10.2, 9.9, 10.1, 18.6.$$

The 18.6 seconds occurred on an off day, when her shoelace got caught on the starting block.

63. Find the mean, median, and mode for the given data.

64. Which measure is the best indication of the athlete's "average" ability that season in this event?

65. Which of the numbers in the set of data values are probably the most significant?

In Exercises 66–68, write lists of numbers having the given characteristics.

66. More values above the mean than below.

67. Half the values above the mean and half below, with mean and median equal.

68. Half the values above the mean and half below, with mean and median unequal.

The **Fearless Fido Frisbee Fetching Fracas** is an annual event in Southern California. To judge the event, five judges score each canine contestant in five skill categories. The mean of the scores in each category is calculated; the total of the five means is the *total artistry score* for a given dog.

The contestants are allowed five runs of Frisbee-catching where the number of yards is significant. The *mean yardage* of runs (counting misses) is calculated.

To determine the world champion Frisbee-catching dog, the *sum* of the total artistry score and the yardage mean is calculated. The dog with the highest number is champion.

69. Calculate the total artistry score for three finalists, Kenilworth, Milquebone, and Yappily, whose scores are in the table for skill categories I–V. (Round all means to the nearest tenth.)

		Kenilworth					Milquebone					Yappily				
I	Sense of direction	8	5	7	7	6	6	7	8	4	5	5	7	7	4	5
II	Depth perception	6	4	5	6	7	4	6	7	7	8	8	4	5	7	7
III	Leaping ability	8	5	7	7	6	6	6	5	6	8	7	7	6	6	8
IV	Level of enthusiasm	9	7	9	8	5	8	7	7	6	5	5	8	8	7	4
V	Level of ability	8	8	7	5	6	4	8	8	7	4	8	6	7	4	7

Find the mean yardage for each finalist. (Round to the nearest tenth of a yard.)

70. Kenilworth: 48 yards; 63 yards; 51 yards; missed on the fourth run; 39 yards

71. Milquebone: 37 yards; 41 yards; 43 yards; 38 yards; 27 yards

72. Yappily: 61 yards; missed; 74 yards; missed; 59 yards

Find the sum of the total artistry score and yardage mean for each finalist.

73. for Kenilworth   **74.** for Milquebone   **75.** for Yappily

76. Which of the three finalists is the champion Frisbee-catching dog?

Some contestants will do anything to improve their scores!

10.3 MEASURES OF DISPERSION

The mean is a good indicator of the middle, or central tendency, of a set of data values, but it does not give the whole story about the data. To see why, compare distribution A with distribution B in the table at the side.

Both distributions of numbers have the same mean (and the same median also), but beyond that, they are quite different. In the first, 7 is a fairly typical value; but in the second, most of the values differ quite a bit from 7. What we need here is some measure of the **dispersion,** or *spread,* of the data.

We shall consider two of the most common measures of dispersion, the *range* and the *standard deviation.*

	A	B
	5	1
	6	2
	7	7
	8	12
	9	13
Mean	7	7
Median:	7	7

Range   The range for a set of data is defined as the difference between the smallest value and the largest value in the set. In distribution A above, the largest value is 9 and the smallest is 5. Thus

$$\text{range} = \text{highest} - \text{lowest} = 9 - 5 = 4.$$

In distribution B, we have

$$\text{range} = 13 - 1 = 12.$$

Dive	Mark	Myrna
1	28	27
2	22	27
3	21	28
4	26	6
5	18	27
Mean	23	23
Median	22	27
Range	10	22

The range can be misleading if it is interpreted unwisely. For example, suppose three judges for a diving contest assign points to Mark and Myrna on five different dives, as shown in the table.

By looking at the range for each diver, we might be tempted to conclude that Mark is a more consistant diver than Myrna. However, by checking more closely, we might decide that Myrna is actually more consistant with the exception of one very poor score, which is probably due to some special circumstance. Myrna's median score is not affected much by the single low score and is more typical of her performance as a whole than is her mean score.

Standard deviation   One of the most useful measures of dispersion, the standard deviation, is based on *deviations from the mean* of the data values. To find how much each value deviates from the mean, first find the mean, and then subtract the mean from each data value.

Example 1   Find the deviations from the mean for the data values

$$32, \quad 41, \quad 47, \quad 53, \quad 57.$$

Solution.   Add these numbers up and divide by 5. The mean is 46. To find the deviations from the mean, subtract 46 from each data value.

Data value	32	41	47	53	57
Deviation	**−14**	**−5**	**1**	**7**	**11**

(To check your work, add the deviations. The sum of deviations for a set of data is always 0.)

To find the measure of dispersion, we might be tempted to find the mean of the deviations. However, this number always turns out to be 0 no matter how much the dispersion in the data is, because the positive deviations will just cancel out the negative ones.

We can get around this problem of positive and negative numbers adding to 0 by *squaring* each deviation. (The square of a negative number is positive.) We take the example above one step further:

Data value	32	41	47	53	57
Deviation from mean	−14	−5	1	7	11
Square of deviation	**196**	**25**	**1**	**49**	**121**

We can now define the **standard deviation:** it is *the square root of the mean of the squares of the deviation.* Luckily, the standard deviation is harder to say than to do.

Continuing our example, we calculate the mean of the squares of the deviations:

$$\frac{196 + 25 + 1 + 49 + 121}{5} = \frac{392}{5} = 78.4.$$

The standard deviation is the *square root* of this number, namely, $\sqrt{78.4}$. The Greek letter σ (sigma) is used to represent the standard deviation.

In summary, for the numbers 32, 41, 47, 53, 57, the standard deviation is

$$\sigma = \sqrt{78.4} = 8.9 \qquad \text{(to the nearest tenth)}.$$

The square root of 78.4 can be found from tables or with a calculator. In this book, we will be satisfied with either $\sqrt{78.4}$ or 8.9; both answers will be given in the answer section at the back of the book.

Example 2 Find the standard deviation of the values

$$7, 9, 18, 22, 27, 29, 32, 40.$$

Solution.

Step 1 Find the mean of the values:

$$\frac{7 + 9 + 18 + 22 + 27 + 29 + 32 + 40}{8} = 23$$

Step 2 Find the deviations from the mean:

Data values	7	9	18	22	27	29	32	40
Deviations	−16	−14	−5	−1	4	6	9	17

Step 3 Square each deviation:

Squares of deviations   256   196   25   1   16   36   81   289

Step 4 Find the mean of the squares:

$$\frac{256 + 196 + 25 + 1 + 16 + 36 + 81 + 289}{8} = \frac{900}{8} = 112.5$$

Step 5 Take the square root of the answer in Step 4. Thus, the standard deviation is

$$\sigma = \sqrt{112.5} = 10.6.$$

One of the main applications of standard deviation comes in working with the **normal curve.** Many different sets of data in the real world lead to graphs which look very much like the normal curve (Figure 10.6). We shall investigate several applications in the exercises of this section and the next.

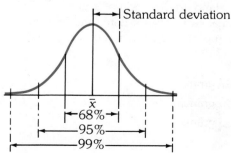

Figure 10.6

It turns out that if a group of data is very closely approximated by a normal curve, then approximately 68% of the data values will lie within 1 standard deviation of the mean. Approximately 95% will lie within 2 standard deviations of the mean; and about 99% will lie within 3 standard deviations of the mean.

Example 3   Suppose an instructor gives a test to 300 students. Suppose further that the grades are closely approximated by a normal curve. Find the number of scores: (a) within 1 standard deviation of the mean; (b) within 2 standard deviations.

(a) As we said above, 68% of all scores lie within 1 standard deviation of the mean. Since there is a total of 300 scores, the number of scores within 1 standard deviation is

$$(68\%) \cdot (300) = (.68) \cdot (300) = 204.$$

(b) A total of 95% of all scores lie within 2 standard deviations of the mean. Since there is a total of 300 scores, the number of scores within 2 standard deviations is

$$(.95) \cdot 300 = 285.$$

10.3 EXERCISES

Find the range and standard deviation for each set of data.

1. 6, 8, 9, 10, 12
2. 12, 15, 19, 23, 26
3. 7, 6, 12, 14, 18, 15
4. 4, 3, 8, 9, 7, 10, 1
5. 42, 38, 29, 74, 82, 71, 35
6. 122, 132, 141, 158, 162, 169, 180

7. 241, 248, 251, 257, 252, 287
8. 51, 58, 62, 64, 67, 71, 74, 78, 82, 93
9. 3, 7, 4, 12, 15, 18, 19, 27, 24, 11
10. 15, 42, 53, 7, 9, 12, 28, 47, 63, 14
*11. 21, 28, 32, 42, 51
*12. 76, 78, 92, 104, 111

Suppose 100 different geology students measure the mass of an ore sample. Due to human error and to limitations in the accuracy of the balance, not all students get the same value. The results are found to be closely approximated by a normal curve. The mean is 37 grams, with a standard deviation of 1 gram. Use the sketch of a normal curve shown here, and find the number of students reporting each reading in Exercises 13–22.

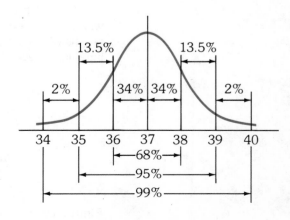

13. more than 37 grams
14. more than 36 grams
15. more than 35 grams

16. more than 38 grams

17. between 36 and 38 grams

18. between 35 and 39 grams

19. between 38 and 39 grams

20. between 36 and 39 grams

21. within 1 gram of the mean

22. more than 2 grams away from the mean

On standard IQ tests, the mean is 100, with a standard deviation of 15. The results are very close to fitting a normal curve. Suppose an IQ test is given to a very large group of people. Find the percent of people whose IQ score is:

23. more than 100

24. less than 100

25. greater than 115

26. between 85 and 115

27. between 70 and 130

28. between 55 and 145

29. less than 55

30. more than 145

Pafnuty Lvovich Chebyshev
(1821–1894) is known mainly for his work on the theory of prime numbers. He and the French mathematician and statistician **Jules Bienaymé** *(1796–1878) developed, independently of one another, an important inequality of probability, known as the Bienaymé-Chebyshev inequality.*

Chebyshev founded the St. Petersburg Mathematical School and taught at the University of St. Petersburg. His mathematical writings covered a range of subjects. His Theory of Congruences made him widely known and was used as a text in Russia for many years.

Another application of standard deviations is given by Chebyshev's Theorem. (P. L. Chebyshev was a Russian mathematician who lived from 1821 to 1894.) This theorem applies to *any* distribution, whether approximated closely by a normal curve or not. It says:

For any distribution of numbers, the fraction of them which lie within k standard deviations of the mean is at least

$$1 - \frac{1}{k^2}.$$

Example: For any distribution of numbers, the fraction of them which lie within 3 standard deviations of the mean is at least

$$1 - \frac{1}{3^2} = 1 - \frac{1}{9} = \frac{8}{9}.$$

Find the fraction of all numbers of a data set lying within the following numbers of standard deviations from the mean.

*31. 2   ***32.** 4   ***33.** 5

In a certain distribution of numbers, the mean is 50 with a standard deviation of 6. At least what fraction of the numbers are between:

*34. 38 and 62?   ***36.** 26 and 74?

*35. 32 and 68?   ***37.** 20 and 80?

In the same distribution, find the fraction of numbers that are:

*38. less than 38 or more than 62

*39. less than 32 or more than 68

*40. less than 26 or more than 74

Carl Friedrich Gauss *(1777–1855), known even in his own lifetime as the "prince of mathematicians," was one of the greatest mathematical thinkers of history. In his masterpiece, Disquisitiones arithmeticae, published in 1798, he pulled together work by predecessors and enriched and blended it with his own into a unified whole. The book is regarded by many as the true beginning of the theory of numbers.*

Of his many contributions to science, the statistical method of least squares is the most widely used today in astronomy, biology, geodesy, physics, and the social sciences. Gauss took special pride in his contributions to developing the method. Despite an aversion to teaching, he gave an annual course in the method from 1835 until he died.

10.4 THE NORMAL CURVE

Many different collections of numbers have distributions that are very closely approximated by normal curves. The normal curve was first developed by Abraham de Moivre (1667–1754), but his work went unnoticed for many years. It was independently redeveloped by Pierre Laplace (1749–1827) and Carl Friedrich Gauss (1777–1855). Gauss found so many uses for this curve that it is sometimes called the *Gaussian curve.*

In the last section, we found that in any distribution which is closely approximated by a normal curve, approximately 68% of all the numbers will lie within one standard deviation of the mean, 95% within two, 99% within three, and so on. What happens if we need to know the fraction of values within $1\frac{1}{2}$ or $2\frac{1}{5}$ standard deviations of the mean? In this case, we must use a table, such as the one in the Appendix at the end of this book.

The table gives the fraction of all scores in a normal distribution that lie between the mean and z standard deviations from the mean. Because of the symmetry of the normal curve, the table can be used for values above the mean or below the mean.

Example 1 Use the normal curve table to find the fraction of all scores that lie between the mean and: (a) 1 standard deviation above the mean; (b) 2.43 standard deviations below the mean.

(a) Here $z = 1.00$ (the number of standard deviations, written as a decimal to the nearest hundredth). Refer to the table. Find 1.0 along the left side, and 0 across the top. The table entry is .341, so that 34.1% of all values lie between the mean and one standard deviation above the mean.

Another way of looking at this is to say that the shaded area in Figure 10.7 represents 34.1% of the total area under the normal curve.

(b) Find 2.4 at the left of the table, and 3 across the top. A total of .492, or 49.2% of all values lie between the mean and 2.43 standard deviations below the mean. This region is shaded in Figure 10.8

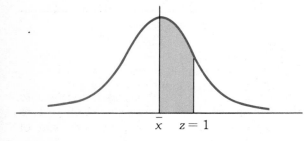

Figure 10.7

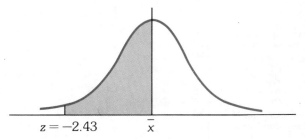

Figure 10.8

Example 2 The average length of a long distance telephone call in a certain city is 4 minutes, with a standard deviation of 2 minutes. What percent of the calls are longer than 8 minutes?

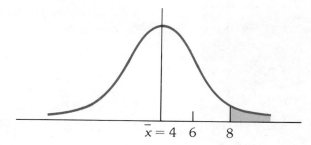

Figure 10.9

Solution.   Here 8 minutes is two standard deviations more than the mean. Thus, we need to find the fraction of all phone calls with length more than two standard deviations above the mean. That is, we need to find the shaded area in Figure 10.9.

From the table in the Appendix, the area between the mean and two standard deviations is .477 ($z = 2.00$). The total area from the mean to the right is .500. We can find the area from $z = 2.00$ to the right by subtracting these numbers: .500 − .477 = .023.   About .023, or 2.3%, of all phone calls exceed 8 minutes.

Example 3   Find the total areas indicated in the following figures: (a) Figure 10.10;   (b) Figure 10.11.

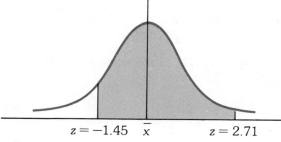

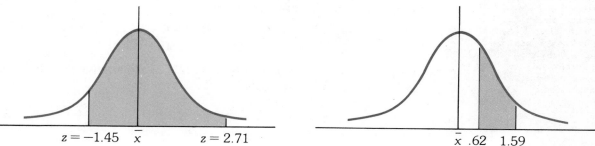

Figure 10.10   **Figure 10.11**

(**a**) Here we need the area from 1.45 standard deviations below the mean to 2.71 standard deviations above the mean. From the table, $z = 1.45$ leads to an area of .426, while $z = 2.71$ leads to .497. The total area is the sum of these, or .426 + .497 = .923.

(**b**) We need to find the area between $z = .62$ and $z = 1.59$. From the table, $z = .62$ leads to an area of .232, while $z = 1.59$ gives .444. To get the area between these two values of z, we subtract the areas here: .444 − .232 = .212.

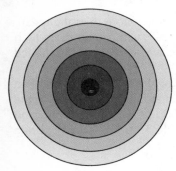

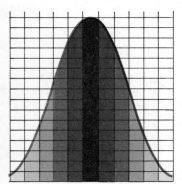

A normal frequency distribution *will occur in the game of darts if the following conditions hold: the player always aims at the bull's-eye; the player tosses a fairly large number of times; the aim on each toss is affected by independent random errors.*

It is not always easy to find the number of standard deviations that one data value is from the mean. This can be found using the formula

$$z = \frac{\text{value} - \text{mean}}{\text{standard deviation}}$$

For example, suppose a normal curve has mean 220 and standard deviation 12. To find the number of standard deviations that a given value is from the mean, we use the formula above. Thus, for 250,

$$z = \frac{250 - 220}{12} = \frac{30}{12} = 2.50;$$

250 is 2.50 standard deviations above the mean. For 204,

$$z = \frac{204 - 220}{12} = \frac{-16}{12} = -1.33;$$

204 is −1.33 standard deviations above the mean (that is, 1.33 standard deviations *below* the mean).

Example 4 The average motorist drives about 1200 miles per month, with standard deviation 150 miles. Assume that the number of miles is closely approximated by a normal curve, and find the percent of all motorists driving: (a) between 1200 and 1600 miles per month; (b) between 1000 and 1500 miles per month.

(a) We need to find how many standard deviations 1600 miles is above the mean. We use the formula given above.

$$z = \frac{1600 - 1200}{150} = \frac{400}{150} = 2.67.$$

From the table, we see that .496, or 49.6%, of all motorists drive between 1200 and 1600 miles per month.

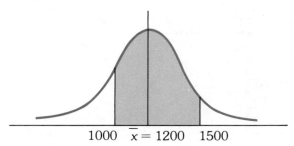

Figure 10.12

(b) As shown in Figure 10.12, we need to find z for both 1000 and 1500.

For 1000: $$z = \frac{1000 - 1200}{150} = \frac{-200}{150} = -1.33;$$

For 1500: $$z = \frac{1500 - 1200}{150} = \frac{300}{150} = 2.00.$$

From the table, $z = -1.33$ leads to an area of .408, while $z = 2.00$ gives .477. Thus, a total of $.408 + .477 = .885$, or 88.5%, of all motorists drive between 1000 and 1500 miles per month.

In the example above, we assumed that a driver could drive any number of miles in a month. A distribution like this that can take on any value is called a **continuous distribution.** Some distributions, however, can take on only a limited number of values. (For example, a die can land with only 1, 2, 3, 4, 5, or 6 up.) These distributions are called **discrete.** We shall see how to handle discrete distributions in the next section. In this section, assume that all distributions are continuous.

10.4 EXERCISES.

Find the percent of area under a normal curve between the mean and each number of standard deviations from the mean.

1.	2.50	**4.**	0.81	**7.**	3.11
2.	1.68	**5.**	−1.71	**8.**	2.80
3.	0.45	**6.**	−2.04		

Find the percent of the total area under the normal curve between two values of z, for Exercises 9–16.

9.	$z = 1.41$ and $z = 2.83$	**13.**	$z = -3.11$ and $z = 1.44$
10.	$z = 0.64$ and $z = 2.11$	**14.**	$z = -2.94$ and $z = -0.43$
11.	$z = -2.48$ and $z = -0.05$	**15.**	$z = -0.42$ and $z = 0.42$
12.	$z = -1.74$ and $z = -1.02$	**16.**	$z = -1.98$ and $z = 1.98$

Find a value of z such that:

*17. 5% of the total area is to the right of z

*18. 1% of the total area is to the left of z

*19. 15% of the total area is to the left of z

*20. 25% of the total area is to the right of z

The Alva light bulb has an average life of 500 hours, with a standard deviation of 100 hours. The length of life of the bulb can be closely approximated by a normal curve. An amusement park buys and installs 10,000 such bulbs. Find the total number that can be expected to last:

21.	at least 500 hours	**27.**	less than 740 hours
22.	less than 500 hours	**28.**	more than 300 hours
23.	between 500 and 650 hours	**29.**	more than 790 hours
24.	between 300 and 500 hours	**30.**	less than 410 hours
25.	between 650 and 780 hours		
26.	between 290 and 540 hours		

The chickens at Colonel Thompson's Ranch have a mean weight of 1850 grams with a standard deviation of 150 grams. The weights of the chickens are closely approximated by a normal curve. Find the percent of all chickens weighing:

31. more than 1700 grams

32. less than 1800 grams

33. between 1750 and 1900 grams

34. between 1600 and 2000 grams

35. less than 1550 grams

36. more than 2100 grams

A box of oatmeal must contain 16 ounces. The machine that fills the oatmeal boxes is set so that, on the average, a box contains 16.5 ounces. The boxes filled by the machine have weights that can be closely approximated by a normal curve. What fraction of the boxes filled by the machine are underweight if the standard deviation is:

37. .5 ounce?   **39.** .2 ounce?

38. .3 ounce?   **40.** .1 ounce?

In nutrition, the Recommended Daily Allowance of Vitamins is a number set by the government to guide an individual's daily vitamin intake. Actually, vitamin needs vary drastically from person to person, but the needs are very closely approximated by a normal curve. To calculate the Recommended Daily Allowance, the government first finds the average need for vitamins among people in the population, and the standard deviation. The Recommended Daily Allowance is then defined as the mean plus 2.5 times the standard deviation.

41. What fraction of the population will receive adequate amounts of vitamins under this plan?

Find the recommended daily allowance for each vitamin:

42. mean need = 1800 units; standard deviation = 140 units

43. mean need = 159 units; standard deviation = 12 units

44. mean need = 1200 units; standard deviation = 92 units

A teacher gives a test to a large group of students. The results are closely approximated by a normal curve. The mean is 74, with standard deviation 6. The teacher wishes to give A's to the top 8% of the students and F's to the bottom 8%. A grade of B is given to the next 15%, with D given similarly. All other students get C. Find the bottom cut-off (rounded to the nearest whole number) for each of the following grades. (*Hint:* Use the table in the Appendix to find the z-value when the area is known.)

*45. A   ***46.** B   ***47.** C   ***48.** D

Depressing data *The girl playing chess has an interest in mathematics equal to that of any boy her age. But what will happen when she enters junior high school and high school? Will she avoid taking more than the minimum requirements in mathematics? In college, will she then lack the mathematics background to major in science, engineering, management science, economics, or in some other male-dominated field?*

Research by sociologists and educators shows that no sex differences vis-a-vis mathematical ability can be detected. However, the socialization of females fosters attitudes that begin to develop about age 13, and usually results in avoidance of mathematics and even anxiety about anything quantitative. Such attitudes are formed because role models are lacking, encouragement at home and in school is withheld, and reference to females in mathematics texts is either stereotyped or absent. Stereotypes reflect common beliefs that women are illogical, inept, and so on—that they work well with people, but not with numbers.

The young woman below seems to have no anxieties about mathematics—she has many options in choosing a career.

Mathematics anxiety *or avoidance is not restricted to women. Any person may feel hopelessness in the face of concepts, exercises, or tests. Anyone may be a victim of early schooling that failed to show the utility of mathematics, failed to build on a student's initial successes, or failed to encourage curiosity.*

To combat mathematics anxiety, colleges around the country have set up special courses, clinics, labs, training workshops, and so on. Here is a partial listing: Wellesley College, *Discovery Course in Elementary Mathematics and its Applications*—Alice Shafer; Wesleyan University, *Mathematics Anxiety Clinic*—Sheila Tobias; Mills College, *Precalculus Laboratory*—Lenore Blum; Manhattan area "Mind over Math" clinics—Stanley Kogelman; University of California at Berkeley, *Math for Girls project*—Nancy Kreinberg, Diane Ressek; University of Missouri at Kansas City, *Introduction to Mathematics sections*—Carolyn MacDonald, Barbara Currier.

Good background information is found in Mathematics and Sex *by John Emest and a research team (University of California, Santa Barbara, 1975) and* Women in Mathematics *by Lynn M. Osen (MIT Press, 1974; in paperback). There are articles by some of the people named above and also by sociologist Lucy Sells.*

*10.5 THE BINOMIAL DISTRIBUTION

A *binomial distribution* is characterized as follows: there are only two possible outcomes (such as *heads* or *tails*, or perhaps *success* or *failure*); repetitions of the experiment are independent of each other; and the probability of a desired result never changes. Examples of binomial distributions include coin tossing, rolling 5 on a die, or selecting a defective radio from a batch produced in a factory.

There are mathematical methods for finding the exact answer to problems involving binomial distributions, but it is usually sufficient to approximate the results using a normal curve.

To see how the normal curve is used, look at the histogram and normal curve in Figure 10.13. This histogram shows the expected results if one coin is tossed 15 times, with the experiment repeated 32,768 times.

Suppose we need to know the fraction of the time that we would get exactly 9 heads on the 15 tosses. We could work this out by the methods of the last chapter. After a huge amount of arithmetic, we would get .153. Geometrically, the answer we desire, the probability of getting exactly 9 heads on 15 tosses, is the fraction we would get by dividing the area of the shaded bar in Figure 10.13 by the total area of all 16 bars in the graph.

The area of the shaded bar is also approximately equal to the area under the normal curve from $x = 8.5$ to $x = 9.5$. The normal curve runs higher than the top of the bar in the left half, but lower in the right half.

To find the area under the normal curve from $x = 8.5$ to $x = 9.5$, we need to find z-values, as we did in the last section. To find the mean and the standard deviation for the distribution, we can use the formulas

$$\text{mean} = n \cdot p, \qquad \text{standard deviation} = \sqrt{n \cdot p \cdot (1 - p)},$$

where n is the number of trials and p is the probability of success on one trial.

In our example of tossing coins, $n = 15$ and $p = \frac{1}{2}$. Using $n = 15$ and $p = \frac{1}{2}$ in the formulas above, we have

$$\text{mean} = 15 \cdot \tfrac{1}{2} = \tfrac{15}{2} = 7.5.$$

$$\text{standard deviation} = \sqrt{15 \cdot \tfrac{1}{2} \cdot (1 - \tfrac{1}{2})} = \sqrt{15 \cdot \tfrac{1}{2} \cdot \tfrac{1}{2}} = \sqrt{3.75} = 1.94.$$

Using these results, we can find z-values for $x = 8.5$ and $x = 9.5$:

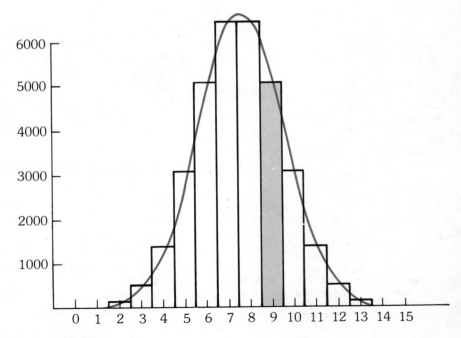

Figure 10.13

$$\text{For } x = 8.5, \quad z = \frac{8.5 - 7.5}{1.94} \qquad \text{For } x = 9.5, \quad z = \frac{9.5 - 7.5}{1.94}$$

$$= \frac{1.00}{1.94} \qquad\qquad\qquad = \frac{2.00}{1.94}$$

$$= .52. \qquad\qquad\qquad\qquad = 1.03.$$

Now we need to find the area under the normal curve from $z = .52$ to $z = 1.03$, a problem similar to those of the last section. From the table in the Appendix, $z = .52$ leads to an area of .199, while $z = 1.03$ leads to .349. To find the result we need, we subtract these two numbers:

$$.349 - .199 = .150.$$

The answer .150 is not far from the exact answer .153 that we stated above.

Example 1   About 6% of the bolts produced by a certain machine are defective. Find the probability that in a sample of 100 bolts: (a) 3 or fewer are defective;   (b) exactly 11 are defective.

(a) We must first find the mean and standard deviation. Here $n = 100$ and $p = 6\% = .06$. Thus,

$$\text{mean} = 100(.06) \qquad \text{standard deviation} = \sqrt{100(.06)(1 - .06)}$$
$$= 6 \qquad\qquad\qquad\qquad = \sqrt{100(.06)(.94)}$$
$$\qquad\qquad\qquad\qquad\qquad = \sqrt{5.64}$$
$$\qquad\qquad\qquad\qquad\qquad = 2.37.$$

As the graph of Figure 10.14 shows, we need to find the area to the left of $x = 3.5$ (since we want 3 or fewer defectives).

$$z = \frac{3.5 - 6}{2.37} = \frac{-2.5}{2.37} = -1.05$$

From the table, $z = -1.05$ leads to an area of .353. Finally, we get the result that we need by subtracting .353 from .500:

$$.500 - .353 = .147.$$

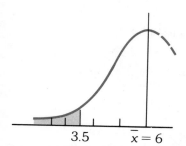

3.5 $\overline{x} = 6$

Figure 10.14

The probability that we will get 3 or fewer defectives in a set of 100 bolts is .147, or 14.7%.

(b) See Figure 10.15. Here we need the area between $x = 10.5$ and $x = 11.5$.

$$\text{For } x = 10.5, \qquad\qquad \text{For } x = 11.5,$$

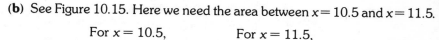

$$z = \frac{10.5 - 6}{2.37} = 1.90. \qquad z = \frac{11.5 - 6}{2.37} = 2.32.$$

Look in the table in the Appendix. The value $z = 1.90$ gives an area of .471, while $z = 2.32$ yields .490. The final answer is the difference of these numbers:

$$.490 - .471 = .019.$$

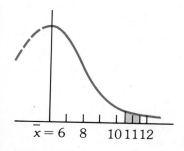

$\overline{x} = 6$ 8   10 11 12

Figure 10.15

There is about a 1.9% chance of having exactly 11 defectives.

10.5 EXERCISES

Suppose 16 coins are tossed. Find the probability of exactly:

1. 8 heads   **2.** 7 heads   **3.** 10 tails   **4.** 12 tails

Suppose 1000 coins are tossed. Find each probability. (*Hint:* $\sqrt{250} = 15.8$)

5. exactly 500 heads   **8.** less than 470 tails

6. exactly 510 heads   **9.** less than 518 heads

7. 480 heads or more   **10.** more than 550 tails

A die is rolled 120 times. Find each probability. (*Hint:* The standard deviation is 4.08.)

11. exactly 20 fives   **14.** exactly 22 twos

12. exactly 24 sixes   **15.** more than 18 threes

13. exactly 17 threes   **16.** fewer than 22 sixes

Two percent of the hamburgers at Tom's Burger Queen are defective. Tom sold 10,000 burgers last week. Find each probability. (*Hint:* $\sqrt{196} = 14$)

17. fewer than 170 burgers were defective

18. more than 222 burgers were defective

A new drug cures 80% of the patients who use it. It is administered to 25 patients. Find the probability that among these patients:

19. exactly 20 are cured   ***22.** no one is cured

20. exactly 23 are cured   ***23.** 12 or fewer are cured

21. all are cured   **24.** between 17 and 23 are cured

10.6 HOW TO LIE WITH STATISTICS

The statement that there are "lies, damned lies, and statistics" is attributed to Disraeli, Queen Victoria's Prime Minister. Other people have said even stronger things about statistics. This often intense distrust of statistics has come about because of a belief that "you can prove anything with numbers." It must be admitted that there is often a conscious or unconscious distortion in many published statistics. The classic book on distortion in statistics is *How to Lie with Statistics,* by Darrell Huff. In the rest of this section we quote some common methods of distortion that Huff gives in his book.

The Sample with the Built-in Bias
A house-to-house survey purporting to study magazine readership was once made in which a key question was: What magazines does your household read? When the results were tabulated and analyzed it appeared that a great many people loved *Harper's* and not very many read *True Story*. Now there were

publishers' figures around at the time that showed very clearly that *True Story* had more millions of circulation than *Harper's* had hundreds of thousands. Perhaps we asked the wrong kind of people, the designers of the survey said to themselves. But no, the questions had been asked in all sorts of neighborhoods all around the country. The only reasonable conclusion then was that a good many of the respondents, as people are called when they answer such questions, had not told the truth. About all the survey had uncovered was snobbery.

In the end it was found that if you wanted to know what certain people read it was no use asking them. You could learn a good deal more by going to their houses and saying you wanted to buy old magazines and what could be had? Then all you had to do was count the *Yale Reviews* and the *Love Romances*. Even that dubious device, of course, does not tell you what people read, only what they have been exposed to.

Similarly, the next time you learn from your reading that the average American (you hear a good deal about him these days, most of it faintly improbable) brushes his teeth 1.02 times a day—a figure pulled out of the air, but it may be as good as anyone else's—ask yourself a question. How can anyone have found out such a thing? Is a woman who has read in countless advertisements that non-brushers are social offenders going to confess to a stranger that she does not brush her teeth regularly? The statistic may have meaning to one who wants to know only what people say about tooth-brushing but it does not tell a great deal about the frequency with which bristle is applied to incisor.

The Well-Chosen Average

A common trick is to use a different kind of average each time, the word "average" having a very loose meaning. It is a trick commonly used, sometimes in innocence but often in guilt, by people wishing to influence public opinion or sell advertising space. When you are told that something is an average you still don't know very much about it unless you can find out which of the common kinds of average it is—mean, median, or mode.

Try your skepticism on some items from "A letter from the Publisher" in *Time* magazine. Of new subscribers it said, "Their median age is 34 years and their average family income is $7,270 a year." An earlier survey of "old TIMErs" had found that their "median age was 41 years. . . . Average income was $9,535. . . ." The natural question is why, when median is given for ages both times, the kind of average for incomes is carefully unspecified. Could it be that the mean was used instead because it is bigger, thus seeming to dangle a richer readership before advertisers?

Be careful when reading charts or graphs—often there is no numerical scale, or no units are given. This makes the chart pretty much meaningless. Huff has a good example of this:

The Little Figures that are Not There

Before me are wrappers from two boxes of Grape-Nuts Flakes. They are slightly different editions, as indicated by their testimonials: one cites Two-Gun Pete and the other says "If you want to be like Hoppy . . . you've got to eat like Hoppy!" Both offer charts to show ("*Scientists proved it's true!*") that these flakes "start giving you energy in 2 minutes!" In one case the chart hidden in these forests of exclamation points has numbers up the side; in the other case the numbers have been omitted. This is just as well, since there is no hint of what the numbers mean. Both show a steeply climbing red line ("energy release"), but one has it starting one minute after eating Grape-Nuts Flakes, the other two minutes later. One line climbs about twice as fast as the other, suggesting that even the draftsman didn't think these graphs meant anything.

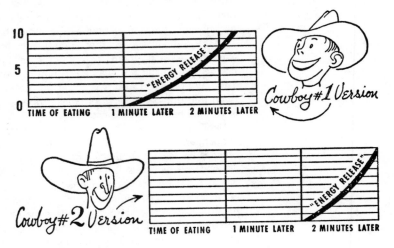

The Gee-Whiz Graph

About the simplest kind of statistical picture, or graph, is the line variety. It is very useful for showing trends, something practically everybody is interested in showing or knowing about or spotting or deploring or forecasting. We'll let our graph show how national income increased ten percent in a year.

Begin with paper ruled into squares. Name the months along the bottom. Indicate billions of dollars up the side. Plot your points and draw your line, and your graph will look like this:

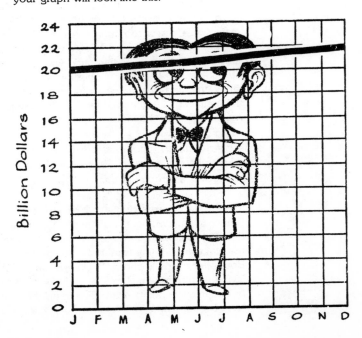

Now that's clear enough. It shows what happened during the year and it shows it month by month. He who runs may see and understand, because the whole graph is in proportion and there is a zero line at the bottom for comparison. Your ten percent looks like ten percent—an upward trend that is substantial but perhaps not overwhelming.

That is very well if all you want to do is convey information. But suppose you wish to win an argument, shock a reader, move him into action, sell him something. For that, this chart lacks schmaltz. Chop off the bottom.

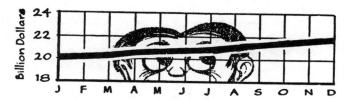

Now that's more like it. (You've saved paper too, something to point out if any carping person objects to your misleading graphics.) The figures are the same and so is the curve. It is the same graph. Nothing has been falsified—except the impression that it gives. But what the hasty reader sees now is a national-income line that has climbed halfway up the paper in twelve months, all because most of the chart isn't there any more. Like the missing parts of speech in sentences that you met in grammar classes, it is "understood." Of course, the eye doesn't "understand" what isn't there, and a small rise has become, visually, a big one.

Now that you have practiced to deceive, why stop with truncating? You have a further trick available that's worth a dozen of that. It will make your modest rise of ten percent look livelier than one hundred percent is entitled to look. Simply change the proportion between the side and the bottom. There's no rule against it, and it does give your graph a prettier shape. All you have to do is let each mark up the side stand for only one-tenth as many dollars as before.

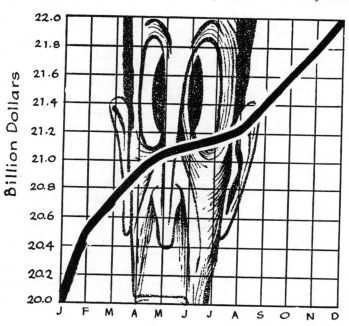

That *is* impressive, isn't it? Anyone looking at it can just feel prosperity throbbing in the arteries of the country. It is a subtler equivalent of editing "National income rose ten percent" into ". . . climbed a whopping ten percent." It is vastly more effective, however, because it contains no adjectives or adverbs to spoil the illusion of objectivity. There's nothing anyone can pin on you.

Suppose Diana makes twice as much money as Mike. One way to show this is with a graph using silver dollars to represent the income of each. If we used one silver dollar for Mike, and two for Diana, we would be fine. But, it is more common to use proportional dollars. It is common to find a dollar of one size used for Mike, with one twice as wide for Diana. This is wrong—the larger dollar actually has four *times* the area, *giving the impression that Diana earns four times as much as Mike. Huff gives another example of this:*

The One-Dimensional Picture
Newsweek once showed how "U.S. Old Folks Grow Older" by means of a chart on which appeared two male figures, one representing the 68.2-year life expectancy of today, the other the 34-year life expectancy of 1879–1889. It was the same old story: One figure was twice as tall as the other and so would have had eight times the bulk or weight. This picture sensationalized facts in order to make a better story. It would be called a form of yellow journalism.

Cause and Effect
Many people often assume that just because two things changed together one caused *the other. The classic example of this is the fact that teachers' salaries and liquor consumption increased together over the last few decades. Neither of these caused the other; rather, both were caused by the same underlying growth in national prosperity. Another case of faulty cause and effect reasoning is in the claim that going to college raises your income. About this Huff says:*

Reams of pages of figures have been collected to show the value in dollars of a college education, and stacks of pamphlets have been published to bring these figures—and conclusions more or less based on them—to the attention of potential students. I am not quarreling with the intention. I am in favor of education myself, particularly if it includes a course in elementary statistics. Now these figures have pretty conclusively demonstrated that people who have gone to college make more money than people who have not. The exceptions are numerous, of course, but the tendency is strong and clear.

The only thing wrong is that along with the figures and facts goes a totally unwarranted conclusion. . . . It says that these figures show that if *you* (your son, your daughter) attend college you will probably earn more money than if you decide to spend the next four years in some other manner. This unwarranted conclusion has for its basis the equally unwarranted assumption that since college-trained folks make more money, they make it because they went to college. Actually we don't know but that these are the people who would have made more money even if they had not gone to college. There are a couple of things that indicate rather strongly that this is so. Colleges get a disproportionate number of two groups of people: the bright and the rich. The bright might show

good earning power without college knowledge. And as for the rich ones . . . well, money breeds money in several obvious ways. Few children of rich parents are found in low-income brackets whether they go to college or not.

Extrapolation

This refers to predicting the future, based only on what has happened in the past. However, it is very rare for the future to be just like the past—something will be different. One of the best examples of extrapolating incorrectly is given by Mark Twain:

In the space of one hundred and seventy-six years the Lower Mississippi has shortened itself two hundred and forty-two miles. That is an average of a trifle over one mile and a third per year. Therefore, any calm person, who is not blind or idiotic, can see that in the Old Oölitic Silurian Period, just a million years ago next November, the Lower Mississippi River was upward of one million three hundred thousand miles long, and stuck out over the Gulf of Mexico like a fishing-rod. And by the same token any person can see that seven hundred and forty-two years from now the Lower Mississippi will be only a mile and three-quarters long, and Cairo and New Orleans will have joined their streets together, and be plodding comfortably along under a single mayor and a mutual board of aldermen. There is something fascinating about science. One gets such whole-sale returns of conjecture out of such a trifling investment of fact.

10.6 EXERCISES

The Norwegian stamp at the bottom left features two graphs.

1. What does the solid line represent?
2. Has it increased much?
3. What can you tell about the dashed line?
4. Does the graph represent a long period of time or a brief period of time?

The Central Bureau of Statistics in Norway *was commemorated by these two stamps on the occasion of its centennial. The stamp at the left pictures respective styles of 1876 and 1976. The stamp at the right shows the growth rate of the Norwegian national product.*

$2.40

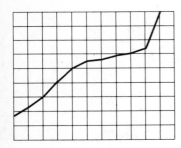

$1.72

The *Newsweek* graph here shows the decline in the value of the British pound for a recent period.

5. Calculate the percent of decrease in the value by using the formula

$$\text{percent of decrease} = \frac{\text{old value} - \text{new value}}{\text{old value}}$$

6. We estimate that the smaller banknote shown has about 50% less area than the larger one. Do you think this is close enough?

Several advertising claims are given in Exercises 7–12. Decide what further information you might need before deciding to accept the claim.

7. 98% of all Toyotas ever sold in the United States are still on the road.

8. Sir Walter Raleigh pipe tobacco is 44% fresher.

9. 8 out of 10 dentists responding to a survey preferred Trident Sugarless Gum.

10. A Volvo has $\frac{2}{3}$ the turning radius of a Continental.

11. A Ford LTD is as quiet as a glider.

12. *Wall Street Journal* circulation has increased, as shown by the graph.

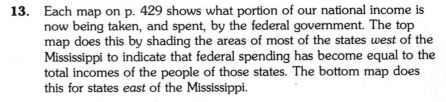

Exercises 13–16 come from Huff's book. Decide how these exercises describe possibly misleading uses of numbers.

13. Each map on p. 429 shows what portion of our national income is now being taken, and spent, by the federal government. The top map does this by shading the areas of most of the states *west* of the Mississippi to indicate that federal spending has become equal to the total incomes of the people of those states. The bottom map does this for states *east* of the Mississippi.

14. Long ago, when Johns Hopkins University had just begun to admit women students, someone not particularly enamored of coeducation reported a real shocker: Thirty three and one-third percent of the women at Hopkins had married faculty members!

15. The death rate in the Navy during the Spanish-American War was nine per thousand. For civilians in New York City during the same period it was sixteen per thousand. Navy recruiters later used these figures to show that it was safer to be in the Navy than out of it.

16. If you should look up the latest available figures on influenza and pneumonia, you might come to the strange conclusion that these ailments are practically confined to three southern states, which account for about eighty per cent of the reported cases.

THE DARKENING SHADOW

(Western Style)

(Eastern Style)

Appendix: Reading Polls

One of the most common uses of statistics in our daily life is in the many polls we see both in newspapers and on television. The book *Lies, Damn Lies, and Statistics* by Michael Wheeler (New York: Liveright, 1976) gives a somewhat biased, but informative, look at polls and pollsters. Wheeler offers the following suggestions for questions to ask yourself when reading a poll.

Healthy skepticism is a necessity in reading polls, not a luxury. What follows is a series of questions which a reader of polls must keep in mind in order to understand what they mean and whether they can be trusted.

Very early in the history of polls, it was discovered that people don't always tell the truth. For example, people often do not give their true ages—young people add to their true age, while older people subtract. (To get around this, the census asks people for the year that they were born. That is harder to fudge on without some thought.)

Sociologists now use polls to help them measure the social acceptibility of certain things. For example, a pollster can ask people about the number of cigarettes they smoke. The answers can be used to get an estimate for the total number of cigarettes smoked in the nation at large. This estimate can be compared to the true number of cigarettes as reported by the various state tax agencies. A few years ago, people were admitting to 73% of the true number of cigarettes. Now, people admit to only 64% of the true number. This decline is used to claim that smoking is not as socially acceptable as it once was.

What do you think would happen if poll takers asked people about their church attendance or the amounts of money they give to charity?

1. Are the numbers right? Americans are beguiled by numbers. We may complain about being digits, not names, in a computerized society, but baseball, celebrated as our national pastime, is less an athletic contest than it is a device for producing statistics like batting averages and runs batted in. Football crowds shout, "We're number one!" The Dow Jones Index measures the pulse of the nation's business in points and fractions.

The careful poll reader must ask two questions: are the numbers right, and what do they mean? As to the first, the reader must know how large the sample was and how it was constructed. For a national poll, a sample of about fifteen hundred should, if fairly drawn, produce results which are within several points one way or the other of what would be obtained if everyone in the country were polled. For a good-sized state, it takes a sample of at least five hundred people to produce the same degree of accuracy.

2. Have the numbers been adjusted? Most of the polls we read in the newspapers are not simple tallies of how many people said yes and how many said no. In most cases, the pollster has adjusted the raw figures in one way or another, but unfortunately that is rarely explained.

Election polls, for example, are often based on interviews with "likely voters." Each pollster has his or her own method for determining who the likely voters are; commonly it involves a series of questions about past voting behavior and interest in the upcoming election. In the end, however, the pollster must make a personal decision about which of the interviews to count in the survey and which to throw out. The pollster must also make a projection of voter turnout on election day. Both of these functions involve a lot more personal judgment than science.

3. How was the poll conducted? Pollsters bicker among themselves as to whether personal interviewing or interviewing by telephone is more reliable.

Both methods have their strengths and weaknesses, and if the poll was conducted by a professional firm, it should not make any significant difference which was used. Amateur polls, those taken by politicians too poorly financed to hire a professional, are often taken by phone because it is cheaper; in such cases, the use of the telephone may be a clue that the poll was a shoestring production.

Postcard polls are never to be trusted. Most congressmen exploit their free franking privilege by sending periodic questionnaires to their constituents. It is rare that more than 10 percent bother to reply, and their responses usually are not representative of the whole. Indeed the fact that they communicated with their congressman marks them as being different from most people.

4. Is it an election poll? Election polls are unique. Unlike all other surveys, they present a clear choice, one that the respondent will shortly have to make as a voter. Regardless of any misgivings, everyone must finally vote one way or the other, or not vote at all. Especially when the election is soon, it is much easier to learn people's true preferences about candidates than their opinions on complex issues.

5. Issue polls: could you answer the question? Polls on issues, be they foreign policy or domestic matters, present an entirely different problem for the pollster. For any given issue, there is a whole range of possible opinions, not just two. The more complex the issue is, the greater the range. Nevertheless, most pollsters try to fit all opinion into the neat categories of agree/disagree, favor/oppose. These simplistic categories make it easy for the keypunch operators to code the results, and they also make for powerful headlines, but they mask the color and depth of public opinion as it truly exists.

6. What do the other polls say? A properly done survey should include many different questions to get at all the nuances and facets of public opinion. All too often the national pollsters rely on only a few questions which give just the barest outlines of opinion. By looking at several polls together, the careful reader can sometimes get a truer picture of public attitudes. Robert Teeter, who has polled for both Nixon and Ford, warns: "Be very suspicious of any poll that claims to reveal new or startling results. If it's out of line with the other polls, it could very well be wrong."

7. What did the pollster really ask? The results of a survey vary significantly with rather inconspicuous changes in wording and format. The December 1975 Gallup Poll that catapulted Reagan into a 40 to 32 percent lead over Ford had presented people with a "laundry list" of ten Republicans, including many who had expressly declared their non-candidacy. The poll had a dramatic effect on press coverage of the campaign; Reagan was transformed from a rather quixotic challenger to the new frontrunner.

In part, however, Reagan's apparent lead may have simply been due to the way that Gallup had posed the question. Just three weeks later, in a far less publicized poll, Gallup asked people to choose between Reagan and Ford. This time the two candidates tied at 45 percent each, with 10 percent undecided. The second study may have been a more realistic look at the then forthcoming primaries, as it pitted the two principal contenders head to head. By contrast, any laundry-list question tends to reduce the percentage of undecided voters; it may also hurt centrist candidates, whose support is chipped away from both sides.

8. Who paid for the poll? There have been flagrant examples of people sponsoring biased polls.

9. Is it really a trend? In late 1975 the *New York Times* headlined, "California Poll Has Ford Losing Ground to Reagan." An examination of the survey, however, shows that conclusion to be virtually groundless. In August the Republican contenders had been in a dead heat, while in November Reagan was reported to have a single-point lead. The sampling error for the survey, however, was plus or minus 5.5 points for each candidate, so it was within the realm of normal chance that Ford actually had a ten-point lead! The newspapers painstakingly keep track of each little dip or rise in a candidate's standings, but such shifts have no real statistical significance.

CHAPTER 10 SUMMARY

Keywords

Statistics	Measure of dispersion
Descriptive statistics	Range
Frequency distribution	Standard deviation, σ
Histogram	Normal (Gaussian) curve
Frequency polygon	Continuous distribution
Class intervals	Discrete distribution
Circle graph	Binomial distribution
Comparison graph	Statistical inference
Central tendency	Hypothesis testing
Mean, $\bar{x}$	Type I error
Median	Type II error
Mode	Significance level

Formulas

$$\text{mean} = \bar{x} = \frac{\text{sum of all numerical values}}{\text{number of values}}$$

$$\text{range} = \text{highest value} - \text{lowest value}$$

$$\text{number of standard deviations from the mean} = z = \frac{\text{value} - \text{mean}}{\text{standard deviation}}$$

Chebyshev's formula For any distribution of numbers, the fraction of them which lie within k standard deviations of the mean is at least

$$1 - \frac{1}{k^2}.$$

Binomial distribution If n is the number of trials, and p is the probability of success on one trial, then

$$\text{mean} = n \cdot p \qquad \text{standard deviation} = \sqrt{n \cdot p \cdot (1 - p)}$$

Standard deviation To find the standard deviation, follow these steps:

Step 1 Find the mean of the data values.
Step 2 Find the deviation from the mean of each value.
Step 3 Square each deviation.
Step 4 Find the mean of the squares (from Step 3).
Step 5 Take the square root of the mean (from Step 4).

The standard deviation, σ (sigma), is defined as the square root (Step 5).

CHAPTER 10 TEST

The following twenty numbers give the number of "Super Whopper Big Burgers" sold in each of the past 20 weeks at Jo's Hamburger Haven.

142 137 125 132 147 129 151 172 175 129
159 148 173 160 152 174 169 163 149 173

1. Complete a frequency distribution for these numbers. Use class intervals 120–129, 130–139, 140–149, 150–159, 160–169, and 170–179.

2. Use the numbers from Exercise 1 to make a histogram.

Find the mean for each of the following.

3. 42, 51, 58, 59, 63, 65, 69, 74, 81, 88

4.

Value	6	10	11	14	19	24
Frequency	7	3	4	2	3	1

Find the median and the mode (or modes) for each set of data.

5. 15, 18, 19, 27, 29, 29, 42   **6.** 41, 39, 45, 47, 41, 39, 51, 38

Find the range and standard deviation for each set of data.

7. 14, 17, 18, 19, 32

8. 26, 43, 51, 29, 37, 56, 29, 82, 74, 93

On standard IQ tests, the mean is 100, with a standard deviation of 15. The results are very close to fitting a normal curve. Suppose an IQ test is given to a very large group of people. Find the percent of these people whose IQ score is:

9. more than 115   **11.** between 70 and 115

10. less than 130

Find the following areas under the normal curve.

12. between $z = 0$ and $z = 1.42$ **14.** to the left of $z = -1.48$

13. between $z = -1.31$ and $z = 2.04$

The average resident of a certain suburb watches television 2.7 hours per evening, with a standard deviation of 1.0 hours. Find the percent of all residents of this suburb who watch:

15. at least 3 hours per evening

16. no more than 2.5 hours per evening

17. between 2 and 4 hours per evening

About 5% of the pills produced by a certain machine are defective. Find the probability that in a sample of 250 pills:

18. 4 or fewer are defective   (*Hint:* $\sqrt{11.875} = 3.4$)

19. exactly 8 are defective

VOICE OF THE PEOPLE
IN FAVOUR OF THE
UNIVERSAL FLY-TRAP
NO SHUT-MOUTH GAME.

11 *Matrices and Their Applications*

Matrix theory as a branch of mathematics is only a little more than a hundred years old, beginning with the work of the English mathematicians Arthur Cayley (1821–1895) and James Sylvester (1814–1897). They were interested in solving systems of linear equations, among other things. We saw one way to solve such systems in Chapter 7, and we shall see another way using matrix methods in Section 3 of this ·chapter.

A matrix is an array of numbers that can be operated on mathematically. The following example shows one source of numbers that lead to a matrix.

Suppose you are running the snack bar at the campus cinema, and need to keep track of the revenue from your various products. For the Monday night showing of ''Casablanca'' with Humphrey Bogart, you decide to list the various dollar amounts in a table:

	Candy	Popcorn	Soft drinks
6:00 show	——	$10	$6.50
8:00 show	$9	$15	$12
10:00 show	$4	$7	$8

If we remember that the rows of this table represent the different shows, while the columns represent the different products, we can write only the numbers (which give the basic information that we want anyway) in a matrix M:

$$M = \begin{bmatrix} 0 & 10 & 6.5 \\ 9 & 15 & 12 \\ 4 & 7 & 8 \end{bmatrix}$$

Hip, hip, array! *Video games are new, but arrays of numbers are not. Chinese mathematicians as early as the first centuries AD used arrays to solve equations. In the West, matrix theory is over a century old. Already its ramifications in physics, engineering, economics, business, and social science are astounding. We discuss only a few applications in this chapter, and not much in detail at that, ending with the use of matrices in game theory. Without matrix theory, video games and other electronic aspects of our civilization would be impossible.*

When numbers are written in a rectangular array like this, and enclosed with brackets or parentheses, the enclosed array is called a **matrix** (plural: *matrices*). The advantage of writing the numbers as a matrix is that the entire array can be treated as a single mathematical entity. A matrix can be named with a single capital letter, as above.

The numbers that make up a matrix are called the **entries,** or *elements,* of the matrix. The entries of matrix M are all numbers, but the matrix itself is not a number, any more than a multiplication table is a number. Various operations can be defined on matrices, including addition, subtraction, and multiplication. For example, we could find $\frac{1}{2}M$, using matrix M from above. This represents the case where all the revenue received from snacks at the Monday night movie is divided into two equal parts, for the two campus clubs sponsoring the sales. We could find the amount that each club gets by multiplying each number in the matrix by $\frac{1}{2}$, as follows.

$$\frac{1}{2}M = \frac{1}{2}\begin{bmatrix} 0 & 10 & 6.5 \\ 9 & 15 & 12 \\ 4 & 7 & 8 \end{bmatrix} = \begin{bmatrix} 0 & 5 & 3.25 \\ 4.5 & 7.5 & 6 \\ 2 & 3.5 & 4 \end{bmatrix}$$

Still remembering that the rows represent different showings of the film while the columns represent different items that were sold, we can look at the matrix on the right and see that each club gets $7.50 from the sale of popcorn at the 8:00 show, for example.

To distinguish a real number from a matrix, real numbers are called **scalars.** Above, we found the product $\frac{1}{2}M$. Here $\frac{1}{2}$ is a scalar, while M is a matrix. Note that the product is also a matrix.

We often classify a matrix by its size, that is, by the number of rows and columns that it contains. For example, matrix M above has 3 rows and 3 columns. In size, it is a 3×3 (read "3 by 3") matrix. A matrix, such as M, that has the same number of rows as columns, is called a *square matrix*. On the other hand, the matrix

$$N = \begin{bmatrix} 2 & 1 & 6 & 3 \\ 4 & -2 & 5 & 1 \end{bmatrix}$$

with 2 rows and 4 columns is a 2×4 matrix. (Give the number of rows first.)

We begin this chapter with a look at some of the basic operations on matrices. We then show how to solve linear systems by matrices. We end the chapter with two other applications—code theory and game theory.

11.1 BASIC OPERATIONS ON MATRICES

In this section we see how to add and subtract matrices, and multiply a matrix by a scalar. Multiplication of a matrix by a matrix is somewhat more complicated and is the subject of the next section.

To see how to add two matrices, we will use matrix M from the opposite page.

$$M = \begin{bmatrix} 0 & 10 & 6.5 \\ 9 & 15 & 12 \\ 4 & 7 & 8 \end{bmatrix}$$

$$T = \begin{bmatrix} 3 & 9 & 8 \\ 1 & 4 & 3 \\ 0 & 2 & 1 \end{bmatrix}$$

Recall that matrix M represents the revenue from the sale of candy, popcorn, and soft drinks at three different showings of the Monday night movie. The rows represent the showings and the columns represent the products. Suppose that on Tuesday night the cinema manager tries to save money by showing an Army training film, Care and Use of Dental Floss, which he advertises as a Japanese anti-modern-world film (and projects out of focus). Revenue from the snack bar on Tuesday is given by matrix T.

How much in total was received from the snack bar on Monday and Tuesday evenings? To find out, we add matrices M and T by adding their corresponding elements:

$$M + T = \begin{bmatrix} \mathbf{0} & 10 & 6.5 \\ 9 & 15 & 12 \\ 4 & 7 & 8 \end{bmatrix} + \begin{bmatrix} \mathbf{3} & 9 & 8 \\ 1 & 4 & 3 \\ 0 & 2 & 1 \end{bmatrix}$$

$$= \begin{bmatrix} \mathbf{0+3} & 10+9 & 6.5+8 \\ 9+1 & 15+4 & 12+3 \\ 4+0 & 7+2 & 8+1 \end{bmatrix}$$

$$M + T = \begin{bmatrix} \mathbf{3} & 19 & 14.5 \\ 10 & 19 & 15 \\ 4 & 9 & 9 \end{bmatrix}$$

From this final result, we see, for example, that a total of $10 was received from the sale of candy at the 8:00 showing, while $14.50 was received from soft drinks at the 6:00 showing.

Example 1 Let $A = \begin{bmatrix} 2 & -1 & 3 & 2 \\ 1 & 0 & 4 & 2 \end{bmatrix}$ and $B = \begin{bmatrix} 5 & 6 & -2 & 5 \\ 3 & -1 & 4 & 7 \end{bmatrix}$.

Find $A + B$.

Solution. Add corresponding elements:

$$A + B = \begin{bmatrix} 2 & -1 & 3 & 2 \\ 1 & 0 & 4 & 2 \end{bmatrix} + \begin{bmatrix} 5 & 6 & -2 & 5 \\ 3 & -1 & 4 & 7 \end{bmatrix} = \begin{bmatrix} 7 & 5 & 1 & 7 \\ 4 & -1 & 8 & 9 \end{bmatrix}$$

The result of adding two matrices is another matrix. Also, by the definition, we add two matrices by adding their corresponding elements. Matrices can have corresponding elements only if they are the same size. Thus, only matrices of the same size can be added together.

Example 2 Let $A = \begin{bmatrix} 2 & 1 \\ 3 & 4 \end{bmatrix}$ and let $B = \begin{bmatrix} 4 & -1 & 0 \\ 2 & 5 & 0 \end{bmatrix}$.

Find $A + B$.

Solution. Since A is 2×2 and B is 2×3, the matrices are not the same size. Thus, we cannot find their sum.

The *difference* of two matrices of the same size is found in a similar way, by *subtracting* corresponding entries.

Example 3 $\begin{bmatrix} 4 & -3 \\ -1 & 5 \end{bmatrix} - \begin{bmatrix} -2 & 4 \\ 1 & 5 \end{bmatrix} = \begin{bmatrix} 4-(-2) & -3-4 \\ -1-1 & 5-5 \end{bmatrix} = \begin{bmatrix} 6 & -7 \\ -2 & 0 \end{bmatrix}$

Example 4   Matrix K shows the weights of four men and four women at the beginning of a diet designed to produce weight loss. Matrix M shows the weights after the diet.

$$K = \begin{bmatrix} 160 & 158 & 172 & 193 \\ 132 & 143 & 119 & 157 \end{bmatrix} \begin{matrix} \text{Men} \\ \text{Women} \end{matrix} \qquad M = \begin{bmatrix} 154 & 148 & 163 & 178 \\ 132 & 154 & 112 & 136 \end{bmatrix}$$

To find the weight loss of the people on the diet, we find matrix $K - M$:

$$K - M = \begin{bmatrix} 160-154 & 158-148 & 172-163 & 193-178 \\ 132-132 & 143-154 & 119-112 & 157-136 \end{bmatrix}$$

$$K - M = \begin{bmatrix} 6 & 10 & 9 & 15 \\ 0 & -11 & 7 & 21 \end{bmatrix}$$

What is your interpretation of the entry -11?

As we said in the introduction to this chapter, a real number is called a *scalar* in distinction to a matrix.

Example 5   Let $A = \begin{bmatrix} 3 & -2 \\ -4 & 6 \end{bmatrix}$ and let b represent the scalar -2, or $b = -2$. Find bA, that is, $-2A$.

Solution.   Multiply each element of A by -2.

$$-2A = -2\begin{bmatrix} 3 & -2 \\ -4 & 6 \end{bmatrix} = \begin{bmatrix} -2(3) & -2(-2) \\ -2(-4) & -2(6) \end{bmatrix} = \begin{bmatrix} -6 & 4 \\ 8 & -12 \end{bmatrix}$$

11.1 EXERCISES

Find, where possible, the matrix sums or differences in Exercises 1–12.

1. $\begin{bmatrix} 3 & 2 \\ 5 & 1 \end{bmatrix} + \begin{bmatrix} 8 & -1 \\ 4 & 3 \end{bmatrix}$

2. $\begin{bmatrix} -6 & 9 \\ 2 & 4 \end{bmatrix} + \begin{bmatrix} -2 & 5 \\ -1 & 3 \end{bmatrix}$

3. $\begin{bmatrix} 2 & 8 & -1 \\ 4 & 0 & 3 \end{bmatrix} - \begin{bmatrix} -1 & 5 & 2 \\ 0 & 4 & 3 \end{bmatrix}$

4. $\begin{bmatrix} -1 & 2 & 4 \\ -2 & 3 & 0 \end{bmatrix} - \begin{bmatrix} 2 & -1 & 4 \\ -5 & 6 & 9 \end{bmatrix}$

5. $\begin{bmatrix} -2 & 3 & 1 \\ 4 & 0 & -2 \\ 5 & -1 & 6 \end{bmatrix} + \begin{bmatrix} -1 & 1 & 1 \\ 4 & 2 & 3 \\ -1 & 4 & 7 \end{bmatrix}$

6. $\begin{bmatrix} 3 & 4 & -1 \\ 7 & -8 & 2 \\ 9 & -1 & 3 \end{bmatrix} - \begin{bmatrix} -4 & 2 & -1 \\ 3 & -1 & 8 \\ 4 & 9 & 0 \end{bmatrix}$

7. $\begin{bmatrix} -4 & 3 & 2 \\ -8 & 0 & 1 \\ 4 & 2 & 1 \end{bmatrix} - \begin{bmatrix} 3 & -4 \\ 2 & -3 \\ 8 & 9 \end{bmatrix}$

8. $\begin{bmatrix} -1 & 6 & 2 \\ 3 & 4 & 7 \\ 9 & 8 & 2 \end{bmatrix} + \begin{bmatrix} -4 & 9 & 2 & 0 \\ -3 & 1 & 4 & 0 \\ 8 & 9 & 7 & 0 \end{bmatrix}$

Determinants came before matrices, *and a brief description will help in understanding how matrix theory developed. Determinants come from arrays made up of numbers arising in systems of equations: numbers multiplying variables (called coefficients) and constants. For example, the system of equations*

$$3x + y = 5$$
$$2x - y = 10$$

contains coefficients 3, 1, 2, and −1; and constants 5 and 10. The four coefficients, for example, can be arrayed and marked with two vertical "fences" to form the determinant

$$\begin{vmatrix} 3 & 1 \\ 2 & -1 \end{vmatrix}$$

This determinant stands for a real number, by definition, the product of the two numbers on the diagonal from NW to SE minus the product of the two numbers on the diagonal from SW to NE. In other words,

$$\begin{vmatrix} 3 & 1 \\ 2 & -1 \end{vmatrix} = 3(-1) - 2(1) = -5.$$

A difference of products such as this comes into play in methods of solving linear systems (see the addition or elimination method in Chapter 7, page 254). Generalizing on this method can do away with trial-and-error in solving any particular system. Determinant theory is just such a generalization.

By the late 17th century, the basic properties of determinants were recognized by Leibniz. In 1750 Gabriel Cramer (1704–1752) published the theory. Determinants played an important role for the next century. Cramer's Rule is still used to find solutions of linear systems via determinants.

9. $\begin{bmatrix} 3 \\ 4 \\ 2 \end{bmatrix} + \begin{bmatrix} -1 \\ 9 \\ 8 \end{bmatrix} - \begin{bmatrix} 2 \\ 1 \\ 3 \end{bmatrix}$

10. $\begin{bmatrix} 7 \\ 8 \\ 9 \end{bmatrix} + \begin{bmatrix} 1 \\ 9 \\ 3 \end{bmatrix} - \begin{bmatrix} -4 \\ 3 \\ 8 \end{bmatrix}$

11. $\begin{bmatrix} -2 & 4 \\ 0 & 9 \end{bmatrix} + \begin{bmatrix} -8 & 1 \\ 3 & 6 \end{bmatrix} - \begin{bmatrix} 4 & 7 \\ -2 & 5 \end{bmatrix}$

12. $\begin{bmatrix} -3 & 2 \\ 4 & 8 \end{bmatrix} + \begin{bmatrix} -9 & -1 \\ 6 & 3 \end{bmatrix} + \begin{bmatrix} -2 & 5 \\ 0 & 8 \end{bmatrix}$

Let $A = \begin{bmatrix} -4 & 0 \\ 3 & -4 \end{bmatrix}$ and let $B = \begin{bmatrix} 2 & -1 \\ -4 & 0 \end{bmatrix}$. Compute the following:

13. $A + B$ **16.** $3B$ *19.** $2A + 3B$

14. $A - B$ **17.** $-4B$ *20.** $-7A + 4B$

15. $2A$ **18.** $-5A$ *21.** $(-1)A + A$

A dietician prepares a diet specifying the allowable amounts of four main food groups: group I, meats; group II, fruits and vegetables; group III, breads and starches; and group IV, milk products. Amounts are given in units, which represent 1 ounce for meat, $\frac{1}{2}$ cup for fruits and vegetables, 1 slice for bread, and 8 ounces for milk.

22. The number of units for breakfast for each of the four food groups respectively are 2, 1, 2, and 1; for lunch, 3, 2, 2, and 1, and for dinner, 4, 3, 2, and 1. Write a 3 × 4 matrix using this information.

23. The amounts of fat, carbohydrates, and protein in each food group respectively are as follows:

Fat: 5, 0, 0, 10
Carbohydrates: 0, 10, 15, 12
Protein: 7, 1, 2, 8

Use this information to write a 4 × 3 matrix.

24. Assume there are 8 calories per unit of fat, 4 calories per unit of carbohydrate, and 5 calories per unit of protein. Write this data in a 3 × 1 matrix.

25. The Jones Furniture Company makes sofas and armchairs in three models of each, A, B, and C. Each month the company sends to each of its warehouses 10 model-A sofas, 12 model-B sofas, 5 model-C sofas, 15 model-A chairs, 20 model-B chairs, and 8 model-C chairs. Write this information as a 2 × 3 matrix. Put sofas in the top row of the matrix.

26. Suppose the New York warehouse of the Jones Company in Exercise 25 had the following stock on September 1.

$$N = \begin{bmatrix} 45 & 35 & 20 \\ 65 & 40 & 35 \end{bmatrix} \begin{matrix} \text{Sofas} \\ \text{Chairs} \end{matrix}$$

with column headings A   B   C

Matrices overshadow determinants *A century after Cramer established the theory of determinants, three mathematicians working in higher algebra contributed to the theory of another kind of array. James Sylvester (see page 441) gave the name "matrix" to any rectangular array from which a determinant could be formed. About this time William Rowan Hamilton (page 443) used arrays to perform linear "transformations" (page 444). He had already conceived of systems without the commutative law for multiplication. (Matrix multiplication is not commutative!) At last, Arthur Cayley (page 440) published the essential theory of matrices, influenced both by the form of matrices and by the brief notation of a system of linear equations in terms of coefficients and constants.*

Sylvester's "matrix" was apt at the time when determinants had supremacy. The word had the general meaning of something in which something else developed or formed. In fact, "matrix" goes back to the Latin mater (for "mother") and a related form of it meaning "womb." Now that matrices are independent of determinants, the term is not so meaningful. A kind of role reversal has taken place, as it were: matrix theory has put determinants in the shadow.

If no stock is sent out of the warehouse during September, how much stock will be on hand October 1? Use the result of Exercise 25 and matrix methods.

27. In Exercise 26, how many model-B chairs were on hand October 1 at the New York warehouse?

28. How many model C sofas?

29. The amount of each model held on September 1 at the Chicago and San Francisco warehouses of the Jones Company are:

$$C = \begin{bmatrix} 22 & 25 & 38 \\ 31 & 34 & 35 \end{bmatrix} \quad \text{and} \quad S = \begin{bmatrix} 30 & 32 & 28 \\ 43 & 47 & 30 \end{bmatrix}$$

Find the total inventory in all three warehouses on September 1.

30. Suppose the Chicago warehouse of the Jones Company shipped the following number of items during September:

$$K = \begin{bmatrix} 5 & 10 & 8 \\ 11 & 14 & 15 \end{bmatrix}$$

Find the stock on hand October 1, taking into account the number of items received and shipped during the month.

Equality of matrices Two matrices are *equal* if corresponding elements are equal. For example, the matrices

$$\begin{bmatrix} -1 & 6 \\ 2 & 4 \end{bmatrix} \quad \text{and} \quad \begin{bmatrix} -1 & 2 \\ 6 & 4 \end{bmatrix}$$

have the same elements, but not in the same positions. Thus, the matrices are not equal. The matrices

$$\begin{bmatrix} x & 3 & 2 \\ 5 & 9 & 1 \end{bmatrix} \quad \text{and} \quad \begin{bmatrix} 8 & y & z \\ 5 & 9 & 1 \end{bmatrix}$$

will be equal only if $x = 8$, $y = 3$, and $z = 2$.

Find the values of all variables in Exercises 31–36.

31. $\begin{bmatrix} 6 & y \\ k & m \end{bmatrix} = \begin{bmatrix} 6 & 3 \\ 8 & 1 \end{bmatrix}$

32. $\begin{bmatrix} 5 & 7 \\ 1 & 8 \end{bmatrix} = \begin{bmatrix} a & b \\ 1 & c \end{bmatrix}$

33. $\begin{bmatrix} x+2 & 5 & 9 \\ 1 & 3 & y \end{bmatrix} = \begin{bmatrix} 2x+1 & z-3 & 9 \\ 1 & 3 & 8 \end{bmatrix}$

34. $\begin{bmatrix} 2 & m+5 & a \\ 1 & n-3 & -1 \end{bmatrix} = \begin{bmatrix} 2 & 2m+6 & 2a \\ 1 & 4n-12 & -1 \end{bmatrix}$

35. $\begin{bmatrix} -2 & 6 \\ k & 2 \end{bmatrix} + \begin{bmatrix} x & y \\ 3 & 5 \end{bmatrix} = \begin{bmatrix} 8 & 2 \\ 7 & 7 \end{bmatrix}$

36. $\begin{bmatrix} 5 & m & 3 \\ 2 & y & 5 \end{bmatrix} - \begin{bmatrix} 3 & 4 & a \\ 1 & 6 & b \end{bmatrix} = \begin{bmatrix} 2 & 5 & 2 \\ 1 & 4 & 1 \end{bmatrix}$

Arthur Cayley (1821–1895) showed such an early affinity for mathematics, that he went to Cambridge instead of entering his father's business. Even while Cayley was practicing law (1848–1862), he wrote numerous theoretical papers on mathematics. During this time he and Sylvester were founding the theory of invariants, out of which grew Cayley's Memoir on the Theory of Matrices (1857). Invariants are algebraic expressions that remain virtually unaltered during certain "transformations," such as flips, rotations, stretches.

In 1863 Cayley became Sadlerian Professor of Pure Mathematics at Cambridge. Except for a lecture series at Johns Hopkins in 1881, his life centered around Cambridge. His work, in contrast, extended to nearly every branch of mathematics. He helped to open the horizons of algebra to include systems that worked even where the commutative law didn't hold. He also helped to free geometry from its three-dimensional Euclidean limitations. Cayley theorized about geometry in n dimensions as early as 1843.

For each of the following statements, give an example to show that it is false, or else give an example illustrating that it is true. For matrices A, B, and C of the same size, and for scalars a and b,

*37. $A + B = B + A$ (*commutative property for addition*)

*38. $A - B = B - A$

*39. $A + (B + C) = (A + B) + C$ (*associative property for addition*)

*40. $A + B$ is a matrix   (*closure property*)

*41.   If 0 is a matrix containing only zero entries, then $0 + A = A$ and $A + 0 = A$.   (*identity property for addition*)

*42.   If $-A = (-1)A$, then $A + (-A) = 0$ and $-A + A = 0$.   (*inverse property for addition*)

*43. $A - B = A + (-B)$, where $-B = (-1)B$.   ***44.** $aB = Ba$

*45. $a(B + C) = aB + aC$ ***46.** $(a + b)C = aC + bC$

*47.   Does the set of all 2×2 matrices and the operation of addition form a mathematical system? (See Chapter 6.)

*48.   Does the system of Exercise 47 form a group? (See Chapter 6.)

11.2 MULTIPLICATION OF TWO MATRICES

Matrix multiplication is not defined in quite so easy a way as was matrix addition. To see how the multiplication of matrices works, let us consider an example.

Suppose a contractor builds two styles of houses, Colonial and Modern, and offers two models in each style. Matrix X shows the number of each type of dwelling the contractor is planning for a new 60-unit subdivision.

$$X = \begin{matrix} & \text{Modern} & \text{Colonial} \\ & \begin{bmatrix} 10 & 15 \\ 15 & 20 \end{bmatrix} & \begin{matrix} \text{Model A} \\ \text{Model B} \end{matrix} \end{matrix}$$

From this matrix, we see that the contractor is planning 10 model-A Modern houses, 15 model-A Colonial, and so on. The number of units of lumber needed for each style of house is shown in matrix Y. (We assume both models in a style take the same amount of lumber.)

$$Y = \begin{bmatrix} 2 \\ 3 \end{bmatrix} \begin{matrix} \text{Modern} \\ \text{Colonial} \end{matrix}$$

From matrix Y, we see that 2 units of lumber are needed for each Modern house, and 3 units are needed for each Colonial house. To find the total number of units of lumber needed for all the model-A houses, we can calculate.

$$10 \cdot 2 + 15 \cdot 3, \quad \text{or} \quad 20 + 45 = 65.$$

That is, 10 Modern model-A's times 2 units of lumber per Modern house, plus 15 Colonial model-A's times 3 units of lumber per Colonial house. This same result could be obtained from the matrices themselves:

$$XY = \begin{bmatrix} 10 & 15 \\ 15 & 20 \end{bmatrix} \cdot \begin{bmatrix} 2 \\ 3 \end{bmatrix}$$

Multiply each entry in the first *row* of X by an entry in the *column* of Y, as the diagram shows.

Multiply: $\begin{bmatrix} 10 & 15 \\ 15 & 20 \end{bmatrix} \begin{bmatrix} 2 \\ 3 \end{bmatrix}$ then $\begin{bmatrix} 10 & 15 \\ 15 & 20 \end{bmatrix} \begin{bmatrix} 2 \\ 3 \end{bmatrix}$

and then add the products: $10 \cdot 2 + 15 \cdot 3 = 65$. The partial product 65 is the entry for row 1, column 1 of the product matrix XY. We can continue by multiplying the entries of row 2 of matrix X by the entries of the column of Y.

Multiply: $\begin{bmatrix} 10 & 15 \\ 15 & 20 \end{bmatrix} \begin{bmatrix} 2 \\ 3 \end{bmatrix}$ then $\begin{bmatrix} 10 & 15 \\ 15 & 20 \end{bmatrix} \begin{bmatrix} 2 \\ 3 \end{bmatrix}$

Add the products: $15 \cdot 2 + 20 \cdot 3 = 90$. The partial product 90 is the entry for row 2, column 1 of XY. The product is:

$$XY = \begin{bmatrix} 10 & 15 \\ 15 & 20 \end{bmatrix} \cdot \begin{bmatrix} 2 \\ 3 \end{bmatrix} = \begin{bmatrix} 65 \\ 90 \end{bmatrix}$$

In the same way, if $\begin{bmatrix} 3 \\ 4 \end{bmatrix}$ is the matrix representing the number of units of brick needed for each model, we have

$$\begin{bmatrix} 10 & 15 \\ 15 & 20 \end{bmatrix} \cdot \begin{bmatrix} 3 \\ 4 \end{bmatrix} = \begin{bmatrix} 10 \cdot 3 + 15 \cdot 4 \\ 15 \cdot 3 + 20 \cdot 4 \end{bmatrix} = \begin{bmatrix} 90 \\ 125 \end{bmatrix}$$

We can combine the separate matrices for units of lumber and of brick into one matrix:

$$Z = \begin{bmatrix} 2 & 3 \\ 3 & 4 \end{bmatrix}$$

The product of matrix X and matrix Z can be found by combining the results from the two separate product matrices above:

	Lumber	Bricks	
$\begin{bmatrix} 10 & 15 \\ 15 & 20 \end{bmatrix} \cdot \begin{bmatrix} 2 & 3 \\ 3 & 4 \end{bmatrix} = \begin{bmatrix}$	65	90	$\end{bmatrix}$ Model A
	90	125	Model B

Hence, 90 units of bricks will be needed for the model-A houses, 125 units for the model-B houses, and so on.

We can use the "criss-cross, multiplying row by column" method of the above example to find the product of any two matrices.

Example 1 Find the product of the matrices

$$A = \begin{bmatrix} -2 & 3 \\ 2 & -1 \end{bmatrix} \quad \text{and} \quad B = \begin{bmatrix} -1 & -3 \\ -2 & 2 \end{bmatrix}$$

James Joseph Sylvester
(1814–1897) was working as an actuary in the early 1850s in London when Cayley was practising law there. They met, and while not collaborators as such, they worked side–by–side on the algebra of invariants. Sylvester thought up "matrix" and a number of other terms such as "invariant" and "syzygy" (look that up!). Sylvester's early career was hampered by English law, which required a religious test (Sylvester was Jewish). Thus he did not get his degree from Cambridge until 1872. Sylvester, in contrast to Cayley, taught at several institutions, including a brief stay at the University of Virginia (1841). He was at Johns Hopkins from 1877 to 1883, where his influence helped to establish mathematical research at the university level. He founded the American Journal of Mathematics.*
Sylvester was also a poet, and published his* Laws of Verse *in 1870, applying principles of "phonetic syzygy."*

Solution.   To find the entry for row 1, column 1 of the product matrix AB, consider only row 1 of A and column 1 of B:

$$\begin{bmatrix} -2 & 3 \\ 2 & -1 \end{bmatrix} \cdot \begin{bmatrix} -1 & -3 \\ -2 & 2 \end{bmatrix}$$

Multiply the entires of row 1 of A by the entries of column 1 of B: $(-2)(-1) = 2$ and $(3)(-2) = -6$. Then add the two products together: $2 + (-6) = -4$. The result, -4, is the entry for row 1, column 1 of the product matrix.

To obtain the entry for row 1, column 2 of the product matrix, we consider row 1 of A and column 2 of B. First, multiply the elements: $(-2)(-3) = 6$ and $3(2) = 6$. Add the products: $6 + 6 = 12$. The result, 12, is the entry for row 1, column 2 of the product matrix.

The entry for row 2, column 1 of the product matrix is found as follows:

$$\begin{bmatrix} -2 & 3 \\ 2 & -1 \end{bmatrix} \begin{bmatrix} -1 & -3 \\ -2 & 2 \end{bmatrix} \qquad \begin{array}{l} \text{Multiply:} \quad 2(-1) = -2 \text{ and } -1(-2) = 2 \\ \text{Add:} \quad -2 + 2 = 0 \end{array}$$

$$\begin{bmatrix} -4 & \square \\ \square & \square \end{bmatrix}$$

$$\begin{bmatrix} -4 & 12 \\ \square & \square \end{bmatrix}$$

$$\begin{bmatrix} -4 & 12 \\ 0 & \square \end{bmatrix}$$

Thus, 0 goes in row 2, column 1 of the product matrix.

Check that -8 goes in row 2, column 2 of the product matrix. The complete product matrix is

$$AB = \begin{bmatrix} -2 & 3 \\ 2 & -1 \end{bmatrix} \cdot \begin{bmatrix} -1 & -3 \\ -2 & 2 \end{bmatrix} = \begin{bmatrix} -4 & 12 \\ 0 & -8 \end{bmatrix}$$

Example 2   With matrices A and B as in Example 1, find

$$BA = \begin{bmatrix} -1 & -3 \\ -2 & 2 \end{bmatrix} \cdot \begin{bmatrix} -2 & 3 \\ 2 & -1 \end{bmatrix}$$

Solution.   Let us find, for example, the entry for row 2, column 1 of the product matrix. From the entries of row 2 of B and column 1 of A, we have

$$(-2)(-2) + 2(2) = 4 + 4 = 8.$$

Thus, 8 is the entry for row 2, column 1 of the product matrix. The complete product matrix is

$$BA = \begin{bmatrix} -4 & 0 \\ 8 & -8 \end{bmatrix}$$

By comparing the product matrices from Examples 1 and 2, we see that the two products are not the same. Matrix multiplication is not commutative! In general, then, for any matrix A and matrix B, provided they can be multiplied together,

$$AB \neq BA.$$

The reason for making the provision in the paragraph above is that we cannot multiply just *any* two matrices. Since we multiply the elements of a *row* of A and a *column* of B to get AB, the number of columns of A must *equal* the number of rows of B. (Recall that we could *add* only matrices of the *same size*.)

Hamilton bridges a theoretical gap   *The year 1843 marked the triumph of William Rowan Hamilton after fifteen years of searching for a way to multiply his quaternions. (A quaternion is a kind of hyper-complex number since it represents a force acting in three dimensions (space) rather than in the plane. Quaternions involve symbols j and k as well as i, the imaginary number discussed in Chapter 5.) Hamilton could add and subtract quaternions, but could not find any product not violating the commutative law AB = BA. It is told that Hamilton was strolling with his wife along the Royal Canal in Dublin when it flashed on him that a mathematical system could be consistent without the commutative law. It happened at Brougham Bridge, and he carved on it the famous formula*

$$i^2 = j^2 = k^2 = ijk = -1.$$

This insight freed algebra, and paved the way for Cayley: as the text indicates, multiplication of matrices is non-commutative!

William Rowan Hamilton
*(1805–1865) had achieved much
by 1843. He was a prodigy as a
linguist and mathematician. His
studies in optics made him famous,
particularly the result that light
refracts as a conical configuration
of rays. He was Ireland's Royal
Astronomer and was knighted in
1835. After 1843 Hamilton devoted
himself to quaternions, which he
believed to have great utility in
science. He elaborated his theories
in the* Lectures on Quaternions
(1853), and his Elements *(1866)
was full of applications. But
applications of quaternions turned
out to be too complicated for the
workaday scientist. Except for
Peter Guthrie Tait (1831–1901),
Hamilton did not rouse any
disciples. After Hamilton's death
a battle of theories broke out
(page 445).*

*In 1943, the fuss was history
when the Irish Academy held its
Quaternion Centenary Celebration,
commemmorated by the above
stamp.*

Example 3 Find each product that exists.

(a)

$$\begin{bmatrix} 2 & -1 & 3 & 1 \\ 0 & 1 & 0 & 2 \end{bmatrix} \cdot \begin{bmatrix} -1 & 2 & 3 \\ 0 & 0 & -1 \\ 1 & 2 & 0 \\ 0 & -1 & 0 \end{bmatrix} = \begin{bmatrix} 1 & 9 & 7 \\ 0 & -2 & -1 \end{bmatrix}$$

The matrix on the left has 4 columns, and the matrix on the right has 4 rows. The number of columns equals the number of rows, a necessary condition for multiplying two matrices.

(b) The product $\begin{bmatrix} -2 & 1 \\ 3 & 4 \end{bmatrix} \cdot \begin{bmatrix} -1 & 2 \\ 0 & 1 \\ 4 & 0 \end{bmatrix}$ cannot be found because there are 2 columns in the matrix on the left, but 3 rows in the matrix on the right. The two matrices cannot be multiplied together.

Another property of matrices is developed in the following example.

Example 4 Let $M = \begin{bmatrix} -8 & 6 \\ 2 & 1 \end{bmatrix}$ and $I = \begin{bmatrix} 1 & 0 \\ 0 & 1 \end{bmatrix}$. Find MI and IM.

Solution. We have

$$MI = \begin{bmatrix} -8 & 6 \\ 2 & 1 \end{bmatrix} \cdot \begin{bmatrix} 1 & 0 \\ 0 & 1 \end{bmatrix} = \begin{bmatrix} -8 \cdot 1 + 6 \cdot 0 & -8 \cdot 0 + 6 \cdot 1 \\ 2 \cdot 1 + 1 \cdot 0 & 2 \cdot 0 + 1 \cdot 1 \end{bmatrix} = \begin{bmatrix} -8 & 6 \\ 2 & 1 \end{bmatrix} = M$$

Here $MI = M$. By multiplying in the same way, we can show that $IM = M$. Matrix I preserves the identity of matrix M for multiplication, and is called the **identity matrix.**

Identity matrices of any square size can be found. For example, the 3×3 identity matrix is displayed below. Notice the similarity between I, the identity for matrix multiplication, and 1, the identity for multiplication of real numbers. As in Chapter 5, $a \cdot 1 = a$ and $1 \cdot a = a$ for every real number a.

$$\begin{bmatrix} 1 & 0 & 0 \\ 0 & 1 & 0 \\ 0 & 0 & 1 \end{bmatrix}$$

11.2 EXERCISES

Find each product in Exercises 1–12 where possible.

1. $\begin{bmatrix} -3 & 2 \\ 4 & 1 \end{bmatrix} \cdot \begin{bmatrix} -2 & 0 \\ 1 & 3 \end{bmatrix}$

2. $\begin{bmatrix} -5 & 2 \\ 1 & 4 \end{bmatrix} \cdot \begin{bmatrix} -1 & 4 \\ 3 & 0 \end{bmatrix}$

3. $\begin{bmatrix} 0 & -2 \\ 5 & 1 \end{bmatrix} \cdot \begin{bmatrix} -3 & 6 \\ 1 & 4 \end{bmatrix}$

4. $\begin{bmatrix} -3 & 5 \\ 1 & 0 \end{bmatrix} \cdot \begin{bmatrix} 8 & -2 \\ 1 & 7 \end{bmatrix}$

5. $\begin{bmatrix} 1 & 3 \\ 4 & 1 \end{bmatrix} \cdot \begin{bmatrix} 2 & 1 & 0 \\ 5 & 2 & 3 \end{bmatrix}$

6. $\begin{bmatrix} -5 & 6 \\ 3 & 4 \end{bmatrix} \cdot \begin{bmatrix} -3 \cdot 5 & 0 \\ 0 & 1 & 5 \end{bmatrix}$

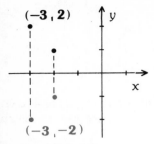

The magic of matrices *The above figure illustrates the idea of a* **linear transformation.** *(This is not a row transformation as in Section 11.3.)*

Take the point (−3, 2), and represent it by the row matrix [−3 2]. Multiply by the 2 × 2 identity matrix:

$$[-3 \quad 2]\begin{bmatrix} 1 & 0 \\ 0 & 1 \end{bmatrix} = [-3 \quad 2].$$

Multiplication by the identity matrix, as expected, leaves the point unchanged; no transformation is involved. Now multiply the row matrix by a matrix with −1 in the row 2, column 2 position:

$$[-3 \quad 2]\begin{bmatrix} 1 & 0 \\ 0 & -1 \end{bmatrix} = [-3 \quad -2].$$

The result is the row matrix that represents the point (−3, −2), shown in color above. The effect is to reflect the point (−3, 2) about the x axis. Transformation has taken place, in this case, a **reflection.** *(Try it with the point (−2, 1).) Multiplying by other matrices, with given patterns of entries, would result in other transformations, such as rotations and shifts.*

7. $\begin{bmatrix} 3 & 2 \\ 5 & 1 \\ 0 & 4 \end{bmatrix} \cdot \begin{bmatrix} 2 & -1 \\ 1 & 3 \end{bmatrix}$

8. $\begin{bmatrix} -1 & 0 \\ 4 & 1 \\ 2 & 0 \end{bmatrix} \cdot \begin{bmatrix} -5 & 0 \\ 4 & 2 \end{bmatrix}$

9. $\begin{bmatrix} -5 & 1 \\ 6 & 2 \end{bmatrix} \cdot \begin{bmatrix} -3 & 2 \\ 4 & 1 \\ 5 & 8 \end{bmatrix}$

10. $\begin{bmatrix} -7 & 3 \\ 9 & 2 \end{bmatrix} \cdot \begin{bmatrix} 4 & -1 \\ 2 & -5 \\ 3 & 6 \end{bmatrix}$

11. $\begin{bmatrix} -5 & 1 & 3 \\ 2 & 0 & 4 \\ 3 & 0 & 2 \end{bmatrix} \cdot \begin{bmatrix} -2 & 0 & 0 \\ 4 & 1 & 0 \\ 0 & 1 & 0 \end{bmatrix}$

12. $\begin{bmatrix} -3 & 4 & 0 \\ 2 & -1 & 5 \\ 0 & 4 & 1 \end{bmatrix} \cdot \begin{bmatrix} 0 & 4 & 0 \\ 2 & 0 & 3 \\ 0 & 5 & 1 \end{bmatrix}$

13. Harry's Donuts, a small neighborhood bakery, sells four main items: sweet rolls, bread, cake, and pies. Matrix A below shows the number of units of the main ingredients needed for each of these items.

	Eggs	Flour	Sugar	Shortening	Milk	
	1	4	$\frac{1}{4}$	$\frac{1}{4}$	1	Sweet rolls
$A =$	0	3	0	$\frac{1}{4}$	0	Bread
	4	3	2	1	1	Cakes
	0	1	0	$\frac{1}{3}$	0	Pies

The cost, in cents per unit, for each ingredient when purchased in large lots or in small lots is given by matrix B:

Cost:	Large lot	Small lot	
	5	5	Eggs
	8	10	Flour
$B =$	10	12	Sugar
	12	15	Shortening
	5	6	Milk

Use matrix multiplication to find a matrix representing the comparative costs per item for the two purchase options.

Suppose a day's orders consist of 20 dozen sweet rolls, 200 loaves of bread, 50 cakes, and 60 pies.

14. Write these orders as a 1 × 4 matrix, and use matrix multiplication to write as a matrix the amount of each ingredient needed to fill the day's orders.

15. Use matrix multiplication to find a matrix representing the costs under the two purchase options to fill the day's orders.

Two matrices A and B are **inverses** of each other if $AB = I$ and $BA = I$. Decide whether or not the following pairs of matrices are inverses of each other by finding their products. (In a later section we see how to actually find the inverse of a given matrix.)

The battle of the vector algebra *After Hamilton's death a 50-year battle got under way involving mathematicians and scientists on both sides of the Atlantic. The importance to science can be seen in the who's who of those pushing their vector algebras to replace quaternions. (A vector represents a force in the plane or in space. Hamilton first used the term — he defined quaternions as quotients (ratios) of vectors. In fact, a row matrix is a vector.)*

Among the first was James Clerk Maxwell, who devised his own vector analysis in the historic Treatise on Electricity and Magnetism (1873). Josiah Willard Gibbs, an American mathematical physicist, produced his own in 1886. Tait said it was a "sort of hermaphrodite monster, compounded of the notations of Hamilton and Grassmann" (see below). Then Oliver Heaviside came out with his own Electromagnetic Theory (1893). By 1900 it was Gibbs vs. Heaviside. Gibbs was particularly influential in the U.S. The battle cooled down in the next decade — and quaternions were out of the picture.

Grassmann discovered at last *In 1844, the year after Hamilton overcame his difficulties, a wider system than Hamilton's appeared. This was the theory of extension by Hermann Grassmann (1809–1877). (Here quaternions are a special case.) Grassmann was a linguist, among other things, and was a scholar of Sanskrit (the classic language of India). Unfortunately for him, Grassmann had a leaning toward philosophy, which tended to distort his meaning for most readers. His work was ignored. Even his revision in 1862 was ignored, although Gibbs expressed his appreciation of the work. In any case, Grassmann gave up on mathematics.*

Grassmann was ahead of his times, and was not appreciated until the early 20th century, when his work was applied to the theory of general relativity.

16. $\begin{bmatrix} 1 & 0 \\ 0 & -1 \end{bmatrix}$ and $\begin{bmatrix} 1 & 0 \\ 0 & -1 \end{bmatrix}$

17. $\begin{bmatrix} 0 & 1 \\ 1 & 0 \end{bmatrix}$ and $\begin{bmatrix} 0 & 1 \\ 1 & 0 \end{bmatrix}$

18. $\begin{bmatrix} 2 & 1 \\ 4 & 3 \end{bmatrix}$ and $\begin{bmatrix} \frac{3}{2} & -\frac{1}{2} \\ -2 & 1 \end{bmatrix}$

19. $\begin{bmatrix} -3 & 2 \\ 6 & -5 \end{bmatrix}$ and $\begin{bmatrix} -\frac{5}{3} & -\frac{2}{3} \\ -2 & -1 \end{bmatrix}$

20. $\begin{bmatrix} 5 & 6 \\ -10 & -13 \end{bmatrix}$ and $\begin{bmatrix} \frac{13}{5} & \frac{6}{5} \\ -2 & 1 \end{bmatrix}$

21. $\begin{bmatrix} 1 & 1 \\ 1 & -1 \end{bmatrix}$ and $\begin{bmatrix} \frac{1}{2} & \frac{1}{2} \\ \frac{1}{2} & \frac{1}{2} \end{bmatrix}$

22. $\begin{bmatrix} 1 & 2 & 0 \\ 0 & 1 & 0 \\ 0 & 1 & 1 \end{bmatrix}$ and $\begin{bmatrix} 1 & -2 & 0 \\ 0 & 1 & 0 \\ 0 & -1 & 1 \end{bmatrix}$

23. $\begin{bmatrix} 0 & 1 & 0 \\ 0 & 0 & -2 \\ 1 & -1 & 0 \end{bmatrix}$ and $\begin{bmatrix} 1 & 0 & 1 \\ 1 & 0 & 0 \\ 0 & -1 & 0 \end{bmatrix}$

24. $\begin{bmatrix} 1 & 3 & 3 \\ 1 & 4 & 3 \\ 1 & 3 & 4 \end{bmatrix}$ and $\begin{bmatrix} 7 & -3 & -3 \\ -1 & 1 & 0 \\ -1 & 0 & 1 \end{bmatrix}$

25. $\begin{bmatrix} -1 & 0 & 2 \\ 3 & 1 & 0 \\ 0 & 2 & -3 \end{bmatrix}$ and $\begin{bmatrix} -\frac{1}{5} & \frac{4}{15} & -\frac{2}{15} \\ \frac{3}{5} & \frac{1}{5} & \frac{6}{15} \\ \frac{2}{5} & \frac{2}{15} & -\frac{1}{15} \end{bmatrix}$

26. $\begin{bmatrix} 4 & 3 & 3 \\ -1 & 0 & -1 \\ -4 & -4 & -3 \end{bmatrix}$ and $\begin{bmatrix} 4 & 3 & 3 \\ -1 & 0 & -1 \\ -4 & -4 & -3 \end{bmatrix}$

27. $\begin{bmatrix} 1 & 0 & 2 \\ -1 & 0 & -2 \\ 1 & 1 & 1 \end{bmatrix}$ and $\begin{bmatrix} 1 & 0 & -2 \\ -1 & 0 & 2 \\ 1 & 1 & 1 \end{bmatrix}$

A **magic square** is a square matrix with the property that the sum of row, column, and diagonal elements are respectively equal. A *pseudomagic square* has at least the sum of its row and column elements equal. Decide whether or not the following matrices are magic or pseudomagic squares.

28. $\begin{bmatrix} 4 & 9 & 2 \\ 3 & 5 & 7 \\ 8 & 1 & 6 \end{bmatrix}$

29. $\begin{bmatrix} 16 & 3 & 2 & 13 \\ 5 & 10 & 11 & 8 \\ 9 & 6 & 7 & 12 \\ 4 & 15 & 14 & 1 \end{bmatrix}$

30. $\begin{bmatrix} 11 & 24 & 7 & 20 & 3 \\ 4 & 12 & 25 & 8 & 16 \\ 17 & 5 & 13 & 21 & 9 \\ 10 & 18 & 1 & 14 & 22 \\ 23 & 6 & 19 & 2 & 15 \end{bmatrix}$

Let $A = \begin{bmatrix} 8 & 1 & 6 \\ 3 & 5 & 7 \\ 4 & 9 & 2 \end{bmatrix}$ and $B = \begin{bmatrix} 6 & 1 & 8 \\ 7 & 5 & 3 \\ 2 & 9 & 4 \end{bmatrix}$. Perform the following matrix operations on matrices A and B. Then decide if the result is either a magic or pseudomagic square.

*31. $A + B$

*32. $2A + 3B$

*33. AB

*34. A^2 (Hint: $A^2 = A \cdot A$)

*35. B^2

*36. $A^2 + B^2$

11.3 MATRIX TRANSFORMATIONS AND SYSTEMS OF EQUATIONS

In Chapter 7, we saw how to solve a system of equations, such as

$$2x - 3y = 10$$
$$2x + 2y = 5.$$

To solve such a system, we need to find a value of x and a value of y that makes both equations true at the same time. We shall use matrix methods to solve this system in this section. These matrix methods involve *row transformations* of a matrix.

To begin the solution of our system, we form an *augmented matrix* using only the numbers from the system. Thus,

$$\begin{array}{c} 2x - 3y = 10 \\ 2x + 2y = 5 \end{array} \quad \text{leads to the augmented matrix} \quad \begin{bmatrix} 2 & -3 & 10 \\ 2 & 2 & 5 \end{bmatrix}$$

The vertical line is used only to separate the numbers from the two sides of the equals signs. Since the numbers in this augmented matrix come from our original system, we can do anything to this matrix that we could do to the original system. In particular, we can do the following three things to the matrix. These three things are called **row transformations.**

(1) Interchange two rows.

(We can certainly interchange any two of our original equations.)

(2) Multiply all elements of one row by a given non-zero number.

(This is the same as multiplying both sides of an equation by a non-zero number.)

(3) Multiply each element of a row of the matrix by some number, and add the results to the corresponding elements of another row.

(This is actually the basis of the algebraic method of solution in Chapter 7.)

$\begin{bmatrix} 2 & -3 & 10 \\ 2 & 2 & 5 \end{bmatrix}$ To see how these row transformations can be used to solve the system given above, we begin with the augmented matrix.

$\begin{bmatrix} 1 & -\frac{3}{2} & 5 \\ 2 & 2 & 5 \end{bmatrix}$ It is usually a good idea first to get the element 1 in row 1, column 1. We do this by multiplying each element of row 1 by $\frac{1}{2}$, getting the new matrix.

$\begin{bmatrix} 1 & -\frac{3}{2} & 5 \\ 0 & 5 & -5 \end{bmatrix}$ Now we should get 0 in row 2, column 1. Do this by multiplying each element of row 1 by -2, and adding the results to the elements of row 2.

$\begin{bmatrix} 1 & -\frac{3}{2} & 5 \\ 0 & 1 & -1 \end{bmatrix}$ Get 1 in row 2, column 2 by multiplying each element of row 2 by $\frac{1}{5}$.

$\begin{bmatrix} 1 & 0 & \frac{7}{2} \\ 0 & 1 & -1 \end{bmatrix}$ Finally, get 0 in row 1, column 2 by multiplying each element of row 2 by $\frac{3}{2}$ and adding the results to the corresponding elements of row 1.

This final augmented matrix leads to the system

$$1x + 0y = 7/2$$
$$0x + 1y = -1,$$

or simply $x = \frac{7}{2}$, $y = -1$. We write this solution as the ordered pair $(\frac{7}{2}, -1)$.

Example 1 Solve the system

$$2x + 3y = 14$$
$$3x - y = 10.$$

Solution. Begin by writing an augmented matrix. If we now use row transformations to change the numbers to the left of the line into the identity matrix, the numbers to the right of the line will become the answers for x and y, as we saw above.

$$\begin{bmatrix} 2 & 3 & | & 14 \\ 3 & -1 & | & 10 \end{bmatrix}$$

$$\begin{bmatrix} 1 & \frac{3}{2} & | & 7 \\ 3 & -1 & | & 10 \end{bmatrix}$$ First, we need 1 in row 1, column 1. Multiply each element of row 1 by $\frac{1}{2}$.

$$\begin{bmatrix} 1 & \frac{7}{2} & | & 7 \\ 0 & -\frac{11}{2} & | & -11 \end{bmatrix}$$ Now get 0 in row 2, column 1. Multiply each element of row 1 by -3 and add the results to row 2.

$$\begin{bmatrix} 1 & \frac{3}{2} & | & 7 \\ 0 & 1 & | & 2 \end{bmatrix}$$ Get 1 in row 2, column 2. Multiply each element of row 2 by $-\frac{2}{11}$.

$$\begin{bmatrix} 1 & 0 & | & 4 \\ 0 & 1 & | & 2 \end{bmatrix}$$ Finally, get 0 in row 1, column 2. Multiply each element of row 2 by $-\frac{3}{2}$ and add the results to row 1.

Read the solution from the rightmost column, $x = 4$ and $y = 2$, written $(4, 2)$.

Example 2 A psychology laboratory which uses rats in experiments has 17 units of wheat and 24 units of barley available daily. One brown rat eats 2 units of wheat and 3 units of barley each day, and one white rat eats 3 units of wheat and 4 units of barley each day. How many of each type of rat should the laboratory keep if all the food must be eaten every day?

Solution. To solve this problem, we must first write a system of equations. Then we can use matrix methods to solve the system. Begin by identifying the variables:

$x =$ number of brown rats   and $y =$ number of white rats

The laboratory has x brown rats. Each brown rat needs 2 units of wheat and 3 units of barley. Thus, the brown rats require a total of $2x$ units of wheat and $3x$ units of barley each day. In the same way, the white rats eat a total of $3y$ units of wheat and $4y$ units of barley daily. Altogether, $2x + 3y$ units of wheat will be needed; the laboratory has 17 units of wheat a day, so that

$$2x + 3y = 17.$$

The total daily amount of barley needed is made up of the barley needed by the brown rats plus the barley needed by the white rats, or $3x + 4y$. Each day, 24 units of barley are available, so that

$$3x + 4y = 24.$$

$\begin{bmatrix} 2 & 3 & | & 17 \\ 3 & 4 & | & 24 \end{bmatrix}$ We must now find the common solution for these two equations. First we write the augmented matrix, and then use row transformations.

$\begin{bmatrix} 1 & \frac{3}{2} & | & \frac{17}{2} \\ 3 & 4 & | & 24 \end{bmatrix}$ Multiply each element of row 1 by $\frac{1}{2}$.

$\begin{bmatrix} 1 & \frac{3}{2} & | & \frac{17}{2} \\ 0 & -\frac{1}{2} & | & -\frac{3}{2} \end{bmatrix}$ Multiply each element of row 1 by -3, and add the results to row 2.

$\begin{bmatrix} 1 & \frac{3}{2} & | & \frac{17}{2} \\ 0 & 1 & | & 3 \end{bmatrix}$ Multiply each element of row 2 by -2.

$\begin{bmatrix} 1 & 0 & | & 4 \\ 0 & 1 & | & 3 \end{bmatrix}$ Multiply each element of row 2 by $-\frac{3}{2}$, and add the results to row 1.

From the rightmost column, read $x = 4$ and $y = 3$. The laboratory needs 4 brown rats and 3 white rats to guarantee that all available food will be eaten.

11.3 EXERCISES

Let $A = \begin{bmatrix} -4 & 0 & 1 \\ 3 & -4 & 2 \\ -1 & 2 & 1 \end{bmatrix}$ and $B = \begin{bmatrix} 2 & -1 & 3 \\ -4 & 0 & 1 \\ 1 & 3 & -3 \end{bmatrix}$.

Apply the row transformations indicated in each case, and write out the resulting matrix. (We work here with a square matrix rather than an augmented matrix just to practice row transformations.)

1. Exchange rows 1 and 3 of A.

2. Exchange rows 2 and 3 of B.

3. Multiply each element of row 2 of A by -3.

4. Double each element in row 1 of A.

5. Multiply each element of row 2 of B by -4, and add the results to row 1.

6. Multiply each element of row 3 of A by 5, and add the results to row 2.

7. Multiply each element of row 1 of A by 3, and add the results to row 3.

8. Multiply each element of row 3 of B by -1, and add the results to row 2.

9. Form matrix $A + B$. Multiply each element of row 2 of $A + B$ by -3, and add the results to row 1.

10. Form matrix $B - A$. Multiply each element of row 1 of $B - A$ by 5, and add the results to row 3.

Some applications of matrices
Matrix algebra is particularly useful in describing phenomena where numerous interconnected relationships exist. Since we are entangled in web upon web of such relationships, physical and economic, for example, the algebra of matrices is pervasive.

In 1925 Werner Heisenberg (1902–1976) introduced matrix algebra into quantum mechanics. This is a branch of physics that studies interaction of atomic particles. (As an electron shifts orbit, energy is emitted in a packet called a quantum.)

Further uses of matrices in physics and engineering were stimulated by a paper on matrix methods in solving oscillation (vibration) problems. This was the work of W. J. Duncan and A. R. Collar, two aeronautical engineers, in 1934.

Linear programming again
Graphs were used in Chapter 7 to solve linear programming problems with two variables. For more than two variables, the simplex method *is an important tool of operations research. The method was devised by George B. Dantzig in 1947. He was concerned with problems of allocating supplies for the U.S. Air Force in the least expensive way. The simplex method, briefly, involves an extended matrix, the* simplex tableau *(a matrix with information in tabular form). Row transformations are used in a systematic way to test for optimum values.*

State which row transformation must be applied to A or B to get the matrices shown in Exercises 11–16.

11. To A; $\begin{bmatrix} 3 & -4 & 2 \\ -4 & 0 & 1 \\ -1 & 2 & 1 \end{bmatrix}$

14. To A; $\begin{bmatrix} -4 & 0 & 1 \\ 3 & -4 & 2 \\ \frac{1}{2} & -1 & -\frac{1}{2} \end{bmatrix}$

12. To B; $\begin{bmatrix} 3 & 2 & 0 \\ -4 & 0 & 1 \\ 1 & 3 & -3 \end{bmatrix}$

15. To B; $\begin{bmatrix} 2 & -1 & 3 \\ 0 & -2 & 7 \\ 1 & 3 & -3 \end{bmatrix}$

13. To A; $\begin{bmatrix} -4 & 0 & 1 \\ 6 & -8 & 4 \\ -1 & 2 & 1 \end{bmatrix}$

16. To B; $\begin{bmatrix} -1 & -10 & 12 \\ -4 & 0 & 1 \\ 1 & 3 & -3 \end{bmatrix}$

Find the sequence of row transformations that will change each of the following matrices into the identity matrix, if possible.

*17. $\begin{bmatrix} 1 & 2 \\ 0 & -1 \end{bmatrix}$

*19. $\begin{bmatrix} -2 & 2 \\ 4 & 1 \end{bmatrix}$

*21. $\begin{bmatrix} 6 & 3 & 0 \\ 0 & 4 & 0 \\ 0 & 0 & 2 \end{bmatrix}$

*18. $\begin{bmatrix} 2 & 1 \\ 1 & -1 \end{bmatrix}$

*20. $\begin{bmatrix} 0 & -1 \\ -2 & 0 \end{bmatrix}$

*22. $\begin{bmatrix} -2 & 2 & 0 \\ 0 & 1 & 0 \\ 1 & 1 & 1 \end{bmatrix}$

Write the augmented matrix for each system. Do not try to solve.

23. $\begin{aligned} 2x + 3y &= 11 \\ x + 2y &= 8 \end{aligned}$

24. $\begin{aligned} 3x + 5y &= -13 \\ 2x + 3y &= -9 \end{aligned}$

25. $\begin{aligned} x + 5y &= 6 \\ 3x - 4y &= 1 \end{aligned}$

26. $\begin{aligned} 2x + 7y &= 1 \\ 5x + y &= -15 \end{aligned}$

27. $\begin{aligned} 2x + y + z &= 3 \\ 3x - 4y + 2z &= -7 \\ x + y + z &= 2 \end{aligned}$

28. $\begin{aligned} 4x - 2y + 3z &= 4 \\ 3x + 5y + z &= 7 \\ 5x - y + 4z &= 7 \end{aligned}$

29. $\begin{aligned} x + y &= 6 \\ 2y + z &= 2 \\ z &= 2 \end{aligned}$

30. $\begin{aligned} x &= 6 \\ y + 2z &= 2 \\ x - 3z &= 2 \end{aligned}$

Use matrix methods to solve the systems of equations in Exercises 31–48.

31. $\begin{aligned} x + y &= 5 \\ x - y &= -1 \end{aligned}$

32. $\begin{aligned} x + 2y &= 5 \\ 2x + y &= -2 \end{aligned}$

33. $\begin{aligned} x + y &= -3 \\ 2x - 5y &= -6 \end{aligned}$

34. $\begin{aligned} 3x - 2y &= 4 \\ 3x + y &= -2 \end{aligned}$

35. $\begin{aligned} 2x - 3y &= 10 \\ 2x + 2y &= 5 \end{aligned}$

36. $\begin{aligned} 6x + y &= 5 \\ 5x + y &= 3 \end{aligned}$

37. $\begin{aligned} 2x - 5y &= 10 \\ 3x + y &= 15 \end{aligned}$

38. $\begin{aligned} 4x - y &= 3 \\ -2x + 3y &= 1 \end{aligned}$

39. $\begin{aligned} 2x - 3y &= 2 \\ 4x - 6y &= 1 \end{aligned}$

40. $\begin{aligned} x + 2y &= 1 \\ 2x + 4y &= 3 \end{aligned}$

41. $\begin{aligned} x + y &= -1 \\ y + z &= 4 \\ x + z &= 1 \end{aligned}$

42. $\begin{aligned} x - z &= -3 \\ y + z &= 9 \\ x + z &= 7 \end{aligned}$

43. $\begin{aligned} x + y - z &= 6 \\ 2x - y + z &= -9 \\ x - 2y + 3z &= 1 \end{aligned}$

45. $\begin{aligned} -x + y \quad &= -1 \\ y - z &= 6 \\ x \quad + z &= -1 \end{aligned}$

*47. $\begin{aligned} 3x + 2y \quad - w &= 0 \\ 2x \quad + z + 2w &= 5 \\ x + 2y - z \quad &= -2 \\ 2x - y + z + w &= 2 \end{aligned}$

44. $\begin{aligned} x + 3y - 6z &= 7 \\ 2x - y + z &= 1 \\ x + 2y + 2z &= -1 \end{aligned}$

46. $\begin{aligned} x + y \quad &= 1 \\ 2x \quad - z &= 0 \\ y + 2z &= -2 \end{aligned}$

*48. $\begin{aligned} x + 3y - 2z - w &= 9 \\ 4x + y + z + 2w &= 2 \\ -3x - y + z - w &= -5 \\ x - y - 3z - 2w &= 2 \end{aligned}$

Use matrix methods to solve each problem.

49. At the Evergreen Ranch, 6 old goats and 5 sheep sell for $305, while 2 old goats and 9 sheep cost $285. Find the cost of an old goat and the cost of a sheep.

50. Linda Ramirez is a building contractor. If she hires 7 day laborers and 2 concrete finishers, her payroll for the day is $346, while 1 day laborer and 5 concrete finishers cost $238. Find the daily wage charge for each type of worker.

51. A biologist wants to grow two types of algae, types A and B. She has available 15 gallons of nutrient 1 and 26 gallons of nutrient 2. A vat of algae A needs 2 gallons of nutrient 1 and 3 gallons of nutrient 2, while a vat of algae B needs 1 gallon of 1 and 2 gallons of 2. How many vats of each type of algae should the biologist grow in order to use all the nutrients?

52. To make his portrait bust of the Maharishi Yoga, Harry bought 2 pounds of dark clay and 3 pounds of light clay, paying $13 for the clay. He later needed one more pound of dark clay and 2 more pounds of light clay, costing $7 altogether. How much did he pay per pound for each type of clay?

53. The perimeter of a triangle is 21 cm. If two sides are of equal length, and the third side is 3 cm longer than one of the equal sides, find the lengths of the three sides.

54. The secretary of the local consumer group bought some decals at 8¢ each, and some bumper stickers at 10¢ each to give to the members. He spent a total of $15.52. If he bought a total of 170 items, how many of each kind did he buy?

*55. The Matrix and the Patrix, a local musical group, is coming to play at a school festival. The reporter for the school newspaper doesn't know how many guitarists and how many members of the rhythm section there are in the group. Janet, one of the guitarists, says that, not counting herself, there are three times as many members of the rhythm group as guitarists. Steve, a member of the rhythm section, says that, not counting himself, the number of members of the rhythm section is one less than twice the number of guitarists. How many of each are there in the group?

11.4 INVERSES OF MATRICES; CODE THEORY

Matrix row transformations, which we used in the previous section to solve systems of equations, can also be used to find the inverse of a matrix, if it exists. Two matrices are **inverses** of each other if their product is I, the appropriate identity matrix. For example, the matrices

$$\begin{bmatrix} 3 & 5 \\ 1 & 2 \end{bmatrix} \quad \text{and} \quad \begin{bmatrix} 2 & -5 \\ -1 & 3 \end{bmatrix}$$

are inverses of each other, since

$$\begin{bmatrix} 3 & 5 \\ 1 & 2 \end{bmatrix} \cdot \begin{bmatrix} 2 & -5 \\ -1 & 3 \end{bmatrix} = \begin{bmatrix} 1 & 0 \\ 0 & 1 \end{bmatrix} \quad \text{and also} \quad \begin{bmatrix} 2 & -5 \\ -1 & 3 \end{bmatrix} \cdot \begin{bmatrix} 3 & 5 \\ 1 & 2 \end{bmatrix} = \begin{bmatrix} 1 & 0 \\ 0 & 1 \end{bmatrix}$$

The inverse of matrix A is written A^{-1} (read "A inverse"). Thus, matrices A and A^{-1} are inverses if $AA^{-1} = I$ and $A^{-1}A = I$. Only square matrices can possibly have inverses. (Why?)

To find the inverse of a given matrix, first combine the given matrix and the identity matrix to form an *augmented matrix*. The identity matrix always goes on the right.

For example, to find the inverse of the matrix

$$A = \begin{bmatrix} 2 & -5 \\ -1 & 3 \end{bmatrix}$$

the first step is to form the augmented matrix:

$$\begin{bmatrix} 2 & -5 & 1 & 0 \\ -1 & 3 & 0 & 1 \end{bmatrix}$$

Then use row transformations so that the given matrix on the left turns into the 2×2 identity matrix. The matrix on the right will then change from the identity matrix to the inverse of the original matrix A.

$$\begin{bmatrix} 1 & -\frac{5}{2} & \frac{1}{2} & 0 \\ -1 & 3 & 0 & 1 \end{bmatrix}$$ To change the matrix on the left into the identity matrix, we first need to get 1 in row 1, column 1. We do this by multiplying the entries in row 1 by $\frac{1}{2}$.

$$\begin{bmatrix} 1 & -\frac{5}{2} & \frac{1}{2} & 0 \\ 0 & \frac{1}{2} & \frac{1}{2} & 1 \end{bmatrix}$$ Now we need 0 in row 2, column 1. To get this 0, multiply each element of row 1 by 1 and add the result to row 2.

$$\begin{bmatrix} 1 & -\frac{5}{2} & \frac{1}{2} & 0 \\ 0 & 1 & 1 & 2 \end{bmatrix}$$ The identity matrix requires 1 in row 2, column 2. We can get 1 there if we multiply each element in row 2 by 2.

$$\begin{bmatrix} 1 & 0 & 3 & 5 \\ 0 & 1 & 1 & 2 \end{bmatrix}$$ Finally, we need 0 in row 1, column 2. We can get an entry of 0 if we multiply each element of row 2 by $\frac{5}{2}$, and add the result to the corresponding entry of row 1.

$$A^{-1} = \begin{bmatrix} 3 & 5 \\ 1 & 2 \end{bmatrix}$$ We have changed the original matrix A into the identity matrix. What had been the identity matrix is now the inverse of A, the matrix A^{-1} at left.

To check, form the products AA^{-1} and $A^{-1}A$. Both these products should equal I, the identity matrix.

In the example above, we found the inverse of a 2×2 matrix. The procedure is the same for a square matrix of any size.

Inverse of a square matrix

Step 1. Form the augmented matrix by placing the identity matrix to the right of the given matrix.

Step 2. Use row transformations to get 1 in row 1, column 1.

Step 3. Get 0's in the rest of column 1.

Step 4. Get 1 in row 2, column 2.

Step 5. Get 0's in the rest of column 2.

Step 6. Continue until the original matrix is changed into the identity.

Step 7. The matrix to the right of the bar is the desired inverse.

If any of these steps are not possible, the given matrix has no inverse. (See Example 2 below.)

Example 1 Find the inverse of $B = \begin{bmatrix} 1 & 3 & 3 \\ 1 & 4 & 3 \\ 1 & 3 & 4 \end{bmatrix}$.

Solution. Form the augmented matrix:

$$\left[\begin{array}{ccc|ccc} 1 & 3 & 3 & 1 & 0 & 0 \\ 1 & 4 & 3 & 0 & 1 & 0 \\ 1 & 3 & 4 & 0 & 0 & 1 \end{array}\right]$$

We already have 1 in row 1, column 1. Thus, we use row transformations to get 0's in the rest of column 1.

$$\left[\begin{array}{ccc|ccc} 1 & 3 & 3 & 1 & 0 & 0 \\ 0 & 1 & 0 & -1 & 1 & 0 \\ 1 & 3 & 4 & 0 & 0 & 1 \end{array}\right]$$

Multiply each element of row 1 by -1; add the results to row 2.

$$\left[\begin{array}{ccc|ccc} 1 & 3 & 3 & 1 & 0 & 0 \\ 0 & 1 & 0 & -1 & 1 & 0 \\ 0 & 0 & 1 & -1 & 0 & 1 \end{array}\right]$$

Multiply each element of row 1 by -1; add the results to row 3.

Now go to column 2. We already have 1 in row 2, column 2, and 0 below the 1. All we need to do now is get 0 above the 1.

$$\left[\begin{array}{ccc|ccc} 1 & 0 & 3 & 4 & -3 & 0 \\ 0 & 1 & 0 & -1 & 1 & 0 \\ 0 & 0 & 1 & -1 & 0 & 1 \end{array}\right]$$

Multiply each element of row 2 by -3; add the results to row 1.

$$\left[\begin{array}{ccc|ccc} 1 & 0 & 0 & 7 & -3 & -3 \\ 0 & 1 & 0 & -1 & 1 & 0 \\ 0 & 0 & 1 & -1 & 0 & 1 \end{array}\right]$$

Multiply each element of row 3 by -3; add the results to row 1.

Read the inverse from the final augmented matrix:

$$B^{-1} = \begin{bmatrix} 7 & -3 & -3 \\ -1 & 1 & 0 \\ -1 & 0 & 1 \end{bmatrix}.$$

To check, show that $BB^{-1} = I$ and $B^{-1}B = I$.

Example 2 Let $C = \begin{bmatrix} 2 & 1 \\ 6 & 3 \end{bmatrix}$. Find C^{-1}.

Solution. Proceed as before:

$\begin{bmatrix} 2 & 1 & | & 1 & 0 \\ 6 & 3 & | & 0 & 1 \end{bmatrix}$ Form the augmented matrix.

$\begin{bmatrix} 1 & \frac{1}{2} & | & \frac{1}{2} & 0 \\ 6 & 3 & | & 0 & 1 \end{bmatrix}$ Multiply each element of row 1 by $\frac{1}{2}$.

$\begin{bmatrix} 1 & \frac{1}{2} & | & \frac{1}{2} & 0 \\ 0 & 0 & | & -3 & 1 \end{bmatrix}$ Multiply each element of row 1 by -6; add the results to row 2.

We cannot go further. There is no way to get 1 in row 2, column 2. Thus, matrix C has no inverse.

Code theory, or *cryptography*, offers an interesting and useful (at least to governments) application of matrix inverses. In the most common type of code constructed by beginners, each letter of the alphabet is replaced with another letter or symbol, such as

$$a = 3 \qquad c = \# \qquad e = +$$
$$b = \$ \qquad d = 9 \qquad f = *$$

and so on. A code like this is very easy for an expert to break. This is because mere substitution of one symbol for another does not change the relative frequency of occurrence of the symbols. Thus, in the substitution of symbols for letters in the English language, $+$ would occur most frequently in any reasonably long message simply because $+$ represents e, and e is the most common letter used in English. (We say "reasonably long message" since in some short messages, such as "summer sessions seldom stimulate" the letter e might not be the most common.) Code breakers have spent a lot of time deciding on the frequency of occurrence of letters, and the following results are now commonly accepted.

Letter	Percent	Letter	Percent	Letter	Percent
E	13	S, H	6	W, G, B	$1\frac{1}{2}$
T	9	D	4	V	1
A, O	8	L	$3\frac{1}{2}$	K, X, J	$\frac{1}{2}$
N	7	C, M, U	3	Q, Z	$\frac{1}{5}$
I, R	$6\frac{1}{2}$	F, P, Y	2		

$\mathcal{A}$ $\mathcal{B}$ $\mathcal{C}$ $\mathcal{D}$ $\mathcal{E}$ $\mathcal{F}$

Aaaaa aaaab aaaba. aaabb. aabaa. aabab.

$\mathcal{A}Aaa$ $\mathcal{B}Bbb$ $\mathcal{C}Ccc$ $\mathcal{D}Ddd$ $\mathcal{E}Eee$ $\mathcal{F}Fff$

Bacon's binary codes *Francis Bacon (1561–1626), philosopher and statesman, was involved in political intrigue and espionage. He invented several binary codes early in the 17th century to hide political secrets carried by messengers between the courts of Europe. His "biliteral cipher" (top line) shows the two letters a and b in combinations of five. The permutations would be $2^5 = 32$, enough for the English alphabet. Bacon's "biformed alphabet" uses two styles of typography to hide a message in a straight text. The decoder deciphers a string of a's and b's, groups them by fives, then deciphers letters and words. Bacon wrote that codes could be made of bells, trumpets, lights, torches, gunshots, and so on, provided that the symbols showed a "twofold difference." In other words, binary codes are not restricted to alphabetic symbols. Centuries later the codes were applied to Shakespeare's plays by people who believed that Bacon was the actual author—they hoped to find messages from Bacon hidden in the texts.*

Governments need to develop more sophisticated methods of coding and decoding messages. Several of the methods depend on matrix theory. To see how, let us assign numbers to letters of the alphabet. For simplicity, let a correspond to 1, b to 2, c to 3, and so on. Ignore punctuation and let 27 correspond to a blank space. Break the message into groups of three letters each. Thus, *mathematics is for the birds* becomes

mat   hem   ati   cs__   is__   for   __th   e__b   ird   s____

where ___ represents a blank. To each group of letters corresponds a matrix of three rows and one column. For example, to the letters m, a, and t, we assign the numbers 13, 1, and 20, respectively, so that

$$\begin{bmatrix} m \\ a \\ t \end{bmatrix} \quad \text{corresponds to} \quad \begin{bmatrix} 13 \\ 1 \\ 20 \end{bmatrix}$$

Our message corresponds to the following sequence of matrices:

$$\begin{bmatrix} 13 \\ 1 \\ 20 \end{bmatrix} \begin{bmatrix} 8 \\ 5 \\ 13 \end{bmatrix} \begin{bmatrix} 1 \\ 20 \\ 9 \end{bmatrix} \begin{bmatrix} 3 \\ 19 \\ 27 \end{bmatrix} \begin{bmatrix} 9 \\ 19 \\ 27 \end{bmatrix} \begin{bmatrix} 6 \\ 15 \\ 18 \end{bmatrix} \begin{bmatrix} 27 \\ 20 \\ 8 \end{bmatrix} \begin{bmatrix} 5 \\ 27 \\ 2 \end{bmatrix} \begin{bmatrix} 9 \\ 18 \\ 4 \end{bmatrix} \begin{bmatrix} 19 \\ 27 \\ 27 \end{bmatrix}$$

Now we choose any 3×3 matrix that has an inverse. Let us choose

$$M = \begin{bmatrix} 1 & 3 & 3 \\ 1 & 4 & 3 \\ 1 & 3 & 4 \end{bmatrix}$$

If we find the product of M and each of the matrices above, we get

$$\begin{bmatrix} 76 \\ 77 \\ 96 \end{bmatrix} \begin{bmatrix} 62 \\ 67 \\ 65 \end{bmatrix} \quad \text{and so on.}$$

The process of translating the message into this series of matrices is called "encoding." The entries of these product matrices can then be transmitted as the message. The recipient of the message divides the numbers into groups of 3, and converts each group of 3 into a matrix. After multiplying these matrices by the matrix M^{-1}, the message can be read.

A code of this type is relatively simple to use, but actually rather hard to decode. For further information on codes, *see The Codebreakers* by David Kahn. Further information on the substitution of symbols for letters is given in the Sherlock Holmes story "The Dancing Men" and in Edgar Allen Poe's "The Gold Bug."

"Shoot from the left eye of the death's head . . ." *In Poe's tale* The Gold-Bug *(1843), the image of a skull on an old piece of parchment starts Legrand, the central character, on a treasure hunt. In the course of it he must decode a message by comparing the frequency of its given symbols with the frequency of letters in the English language. The illustration shows the gold bug (instead of a bullet) dropped from the left eyesocket of the skull.*

11.4 EXERCISES

Find the inverse of each matrix that has an inverse.

1. $\begin{bmatrix} 1 & 0 \\ 0 & -1 \end{bmatrix}$

2. $\begin{bmatrix} 0 & 1 \\ 1 & 0 \end{bmatrix}$

3. $\begin{bmatrix} 2 & 1 \\ 4 & 3 \end{bmatrix}$

4. $\begin{bmatrix} -3 & 2 \\ 6 & -5 \end{bmatrix}$

5. $\begin{bmatrix} 5 & 6 \\ -10 & -13 \end{bmatrix}$

6. $\begin{bmatrix} 1 & 1 \\ 1 & -1 \end{bmatrix}$

7. $\begin{bmatrix} 1 & 1 \\ 1 & 1 \end{bmatrix}$

8. $\begin{bmatrix} -1 & -1 \\ 1 & 1 \end{bmatrix}$

9. $\begin{bmatrix} 2 & 4 \\ 4 & 6 \end{bmatrix}$

10. $\begin{bmatrix} 3 & 5 \\ -1 & 2 \end{bmatrix}$

11. $\begin{bmatrix} 1 & 2 & 0 \\ 0 & 1 & 0 \\ 1 & -1 & 1 \end{bmatrix}$

12. $\begin{bmatrix} 0 & 1 & 0 \\ 0 & 0 & -2 \\ 1 & -1 & 0 \end{bmatrix}$

13. $\begin{bmatrix} 4 & 1 & 2 \\ 0 & 3 & 0 \\ -1 & 1 & 0 \end{bmatrix}$

14. $\begin{bmatrix} -1 & 0 & 2 \\ 3 & 1 & 0 \\ 0 & 2 & -3 \end{bmatrix}$

15. $\begin{bmatrix} 4 & 3 & 3 \\ -1 & 0 & -1 \\ -4 & -4 & -3 \end{bmatrix}$

16. $\begin{bmatrix} 1 & 0 & 2 \\ -1 & 0 & -2 \\ 1 & 1 & 1 \end{bmatrix}$

17. Use the methods of the text to encode the message

arthur is a tree

Break the message into groups of two and use the matrix $\begin{bmatrix} -1 & 2 \\ 2 & -5 \end{bmatrix}$

18. Using the matrix of Exercise 17, encode the message

attack at dawn unless too cold

19. You are a special agent. Your code book contains the following instructions:

Today's matrix is $\begin{bmatrix} -1 & 2 \\ 2 & -5 \end{bmatrix}$. Its inverse is $\begin{bmatrix} -5 & -2 \\ -2 & -1 \end{bmatrix}$.

Use these instructions to decode the following message.

$\begin{bmatrix} -17 \\ 33 \end{bmatrix}$ $\begin{bmatrix} 26 \\ -72 \end{bmatrix}$ $\begin{bmatrix} 53 \\ -133 \end{bmatrix}$ $\begin{bmatrix} 21 \\ -54 \end{bmatrix}$ $\begin{bmatrix} 41 \\ -103 \end{bmatrix}$ $\begin{bmatrix} 35 \\ -97 \end{bmatrix}$ $\begin{bmatrix} 29 \\ -77 \end{bmatrix}$ $\begin{bmatrix} -15 \\ 24 \end{bmatrix}$ $\begin{bmatrix} 39 \\ -98 \end{bmatrix}$

*20. Finish encoding the message given in the text.

*21. The following message is written in a code in which the frequency of the symbols is the main key to the solution.

)?--8))6*+8506*3×6;4?*7*&×*−6.48()6)985)?(8+2:;48)81&?(;46*3)6*;
48&(+8(*598+&8()8=8(5*−8−5(81?098;4&+)&15*50:)6)*;?6;6&*0?−7

Find the frequency of each symbol. By comparing high frequency symbols with the high frequency letters in English (see the text above), try to decipher the message. (*Hint:* Look for repeated two-symbol combinations and double letters for added clues. Try to identify vowels first.)

THE INPUT/OUTPUT STRUC

The whole economy *Wassily Leontief began to investigate connections within entire national economies in the 1930s at Harvard. The chart in the photo shows a portion of the input-output structure of the U.S. economy as of 1963. The coefficient matrix has almost 10,000 cells. Leontief uses computer methods to analyze the U.S. economy and make forecasts on the basis of the matrix model.*

At Cambridge University the model of the British economy under development is a "social-accounting matrix" (SAM) designed also for computer analysis.

*Leontief models   The economist W. W. Leontief, winner of a Nobel Prize in 1973, has developed an input-ouput method for analyzing an economy. This involves the use of matrices. Suppose that an economy produces n commodities. The production of each uses some (perhaps all) of the other commodities in the economy. Thus, the production of oil requires oil, as well as steel, electricity, and so on. The amounts of each commodity used in the production of one unit of each commodity can be written as an $n \times n$ matrix, the *technological matrix* of the system. Suppose the demand of the economy for the net production is known, represented by the $1 \times n$ matrix F. The gross production (the $1 \times n$ matrix P) which is necessary to meet the demand can then be determined. Given the technological matrix A, the following formula can be used to find P:

$$P = F(I - A)^{-1},$$

where I is the identity matrix corresponding in size to A.

In the following exercises, we consider an economy with only two basic products, wheat and oil. Assume that to produce 1 metric ton of wheat requires $\frac{1}{4}$ metric ton of wheat and $\frac{1}{12}$ metric ton of oil. To produce 1 metric ton of oil requires $\frac{1}{3}$ metric ton of wheat and $\frac{1}{9}$ metric ton of oil.

$$A = \begin{bmatrix} \frac{1}{4} & \frac{1}{12} \\ \frac{1}{3} & \frac{1}{9} \end{bmatrix}$$

This leads to the technological matrix shown at the side.

The demand in the economy is 500 metric tons of wheat and 1000 metric tons of oil, so that $F = [500 \quad 1000]$. Go through the following steps to find the necessary gross production of wheat and oil.

*22. Find $I - A$.

*23. Find the inverse of $I - A$.

*24. Find P from the formula $P = F(I - A)^{-1}$. Round off the entries of P to the nearest whole number.

For the economy above, find the gross production required for:

*25. net production of 750 metric tons of wheat and 1000 metric tons of oil;

*26. the net production in Exercise 25, if production of 1 metric ton of wheat consumes $\frac{1}{5}$ metric ton of oil, while $\frac{1}{3}$ metric ton of wheat is required to produce a metric ton of oil.

*11.5 MATRICES AND GAME THEORY

Suppose that you are a citrus farmer. During the winter, you will often be faced with a decision about protecting your crop: you can protect your crop against freezing by burning smudge pots at night at a cost of $1000. If you burn smudge pots, you will be able to sell your crop for a net profit of $14,000, provided that the freeze does develop. If you do nothing, you will either lose $3000 in planting costs if the freeze comes, or will gain $8000 profit if no freeze occurs.

To use mathematics to help with this problem, we first must define the problem clearly. What are the existing conditions, the **states of nature,** under which you may have to operate? There are two in the problem: freezing temperatures or no freeze. Next, we state the *actions,* or **strategies,** that are available to us: use smudge pots or do not use them. The results of each action under each state of nature are summarized in a **payoff matrix,** as shown below. The amounts represent the profit (or loss) for each combination of events.

	States of Nature	
	Freeze	No freeze
Use smudge pots	$14,000	$7000
Don't use them	−$3,000	$8000

There are several criteria which may be used in the decision making. If you are an optimist, you might choose the *maximax* criterion, which selects the action that *maximizes* the *maximum* gain. In our example, if you use smudge pots, your maximum gain would be $14,000, while if you do not use smudge pots, your maximum gain would be $8000. Using the maximax criterion, you would choose the action which gives the larger of the two maximums and use smudge pots.

Economic behavior *In 1944 a new direction in economics was set by the* Theory of Games and Economic Behavior, *containing the famous minimax theorem. It was written by mathematician John von Neumann and economist Oskar Morgenstern. The book's impact relates to the question about which mathematics is proper to economics. Beginning with Antoine Cournot in the 1830s, economists looked to physics for analogies in thermodynamics or mechanics since physics had been so successfully "mathematized." The game-theoretic approach was a break with the past, showing how mathematics other than the calculus could give models of economic behavior. The impact of the book was felt in other social sciences, for example, political science in the study of coalition formation and voting procedures.*

John von Neumann *(1903–1957) achieved much in a life cut short by cancer. Stories told about him made him a living legend. In 1933 he joined the Institute for Advanced Study, newly founded, and was one of its original six mathematicians. By then he had already written important papers on set theory, quantum mechanics, and the minimax theorem. After 1940 he turned to the applications of mathematics, working with Ulam during World War II (Monte Carlo methods, page 386) and extending game theory with Morgenstern. Von Neumann was intensely interested in the theory and use of computers and automata. His interest in artificial intelligence is expressed in his lecture notes published as The Computer and the Brain (Yale paperback).*

On the other hand, you might be a pessimist. In that case, the *maximin* criterion, which *maxi*mizes the *mini*mum payoff, would suit you just fine. The minimum payoffs under the two actions are $7000, with smudge pots, and −$3000 if nothing is done. The larger of these occurs if smudge pots are used, so that this is the appropriate action for this criterion, too.

If you are neither an optimist nor a pessimist, and if you can assign some probability p to the "freeze" state of nature, then you can compromise by choosing the action with the greatest expected payoff. (We discussed expected value in Section 9.5.) Suppose you believe (the weather forecaster has convinced you) that the probability of a freeze is .4. The probability of no freeze would then be .6, with the expected payoffs as follows:

If smudge pots are used: $14,000(.4) + 7000(.6) = \$9800$

If no pots are used: $-3000(.4) + 8000(.6) = \$3600.$

This criterion, choosing the maximum of the weighted payoffs, also indicates that smudge pots should be used. Note that, as your subjective beliefs about the probability of a freeze changes, so would the payoffs. To assign probabilities to such unique events requires a subjective or personal assessment of the event. It works out here that pots should be used as long as you feel the probability of a freeze is greater than $\frac{1}{18}$.

In the example discussed here, all three criteria led to the same action. This is rare, since in a difficult decision, the factors are more closely balanced. If the cost of using the smudge pots were much larger, or the profit when they were used and there were a freeze was much smaller, the situation would look very different. Also, for the payoff matrix to be helpful the effect of all important factors must somehow be included. At the least, the use of a payoff matrix does clarify the problem and permit an orderly approach to considering the alternatives.

The example about the smudge pots is part of a large and growing branch of mathematics called **game theory.** While the name may be somewhat frivolous, the applications are certainly not. Game theory was developed in the 1940's as a means for analyzing competitive situations in business, warfare, and social situations. The theory deals with decision-making in competition with an aggressive opponent.

A **two-person game** between two players A and B is defined by a payoff matrix such as the one shown below. Player A chooses one of the rows, while at the same time, player B chooses one of the columns. A player's choice is called a *strategy.* The payoff is the amount at the intersection of the row and column selected. If the payoff is a positive number, A receives that amount from B; if it is negative, A pays B. In the matrix above, if A selects strategy 2 and B selects strategy II, the payoff is 4, which means that A receives $4 from B. (In practical situations, the payoffs need not be money, but might be goods, real property, or things like "happiness" and "satisfaction.")

A game such as the one above is called a two-player, zero-sum game. A **zero-sum game** is one in which one player must lose whatever the other wins.

$$
\begin{array}{cc}
 & B \\
 & \begin{array}{cc} \text{I} & \text{II} \end{array} \\
A \begin{array}{c} 1 \\ 2 \end{array} & \begin{bmatrix} 2 & -1 \\ -3 & 4 \end{bmatrix}
\end{array}
$$

$$A \begin{array}{c} \\ 1 \\ 2 \\ 3 \end{array} \begin{array}{ccc} & B & \\ I & II & III \\ \begin{bmatrix} -3 & -6 & 10 \\ 3 & 0 & 1 \\ 5 & -4 & -8 \end{bmatrix} \end{array}$$

An $m \times n$ (matrix) game is one in which player A has m possible strategies (rows) and player B has n strategies (columns). For example, a 3×3 game is shown at the side.

From B's viewpoint, strategy II is better than strategy I no matter which strategy A selects, since a gain of $6 is better than a gain of $3; breaking even is better than paying $3 to A; and a gain of $4 is better than a loss of $5. Therefore, B should never select strategy I. In a situation like this, strategy II is said to *dominate* strategy I (or strategy I is *dominated* by II), and strategy I should be removed from consideration, thus reducing the matrix, as shown in the margin.

$$A \begin{array}{c} \\ 1 \\ 2 \\ 3 \end{array} \begin{array}{cc} & B \\ II & III \\ \begin{bmatrix} -6 & 10 \\ 0 & 1 \\ -4 & -8 \end{bmatrix} \end{array}$$

Either player or both players may have dominant strategies. In fact, after a dominant strategy for one player is removed, the other player may then have a dominant strategy where there was none before. For example, in the game at the side, strategy 2 for A now dominates strategy 3, which could be removed.

Which strategies should be chosen by A and B? The goal of game theory is to find *optimum* strategies, those which are most profitable (or least costly) for a player. The resulting payoff, assuming both players select their respective optimum strategies, is called the *value* of the game. We usually find optimum strategies using the *minimax* criterion, in which we *minimize* the *maximum* possible loss. For example, a game for player A and B is shown at the side.

$$A \begin{array}{c} \\ 1 \\ 2 \\ 3 \end{array} \begin{array}{ccc} & B & \\ I & II & III \\ \begin{bmatrix} -7 & 8 & -1 \\ 4 & 7 & 0 \\ -9 & 3 & -6 \end{bmatrix} \end{array}$$

In this game, player A's greatest losses for each of his or her strategies are $7, $0, and $9. To minimize the maximum expected loss, player A should select strategy 2, In the same way, B's greatest losses are $4, $8, and $0, so that B can minimize possible losses by choosing strategy III. Thus, the payoff will be $0 (strategy 2 for A and III for B), which is the value of the game. A game whose value is $0 is called a *fair game*.

The pair of strategies (2, III) determined by the minimax criterion is called the *saddle point* of the game. The saddle point is the pair of strategies for which the entry in the payoff matrix is both the *minimum* in its row and the *maximum* in its column. This fact makes the saddle point easy to determine, if it exists. For example, in the matrix at the side, by the definition above, the saddle point is (4, I) and the value of the game is $3. Whenever there are two or more saddle points, they must all yield equal values.

Unfortunately, not every $m \times n$ game has a saddle point. When there is no saddle point, another approach must be used. Both players will have to *mix* their strategies. If this mixing were to be done in some systematic fashion, the competitor would soon guess it and play accordingly. For this reason, it is best to mix strategies according to previously determined probabilities. For example, if a player has only two strategies and has decided to select them with equal probability, the random choice can be made by tossing a coin.

$$A \begin{array}{c} \\ 1 \\ 2 \\ 3 \\ 4 \end{array} \begin{array}{cc} & B \\ I & II \\ \begin{bmatrix} 2 & 2 \\ 0 & 4 \\ 1 & 6 \\ 3 & 7 \end{bmatrix} \end{array}$$

Heads could represent one strategy; tails the other. Although this would result in the two strategies being used about equally over the long run, on a particular play no player could predetermine which strategy would be used. Some other device, such as a spinner, a die, or random numbers is necessary for more than two strategies, or when the probability is different from $\frac{1}{2}$.

		Nature	
		I	II
You	1	$14,000	$7000
	2	−$3000	$8000

To see how to decide on the probability with which a particular strategy is selected, let us return to the example about the smudge pots. Suppose you must decide several times each season on whether or not to use smudge pots, and you don't really know the probability of a freeze. We need to find the fraction of the time (the probability) that you should use smudge pots. There is no saddle point for this game. We thus let x represent the probability that you will choose strategy 1, with $1 - x$ for the probability that you will choose strategy 2. For Nature's strategy I (a freeze) your expected profit is

$$E_I = 14,000x - 3000(1-x).$$

For Nature's strategy II (no freeze) your expected profit is

$$E_{II} = 7000x + 8000(1-x).$$

You will maximize your profit in the long run if you choose a value of x for which $E_I = E_{II}$. (The reason for this is a little complex, and we shall not go into it here.) To make $E_I = E_{II}$, we solve the following equation.

$$E_I = E_{II}$$
$$14,000x - 3000(1-x) = 7000x + 8000(1-x)$$
$$17,000x - 3000 = -1000x + 8000$$
$$18,000x = 11,000$$
$$x = \tfrac{11}{18}$$

From the Nile to the Mississippi *they have been playing games. The gaming table at right is one of King Tutankhamun's treasures, about 1330 BC. Across the page is a portable backgammon board from the mid-19th century, used by Mississippi riverboat gamblers on the* Natchez.

In Tut's tomb four boards were found, and it is reasonable to think he was fond of games. The board you see was used in the game of chance called senet. It may have been a forerunner of backgammon.

Senet is believed to have been played as follows. Five squares making an L-shaped arrangement were inscribed with hieroglyphics. The object was to get to the square at the corner of the L, which had signs for "happiness." Each player had from five to seven pieces, with moves determined by throwing knucklebones or casting sticks.

Senet is mentioned in the Book of the Dead, an old Egyptian text, as the game the deceased played with the gods. The results were to decide the deceased's fate in the next world.

Thus, the optimum strategy here is to use strategy 1 (use smudge pots) $\frac{11}{18}$ of the time, and strategy 2 (don't use them) a total of

$$1 - x = 1 - \frac{11}{18} = \frac{7}{18}$$

of the time. Your expected gain from this strategy will be found by substituting $\frac{11}{18}$ for x in either E_I or E_II. Using E_I we have

$$14{,}000\left(\frac{11}{18}\right) - 3000\left(\frac{7}{18}\right) = \frac{133{,}000}{18} = \$7388.88.$$

This result, $7388.88, is the value of the game. In general, it comes out the same regardless of which player's optimum strategy and expected gain is considered.

11.5 EXERCISES

In all the following games, assume that player A chooses rows and player B chooses columns.

Eliminate all dominated strategies from each of the following games.

1. $\begin{bmatrix} 4 & 4 \\ 4 & -1 \\ 3 & 5 \\ -4 & 0 \end{bmatrix}$

2. $\begin{bmatrix} 2 & 3 & 1 & -5 \\ -1 & 5 & 4 & 1 \\ 1 & 0 & 2 & -3 \end{bmatrix}$

3. $\begin{bmatrix} 8 & 12 & -7 \\ -2 & 1 & 4 \end{bmatrix}$

4. $\begin{bmatrix} 6 & 2 \\ -1 & 10 \\ 3 & 5 \end{bmatrix}$

Where it exists, find the saddle point and the value of the game for each of the following.

5. $\begin{bmatrix} -6 & 2 \\ -1 & -10 \\ 3 & 5 \end{bmatrix}$

6. $\begin{bmatrix} 7 & 8 \\ -2 & 15 \end{bmatrix}$

7. $\begin{bmatrix} 3 & -4 & 1 \\ 5 & 3 & 2 \end{bmatrix}$

8. $\begin{bmatrix} 2 & 3 & 1 \\ -1 & 4 & -7 \\ 5 & 2 & 0 \\ 8 & -4 & -1 \end{bmatrix}$

9. $\begin{bmatrix} 1 & 4 & -3 & 1 & -1 \\ 2 & 5 & 0 & 4 & 10 \\ 1 & -3 & -2 & 5 & 2 \end{bmatrix}$

10. $\begin{bmatrix} -4 & 2 & -3 & -7 \\ 4 & 3 & 5 & 9 \end{bmatrix}$

Solve each of the following problems:

11. Hillsdale College has sold out all tickets for a jazz concert to be held in the stadium. If it rains, the show will have to be moved into the gym, which has a much smaller seating capacity. The dean must decide in advance whether to set up the seats and the stage in the gym or in the stadium, or both, just in case. The matrix below shows the net profit in each case.

		Nature	
		Rain	No rain
	Set up stadium	−$1550	$1500
Actions	Set up gym	$1000	$1000
	Set up both	$750	$1400

(a) What action should the dean take if he is an optimist?
(b) If he is a pessimist?
(c) If the weather forecaster predicts rain with a probability of .6, what action should the dean take to maximize expected profit?

12. A community is considering an anti-smoking campaign. The city manager has been asked to recommend one of three possible actions: a campaign for everyone in the community over age 10, a campaign for youths only, or no campaign at all. The two states of nature are a true cause-effect relationship between smoking and cancer or no such relationship. The cost to the community (including loss of life and productivity) in each case are shown below.*

	Cause-effect relationship	No such relationship
Campaign for all	$100,000	$800,000
Campaign for youth	$2,820,000	$20,000
No campaign	$3,100,100	$0

(a) What action should the manager recommend if she is an optimist?

(b) If she is a pessimist?

(c) If the Director of Public Health estimates that the probability of a true-cause effect relationship is .8, which action should be recommended to minimize expected costs?

13. The research department of Allied Manufacturing has developed a new process which they feel will result in an improved product. Management must decide whether or not to go ahead and market the new product. The new product may be better than the old, or it may not be better. If the new product is better, and the company decides to market it, sales should increase by $50,000. If it is not better and they replace the old product with the new product on the market, they will lose $25,000 to competitors. If they decide not to market the new product they will lose $40,000 if it is better, and research costs of $10,000 if it is not.

(a) Prepare a payoff matrix.

(b) If management believes that the probability is .4 that the new product is better, find the expected profits under each possible course of action, and find the best action.

14. Suppose Allied Manufacturing of Exercise 13 decides to put their new product on the market with a big advertising campaign. At the same time, they find out that their major competitor, Bates Manufacturing, has also decided on a big advertising campaign for its version of the product. The payoff matrix below shows the increased sales in millions for Allied, which also represent decreases for Bates.

$$\begin{array}{cc} & \text{Bates} \\ & \begin{array}{cc} \text{TV} & \text{Radio} \end{array} \\ \text{Allied} \begin{array}{c} \text{TV} \\ \text{Radio} \end{array} & \begin{bmatrix} 1.0 & -.7 \\ -.5 & .5 \end{bmatrix} \end{array}$$

Find the optimum strategy for Allied, and find the value of the game.

---

*This problem is based on an article by B. G. Greenburg in the September 1969 issue of the *Journal of the American Statistical Association.*

Catching up with imagination
The tense game between Poole and H.A.L. Computer 9000 seemed like pure imagination in 1968, when Stanley Kubrick's 2001: A Space Odyssey came out. HAL won the game, but eventually lost control.

How's this for future shock—in April, 1977, Cyber 176, an ultrafast computer from Control Data Corporation, demonstrated its chess-playing capabilities with the program CHESS 4.5, developed at Northwestern University. Cyber played simultaneous games against ten Expert and Master chess players. It scored 8 wins, 1 draw, and 1 loss!

Shannon's brainchild grows up
In 1949 Claude Shannon (page 73) outlined a "search-and-evaluate" method, the basis of later chess programs. (Computer chess was considered the key problem in the study of artificial intelligence.) John von Neumann proposed that the instructions should be stored inside the computer—an important step in developing computer chess.

In the early 1960s, business people in Europe spent much money in developing a computer program for playing chess at a good level of skill. They hoped to apply the results to computer translation between languages. (Decision trees, as in Chapter 9, are models for both language and chess skills—choosing between alternate meanings and alternate plays are branching phenomena.) The effort failed; the chess tree was too large to be worked out by existing methods.

Later in the 1960s MACHAK was developed at MIT and enabled an IBM computer to play a fairly strong game. In 1972 TECH scored 2 wins and 2 losses in the Ann Arbor tournament, and played "rapid" (5-minute) games with any takers.

But before the spring of '77, the best computers played at a B-class level, with individual games at an A-class level.

Then in April 1977 it was no longer questionable that computers would ever play as well as master players.

In each of the following games, find the optimum strategy for each player, and find the value of the game.

15. $\begin{bmatrix} 5 & 1 \\ 3 & 4 \end{bmatrix}$ **17.** $\begin{bmatrix} 6 & 2 \\ -1 & 10 \end{bmatrix}$

16. $\begin{bmatrix} -4 & 5 \\ 3 & -4 \end{bmatrix}$ **18.** $\begin{bmatrix} 0 & -4 \\ 4 & 0 \end{bmatrix}$

*****19.** In the game of matching coins, two players each flip a coin. If both coins match (both show heads or both show tails), player *A* wins $1. If there is no match, player *B* wins $1, as in the matrix below. Find the optimum strategies for each player, and find the value of the game.

$$\begin{bmatrix} 1 & -1 \\ -1 & 1 \end{bmatrix}$$

CHAPTER 11 SUMMARY

Matrix	Action
Entries	Payoff matrix
Elements	Maximax
Scalar	Maximin
Square matrix	Game
Difference	Strategy
Identity matrix	Zero-sum game
Row transformation	Dominate
Inverses	Optimum strategy
Augmented matrix	Value
Code theory (Cryptography)	Fair
Technological matrix	Saddle point

CHAPTER 11 TEST

Let $A = \begin{bmatrix} 2 & -1 & 3 \\ 2 & 0 & 1 \\ 4 & 1 & 3 \end{bmatrix}$ and $B = \begin{bmatrix} -3 & 0 & 1 \\ 0 & 2 & -1 \\ -1 & 0 & 3 \end{bmatrix}$. Find each matrix:

1. $A + B$ **3.** AB **5.** $2A$

2. $A - B$ **4.** BA **6.** $-3B$

Apply the following row transformations to matrix A from above.

7. Exchange rows 2 and 3.

8. Multiply each element of row 2 of A by -3, and add the result to row 1.

Find each of the following products, when possible.

9. $\begin{bmatrix} -2 & 1 \\ 3 & 4 \end{bmatrix} \cdot \begin{bmatrix} 2 \\ 1 \end{bmatrix}$

11. $\begin{bmatrix} 1 & 6 & 1 \end{bmatrix} \cdot \begin{bmatrix} 3 \\ 2 \\ 1 \end{bmatrix}$

10. $\begin{bmatrix} 3 & 4 & 1 \\ 2 & 0 & 8 \end{bmatrix} \cdot \begin{bmatrix} 1 & 0 & 2 \\ 0 & 0 & 1 \\ 3 & 0 & 2 \end{bmatrix}$

12. $\begin{bmatrix} 8 & 9 & 7 \\ 2 & 1 & 3 \end{bmatrix} \cdot \begin{bmatrix} 2 & 3 \\ 4 & 2 \end{bmatrix}$

Find the inverse of each of the following matrices, where possible.

13. $\begin{bmatrix} 2 & 1 \\ 5 & 3 \end{bmatrix}$ **14.** $\begin{bmatrix} 5 & 10 \\ 2 & 4 \end{bmatrix}$ **15.** $\begin{bmatrix} 1 & 0 & 1 \\ 2 & 3 & 1 \\ 1 & 3 & 2 \end{bmatrix}$ **16.** $\begin{bmatrix} 2 & -1 & 0 \\ 1 & 0 & 1 \\ 1 & -2 & 0 \end{bmatrix}$

Solve each of the following systems of equations using matrix methods.

17. $\begin{aligned} x + y &= 1 \\ 2x - y &= -7 \end{aligned}$

19. $\begin{aligned} x + y + z &= 1 \\ 2x - y + 3z &= 4 \\ 3x + 2y - z &= -5 \end{aligned}$

20. $\begin{aligned} x &= 3 \\ y + z &= 3 \\ x - z &= 2 \end{aligned}$

18. $\begin{aligned} 5x + 6y &= 15 \\ 3x + 7y &= 9 \end{aligned}$

Remove any dominated strategies from the following games.

21. $\begin{bmatrix} 3 & 4 & 4 \\ -1 & -5 & 1 \end{bmatrix}$ **22.** $\begin{bmatrix} 6 & 8 & -3 & 4 \\ 2 & 0 & 2 & -1 \end{bmatrix}$

Find any saddle points for the following games.

23. $\begin{bmatrix} -1 & 0 & 3 \\ 8 & 6 & 9 \\ 4 & 3 & -5 \end{bmatrix}$ **24.** $\begin{bmatrix} -3 & 1 & 4 & 2 \\ 8 & 9 & 6 & 4 \\ 9 & 10 & 15 & 19 \end{bmatrix}$

25. Find the optimum strategy for player A, and the value of the following game.

$$\begin{bmatrix} -3 & -4 \\ -4 & 5 \end{bmatrix}$$

Endgame *Don't feel let down!*

Chapter 12 Computers

Computers have come a long way since scientists working at the University of Pennsylvania unveiled the world's first electronic digital computer, ENIAC (for Electronic Numerical Integrator and Computer) 30 years ago. ENIAC was a hulking, vaguely threatening mechanism. Composed of more than 19,000 vacuum tubes and serviced by dozens of scurrying technicians, it sprawled over 15,000 square feet and weighed 30 tons. But for all its bulk, its memory was limited to just twenty ten-digit numbers, and although its ability to perform 5,000 additions or subtractions a second seemed impressive enough, ENIAC still took days to solve complex problems.

Since then, computers have evolved through three (some say four) generations of technological change, growing ever faster, smaller and more powerful. Today, they are on the threshold of yet another generation of change. Those deceptively nondescript silicon chips are not only stepping up computer speed and power by an entire order of magnitude, they also promise to liberate computers from their prepossessing air-conditioned, dehumidified glass palaces once and for all and dramatically improve cost performance.

"Tomorrow's computers are going to become part of our society at the everyday level," says vice president Neil Gorchow of Sperry Univac, the nation's third-largest computer manufacturer. "By the end of the decade, we're going to be running into them in the grocery store, the drugstore, the lawyer's office and the doctor's office." Computers are already speeding up supermarket checkout lines and the day may not be far off when families use them to keep track of household accounts. What's more, rapidly advancing technical sophistication is leading to a growing simplicity of operation; the coming generation of computers will not require a highly trained cadre of expert programmers to stand as intermediaries between the machines and those who want to use them.

Not everyone subscribes to the notion that a new generation of computers is about to burst on the scene. "We don't refer to generations," says vice president C. A. Conover of Honeywell Information Systems, the nation's No. 2 computer maker. "There isn't any longer a clean break-off point when one generation ends and another one starts." Chairman Frank T.

Cary of International Business Machines Corp, which dominates the electronic-data-processing industry in Olympian fashion, agrees. "What we're seeing is a continuous kind of improvement, rather than the quantum jumps we made the first couple of times around," he says.

Tubes In any case, few in the industry dispute the notion that computer technology has already passed through at least three distinct phases. The first period lasted from about 1953 to 1958 and was characterized by the use of vacuum tubes in computers. Tubes gave way to transistors around 1958 and the second generation was born. That lasted until the early 1960s, when integrated circuits replaced transistors. In the process, computing speed skyrocketed from fifteen calculations a second with the early vacuum tubes to 400,000 a second with the first integrated circuits.

The progress has continued. The basis of the early integrated circuits, solid logic technology, yielded in time to something called monolithic systems technology. MST, as it is known, produced microscopic logic circuits capable of performing well over 1 million calculations a second. In 1971, a young engineer at Intel Corp. took advantage of MST developments and invented the microprocessor. In effect, it was a tiny computer—cheap, reliable, smaller than a sugar cube yet every bit as powerful as the 30-ton ENIAC of 1946. That opened the door to everything from programmable pocket calculators to electronic Ping Pong. Did this mark the beginning of a fourth generation? Some experts say yes. Others say it merely represented further refinement of integrated circuit and semiconductor technology.

In any case, the development of MST chips has laid the foundation for, if not a new generation of computers, at the very least a new wave. Already, the size of computers is shrinking at an astonishing clip. Hewlett-Packard and Wang Laboratories both offer desk-top computers and IBM recently introduced a more powerful portable model of its own that sells for less than $10,000. Even cheaper and more compact models are expected from other manufacturers before long. At the same time, what is known as the architecture of large computer systems—the way their elements are put together—is changing drastically.

Library Traditionally, computer systems have consisted of a central processor (the "number cruncher" where the actual calculations take place), a memory, some control circuits and one or more input-output units. The advent of microprocessors seems likely to transform this setup into what is called a network architecture. A huge memory bank will be the central element. Around the memory bank will be stationed a number of small satellite processors that can handle small problems individually—or team up to tackle big ones. Such network systems promise to be vastly more efficient than present setups in which every problem or operation—no matter how trivial—must be handled by the central computer. "Think of yourself in a library," says editor James Peacock of EDP Industry Report. Instead of a user having to depend on a single librarian to get what he and everyone else using the library needs, he explains, networking allows the user to go and get it himself.

With the present generation of computers that is far easier said than done—especially if the user isn't a highly trained programmer. But the next wave of computers will probably change all that. Because monolithic systems technology allows computer makers to simplify the operation, the coming class of machines will be far more accessible to laymen. "We're doing with computers what the automakers did with automobiles," says IBM's Cary. "When the automobile was first introduced, you practically had to be a mechanic to drive it. But over time, Detroit put the complexity into the system, under the hood, giving us automatic starters and automatic transmissions—making the car simpler and simpler to operate.

Last word   At the same time, scientists are developing new memory chips so small that their intricate workings can be seen only under a scanning electron microscope—and so dense that a one-square-inch wafer can hold up to 5 million bits of information. Entirely new approaches to circuitry may yield tiny logic devices capable of upward of 10 million calculations a second. By the early 1980s, predicts one of the industry's savviest observers, Frederic G. Withington of Arthur D. Little, Inc., the ultimate computer may make its debut. It would be the culmination of computer progress over the last 30 years: a machine that with a minimum of human intervention "can collect, organize and store all existing data (within an organization), then apply them both to conducting routine operations . . . and to supporting management actions."

In other words, a computer system that could run a diversified manufacturing operation by itself—keeping track of raw-material inventories, reordering when stocks get low, scheduling production lines as demand warrants and detecting and correcting whatever mistakes it may have made along the way. A frightening prospect? Perhaps—but then there is no task that even the most advanced computer can do that man can't. What makes them so valuable is simply the fact that they can do it a lot quicker.*

12.1 HOW A COMPUTER WORKS

In this section we shall look at the way that a typical large computer works. Figure 12.1 shows the scheme of a typical computer.

Figure 12.1

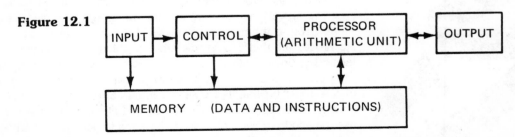

*Introduction and caption from *Newsweek* magazine, "Computers: A New Wave," by Allan J. Mayer with bureau reports, *Newsweek*, February 23, 1976.

The input mechanism is used to enter data and instructions into the machine. This is most commonly done with the familiar punched card (see Figure 12.2). These cards are made up of eighty columns and twelve rows (only ten of the rows have numbers printed on them). The number 6, for example, is represented by a punch in row 6. Letters require two punches in the column directly below them.

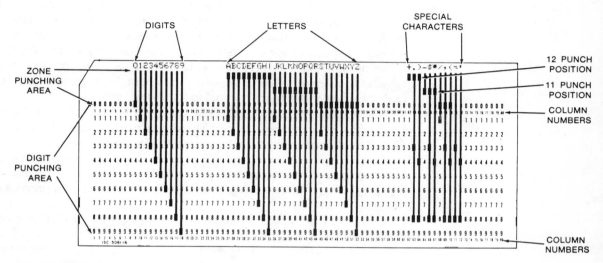

Figure 12.2

As shown in Figure 12.2, the letter A needs a punch in the top row and row 1, while W requires punches in row 0 and row 6. Many computers have input devices that can read over a thousand punched cards per minute. Punched cards are prepared on machines called *keypunches,* which have typewriter-like keyboards. An organization with a large input of data, such as an insurance company or government bureau, may have several hundred keypunches and keypunch operators, all preparing punched cards for one computer.

Paper tape is another common method of inputting information to the computer. Normally faster than cards, paper tape can be used to enter between 80 and 100 thousand characters (numbers of letters) per minute. Much faster is *magnetic tape,* which is very similar to recording tape. Magnetic tape can be used to enter hundreds of thousands of characters per second.

After the information is entered into the computer, it is stored until needed. The binary or base 2 number system is used when storing information in a computer. Recall from Chapter 3 that the base 2 number system is made up of the digits 0 and 1. Any number can be expressed using these two digits. For example,

$$2 = 10_2 = (1 \cdot 2^1) + (0 \cdot 2^0)$$
$$10 = 1010_2 = (1 \cdot 2^3) + (0 \cdot 2^2) + (1 \cdot 2^1) + (0 \cdot 2^0)$$
$$29 = 11101_2 = (1 \cdot 2^4) + (1 \cdot 2^3) + (1 \cdot 2^2) + (0 \cdot 2^1) + (1 \cdot 2^0)$$

To store the number 29 in a computer, we would need five storage locations. The first three would be marked as "on," to represent the 111 of 11101. The next storage location would be marked "off," for the 0 of 11101. Finally, the last location would be marked "on," for the final 1 of 11101.

Each 0 or 1 that is stored is called a **bit** (short for *binary digit*). Letters are stored as special combinations of digits. Bits are stored on *magnetic disks,* made up of several disks coated with magnetic material, stacked one on top of another much like an automatic record changer. A pick-up arm moves back and forth, selecting information from the disk. Disks rotate at a very high speed, but even so, they are very slow when compared to the tremendous speed of the computer. For this reason, *magnetic core* storage is often used. A magnetic core storage is made up of a large number of tiny magnetic doughnuts, called *cores,* which are about the diameter of the head of a pin. Each core can be magnetized to represent 0 or 1.

The supremacy of magnetic core storage is being challenged by new silicon wafer chip memory devices. These new devices offer a combination of speed and economy that is giving cores a run for their money.

Output can be in the same form as input. Computers can produce output as punched cards, paper tape, magnetic tape, or by producing printout on paper. Modern printers can produce over a thousand lines of printout per minute.

Computers often deal with extremely large or extremely small numbers. Usually computers are limited to 8 or 10 digits in one number. To get around the problem of writing large numbers in 8 or 10 digits, we use *exponential notation.* For example, a computer would print 74,000,000,000 as

$$.74 \text{ E} + 11$$

To convert .74 E + 11 to 74,000,000,000, write down the decimal .74, and then move the decimal point 11 places to the right, adding zeros as needed.

Example 1   Convert each of the following.

(a) .9632 E + 15
We write down .9632 and then move the decimal point 15 places to the right. This gives 963,200,000,000,000.

(b) .32572 E + 03
This number leads to 325.72.

If the computer prints out a number with a minus sign after the letter E, move the decimal point to the *left.*

Example 2   Convert each of the following.

(a) .35 E − 03
Write down .35, and move the decimal three places to the left to get the number .00035.

(b) .91624 E − 08
We have .0000000091624.

Computers do all their work internally in base 2. All the instructions given to the computer must be in base 2. A list of these instructions, called a **program,** is made up of a large number of 0's and 1's, and reminds one of a telephone book. Computers would have made little progress if everyone had to "talk" to them in this very artificial language.

The problem of talking with computers has been very largely solved with the introduction of **compilers.** A compiler looks much like a cross between English and algebra, making it relatively easy to learn and use. The first work on compilers was done by Grace Hopper, in 1952. One of the most common compilers in use today is BASIC, which we discuss in the rest of this chapter.

12.1 EXERCISES

Convert each of the following numbers.

1.	.698 E + 02	**9.**	.81375 E + 11
2.	.543 E + 03	**10.**	.2003 E + 05
3.	.8321 E + 04	**11.**	.7629 E − 03
4.	.12897 E + 03	**12.**	.835 E − 01
5.	.82 E + 08	**13.**	.104 E − 04
6.	.173 E + 04	**14.**	.9935 E − 06
7.	.3742 E + 02	**15.**	.414 E − 01
8.	.42974 E + 10	**16.**	.376 E − 04

Write each of the following numbers as they would appear in a computer printout.

17.	48,000,000,000	**21.**	.00009	**25.**	.0000003271
18.	290,000,000,000	**22.**	.0004	**26.**	.0008974
19.	3,800,000,000,000	**23.**	.00000427		
20.	57,000,000,000	**24.**	.00000598		

```
PAY
10 LET R = 74.11
20 LET O = 29.72
30 LET H = 45.26
40 LET T = R + O + H
50 PRINT T
60 END

RUN
149.09
```

12.2 BASIC

One of the most popular languages used by people to talk to a computer is BASIC. This language was developed at Dartmouth University in the early 1960s. Like all languages, BASIC has grammatical rules, which we discuss in the rest of this chapter.

For example, suppose we need to find the total income of a person who made $74.11 last week at the regular pay rate, $29.72 at the overtime rate, and $45.26 at the holiday rate. We could write a list of instructions (a *program*) to find the total pay as in the margin at the left.

PAY
10 LET R = 74.11
20 LET O = 29.72
30 LET H = 45.26
40 LET T = R + O + H
50 PRINT T
60 END

RUN
149.09

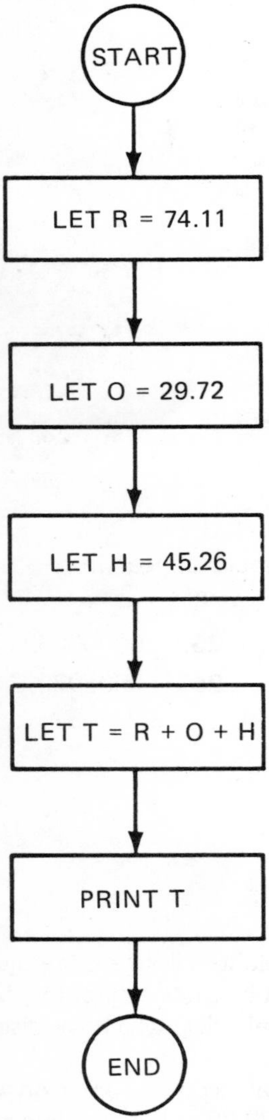

Let's now look at the parts of this program. The word PAY on the first line is the name of the program. We can use pretty much whatever name we like. Line 10 of the program tells the computer to let R take on the value 74.11. Lines 20 and 30 tell the computer the values of O and H. In line 40, we direct the machine to add R, O, and H. The sum is called T. Line 50 calls for the printing of the value of T (the answer) and line 60 tells the computer that the program is completed.

When the entire program has been stored in the computer, we ask it to carry out all the steps by typing the word RUN. As it reaches line 50 it prints out the value of T, which is 149.09. The employee in question earned a total of $149.09 in this particular week.

Notice that every line of a BASIC program must have a number, with the numbers appearing in numerical order. The line numbers make corrections simple. For example, if we wished to change R in the program above from 74.11 to 98.75, we need only type

10 LET R = 98.75

The computer would use this new line 10 to replace the previous line 10.

The letters R, O, H, and T of program PAY are called (*variables.*) In BASIC, a variable name can consist of either a single letter or a single letter followed by a single digit. For example, N, B2, and Z8 are valid variable names, but MEAN, 2A, and Z74 are not.

In many cases, the steps involved in a program are too complicated for us to remember them all. To help, we can draw a **flow chart,** which is a diagram showing the logical sequence of the operations which take place.

A flow chart is necessary for involved problems, and can help with simple ones. A flow chart for program PAY is shown at the side. Circles are used for starting and stopping points. Rectangles are used for inserting and removing data from the computer, and for operations and calculations. Later, we use diamonds when the machine must make a decision.

In the program PAY above, we could change line 50 to

50 PRINT R, O, H, T

This would cause the machine to print out

RUN
74.11   29.72   45.26   149.09

Now we have the original data printed out along with the answers.

We can use quotation marks to have the computer print out words. For example,

50 PRINT "TOTAL SALARY IS"; T   Good

would result in

RUN
TOTAL SALARY IS 149.09

Flow chart (left column):

START

↓

LET R = 74.11

↓

LET O = 29.72

↓

LET H = 45.26

↓

LET T = R + O + H

↓

PRINT T

↓

END

Figure 12.3

In BASIC, multiplication is indicated by an asterisk * and not a cross or dot. Division is indicated by a slash / . Powers are written with an upward arrow ↑ . For example, 3^2 is entered into the computer as 3 ↑ 2, while 5 ↑ 3 represents 5^3.

Powers are evaluated first; then multiplications and divisions from left to right; and finally additions and subtractions again from left to right.

order of operations

Example 1 Find the value of K in the following statement:

$$30 \text{ LET K} = 2 ↑ 3 + 6*4/2$$

Solution. Powers are performed first. Here 2 ↑ 3 means 2^3, or $2 \cdot 2 \cdot 2 = 8$.

$$K = 8 + 6*4/2$$

Now do multiplications and divisions from the left.

$$K = 8 + 24/2$$
$$= 8 + 12$$

Finally, add.

$$K = 20$$

A summary of BASIC symbols is shown at the side.

BASIC	Meaning
=	equals
+	add
−	subtract
*	multiply
/	divide
↑	powers
( )	grouping symbol

One difficulty with the program PAY above is that the data for the employee is actually part of the program. When we need to find the salary for another employee, we would have to reenter the entire program in the computer. To get around this problem, we can use READ and DATA statements. This is shown in the program FANPAY.

```
FANPAY
10 READ N, R, O, H
20 LET T = R + O + H
30 PRINT "NUMBER", "SALARY"
35 PRINT
40 PRINT N, T
50 DATA 3, 74.11, 29.72, 45.26
60 END
RUN
NUMBER   SALARY
   3        149.09
```

The READ statement tells the computer to take values for N (the employee number), R, O, and H from the list of numbers in the DATA statement which comes just before END. Using this method, all the data can be entered in a single line after the rest of the program has been made up. Also, we can easily change the data by retyping line 50 of the program.

Program FANPAY shows the use of a PRINT statement (line 30) to provide a label for the answer. Line 35 contains only the word PRINT. This line causes a line of space to be inserted in the output, for ease of reading.

12.2 EXERCISES

Which of the following are valid variable names in BASIC?

1. Z   **4.** YA   **7.** P6

2. P   **5.** 9   **8.** M9

3. MN   **6.** 12

Evaluate each of the following BASIC statements.

9. $P = 7 + 6*2$

10. $Z = 9*4/2 + 8$

11. $Y5 = 4*2 + 3 - 5 + 2 \uparrow 3$

12. $A2 = 3 \uparrow 4/9$

13. $Q = (6 + 2)*(5 - 1)$

14. $B9 = 2 \uparrow (1 + 3) + 9* 2/6$

15. $H = 2 + 5*3 - 6*2$

16. $J4 = (3 + 6 + 9)*(2 + 9/3)$

Write LET statements that will accomplish the following.

17. $A = X + Y - ZW$

18. $M = 5X - 6K$

19. $Z = P^3M$

20. $Q2 = Y^2X^3$

21. $H5 = \dfrac{8 + P}{2 - Q}$

22. $Z4 = \dfrac{K + Y + P}{M + N}$

23. $K9 = \dfrac{X - 3(Y + R)}{5M + 2N}$

24. $B4 = \dfrac{5P + 2R + M}{7(Y + 2K)}$

```
TIME
2 READ D, R
4 LET T = D/R
6 PRINT T
8 DATA 80
9 END
```

If a READ statement is used in a program, there must also be a DATA line that provides the numbers for the computer to "read." If the computer cannot locate as many data values as you have asked it to read, it will inform you of this fact and stop running the program. See the program TIME at the side, where D, R, and T represent distance, rate, and time, respectively.

25. What is wrong with this program?

26. How should TIME be changed in order to find the time it would take to travel 80 miles at a rate of 35 miles per hour?

Decide what would be printed when each of the following programs is run. If necessary, round answers to the nearest tenth.

27.
```
20 LET A = 3
40 LET B = 5*A
60 LET C = (B + 3)/9
80 LET D = 4 ↑ C
90 PRINT D + 4
```

28.
```
5 LET X = 3.8
10 LET Y = X/2
15 LET Z = (X + 3*Y)/5
20 PRINT "Z ="(;)Z
25 END
```

29.
```
5 READ A, B, C, D
10 LET Q = (A - B)/(C + D)
15 PRINT Q
20 DATA 15, 7, 1, 3
25 END
```

*30.   10 READ X1, X2, X3
        20 LET M = (X1 + X2 + X3)/3
        30 LET D1 = X1 − M
        32 LET D2 = X2 − M
        34 LET D3 = X3 − M
        40 LET S = ((D1 ↑ 2 + D2 ↑ 2 + D3 ↑ 2)/3) ↑ (1/2)
        50 PRINT X1, X2, X3, M, S
        60 DATA 23.5, 28.4, 24.9
        70 END

(*Hint:* An exponent of $\frac{1}{2}$ is the same as taking the square root of the number.)

Make flow charts for each of the following.

31.   Finding M, the mean of the numbers A, B, and C.

⇒ **32.**   Finding Z, the product of Y and R.

12.3 REPETITIVE PROGRAMS

The programs of the previous section showed some of the key ideas of BASIC programming. However, most of the programs of that section could have been worked just as quickly on a pocket calculator (or even with a paper and pencil). The true power of the computer is used when many (perhaps several million) calculations must be made. The computer can be instructed to perform a given operation again and again, even though the instruction is written only once in the program.

One method of doing this is shown in the program SQRT, which is written to find the square roots of the whole numbers from 1 through 5. By the methods of the previous section, this program would have been much longer.

```
SQRT
10 PRINT "NUMBER," "SQUARE ROOT"
15 PRINT
20 READ N
30 LET S = N ↑ (1/2)        (Recall: √N = N^(1/2))
40 PRINT N, S
50 GO TO 20
60 DATA 1, 2, 3, 4, 5
70 END
RUN
```

NUMBER	SQUARE ROOT
1	1.
2	1.41421
3	1.73205
4	2.
5	2.23607

OUT OF DATA IN LINE 20

15 PRINT
20 READ N
30 LET S = N ↑ (1/2)
40 PRINT N, S
50 GO TO 20
60 DATA 1, 2, 3, 4, 5

The new feature in this program is the GO TO statement in line 50. Steps 20, 30, and 40 cause the computer to read the first number from the data list, find its square root, and print both the number and its square root. The GO TO statement in line 50 then sends the computer back to line 20, where it reads the next data item and repeats the same steps (lines 30 and 40) for the new number.

The computer is in a **loop,** which continues until all the data of line 60 has been used. At this point, the computer stops and tells what caused it to stop. In this case, it was "out of data at line 20." It tried to read another number, but found no more in the data list of line 60.

One technique used in many problems is to systematically change the value of a variable. An example of this is shown in the program SQRT1.

SQRT1
10 PRINT "NUMBER," "SQUARE ROOT"
20 PRINT
30 LET N = 1
40 LET S = N ↑ (1/2)
50 PRINT N, S
60 LET N = N + 1
70 GO TO 40
80 END

Infinite loop!!

A loop is formed by lines 40 through 70. Each time through the loop, the value of N is increased by 1. The first time through the program, N has the value of 1 (line 30). In line 40, the machine finds the square root of 1. In line 60, the computer changes N from 1 to $1 + 1 = 2$. It then returns to line 40 and finds the square root of 2. Then N increases from 2 to $2 + 1 = 3$.

The trouble with the program SQRT1 is that there is nothing to make the looping process stop. The computer will continue to calculate and print numbers and their square roots. We can get around this with an IF THEN statement as in the program SQRT2.

SQRT2
10 PRINT "NUMBER," "SQUARE ROOT"
20 PRINT
30 LET N = 1
40 LET S = N ↑ (1/2)
50 PRINT N, S
60 LET N = N + 1
70 IF N<21 THEN 40
80 END

Lines 10 through 60 have the same function as in the program SQRT1. In line 70, the computer checks to see if N<21 (*Recall:* < means "is less than"). If N is smaller than 21, the computer goes to statement 40. If N = 21, or if N>21, the computer continues on to statement 80. Program SQRT2 will result in the printing of the first 20 whole numbers and their square roots. If we wanted the square root of, say, the first 100 whole numbers, we would only have to change 21 to 101 in statement 70.

IF THEN	Meaning
=	equals
<	is less than
>	is greater than
<=	is less than or equal to
>=	is greater than or equal to
<>	is not equal to

ASUM
10 LET S = 0
20 FOR I = 1 to 100
30 LET S = S + I
40 NEXT I
50 PRINT "SUM =," S
60 END

10 READ F, D, N
20 LET T = 0
30 LET T = T + 1
40 LET A = F + (T − 1)*D
50 PRINT A
60 IF T<N THEN 30
70 DATA 3, 4, 4,
80 END

The various symbols which can be used in an IF THEN statement are shown at the side.

Another way to control the number of repetitions of a loop is to use FOR and NEXT statements. To see how these statements are used, look at program ASUM, which is used to find the sum of the first 100 counting numbers, $1 + 2 + 3 + 4 + \cdots + 100$.

Line 10 first sets the value of S at 0. The FOR statement in line 20 causes all steps between there and the NEXT statement (line 40) to be carried out once when I is 1, again when I is 2, when I is 3, and so on until I is 100. At that point, the computer goes to line 50.

The advantage of a FOR and NEXT loop is that we do not need to set up an IF THEN test to tell the computer when to stop looping. We indicate right in the FOR statement how many times the loop is to be carried out.

12.3 EXERCISES

1. Work through the program at the side and write the printout that would be produced.

Work through the program of Exercise 1 again with line 70 changed as follows.

2. 70 DATA 5, 2, 7   3. 70 DATA 8, 5, 6   4. 70 DATA 3, 7, 9

The program MAX below has two places where the computer must make a decision (lines 30 and 40).

MAX
10 READ A
20 READ B
30 IF B<0 THEN 70
40 IF B<=A THEN 20
50 LET A = B
60 GO TO 20
70 PRINT "THE LARGEST ITEM IS"; A
80 DATA 5, 12, 5, 8, 4, 6, 10, 13, 11, 7, −1
90 END

5. Draw a flow chart for the program MAX. (Recall that diamond-shaped boxes are used at decision points.)

6. Work through the program, filling in the values for A and B.

A	B	A	B
5	12	___	___
12	___	___	___
___	___	___	___
___	___	___	___
___	___	___	___

7. Write the printout that would be produced.

8. What would happen if the −1 were left off line 80?

9. Describe in general what the program MAX does.

10. Rewrite MAX so that it will find and print the largest of the following numbers: 25, 21, 32, 26, 28, 30, 26, 33, 29, 31.

11. The following program is the same as ASUM in the text except for new lines 15, 35, and 55. What happens as a result of these new lines?

```
10 LET S = 0
15 LET S2 = 0
20 FOR I = 1 to 100
30 LET S = S + I
35 LET S2 = S2 + I ↑ 2
40 NEXT I
50 PRINT "SUM," S
55 PRINT "SUM OF SQUARES =," S2
60 END
```
semi-colon

Exercises 12–18 refer to the program FCTRL.

```
FCTRL
10 READ N
20 LET P = 1
30 FOR I = 1 TO N
40 LET P = P*I
50 NEXT I
60 PRINT P
70 DATA 5
80 END
```

12. How many times is the loop in the program repeated?

13. What is the value of P after the first time through the loop?

14. After the second time?   15.   The third time?

16. What value of P will finally be printed?

17. Describe the effect (if any) of the following change.

70 DATA 5, 10, 20

18. The program FCTRL prints the value of 5! (recall 5! = 5·4·3·2·1). How could the program be modified to print the value of 8!?

```
10 LET S = 0
20 FOR A = 3 TO 39 STEP 4
30 LET S = S + A
40 NEXT A
50 PRINT S
60 END
```

The program at left shows a slightly different form of the FOR statement. The value of A starts at 3 and increases to 39 in steps of 4, so that successive values of A are 3, 7, 11, 15, and so on.

19. Find the value of S that will be printed.

20. Modify line 20 by changing STEP 4 to STEP 3. Find the values of S that will be printed.

CHAPTER 12 SUMMARY

Keywords			
Keypunch	BASIC	Core	READ
Paper tape	Program	Printout	GO TO
Magnetic tape	Variable	Exponential notation	Loop
Bit	Flow chart	Compiler	IF THEN

BASIC Symbols

=	Equality	( )	Grouping symbols
+	Addition	<	Less than
−	Subtraction	>	Greater than
*	Multiplication	<=	Less than or equal to
/	Division	>=	Greater than or equal to
↑	Exponentiation	<>	Not equal to

CHAPTER 12 TEST

Convert each of the following numbers.

1. .874 E + 08   **2.** .1369 E + 12   **3.** .725 E − 03

Write each of the following numbers as they would appear in computer printout.

4. 389,000,000,000   **5.** .000000725

State whether or not each of the following is a valid variable name in BASIC.

6. MN   **7.** H9   **8.** 8   **9.** M3

Write LET statements for each of the following.

10. $M = 5X + 2Y$

11. $P = \dfrac{8 + Y}{9 + Q}$

12. $Z2 = \dfrac{4M + 2N}{Y - Z - 4}$

13. $Y9 = \sqrt{X + 5M}$

Decide what would be printed as a result of the following programs.

14.
```
10 PRINT X, Y, SUM, POWER
20 PRINT
30 READ X, Y
40 IF X<0 THEN 99
50 LET S = X + Y
60 LET P = X↑Y
70 PRINT X, Y, S, P
80 GO TO 30
90 DATA 2, 3, 4, 2, 3, 4, −1, 0
99 END
```

15.
```
10 READ N
20 IF N<0 THEN 90
30 FOR K = 1 TO 4
40 PRINT K*N
50 NEXT K
60 PRINT
70 GO TO 10
80 DATA 3, 10, 6, −1
90 END
```

A sad ending   **or an up ending?**

For Further Reading You have sampled a great variety of mathematics in this text. What follows is a brief list of books, articles, and films that can give you further insight into what mathematics is all about.

Surveys The Life Science Library includes *Mathematics* by David Bergamini, with splendid illustrations (Time-Life Books 1972). □ *Number: The Language of Science* by Tobias Dantzig will make you enthusiastic (Free Press 1967). □ *The Mathematical Sciences* is a collection of essays on topics at the frontiers of research and application (MIT Press 1969). □ Very readable is Ian Stewart's *Concepts of Modern Mathematics* (Penguin Books 1975). □ The classic survey is by Richard Courant and Herbert Robbins — *What Is Mathematics?* (Oxford University Press 1941). Note Courant's biography below.

History A good general source is *Introduction to the History of Mathematics* by Howard Eves (Holt, 4th ed., 1976). □ Detailed accounts through the mid-19th century are in *History of Mathematics* by David E. Smith (Dover reprint, 2 volumes). □ At the other extreme is *A Concise History of Mathematics* by Dirk J. Struik (Dover 1967). □ The ancient world is discussed in *The Exact Sciences in Antiquity* by Otto Neugebauer (Dover 1969) and *Science Awakening* by Bertel L. Van der Waerden (Oxford U.P. 1961).

Culture Morris Kline gives an overview in *Mathematics in Western Culture* (Oxford U.P. 1964). □ *The Ascent of Man* by Jacob Brownowski offers much insight about Pythagoras, Arabian mathematics, Newton, to name some topics (Little 1970) — you may have seen the TV series narrated by the author. □ *Number Words and Number Symbols* by Karl Menninger is full of lore (MIT Press 1977). □ Mathematics of sub-Sahara Africa is described in Claudia Zalevsky's *Africa Counts* (Prindle, Weber & Schmidt 1973). □ Until a similar book on American Indian mathematics comes along, consult *Picture-Writing of the American Indians* by Garrick Mallory, a government report from 1889 (Dover reprint 1972).

People Short biographies make up *The Great Mathematicians* by Herbert W. Turnbull (Simon and Schuster 1962). □ *Women in Mathematics* by Lynn Osen covers Hypatia to Emmy Noether, along with an essay on the "feminine mathique" (MIT Press 1974). □ Richard Courant and David Hilbert were important influences on 20th century mathematics, and Constance Reid has written biographies of both (Springer, 1976 and 1970 respectively). □ Stanisław Ulam (see page 387) writes about his education in *Adventures of a Mathematician* (Scribner 1976). □ *I am a Mathematician* is the declaration of Norbert Wiener, who popularized cybernetics (MIT 1964). □ Martin Gardner's column in *Scientific American* is called "Mathematical Games," and his many puzzle books are published by Simon and Schuster. Of special interest is his set of filmstrips with audiotapes called *The Paradox Box* (Freeman 1975). □ John Horton Conway, inventor of the computer game "Life," has written a provocative book, *On Numbers and Games* (Academic Press 1976) □ Conway's influence is seen in a unique mathematical novel by Donald E. Knuth, *Surreal Numbers* (Addison-Wesley 1974). □ R. Buckminster Fuller, inventor of the geodesic dome, will give you much to ponder in his *Synergetics: Explorations in the Geometry of Thinking* (Macmillan 1975).

Some Science Fiction   The classic short novel *Flatland* by Edwin A. Abbott tells about life in two dimensions (first published in 1884; Dover reprint 1952). ☐ Logic is featured in Lewis Padgett's story "Mimsy were the borogroves" in *Science Fiction Hall of Fame*, v. 1, Robert Silverberg, ed. (Doubleday 1970). ☐ Mathematical creativity is explored in "Problem child" by Arthur Porges and "Gomez" by C. M. Kornbluth in *Treasury of Great SF*, v. 1, Anthony Boucher, ed. (Doubleday 1959). ☐ Abstract space presents dangers in "The mathenauts" by Norman Kagan in *Year's Best SF*, Judith Merril, ed. (Delacorte Press 1965).

Special Topics   *Thinking and Problem Solving: An Introduction to Human Cognition* by Richard E. Mayer (Scott, Foresman 1977). ☐ *A History of π* (pi) by Petr Beckmann (St. Martin's Press 1971). ☐ "Mathematics of musical scales" by Paul S. Malcolm (*The Mathematics Teacher*, November 1972). ☐ *Art and Geometry* by William M. Ivins, Jr. (Dover 1964). ☐ *Geometry, Relativity, and the Fourth Dimension* by Rudolf v. B. Rucker (Dover 1977). ☐ *The Computer in Art* by Jasia Reichardt (Van Nostrand Renhold 1971). ☐ *How to Take a Chance* by Darrell Huff and Irving Geis (Norton 1957). ☐ *Lady Luck: The Theory of Probability* by Warren Weaver (Doubleday 1963). ☐ *Statistics: A Guide to the Unknown,* edited by Judith Tanur et al. (Holden-Day 1972). ☐ *Tangram: The Ancient Chinese Shapes Game* by Joost Elffers (Penguin 1976). ☐ *The Compleat Strategist* by John D. Williams (McGraw-Hill 1966).

☐ *The Graphic Work of M. C. Escher* (Ballentine 1971) reproduces many creations by Maurits Escher, the artist who fascinates the mathematical world. Two films are available: *Maurits Escher: Painter of Fantasies* and *Adventures in Perception.*

☐ Books about energy include: *The National Energy Plan — 1977* from the Executive Office of the President of the U.S.A. (Ballinger 1977); *Energy and Environment Bibliography,* compiled by Betty Warren (Friends of the Earth 1977); *High Frontier: Human Colonies in Space* by Gerald O'Neill (Morrow 1977).

The World of Mathematics   This is the title of a 4-volume compilation of essays and articles, edited by James E. Newman (Simon and Schuster 1962). Try some of the following: "Infinity," Hans Hahn; "Foundations of vital statistics," John Graunt; "First life insurance tables," Edmind Halley' "Can a machine think?" Alan M. Turing; "Symbolic notation, haddock's eyes and the dog-walking ordinance," Ernest Nagel.

Scientific American Articles   Here are some to start: "Non-Cantorian set theory," Paul J. Cohen and Reuben Hersh (December 1967); "Paradox," W. V. Quine (April 1962); "Symbolic logic," John E. Pfeiffer (December 1950); "Lewis Carroll's lost book on logic," W. W. Bartley III (July 1972); "Geometry" (September 1964) and "Projective geometry" (January 1955), both by Morris Kline; "Topology," Albert W. Tucker and Herbert S. Bailey. Jr. (January 1950); "Chance," A. J. Ayer (October 1965); "What is probability?" Rudolf Carnap (September 1953); "Probability," Mark Kac (September 1964); "The Monte Carlo method," Daniel D. McCracken (May 1955); "Computers," Stanislaw Ulam (September 1964); "Computer logic and memory," David C. Evans (September 1966). Some of these are reprinted in *Mathematics in the Modern World* or *Mathematical Thinking in Behavioral Science* (both published by Freeman, 1968).

Appendix

Table 1 Monthly Payment Necessary to Pay Off a Loan

Amount of Loan	8½% 25 yrs.	8½% 30 yrs.	9% 25 yrs.	9% 30 yrs.	9½% 25 yrs.	9½% 30 yrs.
100	.81	.77	.84	.80	.88	.84
200	1.61	1.54	1.68	1.68	1.75	1.68
300	2.42	2.31	2.52	2.41	2.63	2.52
400	3.22	3.08	3.36	3.22	3.49	3.36
500	4.03	3.84	4.20	4.02	4.37	4.21
10,000	80.53	76.89	83.92	80.46	87.37	84.09
11,000	88.58	84.58	92.32	88.51	96.11	92.50
12,000	96.63	92.27	100.71	96.56	104.85	100.89
13,000	104.68	99.96	109.10	104.60	113.58	109.31
14,000	112.74	107.65	117.49	112.65	122.32	117.72
15,000	120.79	115.34	125.88	120.70	131.05	126.13
16,000	128.84	123.03	134.28	128.74	139.81	134.54
17,000	136.89	130.72	142.67	136.79	148.53	142.94
18,000	144.95	138.41	151.06	144.84	157.27	151.36
19,000	153.00	146.10	159.45	152.88	166.00	159.77
20,000	161.05	153.79	167.85	160.93	174.74	168.17
21,000	169.10	161.48	176.24	168.98	183.47	176.58
22,000	177.16	169.17	184.63	177.02	192.21	184.98
23,000	185.21	176.86	193.02	185.07	200.96	193.41
24,000	193.26	184.54	201.41	193.11	209.69	201.82
25,000	201.31	192.23	209.81	201.16	218.43	210.22
26,000	209.37	199.92	218.20	209.21	227.17	218.63
27,000	217.42	207.61	226.59	217.25	235.91	227.00
28,000	225.47	215.30	234.98	225.30	244.64	235.44
29,000	233.53	222.99	243.38	233.35	253.38	243.85
30,000	241.58	230.68	251.77	241.39	262.12	252.30
31,000	249.63	238.37	260.16	249.39	270.85	260.66
32,000	257.68	246.06	268.55	257.49	279.59	269.08
33,000	265.74	253.75	276.95	265.53	288.32	277.49
34,000	273.79	261.44	285.34	273.58	297.06	285.90
35,000	281.84	269.13	293.73	281.63	305.81	294.37
40,000	322.10	307.57	335.69	321.85	349.50	336.35
45,000	362.36	346.01	377.64	362.10	393.17	378.40
50,000	402.62	384.46	419.61	422.31	436.86	420.43

Table 2 Compound Interest

$$(1 + i)^n$$

i n	1%	2%	3%	4%	5%	6%
1	1.01000	1.02000	1.03000	1.04000	1.05000	1.06000
2	1.02010	1.04040	1.06090	1.08160	1.10250	1.12360
3	1.03030	1.06121	1.09273	1.12486	1.15763	1.19102
4	1.04060	1.08243	1.12551	1.16986	1.21551	1.26248
5	1.05101	1.10408	1.15927	1.21665	1.27628	1.33823
6	1.06152	1.12616	1.19405	1.26532	1.34010	1.41852
7	1.07214	1.14869	1.22987	1.31593	1.40710	1.50363
8	1.08286	1.17166	1.26677	1.36857	1.47746	1.59385
9	1.09369	1.19509	1.30477	1.42331	1.55133	1.68948
10	1.10462	1.21899	1.34392	1.48024	1.62889	1.79085
11	1.11567	1.24337	1.38423	1.53945	1.71034	1.89830
12	1.12683	1.26824	1.42576	1.60103	1.79586	2.01220
13	1.13809	1.29361	1.46853	1.66507	1.88565	2.13293
14	1.14947	1.31948	1.51259	1.73168	1.97993	2.26090
15	1.16097	1.34587	1.55797	1.80094	2.07893	2.39656
16	1.17258	1.37279	1.60471	1.87298	2.18287	2.54035
17	1.18430	1.40024	1.65285	1.94790	2.29202	2.69277
18	1.19615	1.42825	1.70243	2.02582	2.40662	2.85434
19	1.20811	1.45681	1.75351	2.10685	2.52695	3.02560
20	1.22019	1.48595	1.80611	2.19112	2.65330	3.20714
21	1.23239	1.51567	1.86029	2.27877	2.78596	3.39956
22	1.24472	1.54598	1.91610	2.36992	2.92526	3.60354
23	1.25716	1.57690	1.97359	2.46472	3.07152	3.81975
24	1.26973	1.60844	2.03279	2.56330	3.22510	4.04893
25	1.28243	1.64061	2.09378	2.66584	3.38635	4.29187
26	1.29526	1.67342	2.15659	2.77247	3.55567	4.44938
27	1.30821	1.70689	2.22129	2.88337	3.73346	4.82235
28	1.32129	1.74102	2.28793	2.99870	3.92013	5.11169
29	1.33450	1.77584	2.35657	3.11865	4.11614	5.41839
30	1.34785	1.81136	2.42726	3.24340	4.32194	5.74349
31	1.36133	1.84759	2.50008	3.37313	4.53804	6.08810
32	1.37494	1.88454	2.57508	3.50806	4.76494	6.45339
33	1.38869	1.92223	2.65234	3.64838	5.00319	6.84059
34	1.40258	1.96068	2.73191	3.79432	5.25335	7.25103
35	1.41660	1.99989	2.81386	3.94609	5.51602	7.66609
36	1.43077	2.03989	2.89828	4.10393	5.79182	8.14725
37	1.44508	2.08069	2.98523	4.26809	6.08141	8.63609
38	1.45953	2.12230	3.07478	4.43881	6.38548	9.15425
39	1.47412	2.16474	3.16703	4.61637	6.70475	9.70350
40	1.48886	2.20804	3.26204	4.80102	7.03999	10.28572

Table 3   Squares and Square Roots

n	n^2	$\sqrt{n}$	$\sqrt{10n}$	n	n^2	$\sqrt{n}$	$\sqrt{10n}$
1	1	1.000	3.162	51	2601	7.141	22.583
2	4	1.414	4.472	52	2704	7.211	22.804
3	9	1.732	5.477	53	2809	7.280	23.022
4	16	2.000	6.325	54	2916	7.348	23.238
5	25	2.236	7.071	55	3025	7.416	23.452
6	36	2.449	7.746	56	3136	7.483	23.664
7	49	2.646	8.367	57	3249	7.550	23.875
8	64	2.828	8.944	58	3364	7.616	24.083
9	81	3.000	9.487	59	3481	7.681	24.290
10	100	3.162	10.000	60	3600	7.746	24.495
11	121	3.317	10.488	61	3721	7.810	24.698
12	144	3.464	10.954	62	3844	7.874	24.900
13	169	3.606	11.402	63	3969	7.937	25.100
14	196	3.742	11.832	64	4096	8.000	25.298
15	225	3.873	12.247	65	4225	8.062	25.495
16	256	4.000	12.649	66	4356	8.124	25.690
17	289	4.123	13.038	67	4489	8.185	25.884
18	324	4.243	13.416	68	4624	8.246	26.077
19	361	4.359	13.784	69	4761	8.307	26.268
20	400	4.472	14.142	70	4900	8.367	26.458
21	441	4.583	14.491	71	5041	8.426	26.646
22	484	4.690	14.832	72	5184	8.485	26.833
23	529	4.796	15.166	73	5329	8.544	27.019
24	576	4.899	15.492	74	5476	8.602	27.203
25	625	5.000	15.811	75	5625	8.660	27.386
26	676	5.099	16.125	76	5776	8.718	27.568
27	729	5.196	16.432	77	5929	8.775	27.749
28	784	5.292	16.733	78	6084	8.832	27.928
29	841	5.385	17.029	79	6241	8.888	28.107
30	900	5.477	17.321	80	6400	8.944	28.284
31	961	5.568	17.607	81	6561	9.000	28.460
32	1024	5.657	17.889	82	6724	9.055	28.636
33	1089	5.745	18.166	83	6889	9.110	28.810
34	1156	5.831	18.439	84	7056	9.165	28.983
35	1225	5.916	18.708	85	7225	9.220	29.155
36	1296	6.000	18.974	86	7396	9.274	29.326
37	1369	6.083	19.235	87	7569	9.327	29.496
38	1444	6.164	19.494	88	7744	9.381	29.665
39	1521	6.245	19.748	89	7921	9.434	29.833
40	1600	6.325	20.000	90	8100	9.487	30.000
41	1681	6.403	20.248	91	8281	9.539	30.166
42	1764	6.481	20.494	92	8464	9.592	30.332
43	1849	6.557	20.736	93	8649	9.644	30.496
44	1936	6.633	20.976	94	8836	9.695	30.659
45	2025	6.708	21.213	95	9025	9.747	30.822
46	2116	6.782	21.448	96	9216	9.798	30.984
47	2209	6.856	21.679	97	9409	9.849	31.145
48	2304	6.928	21.909	98	9604	9.899	31.305
49	2401	7.000	22.136	99	9801	9.950	31.464
50	2500	7.071	22.361	100	10000	10.000	31.623

Table 4   Areas Under the Standard Normal Curve

The column under A gives the proportion of the area under the entire curve that is between $z = 0$ and a positive value of z.

z	A	z	A	z	A	z	A	z	A
.00	.000	.67	.249	1.34	.410	2.01	.478	2.68	.496
.01	.004	.68	.252	1.35	.412	2.02	.478	2.69	.496
.02	.008	.69	.255	1.36	.413	2.03	.479	2.70	.497
.03	.012	.70	.258	1.37	.415	2.04	.479	2.71	.497
.04	.016	.71	.261	1.38	.416	2.05	.480	2.72	.497
.05	.020	.72	.264	1.39	.418	2.06	.480	2.73	.497
.06	.024	.73	.267	1.40	.419	2.07	.481	2.74	.497
.07	.028	.74	.270	1.41	.421	2.08	.481	2.75	.497
.08	.032	.75	.273	1.42	.422	2.09	.482	2.76	.497
.09	.036	.76	.276	1.43	.424	2.10	.482	2.77	.497
.10	.044	.77	.279	1.44	.425	2.11	.483	2.78	.497
.12	.048	.78	.282	1.45	.427	2.12	.483	2.79	.497
.13	.052	.79	.285	1.46	.428	2.13	.483	2.80	.497
.14	.056	.80	.288	1.47	.429	2.14	.484	2.81	.498
.15	.060	.81	.291	1.48	.431	2.15	.484	2.82	.498
.16	.064	.82	.294	1.49	.432	2.16	.485	2.84	.498
.17	.068	.83	.297	1.50	.433	2.17	.485	2.85	.498
.18	.071	.84	.300	1.51	.435	2.18	.485	2.86	.498
.19	.075	.85	.302	1.52	.436	2.19	.486	2.87	.498
.20	.079	.86	.305	1.53	.437	2.20	.486	2.86	.498
.21	.083	.87	.308	1.54	.438	2.21	.487	2.87	.498
.22	.087	.88	.311	1.55	.439	2.22	.487	2.88	.498
.23	.091	.89	.313	1.56	.441	2.23	.487	2.89	.498
.24	.095	.90	.316	1.57	.442	2.24	.488	2.90	.498
.25	.099	.91	.319	1.58	.443	2.25	.488	2.91	.498
.26	.103	.92	.321	1.59	.444	2.26	.488	2.92	.498
.27	.106	.93	.324	1.60	.445	2.27	.488	2.93	.498
.28	.110	.94	.326	1.61	.446	2.28	.489	2.94	.498
.29	.114	.95	.329	1.62	.447	2.29	.489	2.95	.498
.30	.118	.96	.332	1.63	.449	2.30	.489	2.96	.499
.31	.122	.97	.334	1.64	.450	2.31	.490	2.97	.499
.32	.126	.98	.337	1.65	.451	2.32	.490	2.98	.499
.33	.129	.99	.339	1.66	.452	2.33	.490	3.00	.499
.34	.133	1.00	.341	1.67	.453	2.34	.490	3.01	.499
.35	.137	1.01	.344	1.68	.454	2.35	.491	3.02	.499
.36	.141	1.02	.346	1.69	.455	2.36	.491	3.03	.499
.37	.144	1.03	.349	1.70	.455	2.37	.491	3.04	.499
.38	.148	1.04	.351	1.71	.456	2.38	.491	3.50	.499
.39	.152	1.05	.353	1.72	.457	2.39	.492	3.06	.499
.40	.155	1.06	.355	1.73	.458	2.40	.492	3.07	.499
.41	.159	1.07	.358	1.74	.459	2.41	.492	3.08	.499
.42	.163	1.08	.360	1.75	.460	2.42	.492	3.09	.499
.43	.166	1.09	.362	1.76	.461	2.43	.493	3.10	.499
.44	.170	1.10	.364	1.77	.462	2.44	.493	3.11	.499
.45	.174	1.11	.367	1.78	.463	2.45	.493	3.12	.499
.46	.177	1.12	.369	1.79	.463	2.46	.493	3.13	.499
.47	.181	1.13	.371	1.80	.464	2.47	.493	3.14	.499
.48	.184	1.14	.373	1.81	.465	2.48	.493	3.15	.499
.49	.188	1.15	.375	1.82	.466	2.49	.494	3.16	.499
.50	.192	1.16	.377	1.83	.466	2.50	.494	3.17	.499
.51	.195	1.18	.381	1.84	.467	2.51	.494	3.18	.499
.52	.199	1.19	.383	1.85	.468	2.52	.494	3.19	.499
.53	.202	1.20	.385	1.86	.469	2.53	.494	3.20	.499
.54	.205	1.21	.387	1.87	.469	2.54	.495	3.21	.499
.55	.209	1.22	.389	1.88	.470	2.55	.495	3.22	.499
.56	.212	1.23	.391	1.89	.471	2.56	.495	3.23	.499
.57	.216	1.24	.393	1.90	.471	2.57	.495	3.24	.499
.58	.219	1.25	.394	1.91	.472	2.58	.495	3.25	.499
.59	.222	1.26	.396	1.92	.473	2.59	.495	3.26	.499
.60	.226	1.27	.398	1.93	.473	2.60	.495	3.27	.500
.61	.229	1.28	.400	1.94	.474	2.61	.496	3.28	.500
.62	.232	1.29	.402	1.96	.475	2.62	.496	3.29	.500
.63	.236	1.30	.403	1.97	.476	2.63	.496	3.30	.500
.64	.239	1.31	.405	1.98	.476	2.64	.496	3.31	.500
.65	.242	1.32	.407	1.99	.477	2.65	.496	3.32	.500
.66	.245	1.33	.408	2.00	.477	2.66	.496	3.33	.500
						2.67	.496		

Answers to Odd-numbered Exercises

CHAPTER 1

Section 1.1
1. {12, 13, 14, 15, 16, 17, 18, 19, 20}
3. $\{1, \frac{1}{2}, \frac{1}{4}, \frac{1}{8}, \frac{1}{16}, \frac{1}{32}\}$
5. {17, 22, 27, 32, 37, 42, 47}
7. {1, 2, 3, 4, 5}   9. {1, 2, 3, 4}
11. {4, 6, 8, 10, 12, . . .}
13. {9, 11, 13, 15}
15. finite   17. infinite   19. infinite
21. infinite   23. 5   25. 1000
27. 50   29. well-defined
31. not well-defined
33. not well-defined
35. well-defined; there are none
37. $\in$ 39. $\notin$ 41. $\in$ 43. $\notin$
45. false   47. true   49. true
51. true   53. false   55. true
57. true   59. false   61. false
63. true   65. {1, 2, 3, 4, 5, 6}
67. {HHH, HHT, HTH, THH, TTH, THT, HTT, TTT}
69. {se}, {ve}

Section 1.2
1. $\subset$ 3. $\not\subset$ 5. $\subset$ 7. $\subset$
9. true   11. true   13. false
15. true   17. true   19. true
21. false   23. false   25. false
27. true   29. false   31. 8
33. 32   35. 1   37. 32
39. {2, 3, 5, 7, 9, 10}   41. {2}
43. U 45. $\varnothing$ 47. $\varnothing$, {0}
49. {low cost, high cost, pays off at death, no retirement benefits, retirement benefits}   51. {low cost, no retirement benefits}
53. {high cost, retirement benefits}
55. all five must be present
57. DLA, DLK, DLS, LAK, LAS, DAK, DAS, AKS, LKS, DKS   59. D, L, K, S, A
61. $1 + 5 + 10 + 10 + 5 + 1 = 32$

Section 1.3
1. true   3. false   5. true
7. false   9. true   11. true
13. {3, 5}   15. {2, 3, 4, 5, 7, 9}
17. U, or {2, 3, 4, 5, 7, 9}   19. {7, 9}
21. $\varnothing$ 23. $\varnothing$
25. {2, 3, 4, 5, 7, 9}   27. true
29. true   31. {s, d, c, g, i, m, h}
33. {i, m, h}   35. {s, d, c, g, i m, h}
37. all students studying math and history   39. all students on the G. I. Bill who study math   41. all students who do not study math and are over 25   43. all students who do not study math and study history and are over 25
45. true   47. false   49. true
51. true   53. false   55. true
57. convex   59. convex
61. not convex   63. yes   65. no
67. {3, 9}   69. {18}   71. {3, 9}
73. $A \cap B = \varnothing$

Section 1.4
1.
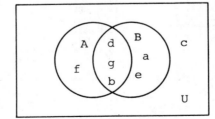

3.
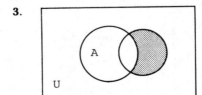

5.
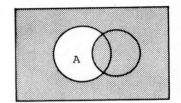

7.
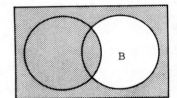

9. $\varnothing$

11.

13. $\varnothing$

15.
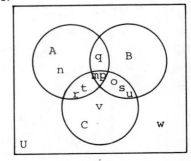

486

17.

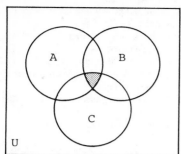

25.

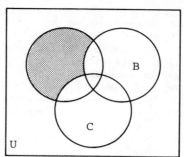

19.

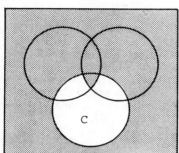

21.

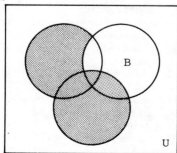

23.

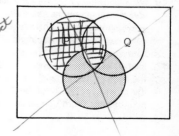

27. true   **29.** false   **31.** true
33. true   **35.** true   **37.** true
39. false
41. normal rate, regular rhythm, p wave precedes t wave   **43.** non-normal rate, regular rhythm, p wave precedes r wave   **45.** normal rate, regular rhythm, p wave precedes r wave   **47.** 4; 8; 16
49. 1, 2, 3, 4, 5, 6, 7, 8, 9, 10, 11, 12, 13, 14, 15 (all except 16)
51. 5; 8; 13   **53.** 2^n

Section 1.5

1. (a) 12   (b) 20   (c) 6
   (d) 10   (e) 48
3. Yes; his data adds up to 142 people
5. (a) A-Negative   (e) B-Positive
   (b) AB-Negative   (f) O-Positive
   (c) B-Negative   (g) O-Negative
   (d) AB-Positive
7. (a) 37   **9.** (a) 40
   (b) 22   (b) 30
   (c) 50   (c) 95
   (d) 11   (d) 110
   (e) 25   (e) 160
   (f) 11   (f) 65

Section 1.6

Exercises 1–4 show only one possible way.
1. {1, 4, 9, 12}
   ↕ ↕ ↕ ↕
   {8, 12, 16, 20}
3. not possible   **5.** 5   **7.** 4
9. 1   **11.** $\aleph_0$ **13.** c
Exercises 15–22 show only one possible way.
15. {5, 10, 15, 20, 25, 30, . . .}
   ↕ ↕ ↕ ↕ ↕ ↕
   {1, 2, 3, 4, 5, 6, . . .}

17. {2, 4, 6, 8, 10, 12, 14, . . .}
   ↕ ↕ ↕ ↕ ↕ ↕ ↕
   {1, 2, 3, 4, 5, 6, 7, . . .}
19. {0, 2, −2, 4, −4, 6, −6, 8, −8, . . .}
   ↕ ↕ ↕ ↕ ↕ ↕ ↕ ↕ ↕
   {1, 2, 3, 4, 5, 6, 7, 8, 9, . . .}
21. {2, 4, 8, 16, 32, 64, 128, . . .}
   ↕ ↕ ↕ ↕ ↕ ↕ ↕
   {2^1, 2^2, 2^3, 2^4, 2^5, 2^6, 2^7, . . .}
   ↕ ↕ ↕ ↕ ↕ ↕ ↕
   {1, 2, 3, 4, 5, 6, 7, . . .}
23.

25. $\aleph_0$ **27.** $\aleph_0$ **29.** c
31. Each guest must move to a room having a number one higher than the current number. This leaves room 1 available for the new guest.

Chapter 1   Test

1. {2, 4, 6}   **2.** {3, 7}
3. {1, 3, 6, 7}   **4.** {2, 4, 6}
5. false   **6.** true   **7.** false
8. true   **9.** true   **10.** false
11. false   **12.** true
13.

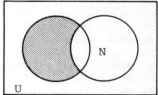

14.

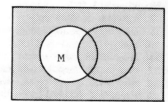

15.

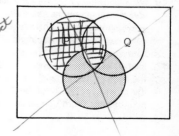

16.

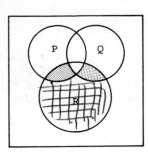

17. {outdoor gaslight, color television, clothes dryer, central air conditioner, water bed heater} **18.** {outdoor gaslight, frostless refrigerator, clothes dryer, central air conditioner, water bed heater} **19.** {coffee maker, dishwasher, manual refrigerator, toaster} **20.** {outdoor gaslight, coffee maker, dishwasher, frostless refrigerator, manual refrigerator, toaster, clothes dryer, central air conditioner, water bed heater}

21. 8 **22.** 27 **23.** 16 **24.** 28 **25.** 43

CHAPTER 2

Section 2.1

1. deductive **3.** inductive
5. deductive **7.** deductive
9. valid **11.** not valid **13.** valid
15. valid **17.** not valid
19. not valid **21.** not valid
23. not valid **25.** not valid
27. not valid **29.** valid
33. They are the same distance from Miami.
35. The two children row across. One stays on the opposite bank and the other returns. One soldier rows across, and the child on the opposite bank then rows back. The two children row across. One stays and the other returns. Now another soldier rows across. This process continues until all the soldiers are across.
37. First balance four against four. Of the lighter four, balance two against the other two. Finally of the lighter two, balance them one against the other.

Section 2.2

1. statement **3.** not a statement
5. statement **7.** statement
9. statement **11.** statement
13. not a statement **15.** statement
17. or **19.** or **21.** not
23. not compound
25. He does not have brown hair.
27. He has brown hair and he is tall.
29. He does not have brown hair or he is tall.
31. He does not have brown hair or he is not tall.
33. It is not the case that he does not have brown hair and he is tall.
35. $p \wedge q$ **37.** $\sim p \wedge \sim q$
39. $(p \vee q) \wedge \sim(p \wedge q)$
41. His mother is not tall.
43. Not all cows were once calves.
45. No people eat pancakes.
47. Not all tax must be paid by April 15.
49. All ground squirrels are happy.
51. Somebody doesn't like Sara Lee.
53. Some people never love anybody.
55. *Literal meaning:* No one can be an expert in mathematics. *Intended meaning:* Not everyone can be an expert in mathematics.
57. *Literal meaning:* I have tried to find a book called *How to Play the Tuba Without Success.* *Intended meaning:* I have tried with no success to find the book.
59. Against birth control

Section 2.3

1. F **3.** T **5.** T **7.** T
9. F **11.** F **13.** F **15.** T
17. F **19.** FTTT **21.** TTFT
23. FTFF **25.** FFTT **27.** TFTT
29. TFTTTFTF **31.** FFFTTTTT
33. FFFFFFFFTFTTFFFF **35.** 16
37. 128 **39.** inclusive disjunction
41. T **43.** F **45.** TFFT
47. FFFT **49.** TFTT **51.**
51. Number the tubes from left to right, across the top of the photograph. One sequence of balls is 1, 2, 4, 5, 8, 11.

54.

H	S	$(H \vee S) \wedge \sim H$
T	T	F
T	F	F
F	T	T
F	F	F

55. row 3 **56.** she did

Section 2.4

Several different wordings possible in Exercises 1–22.
1. If it's Tuesday, then this must be Belgium.
3. If it flies, then it's a bird.
5. If it's Saturday, then Sally goes to town today.
7. If you're a person, then you have a head.
9. If it's a chicken, then it's not a teetotaler.
11. If Napoleon shouts, then Europe trembles.
13. If I study hard, then I pass this lousy course.
15. If my instructor is not weird, then I study hard.
17. If I don't study hard, then my instructor is not weird.
19. If I study hard, then my instructor is weird and I pass this lousy course.
21. If I study hard and I pass this lousy course, then my instructor is weird.
23. $p \to r$ **25.** $r \to \sim q$
27. $r \wedge (q \to \sim p)$ **29.** $r \to q$
31. T **33.** T **35.** F **37.** T
39. F **41.** T **43.** FTTT
45. TTTT (tautology) **47.** TTFT
49. TFTF
51. TTTTTTFT
53. TTTFTTTTTTTTTTTTTTTT
55. equivalent **57.** equivalent
59. not equivalent **61.** not equivalent
63. TTTF **65.** F **67.** F
69. T

Section 2.5

1. If it rains, then I will stay home.
3. If I pass this class, then I did homework.
5. If you are in Raleigh, then you are in North Carolina.
7. If you are elected, then you are against pollution.
9. If the city redevelops Eastlake, then the government approves.
11. If I feel badly, then I lost on the slot machines.
13. If I live in West Virginia, then I like coal.
In Exercises 15–30, first the converse, then the inverse, and finally the contrapositive.
15. If I need a garage, then I buy a Ford. If I don't buy a Ford, then I won't

need a garage.   If I don't need a garage, then I didn't buy a Ford.

17. If you endanger your health, then you smoke cigarettes.   If you don't smoke cigarettes, then you don't endanger your health.   If you don't endanger your health, then you don't smoke cigarettes.

19. If it has fleas, then it's a dog. If it's not a dog, then it doesn't have fleas.   If it doesn't have fleas, then it's not a dog.

21. If they flock together, then they are birds of a feather.   If they are not birds of a feather, then they don't flock together.   If they don't flock together, then they are not birds of a feather.

23. If it's on the north side of the tree, then it's moss.   If it's not moss, then it's not on the north side of the tree.   If it's not on the north side of the tree, then it's not moss.

25. $q \to \sim p; \ p \to \sim q; \ \sim q \to p$
27. $\sim q \to \sim p; \ p \to q; \ q \to p$
29. $(q \lor r) \to p; \ \sim p \to \sim(q \lor r);$
   $\sim(q \lor r) \to \sim p$
31. T   **33.** F   **35.** T   **37.** F
39. contrary   **41.** contrary
43. consistent
45. $b \to \sim m; \ b \lor m; \ m \to b$
47. The butler did it and the maid did it.
48. TTF   **49.** Neither did it; TF
50. The butler

Section 2.6

1. invalid   **3.** valid   **5.** valid
7. invalid   **9.** valid   **11.** valid
13. valid   **15.** invalid   **17.** valid
19. invalid

In Exercises 21–60, the contrapositive of the given answer is also correct.

21. $p \to \sim s$ **22.** $r \to s$
23. $q \to p$
24. None of my poultry are officers.
25. $r \to p$ **26.** $\sim r \to \sim q$
27. $s \to \sim p$
28. Your sons are not fit to serve on a jury.
29. $s \to r$ **30.** $p \to q$
31. $q \to \sim r$
32. Guinea pigs don't appreciate Beethoven.
33. $r \to \sim s$ **34.** $u \to t$
35. $\sim r \to p$ **36.** $\sim u \to \sim q$

37. $t \to s$ **38.** All pawnbrokers are honest.   **39.** $p \to s$
40. $\sim r \to \sim u$ **41.** $t \to p$
42. $s \to \sim q$ **43.** $\sim t \to \sim r$
44. Kittens with green eyes are not willing to play with a gorilla.
45. $p \to q$ **46.** $\sim u \to \sim s$
47. $t \to \sim r$ **48.** $q \to s$
49. $v \to p$ **50.** $\sim r \to \sim u$
51. Opium eaters do not wear white kid gloves.
52. $r \to w$ **53.** $\sim u \to \sim t$
54. $v \to \sim s$ **55.** $x \to r$
56. $\sim q \to t$ **57.** $y \to p$
58. $w \to s$ **59.** $\sim x \to \sim q$
60. $p \to \sim u$
61. I can't read any of Brown's letters.

Section 2.7

1.

3. The statement simplifies to F.

5.

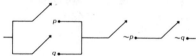

7. The statement simplifies to T.

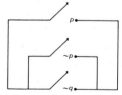

9. p **11.** $\sim p$

Appendix

1. ambiguity   **3.** false cause
5. composition   **7.** ambiguity
9. composition   **11.** emotion
13. experts   **15.** ambiguity
17. false cause   **19.** non sequitur
21. experts   **23.** false cause
25. complex question
27. composition   **31.** $\sim r$
33. $\sim p \lor \sim q$ **35.** s, u **37.** $\sim b$
39. $p \to [\sim j \to (\sim e \land \sim f \land \sim g \land \sim h)]$

Chapter 2   Test

1. invalid   **2.** valid   **3.** statement
4. not a statement   **5.** statement
6. She passes algebra and she does not pass history.
7. If she does not pass algebra, then she will pass history.
8. Not all cluckers are chickens.
9. No spiders eat their mates.
10. T   **11.** T   **12.** T
13. TTTF   **14.** TTTT   **15.** TTTT
16. If he would work, then he would pass mathematics.
17. If Snoopy flies a kite, then Lucy is not around.
18. If you own a bird, then you have yesterday's newspaper.
19. If you are unemployed, then you can get food stamps.
20. If I need mustard, then I'll buy a sandwich.
21. If it's not after 7 P.M., then I drink coffee.
22. If Tom's going, then Sandra's going.
23. Converse: $(q \lor r) \to \sim p$
   Contrapositive: $\sim(q \lor r) \to p$
24. invalid   **25.** valid

CHAPTER 3

Section 3.1

1. simple; 36   **3.** positional; 14
5. simple; 627   **7.** multiplicative; 23
9. simple; 781
11. multiplicative; 80
13. simple; 100   **15.** positional; 390
17. simple; 1375
19. multiplicative; 2375
21. positional; 586
23. positional; 4867
25. ΛΙΙΙ; ⊙Λ⊙Ι; ⊙☺
27. ΛΛΙΙΙΙ; ⊙Λ⊙Ι; ⊙☺
29. Ν ΛΛΙΙ; ⊙Ν⊙Λ⊙Ι; ⊙⊙⊙
31. Ν ΛΛΛΛΙΙ; ⊙Ν⊙Λ⊙Ι; ⊙☺⊙
33. Ν Ν Ν Ι; ⊙Ν⊙Ι; ⊙⊙⊙
35. Μ; ⊙Μ; ⊙⊙⊙⊙
37. ΜΜΝ Ν ΛΛΛΛ; ⊙Μ⊙Ν⊙Λ; ⊙⊙☺⊙
39. Ν Ν Μ; ⊙Ν; ⊙⊙⊙⊙⊙

Section 3.2

1. 231   **3.** 23,201   **5.** 3,400,026
7. 2,005,077   **9.** 82,653
11. 7,170,668

13.

15. 𝄢𝄢

17. 𝄢𝄢𝄢 : 𝄢|||||

19. 𝄢𝄢𝄢 : 𝄢 : |: ||

21. 𝄢𝄢 ၅၅၅ᑎᑎᑎᑎᑎ|||||||

23. 𝄢𝄢𝄢𝄢𝄢𝄢𝄢 ᑎᑎᑎᑎᑎ

25. 𝕐𝕐𝕐𝕐𝕐 ☁☁☁☁☁ 𝄢𝄢𝄢

27. 23   **29.** 86   **31.** 2450
33. 2001   **35.** 21,000
37. 14,000,000   **39.** XII
41. XLVII   **43.** CDLXXIV
45. DCCLIX   **47.** MDCCXXVIII
49. X̄II   **51.** 55   **53.** 394
55. 11,232   **57.** 21   **59.** 37
61. 275   **63.** 50   **65.** 836
67. 764   **69.** 8   **71.** 68
73. 357   **75.** 722   **77.** 533,000

Section 3.3

1. Mayan; 12   **3.** Babylonian; 32
5. Chinese-Japanese; 4036
7. Mayan 242   **9.** Babylonian; 1282
11. Babylonian; 2601
13. Mayan; 2640
15. Mayan; 59,954
17. Babylonian; 80,474
19. Chinese-Japanese; 872

21.   **25.**   **27.**

23.

29. ⟨⟨▷

31. ▼▼▼▼ ⟨⟨ ▼▼▼

33. ⟨⟨ ▼▼▼▼▼ ⟨ ▼▼▼▼

35. ▼⟨⟨ ▼▼▼/▼▼▼ ⟨⟨⟨

37. ⟨▼▼ ▼▼▼▼▼

39. ═   **41.** ⋰   **43.** ═

45.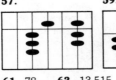

Section 3.4

1. $(5 \times 10^1) + (9 \times 10^0)$
3. $(4 \times 10^2) + (2 \times 10^1) + (6 \times 10^0)$
5. $(1 \times 10^3) + (9 \times 10^2) + (8 \times 10^1) + (4 \times 10^0)$
7. $(2 \times 10^4) + (9 \times 10^3) + (8 \times 10^2) + (4 \times 10^1) + (6 \times 10^0)$
9. $(2 \times 10^6) + (5 \times 10^5) + (0 \times 10^4) + (8 \times 10^3) + (9 \times 10^2) + (0 \times 10^1) + (1 \times 10^0)$
11. $(4 \times 10^3) + (7 \times 10^2) + (1 \times 10^1) + (2 \times 10^0)$
13. $(7 \times 10^6) + (0 \times 10^5) + (0 \times 10^4) + (2 \times 10^3) + (0 \times 10^2) + (0 \times 10^1) + (7 \times 10^0)$
15. 98   **17.** 600   **19.** 7805
21. 57,843   **23.** 8,020,305
25. 77   **27.** 979   **29.** 55
31. 222   **33.** 166   **35.** 825
37. 6910   **39.** 4   **41.** 107
43. 108   **45.** 14   **47.** 208
49. 2992   **51.** 28,742
53. 70,082

55.

57.   **59.**

61. 78   **63.** 13,515

Section 3.5

1. 23   **17.** 214_5 **33.** $2T88_{12}$
3. 44   **19.** 666_7 **35.** 20402_7
5. 43   **21.** 3001_4 **37.** 110_4
7. 124   **23.** 3434_7 **39.** 214_7
9. 113   **25.** 2644_{12} **41.** 2021_8
11. 153   **27.** $1T42_{11}$ **43.** 1727_8
13. 476   **29.** $22E8_{12}$ **45.** 166_{12}
15. 239   **31.** $4TT1_{11}$

Section 3.6

1. 123   **9.** 213   **17.** 214041
3. 1111   **11.** 1202   **19.** 4
5. 1224   **13.** 2003   **21.** 4002
7. 241   **15.** 20221   **23.** 12

25. row 2:   10, 11
   row 3:   10, 12
   row 4:   10, 12
   row 5:   10, 12
   row 6:   6, 10, 11, 12, 13, 14, 15
27. 110
29. 1156
31. 204
33. 4352
35. 354
37. 20343

39.

	0	1	2	3	4	5	6	7	8	9	T	E
0	0	1	2	3	4	5	6	7	8	9	T	E
1	1	2	3	4	5	6	7	8	9	T	E	10
2	2	3	4	5	6	7	8	9	T	E	10	11
3	3	4	5	6	7	8	9	T	E	10	11	12
4	4	5	6	7	8	9	T	E	10	11	12	13
5	5	6	7	8	9	T	E	10	11	12	13	14
6	6	7	8	9	T	E	10	11	12	13	14	15
7	7	8	9	T	E	10	11	12	13	14	15	16
8	8	9	T	E	10	11	12	13	14	15	16	17
9	9	T	E	10	11	12	13	14	15	16	17	18
T	T	E	10	11	12	13	14	15	16	17	18	19
E	E	10	11	12	13	14	15	16	17	18	19	1T

41. 110
43. T73
45. 87E
47. 1036T
49. 9E21

Section 3.7

1. 3   **3.** 7   **5.** 28   **7.** 115
9. 63   **11.** 100   **13.** 1001
15. 10001   **17.** 11110
19. 101111   **21.** 100011110

23.

	0	1
0	0	1
1	1	10

25. 10111
27. 10100101
29. 1000
31. 11011

33.

	0	1
0	0	0
1	0	1

35. 110
37. 1110
39. 1010100
41. 110000001

45.

A	B	C	D	E	F
1	2	4	8	16	32
3	3	5	9	17	33
5	6	6	10	18	34
7	7	7	11	19	35
11	11	13	13	21	37
13	14	14	14	22	38
15	15	15	15	23	39
17	18	20	24	24	40
19	19	21	25	25	41
21	22	22	26	26	42
23	23	23	27	27	43
25	26	28	28	28	44
27	27	29	29	29	45
29	30	30	30	30	46
31	31	31	31	31	47
33	34	36	49	48	48
35	35	37	41	49	49
37	38	38	42	50	50
39	39	39	43	51	51
41	42	44	44	52	52
43	43	45	45	53	53
45	46	46	46	54	54
47	47	47	47	55	55
49	50	52	56	56	56
51	51	53	58	57	57
53	54	54	58	58	58
55	55	55	59	59	59
57	58	60	60	60	60
59	59	61	61	61	61
61	62	62	62	62	62
63	63	63	63	63	63

47. 500,000

Chapter 3   Test
1. simple, 17   **2.** multiplicative, 119
3. positional, 1391
4. Egyptian, 3555
5. Babylonian, 21
6. Babylonian, 40,333
7. Mayan, 2694   **8.** Mayan, 6130
9. Roman, 67
10. $(2 \times 10^1) + (8 \times 10^0)$
11. $(4 \times 10^3) + (6 \times 10^2)$
$+ (9 \times 10^1) + (0 \times 10^0)$
12. 6574   **13.** 117   **14.** 111
15. 490   **16.** 55   **17.** 1202_4
18. 43124_5 **19.** 1529_{12}
20. 1253_8 **21.** 11110010_2
22. 1011_5 **23.** 142_5 **24.** 1232_5
25. 10010_2

CHAPTER 4

Section 4.1
1. 101, 103, 107, 109, 113, 127, 131, 137, 139, 149, 151, 157, 163, 167, 173, 179, 181, 191, 193, 197, 199
3. 10   **5.** 11   **7.** 2 and 3
More than one answer possible in Exercises 9–14.
9. 5 + 7   **11.** 11 + 13
13. 23 + 23   **15.** 17 and 19
17. 71 and 73   **19.** 101 and 103
21. 1, 2, 3, 4, 6, 12
23. 1, 2, 4, 5, 10, 20
25. 1, 2, 4, 13, 26, 52
27. 1, 2, 3, 4, 5, 6, 8, 10, 12, 15, 20, 30, 40, 60, 120
29. 1, 2, 5, 10, 17, 25, 50, 85, 170, 425, 850
31. 1, 2, 7, 11, 14, 22, 77, 154
33. $3 \cdot 5$ **35.** $2^2 \cdot 3^2$
37. $2^4 \cdot 3 \cdot 5$ **39.** $2^3 \cdot 3^2 \cdot 5$
41. $3 \cdot 13 \cdot 17$ **43.** 90
45. 7350   **47.** 5544   **49.** 315
51. N is made up of two parts, $(2 \cdot 3 \cdot 5 \cdot 7 \cdot 11 \cdot 13 \cdots p)$ and 1. Since $(2 \cdot 3 \cdot 5 \cdot 7 \cdot 11 \cdot 13 \cdots p)$ is a multiple of 2, it is divisible by 2. Therefore, the sum $(2 \cdot 3 \cdot 5 \cdot 7 \cdot 11 \cdot 13 \cdots p) + 1$ cannot be.
53. $(2 \cdot 3 \cdot 5 \cdot 7 \cdot 11 \cdot 13 \cdots p)$ is also divisible by 5, so that the sum cannot be.
55. Since it is not divisible by any prime from 2 through p.
57. 807   **59.** 807   **61.** 807

Section 4.2
1. 10   **3.** 60   **5.** 168   **7.** 38
9. 10   **11.** 12   **13.** 6   **15.** $\frac{2}{3}$
17. $\frac{8}{15}$ **19.** $\frac{9}{10}$ **21.** $\frac{4}{5}$ **23.** $\frac{4}{7}$
25. $\frac{50}{91}$ **27.** $\frac{11}{12}$ **29.** 96
31. 140   **33.** 96   **35.** 884
37. 144   **39.** 216   **41.** $\frac{3}{8}$
43. $\frac{43}{48}$ **45.** $\frac{59}{90}$ **47.** $\frac{9}{40}$
49. $\frac{2}{5}$ **51.** We are really finding the prime factorization of each number.
53. 800   **55.** yes   **57.** yes
59. no   **61.** yes   **63.** no
65. no   **67.** 35   **69.** 22
71. yes   **73.** false   **75.** true
77. true   **79.** true

Section 4.3
1. commutative   **3.** associative
5. closure   **7.** identity
9. identity   **11.** distributive
13. identity: no   commutative: no, no   associative: no   closure: no, no
15. yes   **17.** not closed, 2 − 4
19. not closed, 1 + 3   **21.** closed
23. closed   **25.** closed
27. associative, commutative, associative, associative, distributive, identity and associative, distributive
29. 15   **31.** 65   **33.** 130
35. 514
37. (a) 6 (b) 17 (c) 16 (d) 6 (e) 7 (f) 13 (g) 14 (h) 20
39. {m, n, p, q, c, d, e}   **41.** yes
43. Otherwise union is not big enough.
45. 2 + 2 + 2

Section 4.4
1. true   **3.** true   **5.** false
7. true   **9.** true
11.
-4 2 3   5
13. -2 -1 0 1 2
15. -3 -2 -1 0
17. -4 -3 -2 -1
19. <   **21.** <   **23.** >   **25.** >
27. 5   **29.** 2   **31.** -14
33. -23 **35.** 1   **37.** -11
39. -26 **41.** 10   **43.** 19
45. 6   **47.** -13 **49.** -12
51. 18   **53.** -36 **55.** 150
57. -5 **59.** 2   **61.** 3   **63.** -2
65. -36 **67.** 12   **69.** 18
71. 40   **73.** 4   **75.** -11
77. -5 **83.** 26 to 30
85. 44% to 50%

Section 4.5

1. $\frac{27}{20}$ or $1\frac{7}{20}$ **3.** $\frac{1}{5}$ **5.** 3
7. $\frac{3}{10}$ **9.** $\frac{1}{9}$ **11.** $\frac{3}{20}$ **13.** $\frac{5}{12}$
15. $\frac{1}{2}$ **17.** $\frac{7}{40}$ **19.** $\frac{23}{60}$
21. > **23.** > **25.** < **27.** >
29. > **31.** $\frac{13}{10}, \frac{12}{25}$
33. $\frac{7}{9}, \frac{4}{5}, \frac{6}{7}$ **35.** $\frac{3}{10}, \frac{1}{3}, \frac{3}{8}, \frac{3}{7}$
37. $\frac{5}{12}, \frac{1}{2}, \frac{8}{15}, \frac{4}{7}$
We give only one possible answer in Exercises 39–51.
39. $\frac{3}{8}$ **41.** $\frac{19}{30}$ **43.** $\frac{1}{8}$ **45.** $\frac{1}{4}$
47. $\frac{41}{18}$ or $2\frac{5}{18}$ **49.** $-\frac{5}{8}$ **51.** $\frac{1}{4}$
53. $1, \frac{3}{2}, \frac{1}{4}, \frac{1}{2}, \frac{1}{4}, \frac{3}{2}, 1, \frac{3}{2}, 1, \frac{3}{2}, \frac{1}{8}$; 4, 6, 1, 2, 1, 6, 4, 6, 4, 6, $\frac{1}{2}$

Section 4.6

1. 11, 17, 23, 29, 35, arithmetic
3. $-4, -1, 2, 5, 8$, arithmetic
5. $-2, -8, -14, -20, -26$, arithmetic
7. 2, 4, 8, 16, 32, geometric
9. $-2, 4, -8, 16, -32$, geometric
11. 6, 12, 24, 48, 96, geometric
13. $\frac{2}{3}, \frac{3}{4}, \frac{4}{5}, \frac{5}{6}, \frac{6}{7}$, neither
15. $\frac{1}{2}, \frac{1}{3}, \frac{1}{4}, \frac{1}{5}, \frac{1}{6}$, neither
17. arithmetic, 8 **19.** arithmetic, 3
21. geometric, 3 **23.** neither
25. arithmetic, -3 **27.** arithmetic, 3
29. geometric, -2 **31.** neither
33. 70 **35.** 65 **37.** 7
39. -65 **41.** 78 **43.** -65
45. 63 **47.** 174 **49.** 678
51. 183 **53.** 12 **55.** -153
57. 500,500 **59.** 120 **61.** 65
63. 56, 64, B **65.** 17¢
67. $239.98 **69.** $800

Section 4.7

1. 987 **3.** 2584 **5.** 6765
7. 75,025 **9.** 24,157,817
11. 1, 1, 2, 3, 5, 8 **13.** 21
15. 55 **17.** 21 by 13 **19.** 1.6
21. yes **23.** 8 by 5

Chapter 4 Test

1. commutative **2.** associative
3. identity **4.** distributive
5. yes **6.** no **7.** not a prime
8. not a prime **9.** $2^2 \cdot 3 \cdot 5$
10. $5^2 \cdot 13$ **11.** no **12.** yes
13. yes **14.** 5 **15.** 80
16. 12 **17.** 144 **18.** 180
19. $\frac{11}{16}$ **20.** $\frac{57}{160}$ **21.** $\frac{2}{5}$
22. $\frac{3}{2}$ or $1\frac{1}{2}$ **23.** < **24.** >
25. < **26.** 4 **27.** -3
28. -10 **29.** -11 **30.** -6

31. -117 **32.** 36 **33.** 12
34. 4, 10, 16, 22, 28 **35.** 3, 8, 15, 24, 35
36. 6, 4, 2, 0, -2 **37.** 3, 12, 48, 192, 768 **38.** 6, -12, 24, -48, 96 **39.** 129 **40.** 125,250

CHAPTER 5

Section 5.1

1. $.\overline{4}$ **3.** $.\overline{92}$ **5.** $.\overline{469}$
7. $.9134\overline{5}$ **9.** $.083\overline{25}$ **11.** .875
13. .28125 **15.** $.91\overline{6}$ **17.** $.\overline{81}$
19. $.7\overline{3}$ **21.** $\frac{2}{5}$ **23.** $\frac{17}{20}$
25. $\frac{21}{200}$ **27.** $\frac{8}{9}$ **29.** $\frac{6}{11}$
31. $\frac{41}{333}$ **33.** $4\frac{92}{99}$ **35.** $\frac{13}{30}$
37. 1 **39.** rational **41.** irrational
43. rational **45.** rational
47. one answer: $.5438914769\ldots$
49. $.91111\ldots$ **51.** $.\overline{165}$
53. repeating **55.** terminating
57. terminating **59.** true
61. false **63.** true **65.** false
67. false **69.** yes, yes, yes, no, yes
71. no, yes, yes, yes, yes
73. no, no, no, no, yes **75.** yes, no, yes **77.** no, no, no, yes
79. no, no, no, no

Section 5.2

1. 13.5 **3.** 38.9 **5.** 24.7
7. 37.251 **9.** 109.76 **11.** 29.1
13. 24.2 **15.** 28 **17.** $91,150.00
19. $173.49 **21.** $2557.97
23. $147.33 **25.** 21.5
27. 78.4, 78.41 **29.** .1, .08
31. 12.7, 12.69 **33.** 42%
35. 36.5% **37.** .8% **39.** 210%
41. .46 **43.** .08 **45.** 1.59
47. .005 **49.** 166.88
51. 42.48 **53.** 65% **55.** $1612
57. 1650 **59.** 5135 **61.** $11.90
63. $255 **65.** $935 **67.** 40%
69. 25.2%

Section 5.3

1. 104.58 **3.** 24,904.75
5. 1857.78 **7.** 7.68 **9.** 1.20
11. 56.51 **13.** 369 **15.** 25.8
17. 1.89 **19.** .5 **21.** .333
23. .091 **25.** .833 **27.** 1.267
29. 4.917 **31.** 7.733 **33.** .524
35. 1.778 **37.** $855.60
39. $1756.64 **41.** $2740
43. $4336.88 **45.** 407.6

47. 4.36 **49.** .039
51. 1, 2, 3, 4, 6, 7, 8, 9
53. 2, 3, 5, 8, 9

Section 5.4

1. $735, $118 **3.** $560, $140
5. $3088, $225 **7.** $2490
9. $3320 **11.** $7.50 **13.** $4.31
15. $7.24 **17.** $988.36
19. $860.68 **21.** $1,480.31
23. $307.58 **25.** $225.31
27. $88,581.60 **29.** $295.68
31. $423.92 **33.** $236
35. $347 **37.** $38,662
39. $69,538 **41.** $5253
43. $4082 **45.** $5040 **47.** $14\frac{1}{2}$%
49. $15\frac{1}{2}$% **51.** $18, $14\frac{1}{2}$%
53. $1710.34 **55.** $5991.40
57. $22,428.88 **59.** $1378.96
61. $7595.75

Section 5.5

1–7 Answers will vary. **9.** 200
11. .6 **13.** about 83 **15.** 174
17. 67.3 **19.** 58.42 **21.** 40°
23. 280° **25.** 37° **27.** $-40°$
29. 95° **31.** 50° **33.** 5°
35. 57° **37.** 82°
41. not correct **43.** correct
45. .68 **47.** 4700 **49.** 8900
51. .39 **53.** 46 **55.** 976
57. 39.2 **59.** 50.3 **61.** 15.4
63. 1.1 **65.** 307.5 **67.** 1235.7
69. 1.5 **71.** 1861.4 **73.** 85.6

Section 5.6

1. 3.606 **3.** 5.568 **5.** 8.832
7. 1.732 **9.** 3.464 **11.** 7.071
13. 7.810
15. Multiply both sides of $\frac{a^2}{b^2} = 2$ by b^2.
17. $a^2 = 2b^2$, an even number.
19. $(2c)^2 = (2c)(2c) = 4c^2$
21. It is a multiple of 2.
23. The square of an odd number is odd, so a must be even.
25. Since assumption (1) leads to a contradiction.
27. 18.8 sq. m. **29.** 452.2 sq. m.
31. 502.4 cu. cm.

Section 5.7

1. complex **3.** real, complex
5. imaginary, complex
7. real, complex
9. imaginary, complex

Chapter 5 Test
1. .4375 **2.** .45 **3.** 2.$\overline{3}$
4. .41$\overline{6}$ **5.** $\frac{18}{25}$ **6.** $\frac{4}{9}$ **7.** $\frac{58}{99}$
8. $\frac{277}{300}$ **9.** rational
10. rational **11.** irrational
12. 14.68 **13.** 8.275 **14.** 38.7
15. 24.3 **16.** 74.6% **17.** .38
18. 413.84 **19.** 101.5
20. 17.5% **21.** $1215, $159
22. $700, $175 **23.** $4.19
24. $354.05 **25.** $434.05
26. 400 **27.** 900 **28.** 780
29. .297 **30.** 46.7 **31.** 5.5
32. 40.7 **33.** 4.4

Review Exercises
1. yes, yes, yes
2. no, yes, no, yes, yes
3. no, yes, no, yes, yes
4. no, yes, no, yes
5. no, no, yes, yes
6. no, no, yes, no, yes
7. no, yes, no, yes, yes
8. no, yes, no, yes, yes
9. yes, yes
10. no, no, yes, yes, yes
11. yes, yes, yes
12. yes, yes, yes **13.** no, no, yes
14. no, no, yes, yes, yes
15. no, no, no, yes
16. no, no, no, yes **17.** inverse
18. inverse **19.** inverse
20. inverse **21.** commutative
22. commutative **23.** associative
24. closure **25.** distributive
26. distributive **27.** 4 **28.** 12
29. −8 **30.** −30 **31.** 0 **32.** $\frac{1}{2}$
33. −$\sqrt{5}$ **34.** $\sqrt{6}$ **35.** $\frac{3}{2}$
36. $\frac{4}{3}$ **37.** $\frac{11}{9}$ **38.** $\frac{23}{15}$ **39.** $\frac{1}{7}$
40. −$\frac{1}{12}$ **41.** −$\frac{1}{3}$ **42.** $\frac{1}{16}$
43. no reciprocal for 0
44. no reciprocal for 0

CHAPTER 6

Section 6.1
1. 8 **3.** 8 **5.** 3 **7.** 4
9. 1 **11.** 2
13. row 2: 0, 6, 10
  row 3: 9, 0, 9
  row 4: 0, 4, 0, 8, 0
  row 5: 1, 9, 2, 7
  row 6: 6, 0, 0

row 7: 4, 11, 6, 8, 3, 5
row 8: 8, 4, 0, 0
row 9: 6, 3, 9, 3, 9, 6, 3
row 10: 6, 4, 0, 10, 8, 6, 4
row 11: 10, 9, 8, 7, 6, 5, 4, 3, 2

15. 2 **17.** 9 **19.** 11 **21.** 5
23. 11 **25.** 1 **27.** 0 **29.** 5
31. yes **33.** yes **35.** yes
37. 1 **39.** 3 **41.** 1 **43.** 5
45. yes **47.** yes **49.** yes
51. 4 **53.** 2 **55.** 6 **57.** 1900
59. 1500 **61.** yes **63.** yes
65. yes

Section 6.2
1. false **3.** true **5.** true
7. false **9.** true **11.** 5
13. 1 **15.** 4 **17.** 0 **19.** 4
21. 0 **23.** 0 **25.** 2 **27.** 3
29. 7
31. (a) row 1: 2, 0 row 2: 0, 1
  (b) all properties satisfied
  (c) 0 is its own inverse; 1 and 2
    are inverses.
33. (a) row 1: 0 row 2: 0, 1
    row 3: 4, 0, 1, 2 row 4: 0,
    1, 2, 3
  (b) all properties satisfied
  (c) 0 is its own inverse; 2 and 3
    are inverses, as are 1 and 4.
35. (a) 1 (b) all properties satisfied
  (c) 1 is its own inverse
37. (a) row 2: 0, 2 row 3:
  (b) no inverse property
  (c) 1 is its own inverse, as is 3; no
    inverse for 2.
39. (a) row 2: 1, 3, 5 row 3: 2, 1
    row 4: 1, 5, 2, 6, 3
    row 5: 3, 1, 4 row 6: 5, 3, 2
  (b) all properties satisfied
  (c) 1 is its own inverse as is 6; 2
    and 4 are inverses, as are 3
    and 5.
41. prime
43. {7, 16, 25, 34, 43, 52, . . .}
45. {3, 10, 17, 24, 31, 38, . . .}
47. {6, 16, 26, 36, 46, . . .}
49. {2, 9, 16, 23, 30, . . .}
51. {5, 12, 19, 26, 33, . . .}
53. {2, 5, 8, 11, 14, 17, 20, . . .}
55. no answers **57.** identity
59. {4, 9, 14, 19, 24, 29, . . .}
61. 100,000 **63.** 2 **65.** 2, 4
67. 2, 5 **69.** none **71.** none

Section 6.3
1. correct **3.** should be 26,138
5. correct **7.** correct
9. should be 11,793 **11.** correct
13. should be 93,386 **15.** correct
17. should be 258 **19.** correct
21. 23 **23.** 26 **25.** 43
27. no solutions **29.** 1$\overline{6}$1 **31.** 43
33. 86 **37.** 6 **39.** 1, 1
41. 7, 7 **43.** 2, 2 **45.** 8, 8

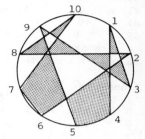

Section 6.4
1. All properties; 1 is its own inverse,
  as is 4; 2 and 3 are inverses.
3. All properties except inverse; no
  inverses for 2, 3, or 4.
5. All properties; 1, 3, 5, 7 are each
  their own inverses.
7. commutative, closure **9.** all
  properties **11.** all properties
13. a **15.** c **17.** c **19.** a
21. row b: d
  row c: d, b
  row d: b, c
23. associative, commutative,
  identity (U), closure
25.

	a	b	c	d
a	a	b	c	d
b	b	a	d	c
c	c	d	a	b
d	d	c	b	a

Section 6.5
1. yes **3.** no **5.** no
7. 468 **9.** 285 **11.** 380
13. 875 **15.** 1496 **17.** 79,992
19. 16p **21.** 14z **23.** 6p
25. 2a **27.** 11r **29.** 10k
31. −6a
33. 5k + 2r can't be further simplified.
35. 5p **37.** −16y
39. 5(3m + 2k) **41.** 6b(2a + 5p)
43. 2b(30ac − 6c + 5a)
45. can't be factored

Section 6.6
1. no−closure   **3.** no−inverse
5. yes   **7.** no−identity, inverse
9. no−closure $(1 + 1 = 2)$
11. yes   **13.** no−closure
15. multiples of 2, of 3, of 4, of 5, of 6, for example
17. S   **19.** R   **21.** V
29. N   **31.** R   **33.** T
35. yes   **37.** yes   **39.** yes
41. no   **43.** no   **45.** no
47. 2   **49.** 3   **51.** 2
53. $4^1 = 4$, $4^2 = 1$, $4^3 = 4$, $4^4 = 1$; not a generator
55. Yes, 3 is a generator.

Chapter 6   Test
1. 5   **2.** 7   **3.** 9   **4.** 8
5. 0   **6.** 3   **7.** 4   **8.** 5
9. 2   **10.** 5   **11.** false
12. false   **13.** true   **14.** 3
15. 3   **16.** 0
17.

+	0	1	2	3	4	5	6	7
0	0	1	2	3	4	5	6	7
1	1	2	3	4	5	6	7	0
2	2	3	4	5	6	7	0	1
3	3	4	5	6	7	0	1	2
4	4	5	6	7	0	1	2	3
5	5	6	7	0	1	2	3	4
6	6	7	0	1	2	3	4	5
7	7	0	1	2	3	4	5	6

18.

×	0	1	2	3	4	5	6
0	0	0	0	0	0	0	0
1	0	1	2	3	4	5	6
2	0	2	4	6	1	3	5
3	0	3	6	2	5	1	4
4	0	4	1	5	2	6	3
5	0	5	3	1	6	4	2
6	0	6	5	4	3	2	1

19. 7, 16, 25, 34, 43, 52, 61, . . .
20. 2, 5, 8, 11, 14, 17, 20, 23, . . .
21. 3, 8, 13, 18, 23, 28, 33, . . .
22. 17   **23.** correct
24. should be 1,540,239
25. correct   **26.** correct
27. correct   **28.** b   **29.** e
30. c   **31.** yes   **32.** yes
33. yes   **34.** yes, c
35. Yes, each element is its own inverse.
36. 616   **37.** 5x   **38.** $4(4k − 3z)$
39. yes   **40.** no

CHAPTER 7

Section 7.1
1. 4   **3.** 15   **5.** 15   **7.** 2
9. −3   **11.** 4   **13.** 7   **15.** 5
17. −2   **19.** 4   **21.** −5   **23.** 7
25. −2   **27.** 18   **29.** −3
31. 8   **33.** −5   **35.** 0
37. 2 hours   **39.** $200   **41.** 10%
43. $67.50   **45.** $2\frac{1}{2}$ hours
47. $4,000 at 5%; $6,000 at 6%
49. 25 mph   **51.** EER = B/W
53. 6   **55.** 2   **57.** 3   **59.** 20

Section 7.2
1. $(2, 13)$, $(−3, −7)$
3. $(2, 17)$, $(−3, −23)$
5. $(2, 4)$, $(−3, −11)$
7. $(2, 2)$, $(−3, −23)$
9. $(2, −12)$, $(−3, 18)$
11. $(2, 4)$, $(−3, 9)$ **13.** $(2, 4)$, $(−3, −1)$
15. $(2, −\frac{2}{3})$, $(−3, 6)$ **17.** yes
19. no   **21.** no   **23.** yes
25. no   **27.** $(1, −2)$ **29.** $(2, 0)$
31. $(6, 2)$ **33.** $(4, 4)$
35. $(−9, −11)$ **37.** $(6, 3)$
39. $(\frac{3}{8}, \frac{5}{8})$ **41.** 9 and 6
43. 10 and 8
45. 22 ten's, 63 twenty's
47. 74 8¢-pieces; 96 10¢-pieces

Section 7.3
1.

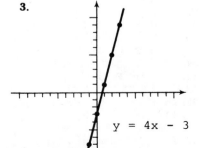

$y = 2x$

3.

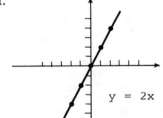

$y = 4x − 3$

5.

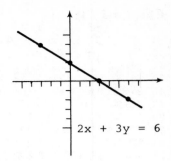

$2x + 3y = 6$

7.

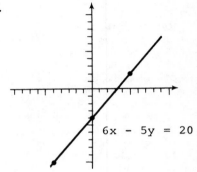

$6x − 5y = 20$

9.

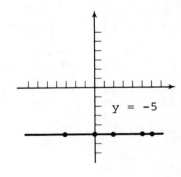

$y = -5$

11.

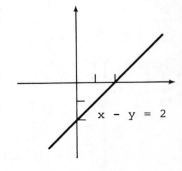

$x − y = 2$

13.

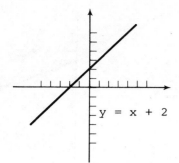

$y = x + 2$

21.

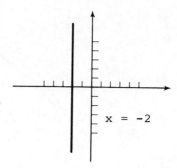

$x = -2$

59. $\frac{1}{4}$ **61.** 65   **63.** 121 pounds
65. 154 pounds

67.

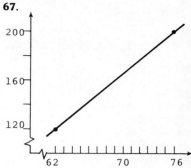

69. 64   **71.** 0   **73.** 0
75. 40   **77.** 80   **79.** (30, 40)

23.

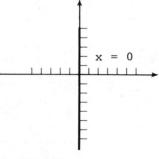

$y = 6$

81.

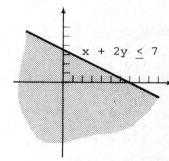

15.

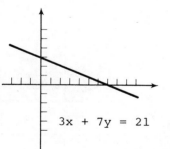

$3x - 2y = 6$

25.

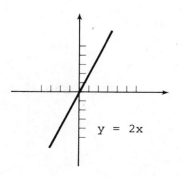

x = 0

Section 7.4

1.

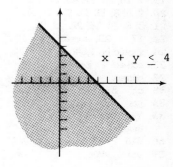

$x + 2y \le 7$

17.

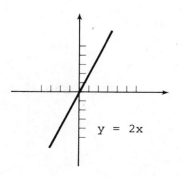

$3x + 7y = 21$

27. -1 **29.** $-\frac{3}{13}$ **31.** $-\frac{5}{2}$
33. 0   **35.** no slope   **37.** -1
39. 3   **41.** 0   **43.** no slope
45. $-2x + y = 7$ **47.** $3x - 4y = 35$
49. $5x + 8y = 43$ **51.** $y = 9$
53. $x = 3$ **55.** $2x + 3y = 17$

57.

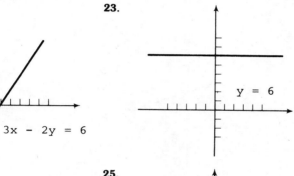

3.

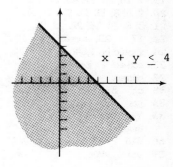

$x + y \le 4$

19.

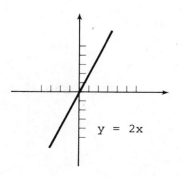

$y = 2x$

5.

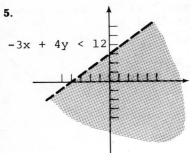

$-3x + 4y < 12$

7.

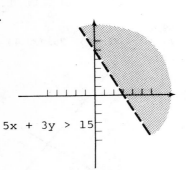

$5x + 3y > 15$

9.

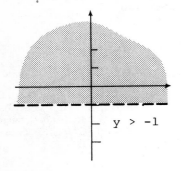

$y > -1$

11.

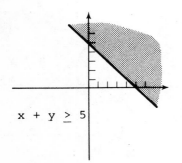

$x + y \geq 5$

13.

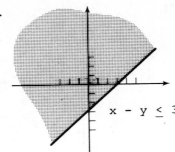

$x - y \leq 3$

15.

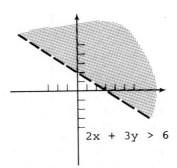

$x + 3y \leq 6$

17.

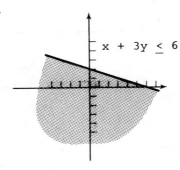

$2x + 3y > 6$

19.

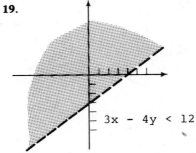

$3x - 4y < 12$

21.

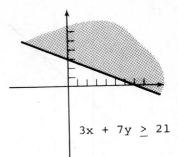

$3x + 7y \geq 21$

23.

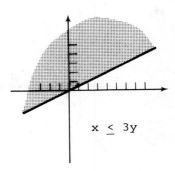

$x \leq 3y$

25.

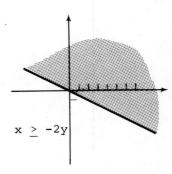

$x \geq -2y$

27.

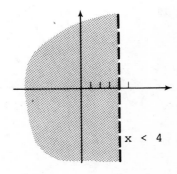

$x < 4$

29.

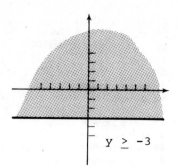

$$y \geq -3$$

31.

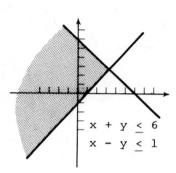

$$x + y \leq 6$$
$$x - y \leq 1$$

33.

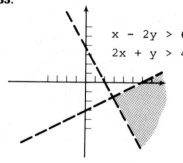

$$x - 2y > 6$$
$$2x + y > 4$$

35.

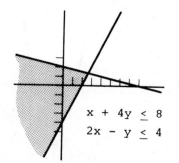

$$x + 4y \leq 8$$
$$2x - y \leq 4$$

37.

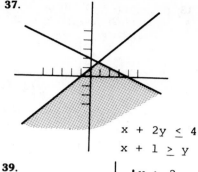

$$x + 2y \leq 4$$
$$x + 1 \geq y$$

39.

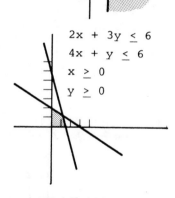

$$x \geq 2$$
$$y \leq 3$$

41.

$$2x + 3y \leq 6$$
$$4x + y \leq 6$$
$$x \geq 0$$
$$y \geq 0$$

Section 7.5

1. Maximum of 65 at (5, 10); minimum of 8 at (1, 1)
3. Maximum of 900 at (0, 12); minimum of 0 at (0, 0)
5. $(\frac{6}{5}, \frac{6}{5})$ **7.** $(\frac{17}{3}, 5)$
9. $(\frac{105}{8}, \frac{25}{8})$
11.

	Maximum	occurs at
(a)	204	(18, 2)
(b)	$117\frac{3}{5}$	$(\frac{12}{5}, \frac{39}{5})$
(c)	102	$(0, \frac{17}{2})$

13. $112, with 4 pigs, 12 geese
15. 8 of #1, 3 of #2
17. 800 bargain, 300 deluxe; maximum profit is $125,000
19. 150 kg of the half and half, 75 kg of the other; maximum profit is $1050

Section 7.6

1. (3, 9); (4, 10); (5, 11); (6, 12); (7, 13) range: {9, 10, 11, 12, 13}; function

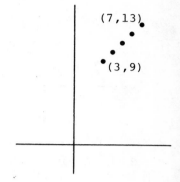

3. (−1, 6); (0, 4); (1, 2); (2, 0); (3, −2); (4, −4) range: {6, 4, 2, 0, −2, −4}; function

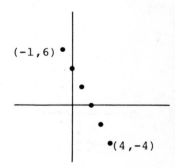

5. (−1, 4); (0, 6); (1, 8); (2, 10); (3, 12); (4, 14); (5, 16) range: {4, 6, 8, 10, 12, 14, 16}; function

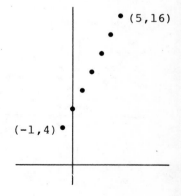

7. $(-4, 3)$; $(-2, \frac{5}{2})$; $(0, 2)$; $(2, \frac{3}{2})$; $(4, 1)$ range: $\{3, \frac{5}{2}, 2, \frac{3}{2}, 1\}$; function

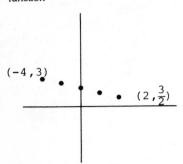

9. $(-2, -4)$; $(-2, -3)$; $(-1, -4)$; $(-1, -3)$; $(-1, -2)$; $(-1, -1)$; $(0, -4)$; $(0, -3)$; $(0, -2)$; $(0, -1)$; $(0, 0)$; $(0, 1)$; $(1, -4)$; $(1, -3)$; $(1, -2)$; $(1, -1)$; $(1, 0)$; $(1, 1)$; $(1, 2)$ range: $\{-4, -3, -2, -1, 0, 1, 2\}$; not a function

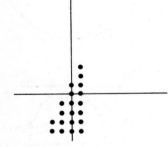

11. $(-1, -1)$; $(-1, 0)$; $(-1, 1)$; $(-1, 2)$; $(0, -1)$; $(0, 0)$; $(0, 1)$; $(0, 2)$; $(0, 3)$; $(1, -1)$; $(1, 0)$; $(1, 1)$; $(1, 2)$; $(1, 3)$; $(1, 4)$; $(2, -1)$; $(2, 0)$; $(2, 1)$; $(2, 2)$; $(2, 3)$; $(2, 4)$; $(2, 5)$;   range: $\{-1, 0, 1, 2, 3, 4, 5\}$; not a function

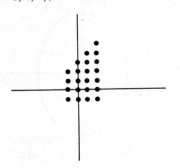

13. $(-4, -2)$; $(-2, -1)$; $(0, 0)$; $(2, 1)$; $(4, 2)$ range: $\{-2, -1, 0, 1, 2\}$; function

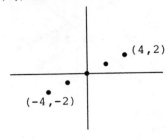

15. $(-9, 3)$; $(-6, 2)$; $(-3, 1)$; $(0, 0)$; $(3, -1)$; $(6, -2)$ range: $\{3, 2, 1, 0, -1, -2\}$; function

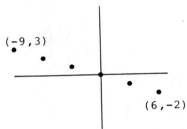

17. $(2, -1)$; $(2, 0)$; $(2, 1)$; $(2, 2)$ range: $\{-1, 0, 1, 2\}$; not a function

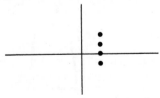

19. $(-2, -1)$; $(-1, -1)$; $(0, -1)$; $(1, -1)$; $(2, -1)$; $(3, -1)$ range: $\{-1\}$; function

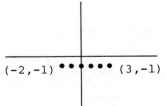

21. (a) -3 (b) 1 (c) 7
23. (a) 13 (b) 9 (c) 3
25. (a) -6 (b) 0 (c) 24
27. (a) 0 (b) -6 (c) 15
29. (a) 0 (b) 2 (c) 20
31. (a) 6 (b) $\frac{8}{3}$ (c) $\frac{11}{6}$
33. (a) no such number   (b) $\frac{3}{2}$ (c) 3
35. function   **37.** function
39. not a function
41. not a function
43. not a function   **45.** function
47. not a function   **49.** $7
51. $13   **53.** $7   **55.** $16
57. $13   **59.** function   **61.** not a function   **63.** no   **65.** no
67. no   **69.** no   **71.** yes

Section 7.7
1.

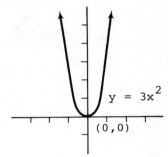

3.

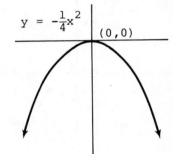

5.

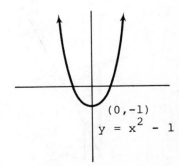

7.

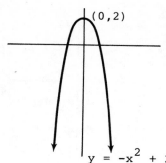

$(0,2)$

$y = -x^2 + 2$

15.

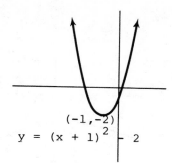

$(-1,-2)$

$y = (x + 1)^2 - 2$

23. circle, center at $(0, 0)$, $r = 4$

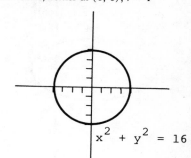

$x^2 + y^2 = 16$

9.

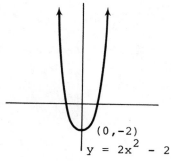

$(0,-2)$

$y = 2x^2 - 2$

17.

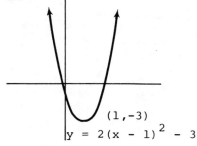

$(1,-3)$

$y = 2(x - 1)^2 - 3$

25. circle, center at $(0, 0)$, $r = 9$

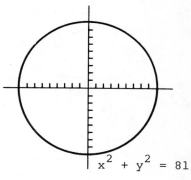

$x^2 + y^2 = 81$

11.

$y = (x - 1)^2$

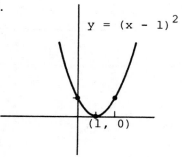

$(1, 0)$

19. $y = -3(x + 2)^2 + 2$

$(-2,2)$

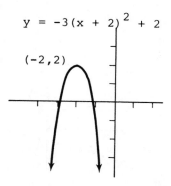

27.

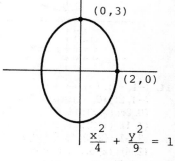

$(0,3)$

$(2,0)$

$\dfrac{x^2}{4} + \dfrac{y^2}{9} = 1$

13.

$(-1,0)$

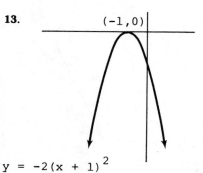

$y = -2(x + 1)^2$

21.

$3y = 2(x - 1)^2 - 6$

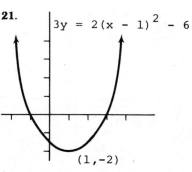

$(1,-2)$

29.

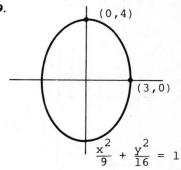

$(0,4)$

$(3,0)$

$\dfrac{x^2}{9} + \dfrac{y^2}{16} = 1$

31.

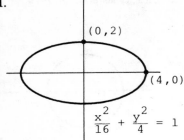

$(0,2)$

$(4,0)$

$$\frac{x^2}{16} + \frac{y^2}{4} = 1$$

33. ellipse, ± 3 on x, ± 1 on y

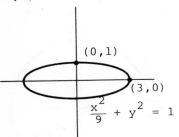

$(0,1)$

$(3,0)$

$$\frac{x^2}{9} + y^2 = 1$$

35.

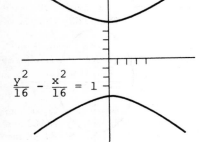

$$\frac{y^2}{16} - \frac{x^2}{16} = 1$$

37.

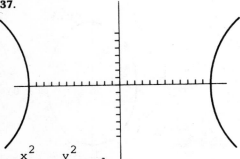

$$\frac{x^2}{144} - \frac{y^2}{49} = 1$$

39.

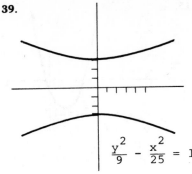

$$\frac{y^2}{9} - \frac{x^2}{25} = 1$$

41. 12 units; minimum cost is $50

Chapter 7   Test

1. $k=1$ **2.** $m=-5$ **3.** 6%
4. $(-2, 3)$ **5.** $(-3, -6)$
6. $(1, 4)$
7.

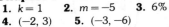

$3x + 2y = 12$

8.

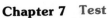

$x - 4y = 8$

9.

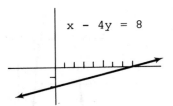

$y + 3 = 0$

10. $y = 2x - 7$
11. $3y = 2x + 26$

12.

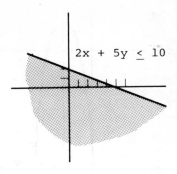

$2x + 5y \le 10$

13.

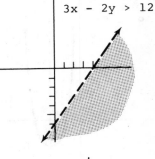

$3x - 2y > 12$

14.

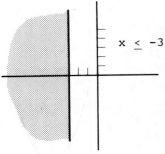

$x \le -3$

15.

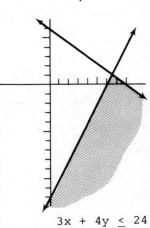

$3x + 4y \le 24$
$2x - y \ge 12$

16.

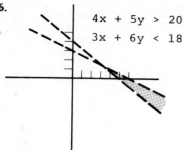

$4x + 5y > 20$
$3x + 6y < 18$

17. maximum of 13 at (2, 1)
18. −15; 3; 39   **19.** −8, 0; −8
20. function   **21.** not a function

22.

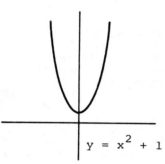

$y = x^2 + 1$

23.

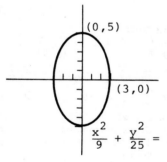

(0, 5)
(3, 0)
$$\frac{x^2}{9} + \frac{y^2}{25} = 1$$

24.

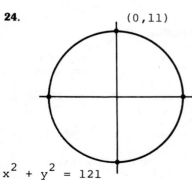

(0, 11)
$x^2 + y^2 = 121$

25.

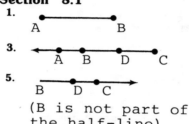

$$\frac{x^2}{16} - \frac{y^2}{25} = 1$$

CHAPTER 8

Section 8.1
1.

A   B

3.

A B   D   C

5.

B   D C

(B is not part of the half-line)

7.

A   B

9.

A   B   C   D

11. 47°   **13.** 90°   **15.** 140°
17.

30°

19.

60°

21.

95°

23. Adjacent angles include *CBD* and *DBA*; *DBA* and *ABE*; *ABE* and *EBC*; *EBC* and *DBC*; vertical angles include *CBD* and *EBA*; *CBE* and *ABD*.
25. Adjacent angles are *YZX* and *XZW*; there are no vertical angles.
27. 43°   **29.** 8°   **31.** 139°
33. 105°   **35.** 70°, 110°
37. 55°, 35°   **39.** 80°, 100°

Section 8.2
1. both   **3.** closed   **5.** closed
7. neither   **9.** both   **11.** closed
13. 5   **15.** 15   **17.** 7   **19.** 96
21. 21   **23.** 16   **25.** 78°
27. 60°   **29.** 77°   **31.** 74°
33. 45°   **35.** acute, scalene
37. acute, equilateral   **39.** right, scalene   **41.** right, isosceles
43. obtuse, scalene   **45.** acute, isosceles   **47.** 4, 4, 6   **49.** 8, 6, 12   **51.** 20, 12, 30

Section 8.3
1. 12 cm²   **3.** 5 cm²   **5.** 8 cm²
7. 2.5 cm² or 2½ cm²   **9.** 418 mm²
11. 8 cm²   **13.** 20 cm²
15. 180 cm²   **17.** four   **19.** 4, sixteen   **21.** $320   **23.** $156
25. $12   **27.** $1,120   **29.** 76.26
31. 80   **33.** 164.48   **35.** 3¾ m³
37. 96 m³   **39.** 168   **41.** 1436 m³
43. 4.19 m³   **45.** 113.04 m³
47. 8   **49.** 64   **51.** $720
53. 3 m
55. $V = 817,000$ mm³; $S = 42,300$ mm²
57. $V = 137,000$ mm³; $S = 12,900$ mm²
59. $V = 1,800,000$ mm³; $S = 84,000$ mm²
61. $V = 251$ cm³; $S = 80.4$ cm²
63. $V = 74.7$ m³; $S = 122.8$ m²
65. $V = 242,000$ m³
67. 108 cm²   **69.** 2 by 14

Section 8.4
1. similar   **3.** not similar
5. similar
7. $P = 78°$; $M = 46°$; $A = N = 56°$
9. $T = 74°$; $Y = 28°$; $Z = U = 78°$
11. $T = 20°$; $V = 64°$; $R = U = 96°$
13. $a = 5, b = 3$ **15.** $a = 6, b = 7\frac{1}{2}$
17. $x = 6$ **19.** 500 m, 700 m
21. 30 m   **23.** 110   **25.** 111.1
27. 320 feet

Section 8.5

In Exercises 17–24, the answer depends on the size of the globe that you use. Our globe is about 16 inches in diameter.

17. about 65°
19. about 105°
21. about 30°
23. about 70°

Section 8.6

1. C   **3.** C   **5.** A, E
7. B, D   **9.** B, D
11. C, G, I, L, M, N, S, U, V, W, Z
D, O
E, F, T, Y
K, X
Q, R
13. I, V
II, IV, VI
III, VII
VIII
X
IX
15. 1   **17.** 2 or 3, depending on the pretzel   **19.** 1   **21.** 2
23. 1   **25.** 0   **27.** 2
29. 6 odd   **31.** 3 even, 4 odd
33. traversible

Start

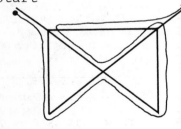

35. traversible

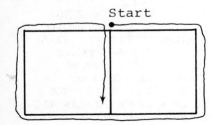

37. not traversible

Chapter 8 Test

1.

29°

2. 172°

3. 106°   **4.** 90°   **5.** both
6. neither   **7.** 34   **8.** 72
9. 60   **10.** 68   **11.** 180
12. $152   **13.** $54,000
14. 904.32   **15.** 864   **16.** 336
17. 1582.56   **18.** $y = 24$, $x = 20$
19. $m = 8$
20. $y = 110°$, $x = z = 33°$
21. 2   **22.** 1

CHAPTER 9

Section 9.1

1. deterministic   **3.** random
5. random   **7.** deterministic
9. $\frac{1}{6}$ **11.** $\frac{2}{3}$ **13.** $\frac{2}{3}$
15. $\frac{1}{13}$ **17.** $\frac{1}{52}$
19. 12 to 1   **21.** 51 to 1
27. classical   **29.** empirical
31. subjective
33. $P(\text{success}) + P(\text{failure}) = 1$

Section 9.2

1. {1, 2, 3, 4}   **3.** $\frac{3}{4}$

5.

	1	2	3	4
1	(1,1)	(1,2)	(1,3)	(1,4)
2	(2,1)	(2,2)	(2,3)	(2,4)
3	(3,1)	(3,2)	(3,3)	(3,4)
4	(4,1)	(4,2)	(4,3)	(4,4)

7. {(1,1), (1,2), (1,3), (1,4), (2,1), (2,2), (2,3), (2,4), (3,1), (3,2), (3,3), (3,4), (4,1), (4,2), (4,3), (4,4)}
9. 16   **11.** $\frac{3}{13}$ **13.** $\frac{9}{169}$
15. Depends on first draw: $\frac{12}{51} = \frac{4}{17}$ if first card was a heart; $\frac{13}{51}$ if first card was not a heart
17. $n_1 n_2$ **19.** $s_1 s_2 / n_1 n_2$ **21.** s_2 / n_2
23. $[n(S) - n(E')]$ to $n(E')$
25. 60   **27.** $\frac{3}{4}$ **29.** $\frac{1}{4}$ **31.** $\frac{1}{4}$

33.

	1	2	3	4	5	6
1	2	3	4	5	6	7
2	3	4	5	6	7	8
3	4	5	6	7	8	9
4	5	6	7	8	9	10
5	6	7	8	9	10	11
6	7	8	9	10	11	12

35. 0   **37.** $\frac{1}{18}$ **39.** $\frac{1}{9}$ **41.** $\frac{1}{6}$
43. $\frac{1}{9}$ **45.** $\frac{1}{18}$ **47.** E' **49.** 12
51. $\frac{1}{4} = 0.25$, considerably less than 0.41
53. 4   **57.** 40°   **59.** 30°
61. Divide the circle into 205 equal parts, 100 representing girls and 105 representing boys.

Section 9.3

1. 4   **3.** 2   **5.** $\frac{1}{4}$ **7.** $\frac{1}{52}$
9. 7 to 6   **11.** $\frac{5}{12}$ **13.** $\frac{1}{4}$
15. .08   **17.** .50   **19.** .13
23. $P(A) + P(B)$ **25.** $\frac{13}{51}$
27. $\frac{13}{52} \cdot \frac{12}{51} \cdot \frac{11}{50} \cdot \frac{10}{49} \cdot \frac{9}{48} = \frac{33}{66,640}$
29. $\frac{33}{16,660}$
35. Any one suit, say hearts, could represent white (W), while the other three suits could represent red (R).

Section 9.4

1. 0   **3.** 1   **5.** $\frac{1}{6}$ **7.** $\frac{4}{17}$
9. $\frac{25}{51}$ **11.** $\frac{1}{17}$ **13.** $\frac{11}{51}$
15. $\frac{13}{52} \cdot \frac{12}{51} \cdot \frac{11}{50} \cdot \frac{10}{49} \cdot \frac{9}{48} = \frac{33}{66,640}$
17. $\frac{9}{48} = \frac{3}{16}$ **19.** $\frac{33}{7840}$ **21.** $\frac{2}{3}$
23. $\frac{1}{4}$ **25.** $\frac{1}{4}$ **27.** $\frac{1}{4}$ **29.** $\frac{9}{25}$
31. One way is to pick 6 red cards and 4 black cards, and then draw from the 10 cards.
35. $\frac{1}{6}$ **37.** 0   **39.** $\frac{2}{3}$

Section 9.5

1. $\frac{21}{6} = 3.5$ **3.** yes
5. $-\$\frac{1}{37} = -2.7¢$ **7.** $\frac{2}{5} = 0.4$
9. No. Expected net winnings are about $-21¢$.
11. $4500   **13.** 118   **15.** no
17. $\frac{7}{10}$ **19.** $\frac{14}{30} = .47$ **21.** $\frac{51}{100}$
23. $\frac{157}{300} = .52$ **25.** $\frac{171}{330} = .52$
27. column 5: 2500, −, 2000, 1000, 25,000, 60,000, 16,000
29. column 7: C, C, C, B, C, A, B
31. 3.73   **33.** 3.28

Section 9.6

1. 12　**3.** 56　**5.** 8　**7.** 24
9. candy bar–lollipop; candy bar–bubble gum; lollipop–licorice stick
11. No. Once an item is chosen, it is no longer available to be chosen again.
13. 6　**15.** Yes. For example, 123 and 231 are different numbers.
17. a　**19.** A first, B second, C third, B first, E second, A third; E first, B second, A third
21. No. A single runner cannot win more than one place.
23. $5 \cdot 4 \cdot 3 = 60$　**25.** 2　**27.** 8
29. $16 \cdot 2 = 32$　**31.** $4 \cdot 5 = 20$
33. 16　**35.** $9 \cdot 10^5 \cdot 1 = 900{,}000$
37. $1^3 \cdot 10^4 = 10{,}000$
39. $26^3 \cdot 10^3 = 17{,}576{,}000$　**41.** $\frac{1}{30}$
43. $C(8, 5)/C(19, 5) = \frac{14}{2907}$
45. $[C(8,3) \cdot C(11,2)]/C(19,5) = \frac{770}{2907}$
47. 0
49. $C(40, 5)/C(52, 5) = \frac{2109}{8330}$
51. $[C(13, 1) \cdot C(13, 2) \cdot C(13, 2)]/C(52, 5) = 507/16{,}660$
53. $C(5, 5) \cdot C(34, 8) = 18{,}156{,}204$
55. $[C(5, 2) \cdot C(4, 1) + C(5, 3)]/C(9, 3) = \frac{25}{42}$
57. $\frac{1}{8}$　**59.** 136　**61.** $\frac{32}{39}$
63. $C(n, r) = P(n, r)/r!$
　　$= n!/[r!(n - r)!]$
65. 85¢
67. 34. The duplicate coins add twenty-five possibilities with repetitions. For example the combinations with repeating dimes are 10–10–10, 10–10–50, 10–10–25, 10–10–5, and 10–10–1. Of the twenty-five new cases there is one duplication of amounts: $25 + 25 + 10 = 60$ and $5 + 5 + 50 = 60$.
69. 3¢

Section 9.7

1. 6　**3.** 6　**5.** 56　**7.** 120
9. $C(5, 3)/C(9, 3) = \frac{5}{42}$
11. $C(3, 3)/C(9, 3) = \frac{1}{84}$
13. $[C(5, 2) \cdot C(3, 1)]/C(9, 3) = \frac{5}{14}$
15. $[C(3, 2) \cdot C(1, 1)]/C(9, 3) = \frac{1}{28}$
17. $C(5, 2) = 10$
19. $C(2, 1) \cdot C(3, 1) = 6$
21. $\frac{15}{64}$　**23.** $\frac{3}{32}$　**25.** $\frac{1}{32}$
27. $\frac{5}{16}$　**29.** $\frac{9}{32}$　**31.** $\frac{1}{2}$
33. $1 - [C(4, 3)/C(6, 3)]$
　　$= 1 - (\frac{4}{20}) = \frac{4}{5}$
35. $\frac{45}{1024}$　**37.** 0

39. $(210 + 120 + 45 + 10 + 1)/1024$
　　$= \frac{386}{1024}$
41. $\frac{252}{1024}$.
45. 1, 1, 2, 3, 5, 8, 13, 21, ... Each element (after the second) is the sum of the preceding two. This is the Fibonacci sequence.
47. Row 8

Chapter 9 Test

1. $\frac{2}{3}$　**2.** $\frac{1}{4}$　**3.** $\frac{1}{26}$　**4.** $\frac{1}{13}$
5. $\frac{8}{13}$　**6.** $\frac{1}{2}$　**7.** $\frac{1}{3}$　**8.** 1
9. 3 to 1　**10.** 25 to 1
11. 11 to 2　**12.** empirical
13. subjective　**14.** classical
15. $\frac{1}{4}$　**16.** $\frac{3}{4}$　**17.** $\frac{1}{2}$　**18.** $\frac{25}{102}$
19. $\frac{77}{102}$　**20.** $\frac{26}{51}$　**21.** $\frac{25}{51}$
22.

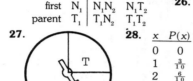

		second parent	
		N_2	T_2
first parent	N_1	N_1N_2	N_1T_2
	T_1	T_1N_2	T_1T_2

23. $\frac{1}{4}$
24. $\frac{1}{2}$
25. $\frac{1}{6}$
26. $\frac{1}{3}$

27.

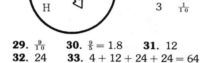

28.

x	$P(x)$
0	0
1	$\frac{3}{10}$
2	$\frac{6}{10}$
3	$\frac{1}{10}$

29. $\frac{9}{10}$　**30.** $\frac{9}{5} = 1.8$　**31.** 12
32. 24　**33.** $4 + 12 + 24 + 24 = 64$
34.

```
            1
          1   1
        1   2   1
      1   3   3   1
    1   4   6   4   1
  1   5  10  10   5   1
```

35. $2^5 = 32$

CHAPTER 10

Section 10.1

3.

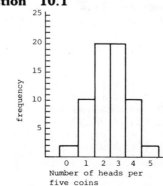

Number of heads per five coins

7. 4　**9.** 6　**11.** 5
13.

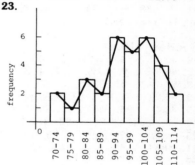

units completed

15. 1　**17.** 2　**19.** 5　**21.** 4
23.

29.

31. 20　**33.** 54　**35.** 15
37. 18　**39.** 10

Section 10.2

1. 14　**3.** 60.3　**5.** 27,955
7. 7.7　**9.** .1　**11.** 300,000
13. $335,000　**15.** 22　**17.** 51
19. 130　**21.** 49　**23.** 562
25. 29.1　**27.** 9　**29.** 64
31. 68 and 75　**33.** none
35. 6.1 and 6.3　**37.** 6　**39.** 3
41. median　**43.** 6.7　**45.** 17.2
47. 118.8　**49.** $41,000
51. $9642.86　**53.** $130,000
55. $64,000　**57.** $7.43
59. $9.50　**61.** $397.40

63. 11.3; 10.2; 9.9
65. all except 18.6
69. 33.2; 31.4; 31.4
71. 37.2   **73.** 73.4   **75.** 70.2

Section 10.3
1. 6; 2
3. 12; $\sqrt{18.3} = 4.3$
5. 53; $\sqrt{407.4} = 20.2$
7. 46; $\sqrt{215.3} = 14.7$
9. 24; $\sqrt{59.4} = 7.7$
11. 30; $\sqrt{111.76} = 10.6$
13. 50%   **15.** 97.5%   **17.** 68%
19. 13.5%   **21.** 68%   **23.** 50%
25. 16%   **27.** 95%   **29.** $\frac{1}{2}$%
31. $\frac{3}{4}$ **33.** $\frac{24}{25}$ **35.** $\frac{8}{9}$ **37.** $\frac{24}{25}$
39. No more than $\frac{1}{9}$

Section 10.4
1. 49.4%   **3.** 17.4%   **5.** 45.6%
7. 49.9%   **9.** 7.7%   **11.** 47.3%
13. 92.4%   **15.** 32.6%   **17.** 1.64
19. −1.03 or −1.04   **21.** 5000
23. 4330   **25.** 640   **27.** 9920
29. 20   **31.** 84.1%   **33.** 37.8%
35. 2.3%   **37.** 15.9%   **39.** .6%
41. 99.4%   **43.** 189   **45.** 82
47. 69

Section 10.5
1. 19.8%   **3.** 12.1%   **5.** 2.4%
7. 90.3%   **9.** 86.7%   **11.** 9.6%
13. 7.6%   **15.** 64.4%   **17.** 1.5%
19. 19.8%   **21.** .3%   **23.** 0%

Section 10.6
1. We have no way of telling.
3. It is going up.   **5.** 28%

Chapter 10 Test
1. frequencies: 3, 2, 4, 3, 3, 5
2.

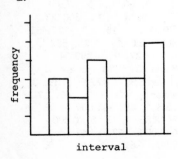

3. 65   **4.** 11.25   **5.** 27, 29
6. 41; 39 and 41   **7.** 18; $\sqrt{38.8}$
   = 6.2
8. 67; $\sqrt{512.2} = 22.6$ **9.** 15.9%
10. 97.7%   **11.** 81.8%   **12.** .422
13. .884   **14.** .069   **15.** 38.2%
16. 42.1%   **17.** 66.1%   **18.** .9%
19. 4.8%

CHAPTER 11

Section 11.1
1. $\begin{bmatrix} 11 & 1 \\ 9 & 4 \end{bmatrix}$ **3.** $\begin{bmatrix} 3 & 3 & -3 \\ 4 & -4 & 0 \end{bmatrix}$
5. $\begin{bmatrix} -3 & 4 & 2 \\ 8 & 2 & 1 \\ 4 & 3 & 13 \end{bmatrix}$ **9.** $\begin{bmatrix} 0 \\ 12 \\ 7 \end{bmatrix}$
7. different sizes
11. $\begin{bmatrix} -14 & -2 \\ 5 & 10 \end{bmatrix}$ **13.** $\begin{bmatrix} -2 & -1 \\ -1 & -4 \end{bmatrix}$
15. $\begin{bmatrix} -8 & 0 \\ 6 & -8 \end{bmatrix}$ **17.** $\begin{bmatrix} -8 & 4 \\ 16 & 0 \end{bmatrix}$
19. $\begin{bmatrix} -2 & -3 \\ -6 & -8 \end{bmatrix}$ **21.** $\begin{bmatrix} 0 & 0 \\ 0 & 0 \end{bmatrix}$
23. $\begin{bmatrix} 5 & 0 & 7 \\ 0 & 10 & 1 \\ 0 & 15 & 2 \\ 10 & 12 & 8 \end{bmatrix}$
25. $\begin{bmatrix} 10 & 12 & 5 \\ 15 & 20 & 8 \end{bmatrix}$
27. 60
29. $\begin{bmatrix} 97 & 92 & 86 \\ 139 & 121 & 100 \end{bmatrix}$
31. $y = 3$, $k = 8$, $m = 1$
33. $x = 1$, $z = 8$, $y = 8$
35. $x = 10$, $k = 4$, $y = -4$
37. true   **39.** true   **41.** true
43. true   **45.** true   **47.** yes

Section 11.2
1. $\begin{bmatrix} 8 & 6 \\ -7 & 3 \end{bmatrix}$ **3.** $\begin{bmatrix} -2 & -8 \\ -14 & 34 \end{bmatrix}$
5. $\begin{bmatrix} 17 & 7 & 9 \\ 13 & 6 & 3 \end{bmatrix}$
7. $\begin{bmatrix} 8 & 3 \\ 11 & -2 \\ 4 & 12 \end{bmatrix}$
9. cannot be multiplied
11. $\begin{bmatrix} 14 & 4 & 0 \\ -4 & 4 & 0 \\ -6 & 2 & 0 \end{bmatrix}$ **13.** $\begin{bmatrix} 47.5 & 57.75 \\ 27 & 33.75 \\ 81 & 95 \\ 12 & 15 \end{bmatrix}$

15. [11,120   13,555]
17. yes   **19.** yes   **21.** no
23. no   **25.** yes   **27.** no
29. magic
31. $\begin{bmatrix} 14 & 2 & 14 \\ 10 & 10 & 10 \\ 6 & 18 & 6 \end{bmatrix}$ magic
33. $\begin{bmatrix} 67 & 67 & 91 \\ 67 & 91 & 67 \\ 91 & 67 & 67 \end{bmatrix}$ magic
35. $\begin{bmatrix} 59 & 83 & 83 \\ 83 & 59 & 83 \\ 83 & 83 & 59 \end{bmatrix}$ pseudomagic

Section 11.3
1. $\begin{bmatrix} -1 & 2 & 1 \\ 3 & -4 & 2 \\ -4 & 0 & 1 \end{bmatrix}$ **3.** $\begin{bmatrix} -4 & 0 & 1 \\ -9 & 12 & -6 \\ -1 & 2 & 1 \end{bmatrix}$
5. $\begin{bmatrix} 18 & -1 & -1 \\ -4 & 0 & 1 \\ 1 & 3 & -3 \end{bmatrix}$
7. $\begin{bmatrix} -4 & 0 & 1 \\ 3 & -4 & 2 \\ -13 & 2 & 4 \end{bmatrix}$
9. $\begin{bmatrix} 1 & 11 & -5 \\ -1 & -4 & 3 \\ 0 & 5 & -2 \end{bmatrix}$
11. Exchange rows 1 and 2.
13. Double each element of row 2.
15. Multiply each element of row 1 by 2, and add the results to row 2.
17. Multiply row 2 by −1. Multiply row 2 by −2 and add to row 1.
19. Multiply row 1 by $-\frac{1}{2}$. Multiply row 1 by −4 and add to row 2. Multiply row 2 by $\frac{1}{8}$. Multiply row 2 by 1 and add to row 1.
21. Multiply row 1 by $\frac{1}{4}$. Multiply row 2 by $\frac{1}{4}$. Multiply row 2 by $-\frac{1}{2}$ and add to row 1. Multiply row 3 by $\frac{1}{2}$.
23. $\begin{bmatrix} 2 & 3 & | & 11 \\ 1 & 2 & | & 8 \end{bmatrix}$ **25.** $\begin{bmatrix} 1 & 5 & | & 6 \\ 3 & -4 & | & 1 \end{bmatrix}$
27.   **29.**
$\begin{bmatrix} 2 & 1 & 1 & | & 3 \\ 3 & -4 & 2 & | & -7 \\ 1 & 1 & 1 & | & 2 \end{bmatrix}$ $\begin{bmatrix} 1 & 1 & 0 & | & 6 \\ 0 & 2 & 1 & | & 2 \\ 0 & 0 & 1 & | & 2 \end{bmatrix}$
31. (2, 3)   **33.** (−3, 0)
35. $(\frac{7}{2}, -1)$ **37.** (5, 0)
39. no solution   **41.** (−2, 1, 3)
43. (−1, 23, 16)   **45.** (3, 2, −4)
47. (−1, 2, 5, 1)
49. Old goats cost $30, sheep cost $25.
51. 4 of A, 7 of B   **53.** 6, 6, 9
55. 6 rhythm, 3 guitarists

Section 11.4

1. $\begin{bmatrix} 1 & 0 \\ 0 & -1 \end{bmatrix}$ **3.** $\begin{bmatrix} \frac{3}{2} & -\frac{1}{2} \\ -2 & 1 \end{bmatrix}$

5. $\begin{bmatrix} \frac{13}{5} & \frac{6}{5} \\ -2 & -1 \end{bmatrix}$ **11.** $\begin{bmatrix} 1 & -2 & 0 \\ 0 & 1 & 0 \\ -1 & 3 & 1 \end{bmatrix}$

7. no inverse

9. $\begin{bmatrix} -\frac{3}{2} & 1 \\ 1 & -\frac{1}{2} \end{bmatrix}$

13. $\begin{bmatrix} 0 & \frac{1}{3} & -1 \\ 0 & \frac{1}{3} & 0 \\ \frac{1}{2} & -\frac{5}{6} & 2 \end{bmatrix}$ **15.** $\begin{bmatrix} 4 & 3 & 3 \\ -1 & 0 & -1 \\ -4 & -4 & -3 \end{bmatrix}$

17. $\begin{bmatrix} 35 \\ -88 \end{bmatrix} \begin{bmatrix} -4 \\ 0 \end{bmatrix} \begin{bmatrix} 15 \\ -48 \end{bmatrix} \begin{bmatrix} -9 \\ 9 \end{bmatrix} \begin{bmatrix} 35 \\ -97 \end{bmatrix}$
$\begin{bmatrix} 53 \\ -133 \end{bmatrix} \begin{bmatrix} 16 \\ -50 \end{bmatrix} \begin{bmatrix} 5 \\ -15 \end{bmatrix}$

19. Santa Claus is fat.

21. Success in dealing with unknown ciphers is measured by these four things in the order named: perserverance, careful methods of analysis, intuition, luck.

23. $\begin{bmatrix} \frac{32}{23} & \frac{3}{23} \\ \frac{12}{23} & \frac{27}{23} \end{bmatrix}$

25. 1565 metric tons of wheat, and 1272 metric tons of oil.

Section 11.5

1. $\begin{bmatrix} 4 & -1 \\ 3 & 4 \end{bmatrix}$ **3.** $\begin{bmatrix} 8 & -7 \\ -2 & 4 \end{bmatrix}$

5. $3 - \mathrm{I}, 3$ **7.** $2 - \mathrm{III}, 2$

9. $2 - \mathrm{III}, 0$

11. (a) set up in the stadium
(b) set up in the gym
(c) set up in both

13. (a)

	better	not
market new	50,000	−25,000
don't	−40,000	−10,000

(b) $5,000 and −$22,000; market new

15. 1: $\frac{1}{5}$, 2: $\frac{4}{5}$, I: $\frac{3}{5}$, II: $\frac{2}{5}$, value $= \frac{17}{5}$

17. 1: $\frac{11}{15}$, 2: $\frac{4}{15}$, I: $\frac{8}{15}$, II: $\frac{7}{15}$, value $= \frac{62}{15}$

19. Both A and B should choose either strategy with a probability of $\frac{1}{2}$. The value of the game is 0, a fair game.

Chapter 11 Test

1. $\begin{bmatrix} -1 & -1 & 4 \\ 2 & 2 & 0 \\ 3 & 1 & 6 \end{bmatrix}$ **2.** $\begin{bmatrix} 5 & -1 & 2 \\ 2 & -2 & 2 \\ 5 & 1 & 0 \end{bmatrix}$

3. $\begin{bmatrix} -9 & -2 & 12 \\ -7 & 0 & 5 \\ -15 & 2 & 12 \end{bmatrix}$

4. $\begin{bmatrix} -2 & 4 & -6 \\ 0 & -1 & -1 \\ 10 & 4 & 6 \end{bmatrix}$ **5.** $\begin{bmatrix} 4 & -2 & 6 \\ 4 & 0 & 2 \\ 8 & 2 & 6 \end{bmatrix}$

6. $\begin{bmatrix} 9 & 0 & -3 \\ 0 & -6 & 3 \\ 3 & 0 & -9 \end{bmatrix}$ **7.** $\begin{bmatrix} 2 & -1 & 3 \\ 4 & 1 & 3 \\ 2 & 0 & 1 \end{bmatrix}$

8. $\begin{bmatrix} -4 & -1 & 0 \\ 2 & 0 & 1 \\ 4 & 1 & 3 \end{bmatrix}$ **9.** $\begin{bmatrix} -3 \\ 10 \end{bmatrix}$

10. $\begin{bmatrix} 6 & 0 & 12 \\ 26 & 0 & 20 \end{bmatrix}$

11. $[16]$ **12.** no product

13. $\begin{bmatrix} 3 & -1 \\ -5 & 2 \end{bmatrix}$ **14.** none

15. $\begin{bmatrix} \frac{1}{2} & \frac{1}{2} & -\frac{1}{2} \\ -\frac{1}{2} & \frac{1}{6} & \frac{1}{6} \\ \frac{1}{2} & -\frac{1}{2} & \frac{1}{2} \end{bmatrix}$

16. $\begin{bmatrix} \frac{2}{3} & 0 & -\frac{1}{3} \\ \frac{1}{3} & 0 & -\frac{2}{3} \\ -\frac{2}{3} & 1 & \frac{1}{3} \end{bmatrix}$

17. $(-2, 3)$
18. $(3, 0)$
19. $(-1, 0, 2)$
20. $(3, 2, 1)$
21. $\begin{bmatrix} 3 & 4 \\ -1 & 5 \end{bmatrix}$ **22.** $\begin{bmatrix} 8 & -3 \\ 0 & 2 \end{bmatrix}$

23. 6 **24.** 9
25. Play row one $\frac{9}{10}$ of the time; row two $\frac{1}{10}$ of the time; value is $-\frac{31}{10}$.

CHAPTER 12

Section 12.1

1. 69.8 **3.** 8321 **5.** 82,000,000
7. 37.42 **9.** 81,375,000,000
11. .0007629 **13.** .0000104
15. .0414 **17.** .48 E + 11
19. .38 E + 13 **21.** .9 E − 04
23. .427 E − 05 **25.** .3271 E − 06

Section 12.2

1. valid **3.** invalid **5.** invalid
7. valid **9.** 19 **11.** 14
13. 32 **15.** 5
17. LET $A = X + Y - Z*W$
19. LET $Z = P \uparrow 3*M$
21. LET $H5 = (8 + P)/(2 - Q)$
23. LET $K9 = (X - 3*(Y + R))/(5*M + 2*N)$
25. No value of R is supplied.
27. 20 **29.** 2

31.

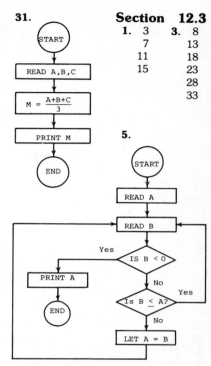

Section 12.3

1. 3   **3.** 8
  7   13
  11   18
  15   23
         28
         33

5.

7. Printout: THE LARGEST ITEM IS 13
9. It identifies the largest number in the data list.
11. The sum of the *squares* of the first hundred counting numbers is also found and is printed below the original sum. **13.** 1 **15.** 6
17. No effect since the computer only reads one data item anyway.
19. 210

Chapter 12 Test

1. 87,400,000 **2.** 136,900,000,000
3. .000725 **4.** .389 E + 12
5. .725 E − 06 **6.** invalid
7. valid **8.** invalid **9.** valid
10. LET $M = 5*X + 2*Y$
11. LET $P = (8 + Y)/(9 + Q)$
12. LET $Z2 = (4*M + 2*N)/(Y - Z - 4)$
13. LET $Y9 = (X + 5*M) \uparrow (1/2)$

14.

X	Y	SUM	POWER
2	3	5	8
4	2	6	16
3	4	7	81

15.

3	10	6
6	20	12
9	30	18
12	40	24

Acknowledgments

Photographs from the collection of Charles D. Miller: pages 2, 7, 59, 171, 195 (bottom), 198, 230, 231, 291 (Courtesy Jack Thornton), 293, 298 (top), 314, 334, 370, 387 (Photograph by Bill Jack Rodgers from Los Alamos Scientific Laboratory), 409

Stamps courtesy of Jory Norballe, Oakland, California, and Frank Warner, New York City

Photographs by Scott, Foresman Staff: 104, 146, 154, 158, 159, 211 (bottom), 340, 342, 347, 358, 374–377, 416 (top), 419 (top)

Cover	From *School of Athens* by Raphael, Scala – Editorial Photocolor Archive
Frontispiece	Alinari – Art Reference Bureau
1 Star map	By courtesy of Yerkes Observatory
4 Chickens	By courtesy of H. D. Hudson Manufacturing Co.
22, 23 Construction	By courtesy of Pace Gallery
28 Cantor	Bavaria-Verlag
30–31 (detail) 6th century B.C. Greek vase	By courtesy of Metropolitan Museum of Art, Rogers Fund, 1914
36 Aristotle	Art Reference Bureau
37 (bottom) Boole	Library of Congress
38 Whitehead	Culver
39 (center) Châtelet	Library of Congress
53 AND/OR game	By courtesy of Ontario Science Center
54 Circuit board	By courtesy of Bell Telephone Laboratories
64 Volunteer	By courtesy of National Institute of Aging, Gerontology Research Center
68 Lewis Carroll	Library of Congress
73 Shannon	By courtesy of Bell Telephone Laboratories
83 Quippu	By courtesy of Musée de L'Homme
84 Marriage tablet	By courtesy of Linden Museen Stuttgart
87 (left) Ishango bone	By courtesy of Prindle, Weber, Schmidt
87 (right) Cartoon	Norman Johnson

88 Tutankhamun   Lee Boltin

91 Constantine   Art Reference Bureau

96 (detail) Yuan Dynasty Kyan Yin   By courtesy of the Art Institute of Chicago

97 Sargon   By courtesy of Iraq Museum

98 Clay tablet   By courtesy of the Trustees of the British Museum

99 Xipe Tótec   By courtesy of the American Museum of Natural History

101 Brahma   By courtesy of the Metropolitan Museum of Art

118 Pythagoras   Art Reference Bureau

121 Prospector   By courtesy of the Minnesota Historical Society

134 Leo Decimus   By courtesy of Staatliche Lutherhalle Wittenberg

137 (left) Franklin   By courtesy of the White House Collection

137 (right) Magic square   By courtesy of Yale University

147 Musician   By courtesy of Staatliche Museum

154 (bottom) Chessboard   By courtesy of Columbia University Library

163 Stevin   Bettman Archive

173 Windmill   By courtesy of Energy Research and Development Administration

179 Centennial problem   From *Mathematical Puzzles of Sam Loyd,* edited by Martin Gardner, © 1959 Dover Publications

181 Hammurabi   Hirmer Photo Archive

182, 183 Town houses   New Century Town Houses, Vernon Hills, Illinois, Bernard E. Ury Associates, Inc.

184, 185 Garbage house   By courtesy of Rensselaer Polytechnic Institute

187 Mastercharge   By courtesy of Continental Illinois Bank

188 Solar car   NASA

189 (bottom) Scanner   By Courtesy of IBM

190 Lagrange   Library of Congress

191 Jefferson   By courtesy of the White House Collection

193 Road sign   Illinois Department of Transportation

194 Bottles   By courtesy of the 7-up Company

195 (top) Bannaker   By courtesy of the Schomberg Collection, New York Public Library

198 (top) Decimal places   By courtesy of the American Mathematical Society

199 Modulor   Howard Kaplan

211 (top) Lasker   Library of Congress

215 Difference Engine   British Crown Copyright, Science Museum, London

233 Galois   Brown Brothers

237 Noether   By courtesy of Professor Gottfried Noether

244 Viète   Library of Congress

252 Poincaré   Library of Congress

262 Agnesi   Library of Congress

268 Circuit pattern   By courtesy of IBM

269 Dantzig   By courtesy of Dr. George Dantzig

274 Windmill   By courtesy of the Central Vermont Service Corporation

275 (left) Catenary windmill   NASA

275 (right) Eggbeater   Canadian Picture Service

282, 283 (left) Smokey Yunick's energy system   Reprinted from *Popular Science* with permission, © 1975 Times-Mirror Magazine, Inc.

283 (right) Wind turbine   NASA

Index

Your Comments

You can help us make future editions of *Mathematical Ideas* more useful by taking a few minutes to answer the following questions. Then tear the page out, fold, seal, and mail it. No postage is required.

Please list the mathematics courses you had before this one (if any).
_____ None _____

Was this course an elective? _____ Required by _____

What is your major (or career goal)? _____

Here is a list of the main topics in the book. Please check those that you covered in class, and check the box which tells how interesting you found the topic.

Chapter	Name	Covered	Interest Little	Interest Some	Interest Much
1	Sets				
2	Logic				
3	Numeration systems				
4	Rational numbers				
5	Real numbers				
6	Mathematical systems				
7	Algebra				
8	Geometry				
9	Probability				
10	Statistics				
11	Matrices				
12	Computers				

Did the topics in the book seem applicable to your daily life or to your career major? Yes _____ No _____

Did the pictures make the book more enjoyable? Yes _____
No _____

Was the material in the captions interesting and useful? Yes _____
No _____

In general, could you follow the instructions of the exercise sets? Yes _____
No _____

Did you find the answers in the back helpful? Yes _____ No _____

Please list in the margin any typographical errors (with the page number) that you may have found in this book.

What did you like about the book?

··· Fold here

What did you *not* like about the book?

School _____ State _____

Fold here

···

First Class
Permit No. 282
Glenview, Ill.

BUSINESS REPLY MAIL
No postage necessary if mailed in United States

Postage will be paid by
SCOTT, FORESMAN AND COMPANY
College Division   Attn: Miller
1900 East Lake Avenue   Heeren MI3
Glenview, Illinois 60025